Wild Mammals of Wyoming and Yellowstone National Park

UNIVERSITY OF WYOMING
Biodiversity
INSTITUTE

The publisher gratefully acknowledges the generous contribution to the production of this book provided by the University of Wyoming Biodiversity Institute (wyomingbiodiversity.org).

Wild Mammals of Wyoming and Yellowstone National Park

STEVEN W. BUSKIRK

UNIVERSITY OF CALIFORNIA PRESS

University of California Press, one of the most distinguished university presses in the United States, enriches lives around the world by advancing scholarship in the humanities, social sciences, and natural sciences. Its activities are supported by the UC Press Foundation and by philanthropic contributions from individuals and institutions. For more information, visit www.ucpress.edu.

University of California Press
Oakland, California

Library of Congress Cataloging-in-Publication Data

Buskirk, Steven, author.
Wild mammals of Wyoming and Yellowstone National Park / Steven W. Buskirk.
pages cm
Includes bibliographical references and index.
ISBN 978-0-520-28689-4 (cloth : alk. paper)
ISBN 978-0-520-96195-1 (ebook)
1. Mammals—Wyoming. 2. Mammals—Yellowstone National Park. I. Title.
QL719.W8B87 2016
599.09787–dc23

2015022452

Manufactured in China
22 21 20 19 18 17 16
10 9 8 7 6 5 4 3 2 1

The paper used in this publication meets the minimum requirements of ANSI/NISO Z39.48-1992 (R 2002) (*Permanence of Paper*).

For Beth, whose curiosity about nature inspires my own

CONTENTS

PREFACE

The near-term stimulus for producing this book was the publication of Douglas Faukner's *Birds of Wyoming* (2010), which was encouraged by Robert Berry and supported by the Wolf Creek Charitable Foundation. I had wanted to write such a book on mammals for over a decade, and had started planning with various possible coauthors, but the research and writing were never started in earnest. Looking at Faulkner's handsome and useful contribution one day with my colleague Carlos Martinez del Rio, he decried the lack of a similar book on mammals, and suggested that I should undertake the task. Nearing the end of my three decades of academic work on mammals, I knew the need, and saw an opportunity. I discussed possible partial funding with another colleague, Craig Benkman, and with Bob Berry. Dr. Fred Lindzey had been a departmental colleague and hunting partner for over 20 years, and was president of the Wyoming Game and Fish Commission at that time. He agreed that the wildlife management community would benefit from such a contribution. Martin Grenier, a former graduate mentee and Nongame Mammal Biologist for the Wyoming Game and Fish Department, wished to see the book supported by the Wyoming Chapter of The Wildlife Society, and with Matt Kauffman encouraged that group to contribute funding to the production of maps.

Several kinds of state mammal faunas have been written about, and deciding what kind of book Wyoming needed and that I could produce was a challenge. A field guide was not needed because so many excellent guides to identification and field sign—for all of North America—were still in print. These guides describe physical traits, tracks, scats, and other field sign as aids to identification. Writing a book that comprehensively reviewed the natural history and ecology of each species was not plausible, because such a book would be too long, and much of that information is already available in print or online. In addition to standard references, book-length species accounts are available for all species of North American ungulates, all large carnivores, and some other mammals that occur in Wyoming: beavers, black-footed ferrets, martens, deer mice, and black-tailed prairie dogs.

I concluded that what was needed was a desk reference on the taxonomy, distributional dynamics, ecology, and conservation status of the mammals of Wyoming, emphasizing knowledge collected in Wyoming and neighboring states, wherever possible. It is written for motivated lay naturalists, wildlife biologists, and mammalogists, which has necessitated the use of vocabulary which I hope will be accessible yet precise to all three audiences. Changes in species' geographic ranges through time are a particular emphasis here. Biogeography lies at the intersection of several disciplines: evolution, ecology, physical geography, and conservation biology. Conservation laws and policies are based in part on the distributions of species; species that are severely or increasingly restricted in distribution are often conservation concerns. The distributions of Wyoming mammals—as depicted in the distribution maps of Hall and Kelson (1959), Long (1965), Hall (1981), and Clark and Stromberg (1987)—have undergone substantial, even drastic change for some species. Some of these changes are due to actual expansions, contractions, or shifts in distributions. Others are due to improved information, which has allowed better depictions of where species occur, or new understandings of what populations fall with a species' boundaries. The systematics of mammals is related to their geographic distributions, because as taxa are split or clumped together, the corresponding distributions must be recast as well. All of these dynamic understandings have implications for conserving and managing Wyoming mammals. This account treats a number of Wyoming species that have been split or combined since earlier accounts, had their species names changed, subspecies confirmed or abandoned as a result of molecular studies, and higher-level phylogenetic relationships re-portrayed based on new understandings.

Identifying Wyoming mammals to species in the field is more problematic that some field guides may suggest, and some traditionally used gross morphological features—size, shape, and color—are now recognized as unreliable. Through binoculars, in hand, or even after cleaning of the skeleton, some pairs of species that occur in Wyoming are very difficult to identify, and the reader would be poorly served if I merely repeated traditional criteria. For those species and others that are challenging, I encourage readers to consult the species identification literature, and refer potentially anomalous or puzzling specimens to museum mammalogists for reliable identification. The literature cited, as nearly specific to Wyoming as possible, represents in some cases much of what is known about a species in and near Wyoming. In other cases, the references cited represent only a small fraction of the available knowledge. In both cases, the references credit the authors of the knowledge presented, and provide the reader with a portal into the body of knowledge about a species or subject, as specific to Wyoming as the literature allows.

Steven W. Buskirk
Laramie, Wyoming
December 15, 2014

ACKNOWLEDGMENTS

Many people and several institutions supported the preparation and publication of this book. Carlos Martinez del Rio provided the initial impetus to undertake the project, and Fred Lindzey and Reg Rothwell were encouraging throughout. Bob Berry was key to providing financial and moral support for the project. Reg did exhaustive reviews of morphology, and traveled to major museums to gather morphological data. Dennis Knight was an indispensable source of ecological and practical advice and encouragement. Bob Lanka of the Wyoming Game and Fish Department supported the project in multiple ways, including providing data on pronghorn and elk abundances across Wyoming. Martin Grenier provided updated results of surveys, as data became available. Gary Beauvais and the staff of the Wyoming Natural Diversity Database were essential in producing distribution maps that were accurate and well documented. Amanda Bowe assumed the onerous task of reviewing online museum records and published accounts for mammal specimens from Wyoming. The online appendices are a record of her diligent work. Doug Keinath helped me to understand the distributions of bats, shrews, and the Wyoming pocket gopher from recent surveys. Shawn Lanning of the Wyoming Geographic Information Science Center lent his technical support to the mapping efforts, and Ken Driese provided biologically informative base maps. The University of Wyoming and College of Arts and Sciences supported my sabbatical leave, which provided the time for most writing and map preparation. Greg Brown made office space available to me in the Berry Center. The newly upgraded University of Wyoming Museum of Vertebrates was an important resource, and I thank Curator Matt Carling and Collections Manager James Maley for providing access. Primary funding was provided by the Biodiversity Institute, University of Wyoming; additional funding was contributed by Wyoming Wildlife – The Foundation, the Wyoming Chapter of The Wildlife Society, and the Office of Research and Economic Development, University of Wyoming.

Kenneth Mills, Department Head of Veterinary Science, provided access to Wyoming State Veterinary Laboratory records for striped skunks, northern raccoons, and red foxes.

Doug Keinath made available locality data for the Wyoming pocket gopher. Jake Ivan of Colorado Parks and Wildlife provided data on movements and field measurements of Canada lynx studied during the Colorado Lynx Reintroduction Project. Dan Stahler tabulated data on field measurements of wolves handled during the Yellowstone Wolf Project. Craig Acres of USDA Wildlife Services provided information on the distributions of Virginia opossums and spotted skunks. Mark Haroldson of the Interagency Grizzly Bear Study Team provided data on the distribution of grizzly bears, and Shannon Podruzny prepared the state distribution map. Michael Bogan and Ernest Valdez evaluated putative *Myotis yumanensis* specimens at the Museum of Southwestern Biology. Kevin McKelvey and Yvette Ortega of the USDA Forest Service, Rocky Mountain Research Station, shared the GIS data that were used to assemble the pre-1999 distribution of Canada lynx. Beth Buskirk drew the illustrations of mouse and bat, and of shrew dentition. Mark Zornes of the Wyoming Game and Fish Department identified a recent geographic range expansion of black-tailed jackrabbits in southwestern Wyoming. Robert Inman, at the time affiliated with the Wildlife Conservation Society, provided the data and prepared the figure of wolverine dispersal. William Zielinski and Ric Schlexer of the US Forest Service Pacific Southwest Research Station assembled and provided field measurements of fishers from the southern Sierra Nevada. Robert Garrott of Montana State University provided data on mountain goat observations from the Greater Yellowstone Area Mountain Ungulate Project, and Carson Butler prepared the GIS layer of mountain goat locations. John Montenieri of the Centers for Disease Control and Prevention provided data on the distribution of sylvatic plague by county, and Kiersten Kugeler prepared the plague distribution map in the species account for black-tailed prairie dog. Matt Hayes prepared data on vehicle collisions with ungulates in the Pinedale area. Thomas Labedz of the Nebraska State Museum compiled data on field measurements of mammals from that collection, and Christy McCain compiled data from the University of Colorado Museum. Anna Barker of The Murie Center prepared and arranged the loan of the image of the Murie brothers.

The continental-scale distribution maps are based on data provided by NatureServe in collaboration with Bruce Patterson, Wes Sechrest, Marcelo Tognelli, Gerardo Ceballos, The Nature Conservancy; Migratory Bird Program, Conservation International; CABS; World Wildlife Fund-US, and Environment Canada-WILDSPACE. Marissa Ochsenfeld conducted literature reviews, and Reid Lawler produced most of the state-scale distribution maps, and modified some continental ones. Elizabeth Ono Rahel of Mariko Design produced the diagrams, histograms, and scatterplots. Sheila Krupp and Shawn Sheen capably managed the various contract accounts associated with this project, and Carol Pribyl oversaw the administrative process. Joy Handley proofread the manuscript and cross-checked citations.

Keith Aubry, Steve Cain, Wendy Estes-Zumpf, Nick Fuzessery, Keith Geluso, Martin Grenier, Matt Kauffman, Douglas Kelt, Marion Klaus, Dennis Knight, Nicole Korfanta, Hayley Lanier, Jason Lillegraven, Charles Long, Bob Luce, Carlos Martinez del Rio, Hall Sawyer,

Doug Keinath, Hannah Griscom, Kelli Trujillo, and Rick Wallen provided constructive reviews of chapters and species accounts. Gary Beauvais, Doug Keinath, Wendy Estes-Zumpf, Mark Andersen, and Ian Abernathy reviewed drafts of distribution maps.

Blake Edgar and Merrik Bush-Pirkle at University of California Press were supportive and efficient in advancing the manuscript toward publication, from its initial submission. Kate Hoffman of UC Press and Mansi Gupta of IDS Infotech Ltd. created an attractive and navigable book from my long and complicated manuscript.

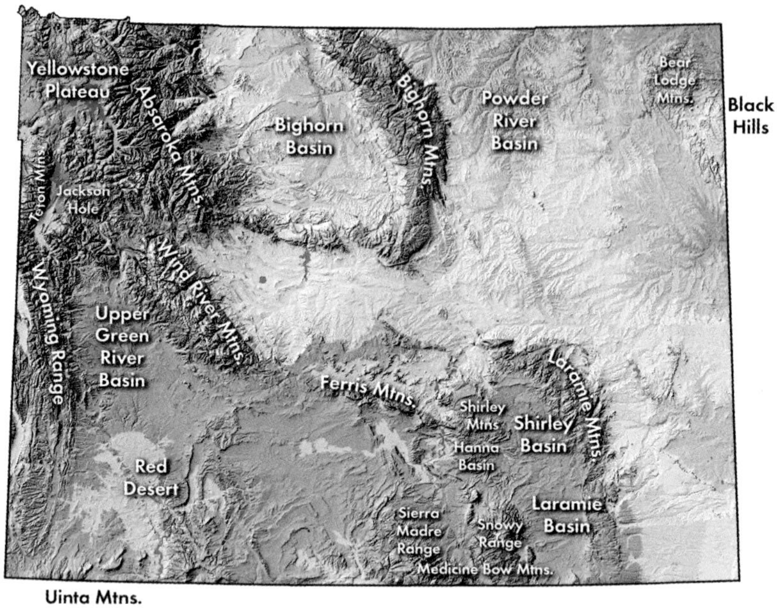

MAP 1.1 The major mountain ranges and basins of Wyoming. The term "northwestern mountains" is nonstandard, and used here to denote the contiguous mountain ranges of the Greater Yellowstone Area, extending to the southern termini of the Wind River Mountains and Wyoming Range.

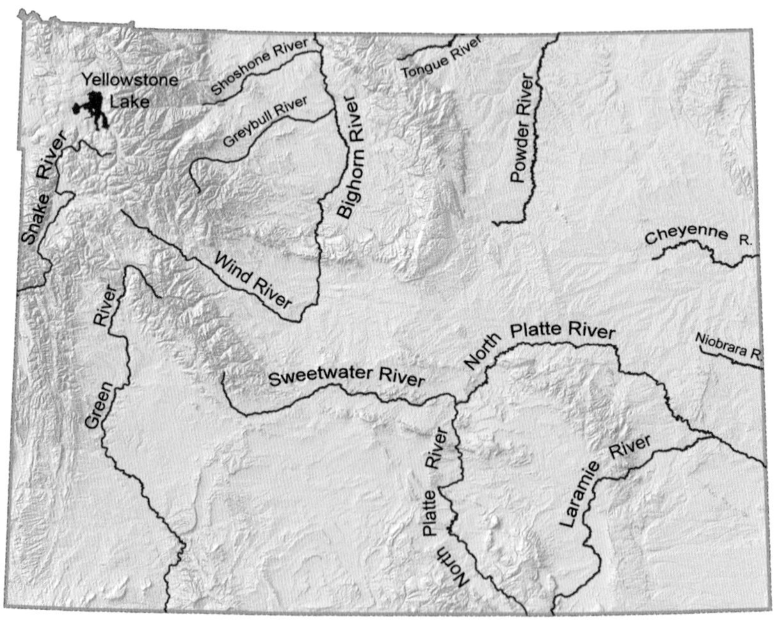

MAP 1.2 The major rivers and tributaries of Wyoming. The Snake River empties into the North Pacific via the Columbia River, the rivers of eastern Wyoming into the Gulf of Mexico via the Mississippi River, and the Green River into the Gulf of California via the Colorado River.

MAP 1.3 The counties of Wyoming, including the part of Yellowstone National Park outside Wyoming.

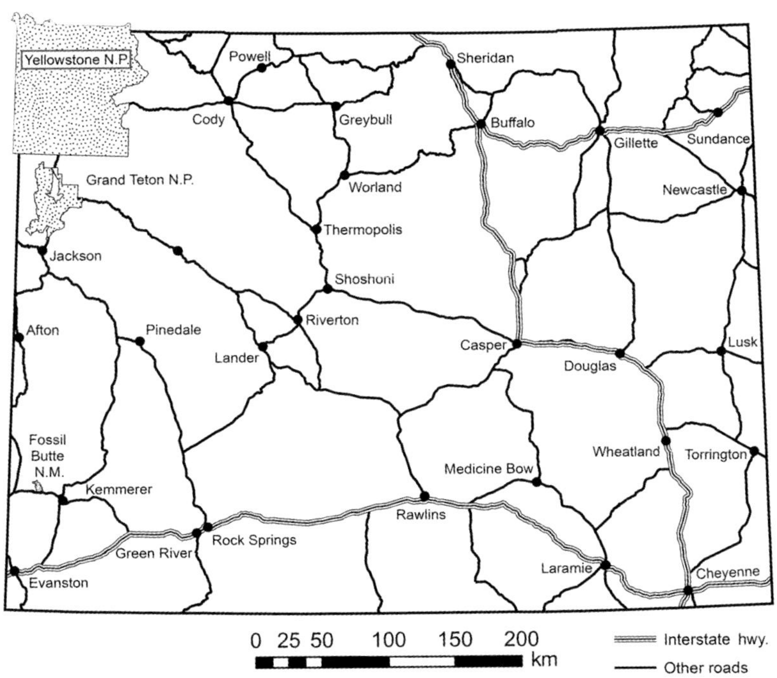

MAP 1.4 The paved highways, cities and major towns, and major units of the National Park System in Wyoming.

PART I

THE WYOMING CONTEXT

Introduction

This book is a review and compilation of the identification, taxonomy, zoogeography, ecology, and conservation of wild mammals native to Wyoming. It includes species accounts for naturalized introduced (or recently range-expanded) North American species, and includes accounts for several species currently in the state but of uncertain residency status, and one species—the fisher—that evidence shows is neither a resident nor a native, but that is worthy of review. The book is intended as a reference resource for mammalogists, wildlife biologists, resource management professionals, and lay naturalists, and the information on which it is based comprises published accounts, museum specimens, online databases of museum holdings, unpublished reports, and unpublished data from the mid-nineteenth century to the present. The unpublished data sources provide some of the best information on current distributions that could be presented, and the generosity of agency and academic scientists who shared these data has been fundamental to the value of this book. The species depicted in these unpublished data sets are diverse—shrews, jumping mice, flying squirrels, pocket gophers, bats, wolves, grizzly bears, wolverines, and mountain goats—and the contributed data have allowed depictions of distributional dynamics of unprecedented detail in Wyoming for a number of species. All state-level distribution maps that feature point data are supported by supplemental online appendices that present data on each locality mapped. Those appendices can be found at http://www.wyomammals.com.

Aside from those presented here, other resources specific to particular taxa will be needed for some problems of identification to species.[1] Tracks or other field signs can be useful in identifying some mammals to species; excellent field guides to animal tracks and field sign cover the geographic area of Wyoming.[2] Keys to genera and species are provided following the relevant order or family introduction, but museum reference collections and help

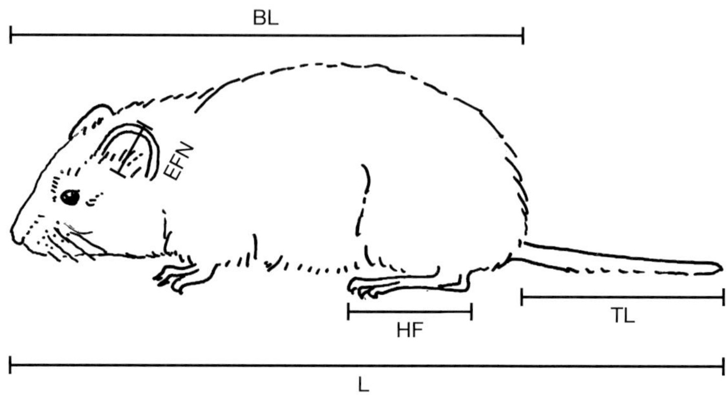

FIG. 2.1 Standard field measurements of a typical small nonflying mammal. L = total length, BL = body length, minus tail, TL = tail length, HF = hindfoot length, EFN = ear length from notch. Tail length is typically measured with the tail held perpendicular to lumbar spine.

from a specialist will be needed to identify such morphologically similar taxa as shrews, some ground squirrels, harvest mice, pocket mice, and chipmunks. Some Wyoming mammals can be reliably identified to species only through analysis of their DNA, and likely some valid biological species in Wyoming have yet to be described using molecular markers. Molecular markers for the identification of a steadily increasing number of Wyoming mammals are available now, and I refer to them where possible.

Field measurements are provided of mammals from Wyoming or as close to Wyoming as possible as an aid in identification, and as an important life history characteristic. If measurements were not available for five females and five males from Wyoming, then measurements were taken from adjacent states, and finally from the nearest area from which such sample sizes could be drawn. Locations for field measurements are given as Wyoming counties or other contributing states. Standard field measurements differ among taxa; the coding used is L = length or total length, BL = body length (length minus tail length), TL = tail length, SH = shoulder height, HF = hindfoot length, EFN = ear length from notch, FA = forearm length (bats), and WT = body weight (fig. 2.1). Sources for data on morphology come from published and unpublished accounts and museum records: University of Wyoming Museum of Vertebrates, Laramie (UWMV); University of Kansas Mammal Collection (KU); Denver Museum of Nature and Science (DMNS), Colorado; Museum of Southwestern Biology, Albuquerque, New Mexico (MSB); University of Nebraska State Museum, Lincoln (USNM); and University of Colorado Museum, Boulder (UCM).

The apparent distributions of many mammalian species in Wyoming have changed over the last 150 years, in some cases dramatically. Some of these perceived changes are due to better information, collected by better-trained biologists with improved survey tools. Other changes are real and have come about as a result of human-caused environmental change, including land use, a changing climate, and altered vertebrate community members. Still other changes have been directly at the hand of humans: mammals have been eradicated, introduced, and reintroduced to Wyoming. Some of these actions have had cascading effects that are well understood, while other effects are only suspected. In the following accounts, the known and hypothesized changes in geographic ranges are presented, and the most plausible mechanisms underlying them identified. Some specialized vocabulary in the areas of anatomy, taxonomy, and biogeography is assumed, and key terms are defined in the glossary.

DISTRIBUTION MAPS

The continental-scale distribution maps presented here are based on maps produced by NatureServe, modified for some species to reflect approximate distributions at the time of European settlement, and to reflect currently recognized species boundaries.[3] The state-scale distribution maps were produced to reflect current distributions, in some cases distributions from earlier in the historical period, and other geographical information. Two kinds of distributional information are presented. The distributions mapped as solid color areas are intended to show an inferred pattern of occurrence of the species. Point data are used to depict localities where a species was collected, observed, or tracked via telemetry. Some species did not permit mapping of their current or past distributions; this was the case for the common gray fox and fisher.

Whether point data or an area is used to depict a distribution is a function of several factors. If the species is not distributed according to known, mapped habitat associations, or if no clear distributional boundary is apparent, and if locality data are available for the few observations available, then the distribution is depicted via points. If the species is either weakly represented in collections or other point data sources, or if its distributional boundaries are known to be sharp and habitat-based, then a smoothed polygon was used to represent its distribution. Using the coyote as an example, identification from a specimen is straightforward, and the species occurs in most or all habitats in the state, but almost no specimens or published data are associated with places where coyotes have been detected in recent decades. Coyote specimens are seldom prepared and preserved in museums. So, the distribution of the coyote is depicted here as statewide, recognizing that the species may be absent from a few sites, for example, alpine zones in winter. The state-scale distribution map for the canyon mouse illustrates the case for the use of points; the species has been collected from only a few locations in southwestern Wyoming, and those locations are documented, but how the species or its habitat is distributed between or beyond those points is not clear.

For a few species for which Wyoming data are scarce, the depicted geographic range was modified from the modeled geographic range presented in the Northwest GAP Species Explorer, a project of the University of Idaho and US Geological Survey.[4] The Northwest

GAP Program produces modeled distributions of vertebrate species, which are representations of where species are predicted to occur based on point data for the species, and environmental information about the sites where the species was observed. The species is predicted to occur where it has been found, and in places with similar environmental conditions—elevation, cover type, topography—nearby. Of course, modeled distributions are only as good as the information entered into the model. For example, a site could be a perfect match for the habitat of a species, but the site could be isolated from the current geographic range by a geographic barrier, have competitors or predators of the species of interest that prevent it from occurring at a site, or harbor diseases that affect the species of interest. In general, modeled distributions tend to be more extensive than the corresponding true species distributions, and the latter is the kind of distribution that I have depicted here, where possible.

Taxonomy and systematics are additional core topics of this book. Taxonomy refers to how taxa (kinds of organisms, such as genera, species, and subspecies) are named, and systematics is the organization of those named types into branching arrangements that reflect phylogeny: patterns of evolutionary divergence. The names and phylogenetic positions of many Wyoming mammals have changed since the seminar work on Wyoming mammals in 1965, and the latest accepted or strongly supported taxonomic and systematic revisions available are used. The standard reference for mammals worldwide is that of Wilson and Reeder (2005), but for common names for mammals north of the Mexican border I have used Bradley et al. (2014). Using standardized common names allows a biologist to communicate about species with a biologist in Eurasia or South America, without ambiguity or reference to the scientific name. In daily use where the regional context is clear, a Wyoming biologist might use such common names as "elk," "beaver," "marten," or "gray fox." However, to communicate with a broader audience, it would be necessary to specify "wapiti," "American beaver," "Pacific marten," or "common gray fox," and I have used the latter conventions here. In colloquial usage in a Wyoming context, the abbreviated common name is customary.

It is important to consider what constitutes a valid record of a species for the purpose of this work. In earlier times, the basic unit of mammalogical investigation was the museum specimen, typically in the form of a stuffed skin and a cleaned skull or skeleton, along with associated field notes. This was the basic tool of the museum mammalogist, who relied on measurements of freshly killed mammals, their cleaned skeletons, and particularly their skulls and dentition to identify them to species, or to recognize and describe new species. This approach has merit in the sense that bones and teeth are important adaptive features, and form a permanent physical record. However, we limit ourselves if we only study prepared bones and skins. First, collecting and preserving mammal specimens has been largely restricted to small-bodied species. Few museum specimens of grizzly bears, wolverines, and porcupines from Wyoming are available for study, especially from a century ago. Secondly, conservation concerns and humane animal treatment require that mammal occurrences be documented without killing them, whenever possible. For some mammal groups, particularly bats, most species can be identified and measured in hand in the field, and the animals released unharmed. For shrews, this is not possible, because few shrews can survive the trauma of capture and examination in hand. Other methods, including remote photography, hair collection, and DNA extraction and identification, are in some cases just as reliable as capture or collection of specimens, and are increasingly used.

The literature review is limited to Wyoming or areas as close to Wyoming as possible, working from the premise that most aspects of the biology and ecology of mammals vary geographically. Still, because there are no intensive field studies of Virginia opossums from Wyoming or adjacent states, it has been necessary to draw from states as far away as Pennsylvania and Florida for basic biological data. This is in spite of the fact that studies from such different ecological settings may tell us little about the life history, population biology, diet, or behavior of the species in Wyoming. The reader is cautioned that the information presented from elsewhere about Virginia opossums—and other species never studied in Wyoming—may not reflect the species' biology here.

The conservation and management of most species are documented to the extent that the literature allows. A great majority of state and federal wildlife management efforts are directed towards a relative few species: ungulates, large carnivores, State Species of Greatest Conservation Need, and threatened and endangered species.[5] For many other species, little knowledge of their conservation status is available, and expressions of conservation concern are often nonspecific and call for further information.

History of Wyoming Mammalogy

For the geographic region in which it lies, Wyoming has an intermediate level of development of mammalogical knowledge—less than that for some neighboring states, more than others. The knowledge base is as large as it is because Wyoming is large in area (10th largest state), was positioned along major nineteenth-century transportation routes, and has had a land-grant university since 1886. Further, the state holds a large amount of publicly owned land (53.6% of total area is owned by the state or federal government), most of it in an undeveloped state and a considerable portion of it native vegetation or part of the National Wilderness System. Public land agencies are required to inventory and monitor biodiversity. These lands include such international conservation icons as Yellowstone National Park (the first national park, 1872), Devil's Tower National Monument (the first national monument, 1906), and large areas of the National Forest System, some of them the earliest federal forest reserves (Shoshone National Forest, originally set aside by the federal government in 1891). Wyoming's large, old, acclaimed nature reserves have been particularly influential in attracting citizen attention, scientific interest, and resources to study questions related to mammalian biodiversity and conservation. Many naturalists and mammalian biologists have made Yellowstone National Park and the surrounding area a travel destination, and studied the fauna and flora of other parts of Wyoming en route. The paved road system of Wyoming was developed during 1910–1935, but the state had been largely accessible by horse and wagon for the preceding 80 years. Wyoming has been the focus of several wildlife conservation issues that have brought national attention and resources to the state; these issues have involved elk, bison, grizzly bears, black-footed ferrets, gray wolves, Preble's meadow jumping mice, and other species. The state of knowledge about these and other mammalian species has benefitted from these initiatives. Finally, several Wyoming ungulates are the focus of an important sport hunting industry, and sustainable management of these species has required monitoring and research.

Limiting the state of knowledge, Wyoming lacks a long history of resident academic mammalogists—university, museum, or government agency scientists who have surveyed for mammals, stored specimens in regional repositories, and published bodies of work on the systematics or zoogeography of mammals of the state. Wyoming has lacked a leadership figure in mammalogy comparable to David M. Armstrong for Colorado or J. Knox Jones Jr. for Nebraska.[1] As a result, many important questions of systematics or zoogeography centered on Wyoming remain unresolved, as will be seen from the accounts that follow. As examples of important unresolved issues, the taxonomic status of the meadow jumping mouse is just beginning to be understood, the systematic and ecological relationships between spotted skunks are a mystery, and the issue of sympatry and introgression between Wyoming and Uinta ground squirrels remains puzzling. The distribution of red foxes in the state is weakly documented and our knowledge of the distribution of bats in Wyoming is in its middle stages.

EARLY ACCOUNTS

Useful information about the distributions and abundances of wild mammals was recorded by early explorers, migrants, and settlers. These accounts appeared in journals, letters, government reports, and newspaper articles, and are referred to in some species accounts that follow. The writers' taxonomic biases were in favor of species that could be killed for food or fur, or were considered dangerous or novel; species that were considered ordinary or not related to food supply or safe travel tended to go unmentioned in their accounts.[2] Through such early accounts we know that grizzly bears and elk roamed the shrub-steppe far from coniferous forest, that wolves were found in Yellowstone National Park at the park's establishment, and that bighorn sheep—referred to by some as "ibex"—were common in the Sweetwater Rocks, just north of the Oregon Trail.[3] But these early historic accounts can also be confusing or misleading: "wolf" can refer to gray wolves or coyotes, the latter at times referred to as "brush wolves." The term "civet" can refer to any of several carnivore species, and "sable" can refer to marten, fisher, or mink, depending on regional usage where the observer lived before. White-tailed deer were called "red deer" in the journals of William Clark and other explorers, but the term is used for a Eurasian ungulate today. In some cases reasonable guesses can be made about the identity of a species mentioned in an early account based on the geographic origin of the writer and the mammalian vernacular names common there.[4] Some writers are known to have fictionalized their accounts for literary effect; Francis Parkman's *The Oregon Trail* has been criticized in this way, but whether his alleged factual liberties extended to observations of mammals is not clear.[5]

Among the earliest of surveys that attempted to systematically document—with specimens—the mammals of the West were those organized and conducted under the direction of the War Department for the purpose of finding the best route for a railroad from the

TABLE 1 Holdings of mammals from Wyoming and some adjacent states in nine national and regional natural history museums that have major holdings in the region surrounding Wyoming, as of November 2011.

Museum	State specimens collected			
	Wyo.	*Colo.*	*Mont.*	*S. Dak.*
University of Kansas Museum of Natural History	9295	6705	3840	5192
National Museum of Natural History, Smithsonian	6367	5987	5980	1094
University of Wyoming Museum of Vertebrates	3838	168	7	43
Museum of Southwestern Biology, Albuquerque	3647	14,152	2425	2827
University of Michigan Museum of Zoology	1891	1368	1930	1935
American Museum of Natural History, New York	1013	2328	604	565
Museum of Vertebrate Biology, Berkeley	751	3479	1211	142
University of Colorado Museum of Natural History	251	9501	118	14
Denver Museum of Nature and Science	73	255	0	7
Totals	27,126	43,943	16,115	11,819

Source: Data for all museums are from online catalogues.

Mississippi River to the Pacific Ocean. These natural history observations were compiled by Spencer Baird, an assistant secretary of the Smithsonian Institution, and extended far beyond any plausible transcontinental route—from Greenland to the recently established border with Mexico.[6] The surveys were spotty over the area that became Wyoming, and the place name "Wyoming" was not yet in use. But such Wyoming place names as Fort Laramie, Bridger Pass, Big Horn River, South Pass, and Medicine Bow River were recorded as the source for various specimens. That book served as the only specimen-based catalogue of mammals of the Wyoming region for over a century, until Charles Long's 1965 monograph was published.

The history of mammal collecting in Wyoming can be glimpsed through the activities of scientists associated with major museums, and the holdings of those museums (Table 1). The number of specimens collected from Wyoming is considerably less than that for Colorado, but greater than for either Montana or South Dakota. The largest holdings of Wyoming mammal specimens are at the University of Kansas Mammal Collection (KU, >9000 specimens), followed by the National Museum of Natural History (NMNH, Smithsonian Institution, >6000 specimens).[7] The major collectors of the specimens at these two museums, and the periods when they were most active were largely nonoverlapping (fig. 3.1). The Smithsonian specimens from Wyoming were mostly gathered during the peak years of the Bureau of Biological Survey (BBS), when such pioneering naturalists as C. Hart Merriam, J. Alden Loring, Edward A. Preble, and Vernon Bailey rode horseback across the West with mouse traps in panniers or wagons, and provided our first deep understandings of mammalian diversity in this region. Their coverage of the area of Wyoming was necessarily uneven (Map 3.1),[8] because of terrain and roads. Their assistants appear prominently in

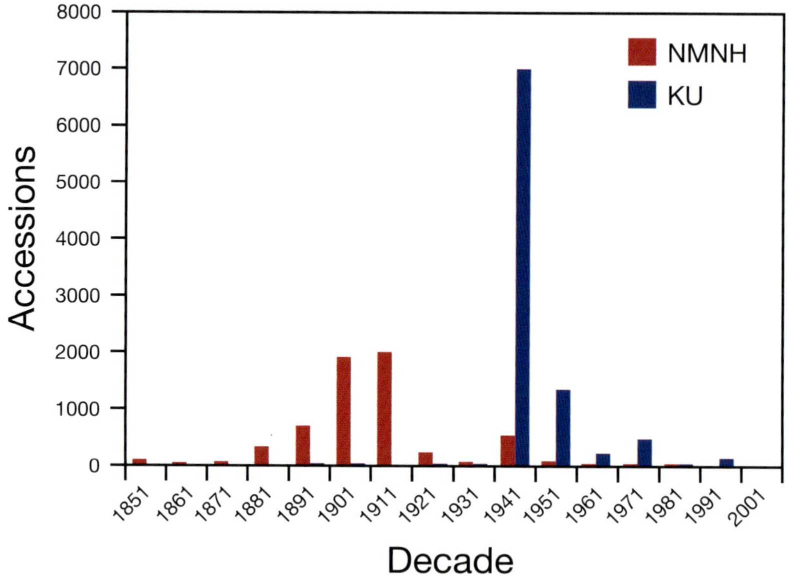

FIG. 3.1 Accessions of mammal specimens into the National Museum of Natural History (NMNH) and University of Kansas Museum of Natural History (KU)—the two largest collections of Wyoming mammals—by decade. The two museums show largely nonoverlapping periods of collection growth. The National Museum specimens were largely collected by personnel of the Bureau of Biological Survey; those at Kansas by faculty, students, and technicians of that institution.

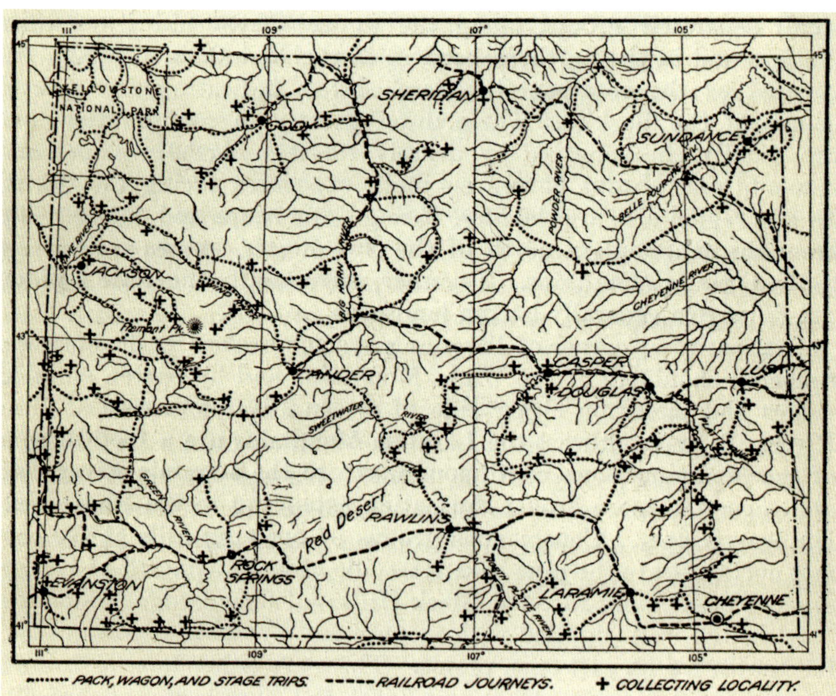

MAP 3.1 Routes traveled by Bureau of Biological Survey naturalists during their peak of mammal collecting activity, 1900–1917. Northeastern Wyoming and Yellowstone National Park received little attention to inventorying mammals.

collection records as well: Alexander Wetmore was an undergraduate at the University of Kansas who collected over 400 specimens from Wyoming—accessioned into the NMNH—under the tutelage of Charles Bunker.[9] Wetmore later earned his PhD at George Washington University, was one of the first to investigate the role of lead shot in causing illness and death in waterfowl, and became the secretary of the Smithsonian Institution.[10]

The single individual who collected the largest number of mammals from Wyoming was Merritt Cary, a BBS naturalist who surveyed Wyoming and collected 1029 specimen—all now held in the NMNH—from Wyoming during 1909–1912. His *A Biological Survey of Colorado* (1911) was an important work for that state, and his *Life Zone Investigations in Wyoming* in 1917 became the first to characterize plants and mammals of Wyoming in the major community types ("life zones") that had been proposed by C. Hart Merriam in several 1890s monographs.[11] Other important contributors to the collection of Wyoming mammals in the Smithsonian collections were Stanley Jewett (527 specimens, 1911–1913) and Olaus Murie (145 specimens, 1927–1938). The BBS merged with the Bureau of Sport Fisheries and Wildlife in 1939, but its survey activities had already waned a decade earlier; the great era of federal inventory of North American faunal resources had ended.

The most productive period of mammal collecting in Wyoming began in the 1940s. The KU was a growing influence in mammalogy, led by such dominating figures as E. Raymond Hall (museum director, 1944–1967) and Rollin H. Baker. This group, including E. Lendel Cockrum and various student assistants, began a period of intensive collecting in Wyoming in the waning weeks of World War II, and extending into the early 1950s. During 1945–1952, the number of mammal specimens from Wyoming in major collections more than doubled (fig. 3.1). This second period of intensive surveys of Wyoming mammals faded during the late 1950s and 1960s. Subsequent efforts at documenting mammal distributions via collections have been sporadic and limited.

THE GROWTH OF MAMMALIAN ECOLOGY

Although Wyoming has lacked major resident scientific leaders in mammalian taxonomy or zoogeography, and small mammal collecting activities peaked decades ago, the state has enjoyed a wealth of productive mammalian ecologists. Dozens of scientists have spent decades or careers studying habitat ecology, population ecology, physiological ecology, and community ecology of various mammalian species and community relationships. These studies differ from purely mammalogical ones in that they investigate processes within and among species, and between mammalian species and the environment, but may contribute only secondarily to our understandings of the diversity and distributions of mammals. The earliest Wyoming mammalogists and the explorer-naturalists who preceded them were keenly observant of the factors associated with where species were seen or collected, but studies of individual species of mammals that resulted in monograph-length accounts did not begin until the 1920s. Since then, graduate students and management and research biologists have investigated a wide range of ecological topics on dozens of Wyoming species. Among the most renowned and highly cited of the ecological studies are those on ungulates and large carnivores, including Douglas Houston's *The Northern Yellowstone Elk*, Adolph Murie's *Ecology of the Coyote in the Yellowstone*, and John Craighead and co-authors' *The Grizzly Bears of Yellowstone*.[12]

THE MURIE BROTHERS

The Murie half-brothers—Olaus and Adolph—deserve special mention, because they personified the transition from mammalogical surveys to ecological studies, contributing to both bodies of knowledge (fig. 3.2). These Minnesota natives first visited Wyoming during the 1920s, began their respective jobs here in the 1930s, and made Wyoming their family homes in the 1940s.[13] The families settled on a former guest ranch near Moose, and through the two men's rather different but complementary careers made key contributions in wildlife science and policy. Both Olaus and Adolph conducted mammalogical surveys in Wyoming early on, both conducted intensive ecological studies focused on individual mammalian species, and both contributed to regional or national conservation debates that resulted in significant advances in conservation policy that affected mammals in Wyoming and beyond.

Olaus began his biological fieldwork after receiving his bachelor's degree from Pacific University in Oregon in 1912. After graduation, he traveled first to eastern Canada, then to northern Alaska in the early 1920s, studying caribou and other mammals with the assistance of his brother Adolph. Upon receiving his MSc degree from the University of Michigan in 1927, Olaus was immediately assigned by the Bureau of Biological Survey to study the Jackson Hole elk herd, which had become problematic. His research on elk in northwestern Wyoming spanned decades, and resulted in an acclaimed 1951 monograph.[14] Adolph earned his PhD from Michigan in 1929 and moved to Wyoming in the 1930s, studying coyote ecology in the Yellowstone area in the wake of the recent eradication of wolves. Adolph's 1940 monograph cast coyotes, and the role they played in the Yellowstone ecosystem, in a neutral scientific light. Such a characterization was coolly received in some quarters, but his study served as the model for his next and highest-profile assignment: studying the ecology of gray wolves, Dall sheep, and other large mammals in Alaska.[15]

Olaus became influential in national conservation policy, leading The Wilderness Society and Izaac Walton League, and advocating for various legislative initiatives—most notably the Wilderness Act of 1964. Adolph focused on predator-related issues at Mount McKinley and other national parks, guiding the National Park Service toward what became the model for predator management in all natural areas of the National Park System.[16] Adolph was drawn into wide-ranging policy discussions of resource management policy in the National Park System; his correspondence records his positions on predator control, fire management, hunting, mining, water impoundment, use of insecticides, use of aircraft, use of snowmobiles, and interpretive signs.[17] As much as any other single person inside or outside government, he influenced what natural areas in the National Park System are like today.

FIG. 3.2 Olaus and Adolph Murie, ca. 1960.

CONTRIBUTIONS BY INDIVIDUALS AND INSTITUTIONS

One of the University of Kansas graduate students who made especially important contributions was Charles A. Long (fig. 3.3), who conducted fieldwork in Wyoming for his PhD dissertation, published in condensed form as *Mammals of Wyoming* in 1965.[18] He provided specimen-based distributional data on the 99 species that he recognized as native to Wyoming, and some morphological data.[19] Long was involved with mammalian taxonomy and distributions throughout his career, most of which was spent at the University of Wisconsin-Stevens Point.[20] Susan G. Clark, who published under the name T. W. Clark during the period 1968–2005, authored at least 30 journal articles on Wyoming mammal distributions and ecology during 1970–1995.[21] Clark and Stromberg's (1987) mammal fauna has served well as a guide to identification and distribution, although it was not specimen-based, and relied heavily on locality data presented by Long (1965).[22]

In addition to the various government agencies and public institutions that have advanced the study of Wyoming mammals, other entities should be mentioned here. One is the Jackson Hole Biological Research Station, now the Jackson Lake Research Station, a facility of the University of Wyoming–National Park Service Research Center.[23] Over its

FIG. 3.3 Charles A. Long, author of the 1965 *Mammals of Wyoming.*

nearly 60-year history, the station has hosted, supported logistically, or helped to fund the fieldwork of many mammalian biologists at a time when support for this kind of work was scarce. Some of these scientists became leaders in their fields. Margaret Altmann, a pioneering ungulate ethologist, worked from the station during summers of the 1950s,[24] as did the Craighead brothers (John and Frank), who began studies of various vertebrate species. Some of the first surveys of the mammals of Jackson Hole were completed there by James S. Findley and coworkers.[25] Kenneth Armitage conducted some of his earliest fieldwork from the station on a species that became a lifelong focus: the yellow-bellied marmot.[26] Douglas Houston used the station while studying moose for his dissertation research in the 1960s.[27] And William Calder used his brief time at the station to conduct his landmark study on the thermal physiology and diving endurance of one of the smallest diving mammals, the Pacific water shrew.[28] This support by the station for mammal researchers continues today.

The Nature Conservancy has become an important nongovernmental contributor to the conservation of biological diversity worldwide, and in Wyoming. Founded as a special committee of the Ecological Society of America, it has played key roles in the development of the Natural Heritage Network, which now has programs in each of the 50 states, including the Wyoming Natural Diversity Database (WYNDD). Natural heritage programs monitor and archive data on species of emerging conservation concern. The Nature Conservancy contributed to the first natural heritage program in Wyoming in the late 1970s, and WYNDD became affiliated with the University of Wyoming—shedding its ties to The Nature Conservancy—in 1998. WYNDD is now the lead state agency for monitoring biological diversity. Many of the recent distributional data on shrew, bat, and jumping mouse species reported here were collected or archived by WYNDD.

Most recently, the Robert and Carol Berry Center for Biodiversity Conservation made its appearance on the Wyoming conservation scene. Opened at the University of Wyoming in 2011, it provided new space, a curator, and a full-time collections manager for the University of Wyoming Museum of Vertebrates (UWMV). This enhancement in collection space and management represents a major advance for the inventory, archiving, and study of mammalian diversity in the region. These and other developments set the stage for a more productive phase of study of Wyoming mammals and other kinds of biodiversity. The Biodiversity Institute will provide resources and leadership for mammalogical study for decades to come.

Zoogeography

Zoogeography is a biological discipline that dates from the Age of Exploration and is concerned with the distributions of taxonomic groups of animals, and what accounts for them. This field of study lies at the intersection of ecology, physical geography, evolution, and population genetics, and focuses on such topics as how well a species of animal disperses, and what kinds of land cover types and geographic features serve as corridors for and barriers to dispersal. Dispersal—the movement of young animals away from where they were born—is a primary consideration in zoogeography, because it is the mechanism by which species expand their geographic ranges and attain the distributions we observe. Zoogeographers tend to understand and depict distributions and their dynamics with maps; they deal with species or population distributions at large scales, rather than the small ones at which individual animals live and die. Charles Darwin was keenly interested in how mammals varied in size and shape across geographic space, as well as through geological time, and used his observations in South America on these patterns to frame his theory of modification with descent.

Systematics is concerned with the evolutionary relationships among kinds of organisms, and the naming of various levels of organization, for example, genera, species, and subspecies—collectively called taxa—according to these understandings. Systematic arrangements of organisms are nested groupings of taxa that mimic their evolutionary histories, with the most closely related taxa most closely positioned on a phylogenetic tree. Zoogeography and animal systematics are closely linked disciplines, because most of the evolutionary variation that we observe originates through geographic separation and genetic differentiation. New lineages of mammals tend to form when geographic barriers interpose between populations, time passes, and the isolated populations become sufficiently

TABLE 2 Factors commonly found to influence dispersal ability among mammal species.

Factor	Weak dispersers	Strong dispersers
Body size	Small	Large
Habitat associations	Specialized	Generalized
Mode of locomotion	Burrowers	Runners, fliers

different—via natural selection or other genetic processes—that they are distinguishable by their phenotypic traits or their DNA. If they diverge sufficiently that, when brought back into reproductive contact, they cannot produce viable and fertile hybrids, then we consider them to have become true biological species. The development of molecular methods, especially DNA markers, has greatly improved our ability to distinguish between lineages of mammals that formerly were evaluated based on their size or shape when alive or when their skeletons had been cleaned. DNA analysis now allows us to infer how recently two or more populations have undergone genetic exchange.

These patterns of genetic divergence or similarity across the landscape are called population genetic structure. A species with strong dispersal abilities, often a habitat generalist, will show little or no genetic structure across an area the size of Wyoming, or perhaps across a much larger geographic range. Examples of such species in Wyoming that readily share genes across long distances of multiple habitat types are the coyote and mountain lion.[1] By contrast, habitat specialists disperse weakly through unsuitable habitats; they tend to remain in a narrow range of environments. These specialists typically show genetic discontinuities at barriers to dispersal, reflecting infrequent or no genetic exchange across such barriers for many generations. Examples of such species from Wyoming include pocket gophers and the American black bear. Pocket gophers must disperse to new areas by tunneling through soil, a slow and energetically costly mode of movement, and bedrock near the surface can completely block dispersal between populations. The black bear has the physical ability to traverse nonforested lands below the elevational limit of trees, but was prevented from doing so over the millennia by the presence of grizzly bears—a dangerous competitor—in habitats away from trees.[2] The factors commonly found to account for dispersal ability in mammal species are summarized in Table 2.

Most species show population genetic structure of a type that is termed isolation-by-distance: their dispersal abilities are intermediate, relative to the permeability of their habitat to movement. These species tend to show progressive genetic differentiation across landscapes that they have occupied for long periods—the farther apart two populations are, the greater the genetic dissimilarity between them. In the case of mammals, humans sometimes facilitate dispersal by translocating animals to establish or augment populations. Such translocations can save a species or population from extinction or help to recolonize formerly occupied geographic range. Examples from Wyoming include the wapiti, bighorn sheep, and black-footed ferret. Humans also interfere with dispersal by constructing barriers to movement, for example, interstate highways.[3] So, human influences are also evident in the population structure of mammal populations, including in Wyoming.

During the late nineteenth and early to mid-twentieth centuries, a major thrust of mammalian systematics was the assigning of new species and subspecies names to populations on the basis of morphological variation, geographic location, or both. During that period, efforts to identify new kinds of mammals took rather extreme forms; for example, mammalogists named over 225 subspecies of the southern pocket gopher (*Thomomys umbrinus*) from central Mexico to the northern Rockies, many of them parapatric populations with no apparent barriers to dispersal between the putative subspecies.[4] This species is now regarded as "overrepresented at the subspecific level."[5] At the species level, a frequently cited example of excessive reliance on morphology in taxonomy is that of the North American brown-grizzly bear. This lineage displays unusually high variation in morphology of the skull, not only among geographic regions, but also among individuals within a region. As a result, many species and subspecies were named during the peak of taxonomic "splitting." C. Hart Merriam concluded that there were at least 86 "forms" of brown-grizzly bears in North America, and that several species could be found in the same place.[6] Over the following decades various biologists expressed reservations about such a taxonomic scheme, until in 1953 Robert Rausch articulated the current view that only one brown-grizzly bear species occupied North America, and that it was the same species as the Old World brown bear, *Ursus arctos*.[7] Even until the 1980s, Merriam's view of brown bear systematics lingered; 77 species were recognized and mapped in 1981; in the Greater Yellowstone Area alone, five species were represented as occurring.[4] Clearly, the morphological species concept used by Merriam was not consistent with the modern concept of biological species of animals that had been articulated by Ernst Mayr beginning in the 1940s.[8] Mayr proposed that biological species were populations that were reproductively isolated from other populations, not simply different in appearance. More recent molecular studies have confirmed that skull morphology is a poor basis for inferring biological species or genetic interchange across brown bear populations.[9]

So, the trend toward splitting species based on morphology had run its course by around 1940, and the adoption of Mayr's biological species concept brought about much consolidation of species during the period 1950–1990. Since then a new countertrend has developed: splitting species based on molecular evidence for population isolation.[10] Several of the new species names introduced since the 1965 and 1987 faunas, including *Perognathus mollipilosus* and *Martes caurina*, reflect this new use of DNA evidence to infer past population interbreeding, and name new species that are scarcely recognizable from their morphology. A further new use for molecular evidence is to infer phylogenetic placement; several new genus names (e.g., *Perimyotis, Callospermophilus*) presented here reflect this important new tool in the study of mammalian evolution.

MAPPING THE DISTRIBUTIONS OF WYOMING MAMMALS

Drawing distribution maps requires that animals of known taxonomy be associated with specific places. This was done by early mammalogists by trapping or shooting mostly small mammals, recording their locations relative to geographic features, and placing the preserved specimens in museums where others could examine them. Large mammals that were observed were recorded in journals, but seldom preserved. Medium-sized mammals—in the size range of raccoons to porcupines—likewise tended not to be sampled by collection, because they were time-consuming to prepare and bulky to transport. Bats began to be sampled alive when mist nets became available in the mid-twentieth century, and more recent electronic devices allow field biologists to hear and record bat biosonar vocalizations. Such ultrasonic vocalizations are useful for identifying some species, but not others, without capture. Here, I have not used acoustic data in preparing the distribution maps for bats, except for the hoary bat and spotted bat, which emit distinctive vocalizations. An important philosophical dichotomy in modern mammalogy involves whether to collect and preserve skins and skulls of mammals. One school of thought holds that such collecting is unnecessary, inhumane, and contributes little information that cannot be collected in less invasive ways. Another view holds that preserved specimens of whole animals are critical to our understanding of taxonomy and zoogeography, and that collecting mammals for this purpose in almost no case represents a population-level impact on the species studied.[11]

Methods began to be developed for single mammalian species or guilds in the second half of the twentieth century, and one of the first species to receive special attention in Wyoming was the black-footed ferret. When a remnant population was discovered in South Dakota in the late 1960s, the ferret required new methods for surveying and capturing: nighttime spotlighting and capture in traps inserted into prairie dog burrows (fig. 4.1).[12] By 1980, survey methods for pygmy rabbits were being developed to take advantage of their distinctive habitats, burrows, and fecal pellets (spherical and small, in piles or windrows). Additional survey methods developed in the last three decades include remote cameras (camera traps), which have become common for surveys of rare or elusive carnivores. Camera traps have been crucial to our improved understanding of the distribution of swift foxes in Wyoming (fig. 4.2) and occasionally record images of species unexpected at a site. Radio telemetry is widely used in habitat and movement studies, but also occasionally provides information on geographic range, for example, when an animal disperses widely. Our understandings of the distributions of wolves and wolverines in the region has been greatly enhanced by telemetry.[13]

One of the fastest-growing methods to understand the distributions of mammals is the use of molecular markers. These methods do not require that the animal be killed, handled, or even observed to collect its DNA.[14] Rather, DNA can be extracted from fecal droppings, or from hairs plucked from the animal by "hair snares." From these snagging devices, DNA is extracted from hair roots and then analyzed for a marker that indicates the species, and in

FIG. 4.1 A black-footed ferret is illuminated by a spotlight, a method developed in South Dakota, and used extensively in Wyoming. The eyeshine is visible for up to 200 m (660 ft) at night.

FIG. 4.2 A swift fox photographed by a camera trap near Rock River, recognizable by its black-tipped tail, black mark on snout, and delicate build. The post holds a chemical attractant; the fox shows facial rubbing behavior typical at such posts.

some cases the individual identity of the animal. Of course, DNA can also be analyzed from carcasses, bones, or stored tissue samples, including specimens stored in museums that are centuries old. Molecular methods will continue to be powerful tools for mammalogists, and now enhance the utility of skins and skulls in museum collections that were collected to be able to measure size, shape, and color.

Depicting the distributions of rare or easily confused species poses special problems. In some cases evidence of occurrence is not disputed, because a species is reported from within its known geographic range, the species is easily detected and identified, physical evidence is gathered, and no other similar species occurs nearby. Complications occur, however, when a species is very rare, difficult to detect, and easily confused with other, more common species, or if no physical evidence is collected. In some cases observations of rare species are reported, even published, by experienced biologists without confirming physical evidence. These reports can then be disputed. This problem can be serious in cases of extremely rare species, for which a single credible observation can have implications for policy and conservation action. This was the case for the reports of an ivory-billed woodpecker in Arkansas in 2004.[15] To evaluate the reliability of evidence on the occurrence of animals, criteria have been proposed by Kevin McKelvey and colleagues. They considered the ranking of evidence, from least to most reliable, to be as follows:

1. visual observations of organism's sign,
2. visual or auditory observations,
3. photographic or recorded auditory evidence lacking diagnostic characteristics,
4. diagnostic track impressions,
5. diagnostic recorded auditory evidence,
6. photographic evidence showing diagnostic characteristics,
7. diagnostic DNA evidence,
8. live or dead specimen (in the possession of someone qualified to identify it).

The use of these criteria does not guarantee that an observation is correctly identified to species, but it is a standard to which mammalogists can refer in considering reports of rare mammals, and I use it in depicting distributions in this book.[16]

DEPICTING MAMMAL DISTRIBUTIONS

Another problematic area for zoogeographers is discerning between the expansions of the geographic range of a species versus a perceived expansion as a result of improved information about distribution. These differences between a perceived and a real distributional expansion can be detected statistically in some cases, but in most cases leave to wonder whether the species was long present, but undetected, or has newly arrived.[17] Likewise, more detailed knowledge of a mammal's actual distribution can result in an apparent distributional contraction. An example of the latter is the Pacific marten, which was mapped by Hall and Kelson (1959), and repeated by Long (1965) as occurring across the western half of Wyoming (Map 4.1), and in the Medicine Bow Mountains.[18] Today we know that the species occurs only in forested areas of western, southern, and northern Wyoming; the reduced area

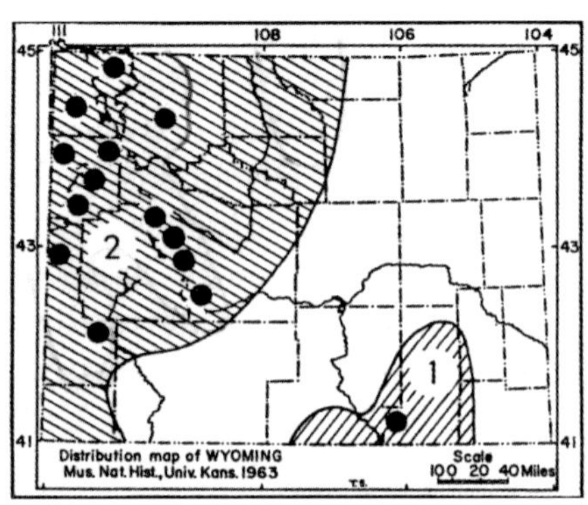

MAP 4.1 The geographic range of American marten depicted by Long (1965), and followed closely by Clark and Stromberg (1987). It is based on the continental-scale distribution map of Hall and Kelson (1959) and makes little reference to the locations of suitable habitats in Wyoming.

of the mapped range in this book is due to greater accuracy in the depiction of sites occupied, rather than any range decline at the scale of the map. Many older distribution maps should be understood to indicate the area within which the species occurs within suitable habitats, rather than the habitats actually occupied. The third kind of problem involving depicting distributions is that very common species can disappear from fairly large areas without the loss being noticed. For example, for ground squirrels, cottontails, jackrabbits, and many species of mice and shrews, substantial reductions in distribution can occur undetected, so easily are the species overlooked.

ZOOGEOGRAPHIC REGIONS OF WYOMING

Wyoming is a meeting point for several diverse biotic provinces: eastern North American, Great Plains, Colorado Plateau, Great Basin, Cordilleran, boreal forest, and alpine tundra habitats all extend to within the state borders.[19] Wyoming's mammal fauna contains representatives of all these provinces, and their current distributions reflect earlier distributions and dispersal abilities. Some species associated with the Great Plains—the American perimyotis, Nebraska pocket gopher, and eastern cottontail—extend westward to the high plains and Front Range and advance no farther. Mammals associated with the Colorado Plateau, some of them originating from the Chihuahuan Desert of northern Mexico, colonize from the south, several species specifically along the Green River corridor. A few species—Yuma myotis, cliff chipmunk, pinyon mouse, canyon mouse, and ringtail—reach their northern limits in the interior portion of their geographic range near the towns of Green River or Rock Springs. Boreal forest species occur on forested mountains today, but are remnants of peninsular Pleistocene forests that extended much farther south and to lower elevations than today. For these species, including the snowshoe hare, red squirrel, northern flying squirrel, southern red-backed vole, Pacific marten, and Canada lynx, their current distributions on mountains are relicts of the contraction and natural isolation of Late Pleistocene geographic ranges. Two montane species specifically associated with the Cordillera likewise arrived here during the late Pleistocene and are now isolated from the Canadian cores of the species' distributions: American pika and bighorn sheep. Great Basin mammals have pushed into Wyoming from the west, some of them to a point east of the eastern boundary of the Great Basin itself; these species include the pygmy rabbit, bushy-tailed woodrat, and sagebrush vole.[20] Thus, a map of zoogeographic regions of Wyoming would not show discrete fingers that extend from the periphery, but a set of overlapping, or isolated, or ribbon-like zones that would show varying degrees of connectivity with the region of their origin.

Two kinds of corridors are of particular interest to mammalian zoogeographers of Wyoming. The first is riparian zones that follow rivers into and across the state. These riverine corridors and their trees, shrub meadows, and mesic herbaceous vegetation provide movement corridors for many mammalian species between basins, between mountain ranges, and into Wyoming from the south and east. Examples of species that have used riparian corridors for dispersal include the Virginia opossum, red squirrel, Nebraska pocket gopher, and northern raccoon. The second kind of dispersal corridor is coniferous forest. Although both Wyoming and Montana help form the spine of the Cordillera, both states also have isolated or peninsular coniferous forests surrounded by a matrix of shrub-steppe. Boreal forest species such as the red squirrel, southern red-backed vole and Pacific marten use—now or in the past—mountain topography or vegetation to colonize new areas. However, how mammalian species perceive montane corridors is not the same for all species. For example, the Bighorn Mountains seem to be an extension of the main Cordillera from the perspective of bighorn sheep, but are more closely allied with the Black Hills when considering genetic structure among red squirrel populations.[21] Riparian zones have provided connections between forested mountain ranges through the millennia; these ribbons of forest likely were wider and more easily navigated by forest mammals during the Late Pleistocene than today. Thus, the Wind/Bighorn River provided a forested corridor from the northwestern mountains to the Bighorns, the North Platte and Laramie Rivers connected the Medicine Bow and Laramie Mountains, and the Cheyenne River provided an avenue between the Black Hills to the Hat Creek Breaks. The Hat Creek Breaks, between the Niobrara and Cheyenne Rivers, represents a ribbon of coniferous forest extending from Pine Ridge of southern South Dakota westward to western Niobrara County. As a result of such connections as these, the forests today restricted to mountain ranges were, during past periods of cooler temperatures, broader and more contiguous corridors for mammalian dispersal.[22]

The importance of forested corridors to the zoogeography of montane mammals of the Colorado Rocky Mountains was recognized over 50 years ago by Findley and Anderson (1956). They identified species that—to varying degrees—had been able to colonize Colorado from forested habitats to the north.[23] The principal barrier to dispersal of mammals southward along the Cordillera at temperate latitudes has been the Wyoming Basin, an arid, treeless steppe.[24] Some northern species that likely could thrive in Colorado forests, but have been unable to traverse the Wyoming Basin from the north include the yellow-pine chipmunk and American water vole. The moose crossed the barrier only with the assistance of wildlife managers. Other species occur both north and south of the Wyoming Basin, some as separate, others as the same nominal subspecies: masked shrew, yellow-bellied marmot, golden-mantled ground squirrel, red squirrel, North American porcupine, and Pacific marten.

The importance of the large, arid center of Wyoming as a zoogeographical barrier that affects species differently is illustrated by two large carnivores. The mountain lion and black bear both occur widely across the state, and both species are known to occasionally cross wide unforested areas. Has this always been the case, so that populations among the various mountain ranges are closely related? Charles Anderson and colleagues tested this hypothesis for mountain lions, sampling animals from five locations surrounding the unforested center of the state. They found that mountain lions in Wyoming formed a single interbreeding population, with little evidence that the Wyoming Basin had been any barrier to movement over millennial time periods.[25] By contrast, Michael Cachelin and David McDonald found strong genetic structure in American black bears likewise sampled from the major mountain ranges ringing the state. They found large genetic distances between these black bear populations; those across the gap between the Teton and Bighorn Mountain (275 km [171 mi]), and between the Bighorn and Medicine Bow Mountains (220 km [136 mi]) were around 10 times the values observed for black bears across comparable distances in continuous forest of Canada.[26] What could have deterred black bears from readily crossing broad shrub-steppe areas between mountain ranges of Wyoming over the millennia? Cachelin and McDonald hypothesized that the commonness of grizzly bears in treeless shrub-steppe before Europeans arrived posed a deadly threat to black bears that might try to disperse among forested mountain ranges. It is well known that American black bears restrict themselves to forested habitats in regions where grizzly bears are common, and risk being killed by grizzlies when they venture away from trees that they can climb. This interpretation is consistent with what is known about presettlement conditions, and with the behaviors of black and grizzly bears. Here, evidence from DNA sheds light on how two species have perceived corridors and barriers to dispersal differently in the prehistoric past.

Another species that displays the genetic evidence of its past dispersal patterns is the American water vole. This remarkable species lives along forest streams of the Cascades and Rocky Mountains to as far south as southern Utah, using mountain streams and rivulets to traverse its home range. Working in northern Wyoming and adjacent Montana, Marion Klaus posed the question whether water voles were more genetically connected by land or by water. That is, are water voles at a site most closely related to animals close to them by land distance, or by stream distance? Klaus and her coworkers found that water voles were very effective at traversing upland areas to disperse among streams; distance by water was not a strong predictor of genetic distance among populations.[27]

One last example illustrates the disparity between how we might think mammals disperse based on their natural history and how their DNA tells us they actually travel. The pygmy rabbit is a small, burrowing dweller of sagebrush steppe. Distributed across the Great Basin eastward as far west as California, the species has long been described as closely associated with tall, old big sagebrush.[28] Because pygmy rabbits and the tall, old stands of big sagebrush that they occupy are so discontinuously distributed across southern Wyoming, it was long assumed that they would exhibit strong genetic structuring among sites; their genes were expected to show evidence of isolation among many small subpopulations. But when Amanda Thimmayya analyzed pygmy rabbit DNA from the Wyoming part of the geographic range, she found almost no genetic structure; pygmy rabbit subpopulations were almost identical over hundreds of miles of distance. So, pygmy rabbits have been able to disperse frequently among many sites in Wyoming over the millennia, and show no particular evidence of isolation.[29]

Prehistory

The earliest mammals are defined as those having most of a set of skeletal features now considered mammalian. Principal among these is a jaw joint composed of the dentary and squamosal bones, as opposed to various other joint components of our reptilian ancestors.[1] The earliest-known radiation of recognizable mammals took place in the Late Triassic to Early Jurassic Period (fig. 5.1), around 200 Ma (1 Ma = one million years or one million years ago, depending on context). One of the earliest branches in mammalian phylogeny was that forming the two major lineages represented in modern mammals: one lineage gave rise to modern monotremes, including the duck-billed platypus, and the other led to the lineage comprising modern marsupials and placental mammals. Most members of both lineages, and some other side branches as well, went extinct; modern mammals are the survivors of over 180 Ma of diversification and extinction.[2] From the time of their origins until the end of the Mesozoic Era, mammals lived in the ecological shadow of the much larger-bodied dinosaurs. Dinosaurs are known from their fossils to have been morphologically and ecologically diverse; they included herbivores as well as aggressive predators. The latter are generally thought to be the reason that mammals remained mostly small-bodied and nocturnal for two-thirds of their evolutionary history. Mammals also were mostly uniform in their body plan—the shape of today's rats, although considerable morphological diversity has been discovered in the last 20 years. Species have been found that were predators of vertebrates, semiaquatic, or as large as modern badgers.[3]

The mammalian fossil record from the Mesozoic is highly incomplete, so to infer patterns of evolution over this era we must resort to the known fossil sites, wherever they may be. Mammal fossils from Wyoming have played a disproportionately large role in the reconstruction of mammalian history, including from the Mesozoic. For example, marsupials

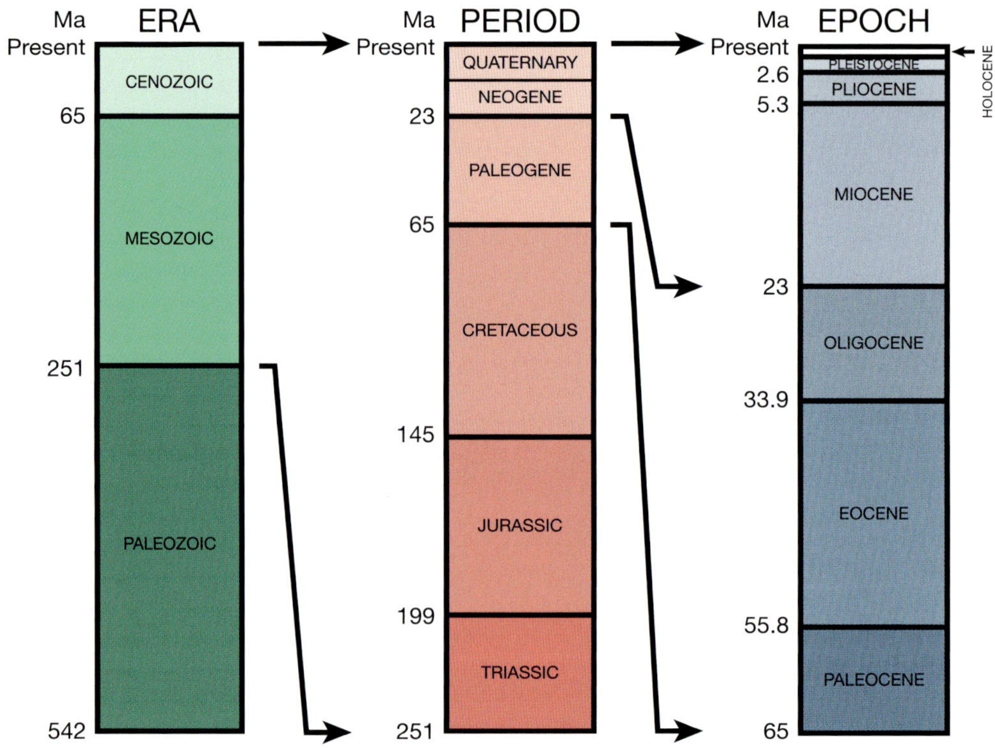

FIG. 5.1 The geologic timescale used in this book. The Holocene is generally assumed to have begun around 10,000 years ago.

were the first mammals described from the Cretaceous, discovered in the Lance Formation of Wyoming by Edward D. Cope in 1892.[4] Wyoming's mammalian record since then has been extended back into the Jurassic and the area around Medicine Bow, including at Como Bluff Quarry 9, holds the richest concentration of Late Jurassic mammal assemblages in the world.[1] The earliest true placental mammals have been found in deposits from the Early Cretaceous, around 125 Ma; placentals remained uncommon in the North American fossil record until around 84 Ma. Late Cretaceous mammalian faunas from North America usually contain many multituberculates and marsupials. The former represented a group of rodent-like mammals that first appeared in the fossil record around 155 Ma, and persisted until the earliest parts of the Oligocene Epoch, around 35 Ma. Mammals from the latest Cretaceous are well represented in Wyoming fossils, with documented sites in eastern Niobrara County, the northern Bighorn Basin, the Rock Springs Uplift, and Hanna Basin.[1] These fossils are almost entirely teeth and incomplete jawbones.

One geological process that was central in Wyoming's mammalian prehistory was a mountain-building episode—the Laramide Orogeny—that began during the late Cretaceous, by around 75 Ma. This period of uplifting continued until at least 45 Ma and created most of the major topography of the modern Rocky Mountains, including almost all of

Wyoming's mountains and basins. The volcanic Absarokas and the Teton Mountains are exceptions, having been formed more recently. Within Wyoming, from the Wind River Mountains to the Black Hills and from the Bighorn Basin to the Laramie Basin, the Laramide Orogeny accounted for the mountain-basin topography that accounts for so much of Wyoming mammalian diversity today. This prolonged period of mountain raising also had the effect of finally draining the Western Interior Seaway, although minor incursions of the seaway persisted in southern Wyoming until as late as the early Paleocene. Beginning late in the Eocene (around 38 Ma), after the end of the Laramide mountain raising, multiple explosive volcanic events shook the Rocky Mountains to the west. This volcanism continued for 30 Ma and deposited thousands of vertical feet of initially airborne clastic material into the basins and across the mountain slopes. The slow basin-filling process, in some cases, eventually overtopped the surrounding mountains, so that Wyoming's topographic relief was much reduced. Episodically, entire assemblages of land animals were suddenly buried under dense clouds of volcanic debris, and some of these events resulted in extraordinary fossil records of mammalian life, such as is seen in Badlands National Park of South Dakota. The Wind River and Bighorn Basins were mostly filled with this highly erodible air-deposited material, which protected the more durable rocky flanks of the mountains of Laramide origin. As a result, the Laramide mountains have persisted much longer than had they been exposed fully to erosive processes since their formation. Small pockets of volcanic material persist as remnants high on the slopes of some mountains, and larger patches are found in the basins, especially in the eastern part of the state.[5]

THE AGE OF MAMMALS

The Cenozoic Era has been called the Age of Mammals with good justification. Its beginning was violent, marked by a bolide impact in northern Yucatán, Mexico, around 66 Ma. This event, which marks the Cretaceous–Paleogene (K-Pg) boundary, occurred close to the time of the extinction of dinosaurs, although current evidence does not allow us to conclude that the two events coincided exactly.[6] Based on studies in Montana and other places, we know that most mammalian species became extinct then. But, importantly, some small-bodied and aquatic-dwelling mammalian lineages species persisted, and by 600,000–700,000 years after the event, mammalian diversity had expanded; in the absence of dinosaurs, a major radiation of mammalian groups was underway.[7] The Cenozoic became distinguished by the dominance of mammals over other land vertebrates in their numerical abundances, body sizes, morphological diversity, and ecological influence. Throughout the Paleocene and Early Eocene, mammalian orders continued to radiate at around the same rate as they had in the Late Cretaceous, so that by 50 Ma, all of the extant orders were leaving recognizable evidence in the fossil record.[8] At the level of genera and species, mammalian diversity expanded rapidly during the Paleocene and Eocene. And although mammals remained small-bodied throughout the Mesozoic, the average body size of the largest mammals began to increase early in the Cenozoic. While small mammals (10–100 g) continued to make up a large proportion of mammalian species, large-bodied mammals (>44 kg) became larger over the course of the Cenozoic, by a factor of 10–100.[9] By Oligocene time very large-bodied mammals were common, and the largest-bodied mammals approached the size of some large herbivorous dinosaurs of 50 Ma earlier. For example, *Paraceratherium* was a hornless rhino-like herbivore of Eurasia; some individuals are believed to have stood 8 m (26 ft) at the shoulder and weighed 17 metric tons (36,000 lb).[10] Mammals became ecologically dominant on land, accounting for far more biomass and energy consumption than other warm-blooded vertebrates. Even after major die-offs of large mammals on three continents in the Late Pleistocene, mammals of all body sizes today are 4–100 times more numerically abundant across the land surface of the world than are birds of the same size class.[11] In addition to attaining large body sizes and becoming numerically abundant within all body-size categories, mammals came to occupy a wide range of niches: burrowing, flying, swimming, running, grazing, browsing, and filter-feeding in the ocean. Most of these niches were filled by mammals by the Middle Eocene.

An especially important period for mammalian evolution was the Paleocene–Eocene Thermal Maximum (PETM), a period 55 Ma when mid-latitude atmospheric temperatures increased by 4–5°C (7–9°F). This warm period is believed to have had a sudden onset, and to have lasted for around 170,000 years.[12] The most detailed studies of the mammalian fossil communities of that time have been conducted on Polecat Bench in the northern Bighorn Basin. Coinciding with the onset of the PETM, three mammalian orders important today appeared synchronously: Artiodactyla, Perissodactyla, and Primates. That three major taxa could so suddenly and simultaneously appear in the fossil record has been a challenge to explain, but the fact that this was a period of rapid environmental change makes this synchronous appearance plausible. Noteworthy also is that the body sizes of at least some mammals became smaller during the PETM, in the case of one species of hoofed mammal, *Ectocion parvus*, by 50%. When the PETM ended and atmospheric temperatures returned to their long-term average, the body size of *Ectocion* returned to its long-term average as well.[13] An inverse relationship between environmental temperature and body size of mammals is predicted from physiological principles, expressed as Bergmann's rule. However, it is plausible that the changes in body size were facilitated or caused by dispersal of *Ectocion*, particularly across latitudes, so that smaller-bodied populations shifted their geographic ranges in response to climate change.

The Cenozoic was an era of gradual temperature decline, punctuated by a major cooling event in the earliest Oligocene, around 33 Ma. This event lasted around 400,000 years, and saw ambient mean temperatures in mid-continental North America drop by around 8°C (14°F). Vegetation in what is now Wyoming shifted from dense forests to open wooded grasslands similar to those of today. Remarkably, the mammalian assemblage was little affected by this major environmental shift; most species studied from the badlands of South Dakota show little or no change in size over the course of the event.[14] Much of the remainder of Cenozoic mammalian evolution in Wyoming is obscured by the dearth of surface deposits representing the Oligocene, Miocene, and Pliocene Epochs (Map 5.1), due to the erosion of

MAP 5.1 Surface formations of Wyoming by geological age, from the Precambrian to the Quaternary. Deposits of Paleocene and Eocene ages are widespread (37% of area), fossil rich, lightly vegetated, and renowned for their contributions to the study of mammalian evolution. Those of Oligocene to Pliocene age account for only around 12% of Wyoming's surface, and are relatively fossil poor. Source: Gingerich (2003, 2006).

the clastic volcanic deposits of that part of Wyoming's history. Those epochs, representing nearly half of the last 66 million years, are represented by only 11% of Wyoming's surface geological formations today, and not highly fossil-rich ones at that. By contrast, the Paleocene and Eocene periods represent a comparable time span earlier in the Cenozoic, but deposits of those ages cover over 36% of Wyoming's area, and include some of the richest fossil-bearing deposits per unit of land area of these ages in North America.[15] As a result, we know more about mammalian diversity in Wyoming during the time 66–32 Ma than that 32–2.6 Ma.

THE PLEISTOCENE EPOCH

The beginning of the Pleistocene Epoch (2.6 Ma) is a useful point of reference for studying modern Wyoming mammals, for several reasons. First, the geographic setting (latitude, longitude, elevation) of Wyoming had converged on that of today. Second, most of the mammalian genera of modern North America were present on the continent by then, although perhaps distributed differently. A number of other genera were in Eurasia or northern South

America, about to cross over the respective land bridges from those continents into North America. Third, the climate and mammalian fauna were beginning to undergo a dynamic period; the Pleistocene, in contrast with other late Cenozoic epochs, was a time of high turnover rates for mammals, meaning that extinctions and appearances of species in the fossil record occurred at higher rates than for other vertebrate groups, and higher than for mammals during the Cenozoic before the Quaternary.[16]

The agents of this dramatic change included two routes for immigration: the Panamanian Isthmus, connecting North and South America, and the Bering Land Bridge, connecting Asia to northern North America. The former became a dry route in a complex pattern, with the earliest evidence of a land connection in the middle Miocene. Mammalian colonizations across the isthmus become apparent at 3 Ma.[17] The joining of the land connection between the two Americas was the result of tens of millions of years of northward shifting of the tectonic plate on which South America rests, uplifting of the seafloor in the area between the continents, stimulating volcanism, and resulting sea level fluctuations. Climate was also a significant factor in causing the Panamanian Isthmus intermittently to have more grasslands and less forest than today, allowing the passage of steppe-adapted taxa, such as camels.[16] Biologically, the land connection allow dispersal of plants and animals between the two American continents and accounted for substantial changes in the North American mammal fauna; it caused even greater changes in South America. This exchange followed nearly 90 Ma of near-total isolation from other continents of the South American land flora and fauna. By early in the Miocene Epoch, the southern continent had evolved a diverse set of endemic mammalian orders, some of which seem curious today: marsupials were morphologically and ecologically diverse, and included large-bodied, saber-toothed predators. Another group, the xenarthrans, represented by modern sloths and armadillos, included large-bodied herbivores with elaborate armored plates covering them from head to tail. Ungulate orders now extinct—Litopterna and Astrapotheria—were widespread and varied. Mammalian species dispersed in both directions, from the north and south, but the salient effect of GABI for western North America at Wyoming's latitude was to allow several species of South American lineages to arrive here over the last few million years. Some went extinct later in the Pleistocene, as is discussed later, but the nine-banded armadillo still occupies the southern tier of the United States today, and the Virginia opossum and North American porcupine—both descendants of South American migrants—occur in Wyoming. Some lineages are understood to have migrated to South America from the north, undergone adaptive change during the 1–2 million years, and then returned to North America in their respective australly modified forms; the long-tailed weasel and mountain lion are such species.[18]

The most recent avenue to enrich the North American and Wyoming mammalian faunas was the Bering Land Bridge, which opened and closed episodically during the Pliocene and Pleistocene coincident with major continental glaciations. The Bering Sea was shallow throughout that time and during the most extreme glacial advances sufficient water was bound in continental ice sheets that Beringia became dry land. At its most extensive, the Bering Land Bridge measured nearly 1000 km (600 mi) from north to south, and was vegetated with tundra or steppe vegetation. These land connections have been found to have occurred as early as 7.4 Ma; the most recent glacial peak was 20,000–18,000 years ago and the last emergence of a watery Bering Strait occurred around 11,000 years ago.[19] During these periods of bridging, mammal lineages passed between Eurasia and North America; large mammalian species in Wyoming today that arrived via Beringia during the Pleistocene include the grizzly bear, wapiti, both deer species, moose, bison, and bighorn sheep.[20] The passage from Eurasia to temperate North America was complicated by the fact that at the time of the greatest drops in sea level, all of Canada was covered by an ice sheet over 3000 m (9900 ft) thick in places. This ice sheet blocked animal movements by land from Beringia to temperate North America. An ice-free refugium in Interior Alaska, contiguous with the Bering Land Bridge, was steppe-covered through even the most extreme glacial events, so that mammals arriving from Siberia persisted in Alaska until the ice-free corridor opened and movement southward became possible, after the Bering Land Bridge had been inundated. Thus, dispersals between the continents in either direction were achieved in pulses, separated by glacial and interglacial periods.

At around 3 Ma, late in the Pliocene, 18 genera arrived in North America from Eurasia and South America.[15] At around 2.5 Ma, in the earliest Pleistocene, 11 more genera arrived via the Bering Land Bridge alone, including jackrabbits, two kinds of bog lemming, and now-extinct types of Bovidae, Cervidae, a lesser panda, the flying squirrel, and a muskrat. Around the same time, from South America, 17 genera immigrated northward, including the ground sloth *Glossotherium*, several armadillo-like mammals, a now-extinct capybara, and the earliest North American porcupine (*Erethizon*). No other period during the last 40 Ma of North American prehistory records such a surge of mammalian immigration.[21] These immigrations were not, however, reciprocated by similar movements to neighboring continents: North America contributed only around 20% as many mammalian genera to Siberia as it received, whereas South America became flooded with 29 North American genera.[22] Beginning around 1.8 Ma, North America received further immigrants from South America in the form of a now-extinct ungulate the size of a rhinoceros, large and small-bodied sloths, giant anteaters (*Myrmecophaga*), and vampire bats. Immigrants from Siberia around that time included mammoths, two genera of muskox, caribou, the jaguar, voles of the genera *Microtus* and *Myodes*, lemmings, the modern gray wolf, the short-tailed weasel, and an otter of the genus *Lutra*. At around 240,000 years ago, a recognizable bison made its first appearance via Beringia.[23] As a result of these profound changes, North America, including Wyoming, became a zone of contact for diverse mammals recently arrived from a vast arc extending from South America to Siberia, as well as forms native to North America for millions of years.

The environment in Western Interior North America during the last glacial advance featured ambient summer temperatures 7.0–8.5°C (13–15°F) cooler than today; precipitation was lower as well.[24] The vegetation at 18,000 years ago is difficult to infer, because so little fossil pollen from Wyoming is available for times before 12,000 years ago. Probably, however, it included more sagebrush, grasslands, and conifers than today. Some areas of northwestern Wyoming were under the climatic influence of the large glacier that covered the Yellowstone Plateau.

Mammals of Wyoming and northern Colorado from the Late Pleistocene are known from several well-studied sites. Natural Trap Cave is a karst sinkhole in the northern Bighorn Mountains that has produced over 30,000 mammal specimens dated to as early as 110,000 years ago and as late as the early Holocene. The now-extinct mammals that have been found in Natural Trap Cave include the dire wolf (*Canis dirus*), American lion (*Panthera atrox*), American cheetah (*Miracinonyx trumani*), short-faced bear (*Arctodus simus*), a mammoth (*Mammuthus* sp.), four types of now-extinct horse (*Equus* spp.), American camel (*Camelops* sp.), woodland muskox (*Symbos cavifrons*), a bison (*Bison antiquus*), and a bighorn sheep (*Ovis catclawensis*). A few other species represented in cave specimens do not live near the cave today; noteworthy among them are the arctic hare (*Lepus arcticus*) and a collared lemming (*Dicrostonyx* sp.). Both of these taxa are known today as arctic mammals, limited to tundra regions thousands of kilometers north of Wyoming. Still other species detected in the deposits are found near the cave today: cottontails, jackrabbits, yellow-belled marmots, ground squirrels, and deer.[25] The apparent coexistence in the Late Pleistocene of mammal species—for example, jackrabbits occurring with collared lemmings—that are not sympatric anywhere today is consistent with the idea that climate change can result in environments for which there is no current analogue, and the mammal communities that occupied these environments likewise lack modern analogues. The observation of past no-analogue climates and communities suggests that mammal communities in Wyoming may in the future shift markedly in the face of our rapidly changing climate.[26]

Two sites farther south, Little Box Elder Cave on the north side of the Laramie Mountains and Chimney Rock animal trap in northernmost Colorado held many of the same and some additional Late Pleistocene mammalian species: red fox, striped skunk, a type of lynx, wapiti, and mountain goat.[27] The Colby site, near Ten Sleep, dates from the Clovis period—13,500–13,000 years ago—and while it did not hold a rich mammalian fauna, it yielded one especially important finding—a Columbian mammoth (*Mammuthus columbi*) in association with human projectiles and tools of Clovis design.[28] This mammoth occurrence is renowned, because it is the only Clovis-age kill site of mammoth from Wyoming, and one of only two ever found in North America.[29]

THE END OF THE PLEISTOCENE

The last major glaciation peaked around 20,000–18,000 years ago, when glacial ice covered North America down to around 38° N latitude in the central states, and sea level was around 130 m (430 ft) lower than today. Since then, continental and mountain glaciers have melted at a fairly steady rate until the last two centuries, when the rate of melting accelerated sharply. The Pleistocene ended formally around 10,000 years ago, but the ice-free corridor between Alaska and temperate North America opened around 1000 years later.[30]

It has been known for over a century that a number of mammalian lineages disappeared from North America around the time of the end of the last glaciation. But why did they do so, and exactly when? Debate has raged for decades over the timing and cause of the extinctions, with two main single-cause explanations in contention. The first, shifting climate, posited that the extinctions were not synchronous with each other or with the arrival of humans and that rapidly warming climate and its cascading effects could easily account for asynchronous disappearance of many herbivores and some predators.[31] The second explanation stated that the hunting behaviors of newly arrived humans overwhelmed large mammals that were naïve to human predators and stood unalarmed as hunters approached them and hurled their pointed weapons from short distances. According to this hypothesis, the large mammals and their obligate predators disappeared fairly quickly and immediately after humans attained population densities that could be ecologically influential. The hypothesis that human hunting had driven some Pleistocene large mammals to extinction had been implied in the eighteenth century, but was explicitly articulated and advocated in the nineteenth century. It began to receive increased empirical support in the late nineteenth and early twentieth centuries, but many still doubted the ability of humans to kill enough members of a species to cause its extinction.[32]

A crucial event in the development of the human-hunting hypothesis occurred in 1926, when an arroyo near Folsom, New Mexico, was excavated by a team led by Jesse Figgins, an archeologist affiliated with the Colorado Museum of Natural History (now the Denver Museum of Nature and Science). It was found to hold 23 individuals of a now-extinct species of bison that had been killed using a distinctive weapon that today carries the town's name: Folsom point. The site was excavated methodically and with many independent observers, leaving no doubt as to its authenticity. This effort left the clear impression with archeologists that humans had lived in North America at least as early as 9000–8000 years ago, and had hunted now-extinct large mammals with technology recognizably different from that used by later native peoples.[33] The Folsom Site presented the first credible evidence uncovered in North America of humans coexisting with extinct kinds of mammals, and the earliest dated evidence of humans in the New World. Within two decades, an even older site was excavated near Clovis, New Mexico. It produced another distinctive type of stone projectile—the Clovis point—that was now believed to represent the technology of the earliest Americans, a view that persisted for decades. Today, the arrival of the first humans in temperate North America is thought to have been around 15,000 years ago, while Clovis culture is limited to the period 13,500–13,000 years ago. Few cultural sites predate Clovis ones in temperate North America, but this may be because of low human population densities in interior areas until Clovis time.[34]

The chief advocate for the human-hunting explanation for the sudden disappearance of so many North American large mammals became Paul Martin of the University of Arizona, who over a 40-year career assembled the evidentiary pieces of what became known as the "overkill hypothesis":

1. Considered worldwide, the disappearance of large mammals and flightless birds coincided with the times and places of arrivals of humans. When humans appeared in new areas, vulnerable species of large mammals and flightless birds quickly disappeared.

2. Extinctions among small-bodied mammal species at the same places and times were much less common. Thus, mammals over 44 kg (100 lb) adult body weight were affected to a much greater degree than smaller mammals. This is thought to reflect the preferences by humans for large food animals, and the low population densities of those species, relative to small mammals.
3. No major extinction event has been observed in Africa coincident with the ones in the Americas, Australia, New Zealand, the Mediterranean islands, or Cuba-Hispaniola. Such a lack of African extinctions of large mammals is consistent with the fact that humans and large mammals coevolved there for millions of years; thus, the relationship between humans and large African mammals developed gradually, before 50,000 years ago.[35]

The debate over which narrative better accounted for the extinction event was prolonged and passionate, and came to hinge on evidence for when and how quickly the event occurred. The overkill hypothesis predicted a sudden event over just a couple of millennia, whereas the climate hypothesis predicted a more gradual or staged disappearance. Further, the climate hypothesis allowed for large mammalian extinctions before the arrival of humans, as each species in turn reached an environmental threshold below which it perished. Dating methods improved markedly through the years and by 2009 it could be demonstrated that most of the North American large mammal extinctions occurred over a 500-year period not long after the arrival of humans in sufficient numbers in the New World, when 30 genera of large mammals suddenly disappeared. South American extinctions lagged by several hundred years and affected an even larger proportion of the large mammal community.[36] These findings are consistent with the overkill hypothesis, but not the climate narrative. Whether these new, synchronous dates will lay to rest the debate about what caused the late Pleistocene extinction spasm is unclear, but it is certain that most genera of North American large mammals suddenly disappeared at about the same time, a few centuries after the first arrival of humans in numbers. This change vastly altered the appearance and functioning of mammal communities of Wyoming. In the 9000 or so years since the extinction spasm, the changes to the mammal assemblage have consisted mostly of rearrangements of geographic ranges in response to changing environments and effects attributable to humans since the arrival of Europeans.

Vegetation as Habitat

Habitat is one of the most important and pervasive concepts in mammalian ecology, with habitat use, preference, and quality frequently reported in ecological studies. It is also one of the most ambiguous concepts, subject to various definitions and interpretations.[1] Here, I use "habitat" to denote the more or less permanent features of the landscape that are associated with the presence, abundance, or fitness of a kind of animal. In referring to permanence, I mean that some environmental factors that can be important to mammals are not necessarily included beneath the rubric of "habitat." One such transient factor is disease, which can affect the distribution and abundance of a species strongly, but can increase in incidence within days or weeks, and disappear as quickly. Disease does not necessarily affect the physical properties or resources of an area of land, although it can be an important factor in distribution and abundance of animals. Likewise, predators and competitors of an animal species can influence its distribution or abundance, but are not generally considered a feature of habitat. For example, if wolves were to limit the distribution or abundance of a smaller carnivore, their influence would not usually be considered habitat-mediated. Habitat in the usual sense includes such attributes as climate, soil attributes, dominant plant species, successional or structural stage of vegetation, availability of free water, slope, aspect, and elevation—all fairly permanent features of the landscape, at least within a season. Examples of habitat attributes important to Wyoming mammals include soil type (e.g., pocket gophers), dominant plant species (e.g., Abert's squirrel), successional stage (e.g., most herbivores), amount and characteristics of coarse woody debris (e.g., Pacific marten), proximity of free water (e.g., moose), presence of roads or other cultural features (e.g., wolverine), and physical characteristics of snow (e.g., Canada lynx).

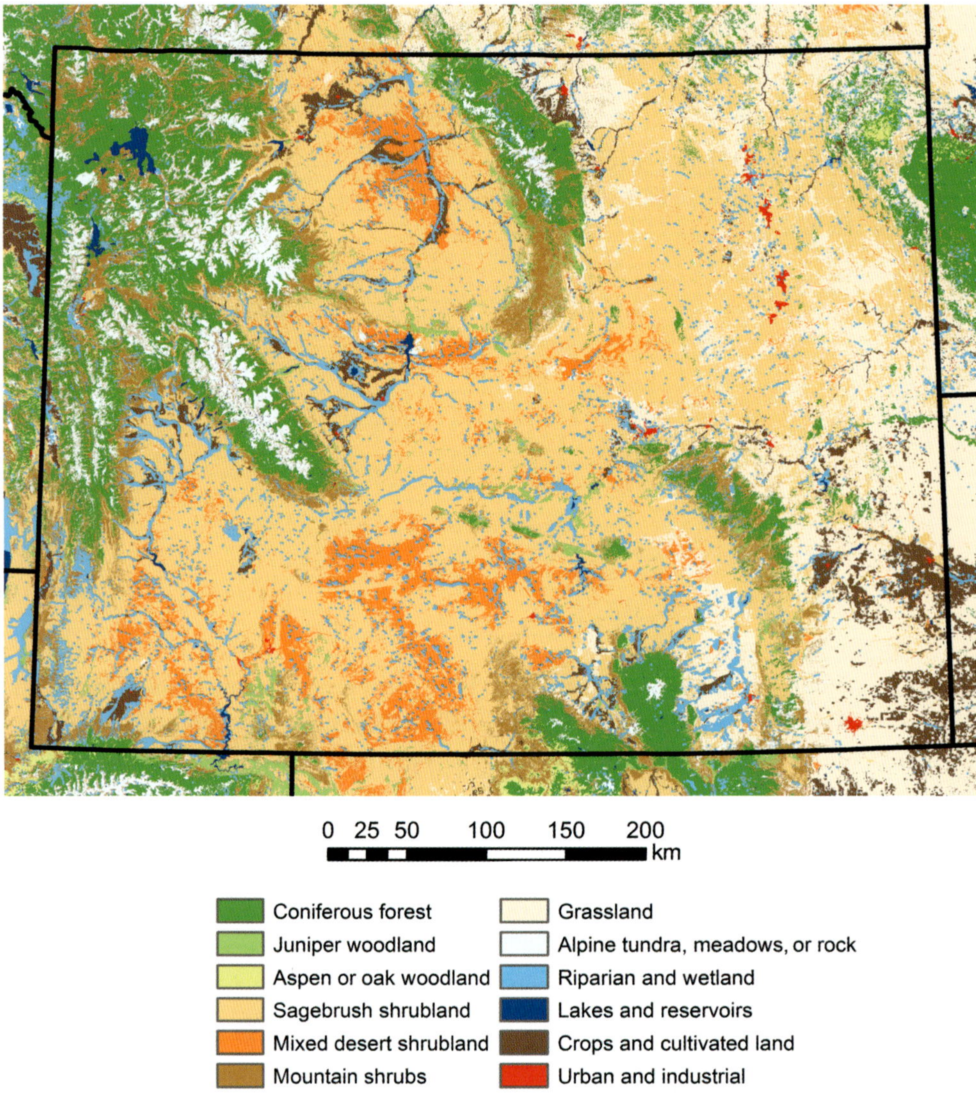

MAP 6.1 Simplified vegetation map of Wyoming and adjacent lands.

Mammal species differ in the specificity of their habitat associations. Habitat generalists have broad habitat tolerances, and are not limited by many naturally occurring features. The coyote is such a species. Other species, for example, the black-footed ferret, are highly sensitive to one or a few environmental attributes, and might not occur at all where an attribute is inconsistent with its needs. Some species—raccoons among them—are broadly tolerant of the presence of humans and human developments, using foods of human origin, and sheltering in structures created by humans. Others are true wilderness species, and tolerate little contact with humans or cultural features. Many of the species treated here perceive habitat at multiple spatial and temporal scales that are difficult for biologists to understand or measure. For example, it is difficult for biologists to explain the habitat factors that cause a species of shrew to be found in some sites, but not others. Scientists' senses and instruments do not function well at the scales at which a shrew lives out its daily life. Likewise, it has long been understood that wolverines are dwellers of wilderness, but the specific attributes of wilderness that are important to wolverine survival and reproduction remain puzzling. As a result of these limitations, the habitat associations of some mammalian species of Wyoming are largely unknown. It is known where the species occurs, but not why it occurs there.

Vegetation type (Map 6.1) or condition is the habitat attribute most commonly referred to as important to mammals. Wyoming vegetation can be divided most broadly into forested and non-forested types, and these major types are further subdivided. Here, I describe six major vegetation types, their attributes as related to wildlife habitat, and the mammalian species that are commonly associated with each.

GRASSLAND

Grass-dominated prairies are concentrated in the eastern one-fourth of the state, particularly east of the Laramie Mountains to as far north as northern Weston County, in the Laramie Basin, and along the eastern flank of the Bighorn Mountains (Map 6.2).[2] This type occurs where summer precipitation is sufficient to provide moisture to the relatively shallow root systems of grasses and forbs. Grasslands in Wyoming are roughly divided into short-grass and mixed-grass types. Short-grass prairie is concentrated in the southeastern corner of the state, and is characterized by the presence of blue grama and buffalograss. These species tend to form isolated clumps, rather than continuous sod. Mixed grass is more widely distributed across eastern Wyoming, occurs in lower elevations, and is characterized by needle-and-thread grass, blue grama, western wheatgrass, Sandberg bluegrass, and other species. Mixed-grass types are more likely to form continuous sod-like cover. On more alkali soils, grasslands tend to include alkali sacaton, four-wing saltbush, and greasewood. Sandy soils have more Indian rice grass and prairie sandreed.[3]

The mammals that are closely associated with grassland habitats include many Great Plains species: spotted ground squirrels, black-tailed prairie dogs, pocket mice, Nebraska pocket gophers, prairie voles, swift foxes, black-footed ferrets, American badgers, pronghorns, and bison. Bison attained their highest densities in presettlement times on the grasslands of the eastern part of the state, and the highest population densities of pronghorns today likewise occur at the edges of the Great Plains.[4] The features that are important to mammals on grasslands include the abundant, easily digestible herbaceous biomass, much of it belowground. Species that prefer open areas with good visibility—black-tailed prairie dogs, swift foxes, and pronghorns—find security in such sites, because they detect their mammalian predators by vision. But species that require physical structure to escape predators or extreme temperatures find little cover on grasslands. Trees, logs, shrubs, talus

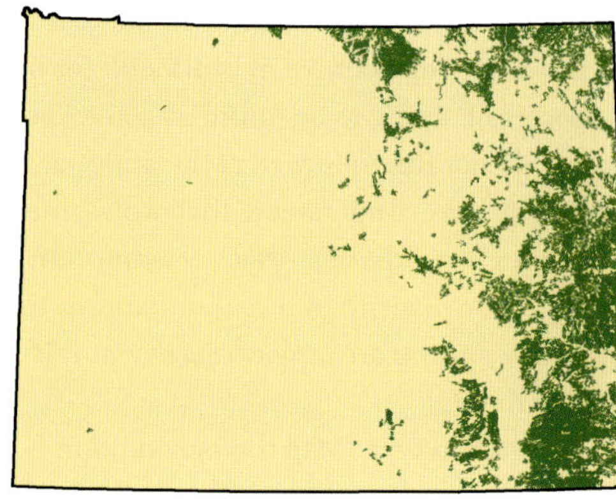

MAP 6.2 Grassland vegetation in Wyoming and Yellowstone National Park is concentrated in the eastern one-half of the region, comprising mixed grass in the lower-elevation areas and short grass in the higher ones.

fields, rock outcrops, and other structures are scarce or absent. So, some of them must dig their own burrows or use burrows dug by other mammals. In this way, burrowing mammals are particularly ecologically influential in grasslands. Ground squirrels, prairie dogs and badgers are key to the occurrence of many other species of land vertebrates, by nature of their digging and then abandoning burrows.[5] These burrows also increase water percolation to underground stores. Grasslands are a conservation concern because they have been so extensively converted to other land uses, particularly crop agriculture. Where crop production or livestock grazing is the primary land use, competition from wild mammals may be problematic for producers, and aggressive mammal control may be employed.

SHRUB-STEPPE

Shrub-steppe is the most widespread vegetation type in Wyoming, covering a wide swath from the southwest to northeast corners (Map 6.3). The main shrub species of Wyoming's basins and playas include several species of sagebrush, winter fat, shadscale, Gardner's saltbush, greasewood, and rabbitbrush. Of these, sagebrush in its various forms is one of the most widespread and prominent features of the Wyoming landscape, occurring as a genus nearly statewide. Big sagebrush is arguably the most important shrub species as mammal habitat, by nature of its great extent.[6] One subspecies, Wyoming big sagebrush, occupies middle elevations in the state—1700 to over 2700 m (5600 to over 8900 ft)—and other subspecies occur higher on mountain slopes, or in ravines or riparian settings. Big sagebrush tends to occur where moisture is more abundant than it is in grasslands, and arrives primarily via snowmelt, but is limited in availability by late summer. This species of sagebrush produces abundant stout woody stems that can reach 2 m (7 ft) above ground, but are more typically 0.5 m (1.4 ft) tall. The non-deciduous leaves on the plant over winter are important browse for pygmy rabbits, jackrabbits, pronghorns, and mule deer. These species have the ability—not universal among herbivores—to detoxify sagebrush leaves. Sagebrush leaves contain high concentrations of terpene-like compounds, which are toxic to most herbivores and the bacterial floras that they carry in their gastrointestinal tracts.[7] Other sagebrush

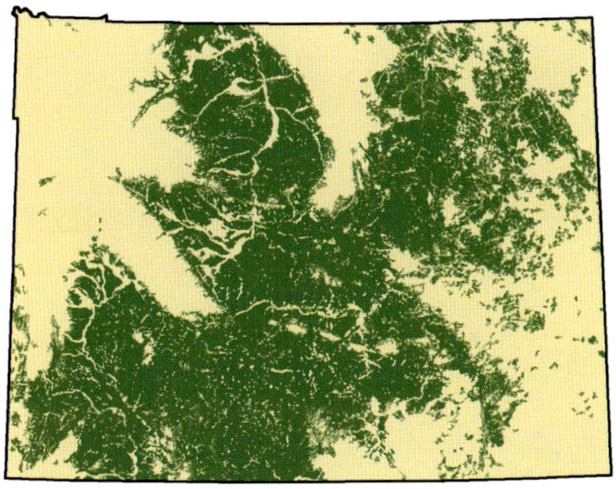

MAP 6.3 Shrub-steppe vegetation is the dominant type in most of Wyoming and Yellowstone National Park, except for the northwestern mountains and southeastern corner.

species include low sagebrush, black sagebrush, and silver sagebrush, which have their own moisture and temperature tolerances, and are distributed less broadly than big sagebrush.

In the hottest, driest basins of Wyoming are shrub-dominated communities that differ from sagebrush-dominated steppe in value as wildlife habitat. These communities are variously called desert shrublands and playa shrublands, and collectively vegetate around 8% of Wyoming's area.[8] Desert and playa shrublands experience especially hot, dry conditions in late summer, and in many areas are characterized by saline or alkali soils. The salts involved are sulfates, chlorides, and bicarbonates of magnesium, sodium, and calcium, which are released by the weathering of rock, but are not drained by runoff or percolation through soil. As a result, they tend to be concentrated near the surface, leaving a white surface powder in extreme cases. The plant species typically associated with these sites are Gardner's saltbush, birdsfoot sagewort, winter fat, shadscale, and greasewood. In contrast to sagebrush-dominated steppe, desert and playa shrublands have low productivity, provide relatively little aboveground structure that can shelter small mammals, and are of limited value as mammal habitat. Pocket mice, some ground squirrel species, Wyoming pocket gophers, jackrabbits, and coyotes occur in this type, but it has limited value as ungulate winter range.

RIPARIAN AND WETLANDS

Riparian zones are areas near flowing water, and contain vegetation and soils that are influenced by the presence of shallow groundwater. Wetlands are habitats characterized by at least seasonal presence of standing surface water. The wetlands of Wyoming are widely dispersed, but most Wyoming wetlands are dry for part of the summer, and some are the product of human activities: leakage from irrigation canals, flood irrigation practices, and water discharged from natural gas wells.[7] So, the distribution of surface water in late summer is far less extensive than suggested by Map 6.4. The vegetation pattern of riparian zones is highly variable, from the presence of a few scattered plains cottonwoods along stream courses in basins and the Great Plains to dense associations of woodland, willows, and other shrubs that can exceed a mile in width in Jackson Hole, the Wind River Basin, and other

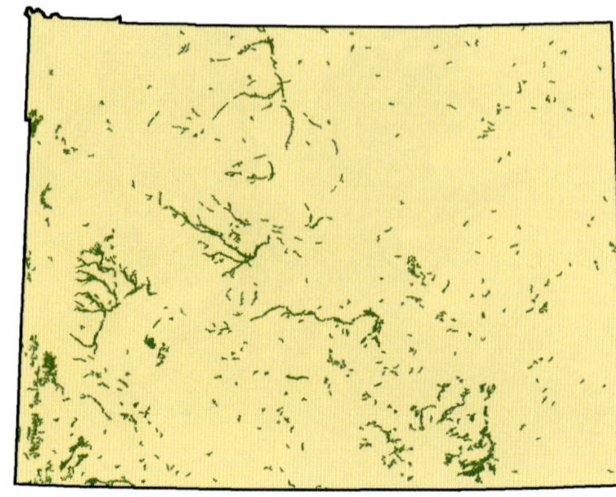

MAP 6.4 Riparian and wetland vegetation is limited in extent, but highly dispersed across Wyoming and Yellowstone National Park. It is disproportionately important to mammals.

basins. Tall willow thickets are found on some sites, including such species as Booth's willow, Drummond's willow, and Geyer's willow. Invasive nonnative tamarisk and Russian olive occur near water in lower-elevation riparian zones. Where willow thickets do not occur, gravel bars may be vegetated with sandbar willow, silver buffaloberry, silver sagebrush, wild rose, forbs, and grasses. Riparian zones are one of the most important habitat types to Wyoming mammals, disproportionate to their areal extent, which is around 1.2% of the state.[9] Many large-bodied mammals that occupy arid upland habitats can do so only because they have access to riparian water for drinking; small rodents are much better at producing concentrated urine and living without access to free water.[10] Also, riparian vegetation supports many more kinds of mammal than adjacent arid uplands: shrews, tree squirrels, jumping mice, American water voles, northern raccoons, skunks, Pacific martens, white-tailed deer, and moose occur in riparian zones surrounded by arid habitats that are inhospitable to them.[11] These ribbon-shaped habitats also provide dispersal corridors that allow certain mammal species to colonize or recolonize forested or mesic habitats that would be otherwise inaccessible in Wyoming's arid landscape.

The major kinds of wetlands of habitat value to mammals are marshes, playa wetlands, and meadows. Marshes are easily recognizable by their standing water and herbaceous vegetation emerging from the water surface: grasses, sedges, rushes, and some broadleaved plants. Cattail, common spikerush, sedges, and bulrushes are typical marsh plants. Marshes are highly productive for mammalian diversity and biomass, and the species that occur there include voles, American beaver, common muskrats, northern raccoons, American mink, northern river otters, and moose. Perhaps muskrats have the strongest association with marshes among Wyoming mammals, by nature of their diet of herbaceous aquatic plants. They can overwinter in marshes that do not freeze to the bottom, and build structures up to several feet in height that protect them from cold and predators. Wetlands are likewise important for shrews, not because most shrews need open water per se, but because they associate with the moist vegetation and rich invertebrate fauna at the margins of permanent standing water. Playa wetlands are those that remain after evaporation has removed most standing water from shallow basins in late summer. These sites tend to have little surface vegetation because of the high salinity of the soil; red swampfire is a common late summer plant of the margins of playa wetlands. The value of this habitat type to mammals is limited, although the water itself attracts some species. Wet meadows are characterized by standing water in early summer, which recedes below the surface by late summer. This habitat features grass-like plants and forbs, and shrubby cinquefoil. It has high value for grazers, and the amount of physical structure favors small rodents and shrews.

MOUNTAIN SHRUB

Mountain shrub communities are diverse, comprising several plant associations that have in common that they include woody plants shorter in stature than trees, and occur near the lower elevational limit of coniferous forest. This community type tends to ring the major mountain ranges of the state, extending into the mountains in places (Map 6.5). Its extent is limited in Wyoming, but it has disproportionately great significance as winter and transitional habitat for elk, mule deer, and pronghorns. The key features of this community are varied topography and important winter browse plants eaten by ungulates and other mammals. The varied topography causes snow to drift into the lees of ridges and vegetation patches, creating places for ungulates to shelter and snow drifts that are water sources in early summer. Shrub species are patchily distributed across this heterogeneous moisture mosaic and include mountain-mahogany, juniper, skunkbush sumac, antelope bitterbrush, mountain big sagebrush, serviceberry, and snowberry. Over limited areas of the state, these species are admixed with aspen, oak, or pinyon pine.[2]

Mountain-mahogany is particularly important and comprises two allopatric species, alderleaf mountain mahogany of the eastern and southern parts of the state, and curl-leaf mountain mahogany of the Bighorn Mountains and western edge of the state. Both species are nitrogen-fixing and tend to occur where soils are shallow and rock near the surface concentrates soil moisture. Both species are heavily used as winter browse for deer, and respond to heavy browsing by producing protective spines and lateral branches. The tree-like junipers likewise comprise two Wyoming species, Rocky Mountain and Utah juniper, which are

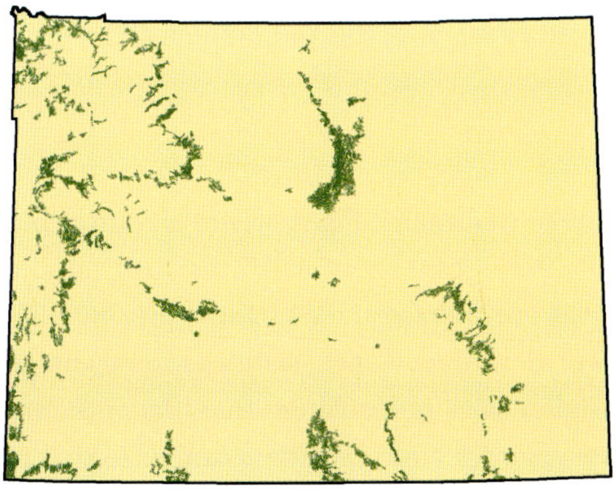

MAP 6.5 Mountain shrub vegetation is limited to the lower slopes of mountain ranges in Wyoming and Yellowstone National Park, below the coniferous forest zone. It is especially important for wintering ungulates.

found in foothill habitats. Rocky Mountain juniper occurs in foothill and escarpment settings almost statewide, whereas Utah juniper is confined to similar sites in the central and southwestern regions of the state. Both juniper species produce fruits and seeds that are eaten by a range of small mammals, and deer use junipers for shelter and winter browse. Serviceberry comprises two Wyoming species as well, Saskatoon serviceberry and Utah serviceberry, both of which are found on fairly protected sites over most of the state. These species are taller than mountain-mahogany, in places assuming a small tree form, and provide visual cover for deer and elk, as well as palatable browse. Antelope bitterbrush is much shorter and occurs in all but the northeastern corner of the state. Its seeds are eaten by pocket mice and other species, and the leaves and stems are highly palatable to deer and pronghorns. It can be severely overbrowsed or eliminated under heavy browsing. Oaks of two species are locally important mammal habitats. Gambel's oak grows in the Little Snake River region along the south-central border of the state, whereas bur oak grows in the hardwood draws leading to and within the Black Hills and Bear Lodge Mountains. Both species have value for the dense cover they provide, as well as their acorns—important foods for rodents and black bears. Chokecherry occurs nearly statewide, and is cultivated as well; it produces fruits that are highly palatable to many mammalian species: squirrels, mice, coyotes, foxes, and bears. Stems are browsed by deer in winter. Quaking aspen forest must be mentioned as a foothill habitat type, although the species occurs interspersed with conifers at higher elevations. Aspen has the typical form of a tree, although where snow drifts deeply, it can take the shape of a twisted midsized shrub. Aspen has value as browse for elk; elk and moose also scrape the bark with their incisors and eat the cambium. Beavers find aspen especially palatable and fell the trees to gain access to stems and leaves above the reach of ungulates. These plant parts are not heavily defended by plant chemicals, and are highly palatable to browsers. Aspens can be eliminated above ground from a riparian zone by a beaver colony, after which the colony typically abandons the site and disperses to another stream reach. The relationships among aspens, elk, wolves, and hunting have been debated for decades, with little resolution to basic questions about what allows aspen to regenerate and grow to a height above browsing level. The key questions in this discussion are whether the high population densities of elk desired by hunters are consistent with aspen regeneration, particularly near migration routes and feed grounds, and whether the return of wolves to Yellowstone has had an observable effect on the distribution or growth form of aspen in the Yellowstone area.[10]

CONIFEROUS FOREST

Coniferous forests in Wyoming are largely limited to mountain slopes, where elevation and topography create a range of environmental conditions consistent with tree survival and reproduction (Map 6.6). In general, forested slopes are characterized by more precipitation but cooler growing season temperatures than lower elevations. Aspect serves as somewhat of a proxy for elevation; south-facing slopes have higher rates of insolation than flat ones, dry out sooner after precipitation, and have moisture regimes and vegetation similar to that of

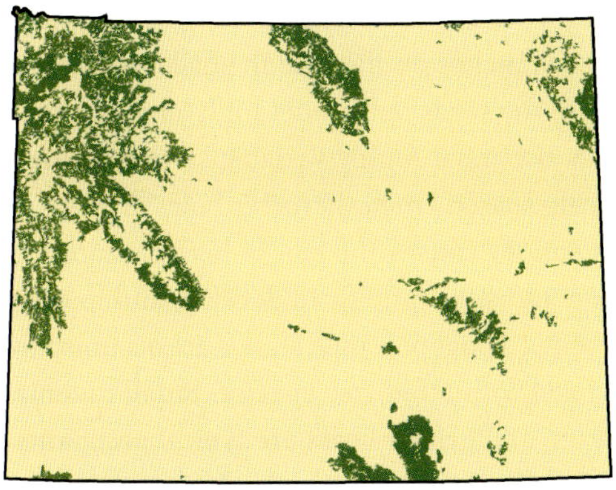

MAP 6.6 Coniferous forest vegetation occupies the mid-slopes of the mountain ranges of Wyoming and Yellowstone National Park.

lower-elevation sites. North-facing slopes are the reverse. The major coniferous tree species of Wyoming forests comprise ponderosa pine, Douglas-fir, lodgepole pine, Engelmann spruce, subalpine fir, and whitebark pine.[7] Limber pine and aspen tend to occur as admixtures with these major types, especially near the lower elevation limit of forest. A number of mammalian species can be considered forest obligates, in the sense that their distributions do not extend far from the forest edge, and they commonly have a behavioral aversion to open areas. Such species include the pygmy shrew, yellow pine chipmunk, golden-mantled ground squirrel, red squirrel, northern flying squirrel, western heather vole, southern red-backed vole, western jumping mouse, snowshoe hare, Pacific marten, and American black bear.

Ponderosa pine occurs near the lower limit of forest in the eastern half of Wyoming; it is rare or absent over the western half of the state. In various places, it interfaces with grassland or mountain shrubs. Because they occur on dry sites and typically have fairly open canopies, ponderosa-pine-dominated stands tend to lack abundant herbaceous growth. Ponderosa pine is fire-adapted by nature of its thick bark and lack of branches on the lower bole. Thick bark protects the cambium from low-intensity fire, and the lack of branches on the lower bole slows the spread of fire to the canopy. Ponderosa stands also tend to have even-aged population structure, so that stands with tall canopies tend to lack young trees, unless the stand has been protected from fire for decades. In short, ponderosa-dominated stands tend to lack abundant herbaceous vegetation, coarse woody debris, or live branches near the ground—important structures to various mammals for food or shelter. Of course, under intensive forest management these attributes may vary. Ponderosa pine cones produce a fairly fleshy seed, and these seeds and other tree parts are important to Abert's squirrels, a ponderosa pine obligate. Red squirrels and northern flying squirrels, however, are only loosely associated with pure stands of ponderosa pine, preferring more mesic stands with at least some spruce or subalpine fir.[10] Porcupines frequently eat the cambium of ponderosa pine, and cause locally intensive damage. Douglas-fir has similar environmental tolerances to ponderosa pine and tends to occur admixed with it.

The most widespread Wyoming conifer is lodgepole pine, which grows at higher elevations than ponderosa pine on sites with intermediate moisture regimes. It occurs admixed with other Wyoming conifers, particularly Engelmann spruce and subalpine fir. Lodgepole pine is characterized by a tall, slender shape, lack of lower branches on the bole in shaded sites, and cones that are variably serotinous, opening to release the seeds only under high temperatures, such as are generated by forest fires. This species occurs at highly variable stem densities, including well-spaced stands and those with closely spaced small-diameter boles, known as "dog-hair." On some sites, the development of a canopy causes the soil to become moister, which allows an understory of Engelmann spruce and subalpine fir to develop, and to even overtop the lodgepole pine if no disturbance intervenes. Lodgepole pine stands are occupied by a wide range of mammals, especially where admixed with other tree species. Small rodents, bats, martens, wapiti, both deer species, black bears, and grizzly bears are typical occupants. Lodgepole pine cones, especially nonserotinous ones, are easily dismantled by red squirrels, and the seeds eaten. Northern flying squirrels and southern red-backed voles occupy these stands, feeding on the fruiting bodies of fungi that are mutualistic with conifers.[12] Pacific martens are common in lodgepole pine during most of the year, tending to leave it for late-successional spruce–fir stands during the coldest weather of midwinter.[13] Black bears feed on the cambium of lodgepole pine. There are no mammalian species that appear to have an obligate association with lodgepole pine; all mammalian species that occur in this type also occupy, and some show a behavioral preference for, stands dominated by spruce and subalpine fir.

Engelmann spruce–subalpine fir forests tend to occur at higher elevations than lodgepole pine, on north-facing sites within the elevation range of lodgepole pine, and in streamside areas as low as the ponderosa pine zone. In the Medicine Bow Mountains and the mountains around Jackson Hole, blue spruce is locally common and ecologically similar to Engelmann spruce. Spruces and subalpine fir are found on moist sites with infrequent fires, growing to large diameter and old age in the absence of disturbance. They have lower branches on the boles that retain their leaves even when shaded, reproduce vegetatively when branches are pressed to the ground, and occur in uneven-aged populations, where seedlings and saplings grow in the shade of old standing trees. These old-growth stands of spruce–fir have abundant physical structure near the ground, in the form of live tree branches and coarse woody debris of large diameter. In such stands, death of the largest trees opens the canopy and allows the growth of small areas of herbaceous plants and mast-producing shrubs that are important foods for many mammalian species from mice to moose. The moist conditions of these sites encourage the growth of fungi that are important foods to red squirrels and northern flying squirrels. The seeds of spruces are readily accessible beneath the papery cone bracts, and red squirrels prefer them over other Wyoming conifers; squirrel middens are best developed beneath large, old spruces. Elk use spruce–fir stands for sheltering during hunting seasons, and moose browse the leaves of subalpine fir in winter.[10]

More than any other habitat type considered here, coniferous forests of Wyoming are undergoing rapid, potentially transformative change. These changes are the result of insect outbreaks by several forest species: mountain pine beetle infests ponderosa pine, lodgepole pine, limber pine, and whitebark pine. Engelmann spruce is infested by spruce beetle, Douglas-fir by Douglas-fir beetle, and subalpine fir by western balsam bark beetle. Each of these insect species has undergone explosive rises in abundance during the last two decades, somewhat asynchronous across geographic areas and insect species. Multiple causes have been invoked for these outbreaks, including climatic warming.[14] The effect on coniferous trees is dramatic: in the Sierra Madres of southern Wyoming after 5–10 years of epidemic, 90% of lodgepole pine over 30 cm (12 in) diameter at breast height (dbh) were killed, 80% of Engelmann spruce over 40 cm (16 in) dbh were killed, and 40% of subalpine fir over 40 cm (16 in) dbh were killed by insects.[15] In the short term, expected effects on mammals include sharp reductions in abundances of red squirrels, northern flying squirrels, and southern red-backed voles. In the intermediate term, increases in sunlight reaching the forest floor could result in lush herbaceous growth, with implications for many herbivores. In the long term, it is difficult to assess whether pre-outbreak vegetation conditions will be reestablished or whether new vegetation communities will replace the coniferous forests of the central Rocky Mountains.

ALPINE TUNDRA, MEADOW, AND ROCK

The upper elevational limit of coniferous forest marks the lower limit of alpine vegetation types (Map 6.7).[16] It includes such dissimilar features as meadows, dry ridges, and rocky areas with little or no vegetation. Alpine meadows exhibit a range of moisture regimes, typically fine soils, and perennial plants that include grasses and sedges, forbs, and subshrubs. Typical alpine meadow plants include alpine bistort, high northern buttercup, and alpine forget-me-not. Tundra occurs on variably well-drained slopes or ridgetops, and features perennial subshrubs, cushion plants, and some perennial forbs. Cushion plants occur throughout the range of North American arctic and alpine tundra, and some species are found from northern Canada to the southern Rocky Mountains: moss campion, mountain sorrel, and purple mountain saxifrage.[17] The elevations at which these species occur are perhaps 2700 m (9000 ft) lower in northern Canada than in Wyoming, and their presence so far south is a relic of much colder climates, when the distributions of these species extended in peninsular fashion from Pleistocene glacial margins to the central Rocky Mountain ranges where they are found today. Tundra plants that are not buried beneath snow in winter are exposed to extreme physical environments, including high wind, strongly varying daily temperatures, and high rates of evaporative water loss.[7] The third cover type treated here is rock, which I mean to include fell-field, scree, talus, and exposed bedrock. These types become increasingly prevalent as one ascends in elevation above the limit of trees. Fell-field is a talus- or scree-strewn type that is strongly influenced by freeze–thaw cycles and wind, so that cushion plants are limited to interstices between frost heaves and in lees formed by projecting rocks. Scree is rock debris of small size, typically on marginally stable slopes. It has interstices too small to admit any mammals, and almost no vegetation. Still, it has value as wildlife habitat; bighorn sheep use scree slopes to escape predators. Talus is rock debris of larger size than scree, typically larger than 20 cm (8 in) in diameter. It can be over 2 m (7 ft) in diameter, and has interstices large enough to allow the movement of many kinds of small mammals: voles, pikas, marmots, weasels, and martens. Large-diameter talus fields can

MAP 6.7 Alpine tundra, alpine meadow, and rock vegetation types occur above the elevational limit of trees in Wyoming and Yellowstone National Park.

extend continuously for miles, for example, in the Wind River Mountains, providing avenues for mammals to cover their home ranges beneath the rock surface, or to gain access to areas above tree line seasonally. The crevices beneath and between rocks protect these mammals from avian and large mammalian predators. Martens and weasels commonly climb to elevations over 3600 m (12,000 ft) in summer in this way.[18] Wolverines select crevices between large alpine boulders to give birth; they typically dig through snowdrifts to access them in late spring.[19] Exposed bedrock above the elevational limit of trees may have little vegetation with food value for mammals, but it is essential as a predator-escape feature for mountain goats and bighorn sheep, particularly where large predators are common.

A larger number of Wyoming mammalian species use alpine zones seasonally than is commonly appreciated. While seasonal migrations of wapiti and deer to subalpine zones from lower elevations are well documented,[20] reported migrations from forest or plains to alpine zones by badgers, martens, white-tailed jackrabbits, white-tailed deer, bison, and other species are less so, but are often reported and doubtless occur. Few mammals can survive above the elevational limit of coniferous forest in winter, and those that do mostly employ thermal insulation, sheltering in protected microenvironments, hibernation, or combinations of these strategies to survive.[21] Among mammals too large to shelter beneath the snow, gray wolves, wolverines, bighorn sheep, and mountain goats are alone in being able to withstand for sustained periods the convective heat losses experienced in alpine zones in midwinter. Bighorn sheep used alpine zones much more commonly in presettlement times than today. The decline in abundance of bighorns in alpine zones over the last 120 years results from two factors. First, lower-elevation winter ranges and migration routes have been lost to agriculture and other land uses. Second, domestic sheep grazing in alpine zones, which reached its peak in the 1930s, facilitates disease transmission to bighorns, reducing their populations drastically. Still, alpine zones of Wyoming and Colorado that are usable as summer bighorn sheep habitat are much more extensive than is true for Montana and Idaho.[22] With decreases in use of alpine areas for domestic sheep grazing over the past few decades, these habitats have the potential to experience some recovery of the former abundances of bighorn populations.

Conservation and Management: Yellowstone and Beyond

The mammalian fauna of Wyoming and Yellowstone is remarkable for its intact and natural state. Of the species that have been introduced or reintroduced to sites in the region, most (e.g., bison, wapiti, black-footed ferret, moose, bighorn sheep, American beaver) were translocated from populations within the state's borders. Among mammals, only the wolf—extirpated in Wyoming by the late 1920s—occurs as a result of translocation from a distant site, Alberta. Moose occur in several parts of Wyoming as a result of introduction, but that species originally arrived in Wyoming without human help, and most of the moose translocated to the state originated from populations in or near Wyoming. In addition to translocations of moose from Jackson Hole to the Bighorn Mountains in the 1940s–1980s, moose migrated into Wyoming from Colorado, where they had been introduced from Wyoming in the 1970s. A similar narrative applies to wild bison in Yellowstone National Park, which descended from wild animals never taken into captivity mixed with recently captive animals the geographic source of which is unclear, but near Yellowstone. Thus, nearly all species of mammal present in Wyoming 200 years ago are represented today by populations descended from Wyoming ancestors or sites near Wyoming, adapted to Wyoming environments.

Another important asset to Wyoming's natural heritage is the absence of feralized domestic mammals of species not native to North America. Wyoming has for decades enforced rules limiting captive possession or introduction of native or nonnative big game and trophy game mammals. State lawmakers prohibited game farming of these species beginning in 1939[1] and the rules were stiffened in 1992, when the Game and Fish Commission, in response to litigation over an exotic game ranch in Crook County, strengthened its position on farmed game.[2] As a result, Wyoming today has almost no free-ranging nonnative mammals. The exceptions are the small populations of mountain goats and eastern gray squirrels that have moved across the state lines from Montana and Idaho, both of them representing range extensions of less than 300 km (180 mi). The mountain goats numbered fewer than 300 animals as of 2012, and the eastern gray squirrels apparently are restricted to urban and nearby riparian habitats in the Sheridan–Buffalo–Gillette area. By contrast, some other western states have few rules regarding exotics and harbor many species of nonnative ungulates on both private and public lands. Of course, domestic horses have lived in a feral state in Wyoming since at least the eighteenth century, and their populations have increased and declined in response to economic and regulatory circumstances. As of 2012, around 5300 feral horses roamed the Red Desert and other areas of Wyoming, mostly on public land.[3] Other feralized nonnative species (e.g., domestic dog, domestic cat, wild boar, domestic goat, ass) that are widespread and in some cases ecologically problematic in other states are either rare outside of captivity or completely absent from Wyoming. In the context of recent developments in wildlife diseases, particularly chronic wasting disease in domesticated cervids, the long-standing state policy limiting the possession or introduction of big game seems wise and foresighted.[4]

Two additional factors serve to reduce the prevalence of nonnative mammals: first, Wyoming environments are sufficiently harsh that few of the nonnative species widespread elsewhere in temperate North America can persist or increase here. Cold, windy winters and the lack of vegetative structure limit the kinds of exotic species that can take hold and spread. The climate and landscape also account for the low population density of humans—an average of $2.3/km^2$ ($6/mi^2$) statewide. The harsh climate also dictates that a small fraction of Wyoming's land area is in crop cultivation; wildland—land with native vegetation—is the predominant land cover type. Further, large holdings of public lands (57% of Wyoming's total area) and statutory mandates for their stewardship mean that federal and state agencies are attentive to any occurrences of nonnative mammals and their environmental effects. The result is the infrequent occurrence of free-ranging nonnative mammal species in Wyoming, with the exception of feral horses, which are protected by federal law.[5]

Although extinctions of native mammals and introductions and reintroductions of nonnatives have been rare in the state during the historical period, range expansions into the state facilitated by human alterations of the environment are more common. Virginia opossums, eastern fox squirrels, and northern raccoons apparently were absent from the state in presettlement times. Raccoons occur in all counties today, their spread attributed to the reduction in abundance of wolves, which occurred decades before the final extinction of wolves in around 1929. Virginia opossums now occur only along the most downstream 48 km (30 mi) stretch of the North Platte River in the state. Oral histories suggest that skunks of all three species were rarer and more local in distribution before settlement; it is possible that settlement allowed skunks to move upstream along various riparian zones into and through the state. And black-tailed jackrabbits clearly have expanded their range in the state, possibly as a result of a warming climate, or of recent drought.[6] The current widespread distribution of the eastern fox squirrel is the result of human alteration of the environment, as well as direct introductions of animals to new areas. So, humans have had little effect on the mammal species present in the state, but considerable influence on their distribution around the state.

YELLOWSTONE AND CONSERVATION

Over the last 150 years, the outdoor recreation enterprise has been central to the conservation of Wyoming's mammalian fauna. This enterprise has included wildlife watching, hunting, fishing and trapping, and all of the ancillary regulatory, social, and economic infrastructure to protect wildlife resources. Of course, the nineteenth and early twentieth centuries saw intensive hunting and trapping for subsistence, markets, and recreation, some of it at unsustainable levels. But since the early twentieth century, outdoor recreation in its various forms has been a net asset to the conservation of Wyoming mammals. The first manifestation of the alignment between recreation and conservation was the movement to establish Yellowstone National Park in the early 1870s. This was the beginning of a long-term trend toward the creation of protected nature reserves around the world. How Yellowstone was discovered and designated as the world's first national park is foundational in the conservation literature.[7] The park has also been highly influential in shaping national sentiments regarding wildlife in Wyoming.

In addition to having symbolic importance for 140 years, Yellowstone served as an important refugium for several mammal species during the nadir of mammalian abundance in the late nineteenth and earliest twentieth centuries. During that time, bison, grizzly bears, pronghorns, and wapiti underwent well-documented declines throughout the Rocky Mountain West, but persisted in Yellowstone National Park due to protections there. Some accounts specifically suggest that wolverines and river otter populations likewise benefitted from Yellowstone's protections, so that those species were able to recolonize adjacent areas of Wyoming decades later.[7] The wolf—extirpated from all of Wyoming including Yellowstone—could not have been returned to northwestern Wyoming had not Yellowstone National Park and surrounding public lands served as a core protected area for the reintroduced population. After the establishment of Yellowstone National Park and continuing to around 1920—much later in some areas—the exploitation of various big game populations continued to the point of depletion across Wyoming. Bison were nearly exterminated by around 1890, but elk reached their low ebb 20–30 years later, and Yellowstone was one of the few protected reservoirs from which restoration of those two species could proceed.

Statewide Mammal Conservation

On the national stage, the assembly of the legal, regulatory, and economic infrastructures of conservation that we now take for granted began during the Reconstruction era, when national attention turned west, settlers crossed the plains by the hundreds of thousands, explorers located gold deposits and railroad routes, and military expeditions increased in frequency and size. During this period sporting organizations sprang up across the eastern, populated parts of the country, and periodicals began to popularize and dramatize outdoor adventures in the West. Outdoor writers also argued for wildlife conservation, particularly in Yellowstone. These periodicals featured outdoor adventure stories and opinion pieces by the early leaders of American conservation: Theodore Roosevelt, George Bird Grinnell, and C. Hart Merriam. Each of these men traveled through Wyoming and made well-documented pilgrimages to Yellowstone. The public responded with calls for greater state and federal protections. By the 1900s, wolves and other predators notwithstanding, effective wildlife conservation and restoration were gaining traction within Yellowstone National Park.

Beyond Yellowstone's borders, the implementation of wildlife conservation goals was slower and more challenging, for good reason. Whereas Yellowstone was a relatively small tract of land managed by a single federal agency, holding exclusive jurisdiction, under a single management philosophy, the other 97% of Wyoming became a mosaic of private and public lands under various land-use regimes and legal jurisdictions during the period 1875–1910.[8] These included lands where hunting was allowed or forbidden, croplands were tilled or wilderness prevailed, public access was permitted or restricted, military reservations were established and abandoned, and reservations owned and managed by native peoples were created, and their governance developed. The earliest manifestation of mammal conservation policy at the state level occurred in the 1880s—before statehood in 1890—when the territorial legislature passed laws intended to set hunting seasons and bag limits, limit shipments of untanned hides across state lines, and restrict the seasons when game meat could be offered for sale. An 1886 law prohibited capturing live or holding in captivity wild mammals for the purpose of domestication or for scientific purposes, the earliest statutory recognition that mammals could not be taken alive from the wild without authorization.[9] But these laws generally lacked a statutory basis for their enforcement, so the mechanisms of implementation were absent. In light of the exhaustive overharvest of most big game animals, and with the potential for overharvest of other species, many people who had not participated in the big game slaughter—and some who had—now saw wisdom in the protection of remaining populations. By the time of Wyoming statehood in 1890, the idea of wildlife conservation was in the minds of many, but the laws and infrastructure for enforcing them were yet to be realized.

The new State of Wyoming enacted various laws for wildlife protection, as well as the enabling legislation for enforcing existing laws. But the laws continued to be inconsistently enforced, and ran afoul of the view held by some settlers and tribal members that subsistence use of game—particularly in remote areas—was a basic personal right. This right was specifically granted to certain tribes by treaty provisions, and tribal members exercised their right to hunt for food and hides on public land. Such subsistence uses led to conflicts—in some cases violent—between white settlers and Native Americans. One such incident led to a US Supreme Court case that tested whether federal treaties with tribes preempted newly established state ownership, via the Wyoming Statehood Act, of wildlife outside of Yellowstone National Park. With that case, the right of the State of Wyoming to manage the wild mammals within its borders—aside from Yellowstone National Park and tribal lands—was firmly established, and the stage was set for state jurisdiction over wildlife population management on over 90% of Wyoming's lands for nearly a century.[10]

Public attitudes and state legislative action were now trending rapidly in the direction of complete protection for some species. In 1895 a state law was enacted making the killing of bison a felony, punishable by up to 10 years' imprisonment. The same legislature designated the existing Fish Commissioner as State Game Warden. Fish Commissioner Gustave Schnitger assumed these new duties, but soon became discouraged by the magnitude of his new responsibilities, and the dearth of resources available to fulfill them. He deputized 20 assistant wardens, who went about their volunteer, uncompensated enforcement activities. Some arrests were made and fines levied, but some of the cases brought by deputy wardens were dismissed, and the wardens were soon disheartened. The big game hunting season was now open September 1–November 30, and Commissioner Schnitger enforced it as well as circumstances allowed. A move to delegate the responsibility for hiring—and paying—local wardens to the counties was met with little local interest. So, Schnitger began to press for a separate state game office.[11]

The legislature granted his request in 1899. In his new capacity Warden Schnitger could hire assistant wardens for pay, and they were to be bonded and held culpable for failing to perform their duties. Forest rangers—a new arrival on the Wyoming scene—could be deputized and paid $1 per day while working on game cases. Hunting of moose and swans and trapping of beaver were suspended, and bag limits were set for elk, deer, and antelope.[11] During this period when wildlife laws and enforcement bodies were nominally in place, game wardens still had difficulty making criminal cases because poachers were so effective at quickly transporting game meat across state lines. Wardens from adjacent states had no jurisdictional powers over violations in Wyoming and the meat could legally be sold there. Throughout the West, market hunting became an exercise in transporting big game carcasses across state lines quickly. Then, in 1900, a federal law was enacted with profound implications for wildlife law enforcement in all states. The Lacey Act, introduced by Rep. John Lacey of Iowa, made it a federal offence to transport across state lines any wildlife parts taken in violation of the state law where it was killed. This law was viewed at the time as a straightforward exercise of the federal government's constitutionally granted commerce powers, and had the prompt effect of drastically curtailing illegal market hunting nationwide.[12] Any fresh game meat found by state game wardens outside the autumn hunting seasons meant that the animals had been poached, and proof of geographic origin was not at issue.

The evolution of Wyoming's wildlife conservation policies continued, with the requirement that hunting guides be certified, be deputized as assistant game wardens, and accompany any nonresident hunter. Transportation of wildlife parts across state lines required a permit from the local justice of the peace. The hunting season on pronghorns was closed from 1909 to 1915.[13] The use of dogs was prohibited for hunting deer, elk, and bighorn sheep. One of the earliest and most effective game wardens was Daniel Nowlin (fig. 7.1). In his first report to the governor in 1902, he described his work progress, mentioning the considerable amount of money that nonresident hunters had spent in the state during their

FIG. 7.1 Daniel C. Nowlin, the first effective game warden in Wyoming.

hunting trips. This was surely the first attempt to estimate and emphasize the economic activity generated by nonresident hunting in Wyoming. Warden Nowlin diligently monitored and reported to the governor conditions regarding game and its scarcity across Wyoming, and his reports contained many anecdotes of egregious overharvest and waste. He recommended a wide range of corrective measures: accountability for taxidermists and guides, more restrictive hunting seasons, licenses for all ungulate hunting, and better salaries for wardens. The 1909 legislature was particularly active in the area of wildlife management, and an entirely new game code was enacted, including many provisions based on Nowlin's recommendations.

While big game was still scarce in some areas, Wyoming was now confronted in the Yellowstone area with a new challenge—the perceived overabundance of elk. This species had undergone rapid population growth in response to mild winters, the curtailment of overhunting, and reduced populations of large predators.[11] Further, traditional migratory routes were now being blocked by intensive hunting at the Yellowstone National Park boundary, and by settlement in southern Jackson Hole. As a result, elk were wintering in places that they had not previously, and deaths from poor forage on these areas were shocking to those who observed and heard reports of them. Jackson Hole soon became the site of the next major conservation action targeting mammals when, at the request of the state, the federal government first appropriated funds for hay to feed elk, then funds to buy land to raise hay, and then declared the area to be part of the National Wildlife Refuge System: the National Elk Refuge.

Steadily the state improved its wildlife conservation system, including the appointment of a Game and Fish Commission, more detailed regulations, population monitoring, and license fee increases. Revenue sources to operate the Game and Fish Department were limited then, as they are today, because the primary funding source has always been sales of hunting and fishing licenses. During years when ungulate populations were high and many sportsmen bought licenses, revenues were good. But when major die-offs occurred, and when interest in hunting waned, agency resources declined in step. An important new source of revenue became available when, in 1937, a Federal Aid program—the Pittman–Robertson Act—was enacted to place excise taxes on hunting and shooting equipment and supplies, and to share the revenues raised with states, using a formula based on land area and number of licensed hunters. With much land and a moderate number of hunters, Wyoming's mammal conservation programs benefitted greatly.

The Modern Era

The period from 1960 to the present saw a number of changes at the national level that affected wildlife management in Wyoming. These were catalyzed by a surge in environmentalist views and activity, which led to such federal laws as the National Wilderness Act of 1964; a succession of three endangered species laws, the last of which was the Endangered Species Act of 1973 (E.S.A); and the National Environmental Policy Act of 1970. The conservation of mammalian predators was particularly influenced by this phase of history, because of growing recognition that predators served legitimate and important functions in wildlands, and the further recognition that traditional methods of predator removal had severe unintended effects on nontarget species. The management of wapiti in Yellowstone National Park had become particularly controversial, and in response to this and other issues the National Park Service commissioned a review of wildlife management programs in areas under its jurisdiction. The review panel was chaired by A. Starker Leopold, who led the first outside committee invited to comment on National Park Service policies.[14] The resulting report recognized that management of Yellowstone's elk population had been ineffective, that protection of habitat was as much an agency goal as was protection of elk themselves, and that predators served important ecological functions in natural areas. Another national policy development with important implications for conservation of carnivores was the Executive Ban on the use of wildlife toxicants, signed by President Richard Nixon in 1972.[15] The order applied to all federal lands and federally funded programs, and led to the Environmental Protection Agency banning the use of three wildlife toxicants that had been used broadly: sodium monofluoroacetate (Compound 1080), sodium cyanide, and strychnine. It is difficult to evaluate the effect of these toxicants on predator communities in Wyoming before their ban, but it is clear that they were placed in baits over wide areas of the West, and likely had population-level effects on all carnivores that ingested them—scavengers and predators—for most of a century. In 2013, strychnine remained banned, and sodium monofluoroacetate and sodium cyanide were allowed only under limited and closely regulated conditions.

TABLE 3 Mammalian species in Wyoming listed or reviewed for listing under the Endangered Species Act during 1995–2012. Other species have been petitioned for listing but were not reviewed.

Species	Listing status (year)	Federal register reference
Black-footed ferret	Endangered (1967), experimental population, nonessential (1991)	Goble et al. (2005) for original listing; 56 FR 41473–41489 for experimental, nonessential status
Canada lynx	Threatened (2000)	65 FR 16052–16086
Swift fox	Found not warranted (2001)	66 FR 1295–1300
Preble's meadow jumping mouse	Threatened status reinstated (2008)	73 FR 39790–39838
Black-tailed prairie dog	Found not warranted (2009)	74 FR 63343–633366
American pika	Found not warranted (2010)	75 FR 6438–6471
Grizzly bear	Threatened status reinstated (2010)	75 FR 14496–14498
Pygmy rabbit	Found not warranted (2010)	75 FR 60516–60561
White-tailed prairie dog	Found not warranted (2010)	75 FR 30338–30363
Wolverine	Warranted but precluded (2010, 2012) Withdrawn (2014)	75 FR 78030–78061, 77 FR 69994–70059, 79 FR 47522–47545
Wyoming pocket gopher	Found not warranted (2010)	75 FR 19592–19607
American bison	Found not warranted, 2011	76 FR 10299–10310
Common gray fox	Under review (2012)	77 FR 71759–71771
Eastern spotted skunk	Under review (2012)	77 FR 71759–71771
Wolf	No longer threatened (2012)	77 FR 55530–55604
Little brown myotis	Under review (2014)	None[a]
Northern long-eared myotis	Threatened (2015)	80 FR 17974–18033

a. The little brown myotis was not petitioned for listing and has followed a nonstandard regulatory route.

E.S.A is now recognized as one of the most influential US environmental laws of the late twentieth century, and it has been especially influential in such states as Wyoming and California.[16] Not only have a number of species found in Wyoming been listed under the act, but also the mere possibility of protection of species perceived as rare or at risk has been a powerful incentive for various agencies—state and federal—to monitor their distributions and abundances. These monitoring efforts have increased our knowledge of the distributions of Wyoming mammals markedly, and led to some important conservation actions. Just since 1995, four mammalian species from Wyoming have been newly listed or carried over from earlier listings, and 13 others have been reviewed for listing, but found not warranted, or warranted for listing but precluded by limited agency resources (Table 3). Additionally, some listed populations have undergone delisting or downlisting following improvement of their conservation status, or better understanding of their distribution and abundance. E.S.A.-related decisions are often controversial, because of perceptions that a listing decision will affect the survival of a population, subspecies, or species. Alternately, some perceive E.S.A. listing as an unnecessary restriction on the rights of people to live and work on the land, including private land. Such controversies are exacerbated by the fact that candidate or listed species are often poorly understood, not surprising considering their rarity. When species of concern receive more intensive study, the level of concern about their conservation often declines. For example, in the case of the grizzly bear it was unclear during the early 1980s how many bears still occupied the Greater Yellowstone Area; the development of improved surveying and modeling methods later demonstrated a larger population than had been supposed.[17]

Regulations and programs put in place in response to E.S.A. have stimulated surveys for rare mammals or their habitat and have resulted in studies of possibly limiting factors, and models of life histories. For example, our understandings of the distribution, abundance, and ecology of prairie dogs in the state are stronger because of the E.S.A. listing of the black-footed ferret. Likewise, the reintroduction of wolves to parts of Wyoming has enabled studies of how wolves and beavers affected the communities from which they had been eliminated. So, E.S.A. has had diffuse and positive effects on what is known about the distribution, abundance, and life histories of a wide range of Wyoming mammals, not only those that have been listed.

Beginning in the early 2000s, a new federal program was established that emphasized state-based programs for the conservation of all wildlife, not just game fish and mammals. State Wildlife Grants provided funds to states to study or conserve so-called "non-game wildlife," with funds originating from those set aside from federal excise taxes on hunting, angling, archery, and shooting equipment and supplies. Participating states were required to write State Wildlife Action Plans, within which State Species of Greatest Conservation Need were required to be identified and revised periodically. The action plan for Wyoming provides an overall framework for anticipating conservation issues concerning all vertebrates, rather than merely reacting to crises as they arise. As of 2010, Wyoming had identified 46 mammalian State Species of Greatest Conservation Need, which are priorities for receiving State Wildlife Grants.[18] Some of the information presented here has resulted from these grants to investigators working in Wyoming, but the total amount of funding available is small compared to the number of species about which little remains known. A means to understanding the ecology and conservation status of most mammals in the state is still beyond our reach.

A means to provide funding to study and manage all wildlife, not just hunted species, has long been a national conservation goal, one that has not yet been reached. The Fish and Wildlife Conservation Act of 1980 articulated the intent to provide federal aid to the states for nongame management, but a sufficient revenue source to fund the act has never been found.[19] Various funding models have been proposed; a national excise tax on outdoor recreation equipment would parallel the taxes on equipment used in hunting, fishing, and shooting. Such a tax would provide revenue to the states using the successful Pittman–Robertson model, and would turn nonhunters into contributing stakeholders in the wildlife conservation enterprise. And the tax rate could be low; the annual economic activity generated by all outdoor recreation in the United States is $646 billion,[20] and a tax rate of less than one-tenth of 1% on that activity would go far to fund a wide range of nongame programs nationwide.

CONSERVATION STORIES FROM WYOMING

The intactness of the Wyoming mammal fauna speaks to the fact that successes outnumber failures among conservation stories in the state. Some of these stories are told in detail in the species accounts. But not all conservation battles have been won, not all can be characterized as successes or failures, and some involve multiple species and deserve special consideration.

Wapiti, like bison, had a nearly continent-wide distribution at temperate latitudes when Europeans arrived. They were dwellers of New England forests, Gulf Coastal savannahs, Chihuahuan Desert in northern Mexico, and the California Coast Ranges. Like bison, elk were overhunted to extinction over the entire area east of Wyoming, south of the Mexican border, and in most of the Pacific states. By the 1890s, the need became apparent to protect some remaining wild animals in captive settings, and elk were translocated from Yellowstone for that purpose. As the conservation infrastructure was assembled nationwide, and free-ranging elk populations became able to survive exploitation, animals from Wyoming, or their descendants, were translocated to most of the states from which elk had been hunted to extinction (Map 7.1).[21] To be sure, the species was never at risk of extinction, because of populations in western Canada, but Yellowstone elk facilitated the process of returning the species to at least some locations in almost every state in the presettlement range, where they now provide recreation and esthetic values.

A less well-known story involves one of the most important single actions to have implications for mammalian diversity in Wyoming: the construction of the Flaming Gorge Dam, with its associated impoundment. The dam was proposed in the 1930s and 1940s and became an alternative to the controversial Echo Park Dam in Utah, which was abandoned in the face of fierce opposition by environmental groups led by David Brower.[22] Construction of the Flaming Gorge Dam began in 1958, with filling of the impoundment zone starting in 1962. Although the area of the impoundment was not large, by regional standards, the habitats flooded were rare ones so far north, with limited extent in Wyoming. These habitats were Colorado Plateau types characterized by riparian woodlands and shrub meadows, surrounded by steep canyon walls (Map 7.2). The inundated escarpments featured foothill shrublands with various mountain shrub species, sagebrush, and grasses. On adjacent higher slopes, juniper woodlands with pinyon were found. Several mammalian species with extremely limited distributions in the state had their geographic ranges within Wyoming markedly reduced. The cliff chipmunk, pinyon mouse, and canyon mouse were all recorded in the impoundment zone before flooding, and all have Wyoming distributions now restricted to the neighboring side canyons and benches.[23] Bighorn sheep occurred in the steep canyon country, in what were likely the warmest, steepest habitats with access to water that bighorns occupied in

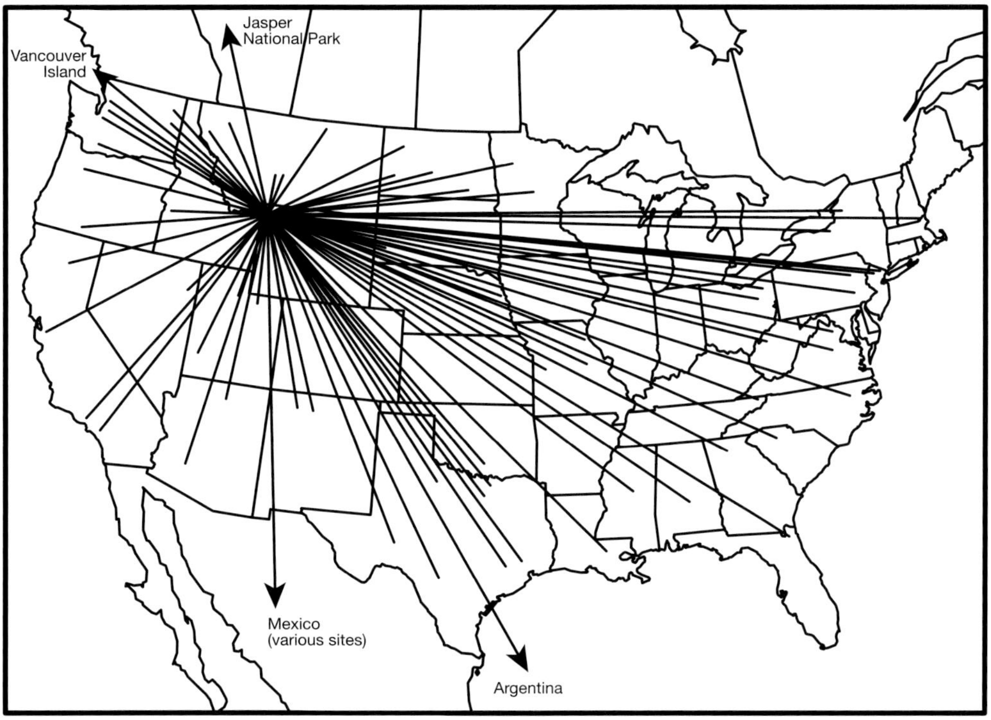

MAP 7.1 Wapiti were shipped from Yellowstone National Park to many states and Canadian provinces during 1889–1967. Some shipments from Wyoming were made to zoos and private reserves, but most were to restore free-ranging populations that were known earlier in the historical period. Secondary translocations were made from these destinations to others not mapped, so that Wyoming has been key to the repopulation of the elk throughout much of its former geographic range in the United States.

the state.[24] Ringtails are uncommon in the area around the reservoir today, but likely would have occupied the riparian woodland and shrub meadows along the Green River before impoundment (Map 7.2). The Flaming Gorge Dam may have produced benefits for water resources and water-based recreation, but a finger of the Colorado Plateau biotype was nearly eliminated from Wyoming, along with at least four species rare in or absent from southwestern Wyoming today. This story illustrates how shifting baselines lead to under-appreciation of the mammalian diversity and habitats that Wyoming held just 60 years ago. It also shows that what exactly was lost cannot be known, because the requirements for predevelopment monitoring and documentation were not yet in place when Flaming Gorge was impounded. It also argues for personal contact between mammalian biologists and naturalists with all parts of Wyoming—not just the famous destination sites—so that the diversity of mammals and habitats in the state are better understood and appreciated.

A third conservation narrative is one familiar to most Americans—the return of the gray wolf to Yellowstone National Park and the surrounding area. How thoroughly publicized this issue has been is suggested by the fact that the University of Wyoming library catalogue

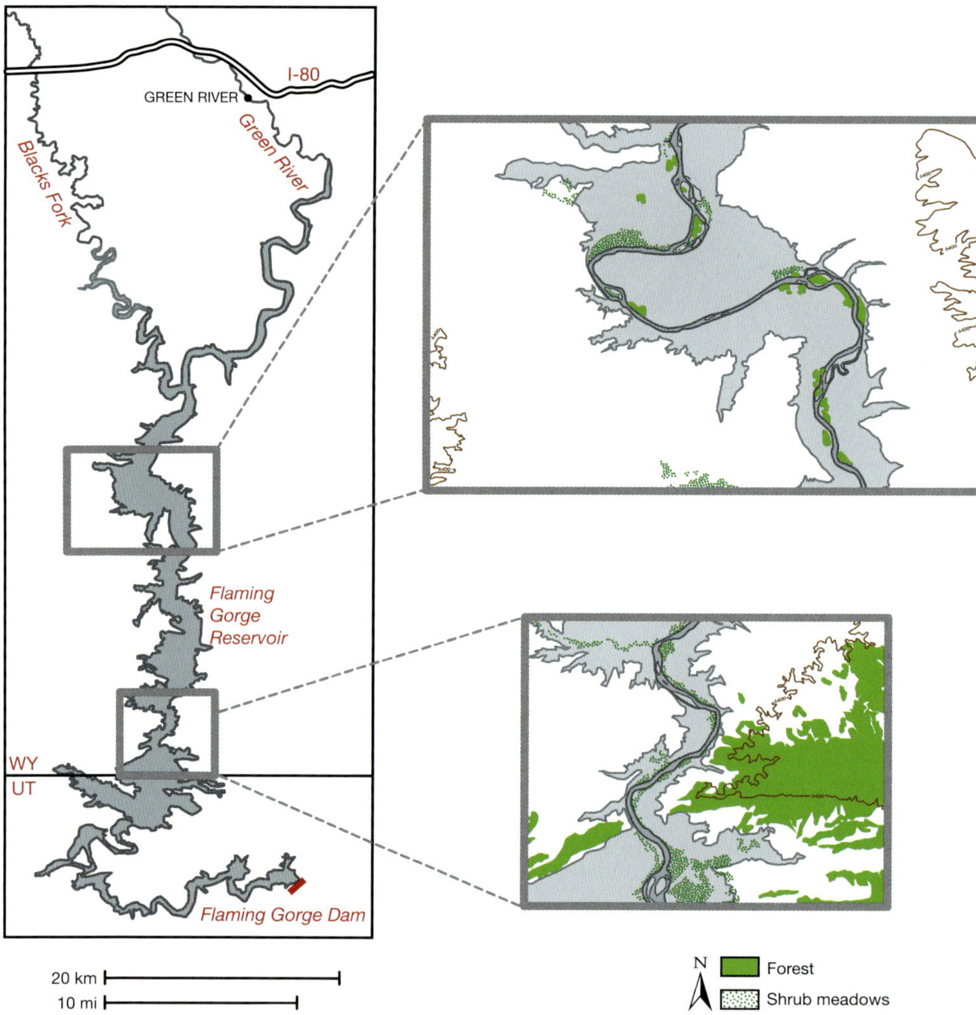

MAP 7.2 The lower Green River in Wyoming before and after the impoundment of Flaming Gorge Reservoir (gray). Insets show the distribution of steep-walled canyons, coniferous (juniper–pinyon) forest, and riparian shrub meadows. Several species now rare and restricted in Wyoming likely were common along the river or in side canyons before impoundment.

lists over 100 books and other media with "Yellowstone + wolf" as keywords, and the online index Web of Knowledge lists over 300 scientific journal articles with the same keywords. The number of popular articles on the subject likely runs into the thousands.[25] The identification of the possible need for the reintroduction of the wolf to Yellowstone was implied by the 1964 Leopold Report and more explicitly identified by later studies, but the statutory authority came from the E.S.A., and the implementation of the reintroduction was discussed, debated, and litigated in innumerable venues.[26] The reintroduction was conducted under the auspices of section 10(j) of E.S.A., whereby experimental, nonessential

FIG. 7.2 Wildlife overpass completed over State Highway 191 west of Pinedale in 2012. This and another overpass help pronghorns, mule deer, and other species to cross busy state highways, keeping intact an important traditional migration route, and reducing vehicle collisions.

populations can be established and managed—including by the use of lethal removal—under terms not otherwise allowed under E.S.A. The action cannot be said to have saved the wolf, as a species, from extinction. Indeed, *Canis lupus* is one of the most widely distributed wild mammal species in the world, with extensive areas of distribution in Eurasia, Alaska, and Canada.[27] It was the ecological consequences of the reestablishment that were of primary interest, and so far those have been well studied, but still uncertain. Effects of the expanding wolf population on elk numbers, woody vegetation, grizzly bears, other carnivores, and other prey have been investigated, discussed and debated, with some assertions of ecological effects of wolves being rebutted by other scientific reports.[28] One reason for the lack of clarity in the outcome of the before–after "experiment" presented by the wolf reintroduction is the fact that other confounding events occurred at around the same time (1995) as the wolf reintroduction. Just considering possible factors that could affect elk numbers, these include vegetative recovery from the 1988 fires, recovery of grizzly bears from their population lows of the 1970s, the beginning of a drought of nearly a decade in duration, and the introduction of lake trout to Yellowstone, which may have affected elk populations by an ecological path that was only recently identified.[29] The wolf reintroduction will be a decades-long experiment, and its ecological results will never be known with certainty, because the Yellowstone ecosystem is subject to many other forces, besides the effects of wolves. But the program has been intensively, if unevenly documented, and future generations will not lack for information about the wolf reintroduction to Yellowstone.

The last narrative involves our emerging understanding of the importance of traditional migration routes to ungulates. Ungulate migration is mentioned as important in the ecology of various ungulate species,[30] but the need to mitigate the loss of migration routes due to human developments was recognized traditionally only for wapiti, and to a lesser extent, mule deer. The recognition in the early 2000s of the length and longevity of the migration taken by pronghorns that move seasonally from Jackson Hole to the Upper Green River Basin was a surprise to many. Concern was expressed about potential interruption of the migration by major energy development and exurban residential development in that area. A further concern involved increasingly heavy traffic on State Highways 189 and 191, which cross the primary routes of travel of pronghorns and deer, producing many vehicle–wildlife collisions.[31] The worsening situation, from the standpoint of pronghorns, spurred a major initiative to mitigate these impacts via two major wildlife overpasses and six underpasses along a 12 mi stretch of highway west of Pinedale (fig. 7.2). These structures quickly saw heavy use by pronghorns and mule deer in their first few months of use; over 2400 pronghorns and over 6400 mule deer traversed the structures in fall 2012.[32] Traffic deaths of ungulates on the 12 mi stretch for the period September 1–December 31, which had averaged 46/year for the period 2006–2011, declined to 9 for the same period in 2012.[33] What is not clear is whether the overpasses and underpasses will mitigate the broader effects of dispersed development on pronghorns, so that the traditional migrations will continue.

Notes for Part I

INTRODUCTION

1. Jones and Manning (1992) is useful for genus-level identification. The Mammalian Species accounts published by the American Society of Mammalogists (available online) provide summaries of diagnostic traits. Hoffmeister (1986) and Armstrong et al. (2011) provide the most detailed and geographically relevant taxonomic keys.

2. Murie (1954) has been in press continuously for over 50 years; Elbroch (2003).

3. NatureServe is a program of the Natural Heritage Network (Patterson et al. 2007); its maps are used here by permission.

4. University of Idaho and US Geological Survey Gap Analysis Programs. http://www.gap.uidaho.edu/NorthwestGAPSpeciesviewer/, accessed 2012–2013.

5. See chapter on "Conservation and Management" for a discussion of State Species of Greatest Conservation Need.

HISTORY OF WYOMING MAMMALOGY

1. See Armstrong publications cited in Armstrong et al. (2011). J. N. Jones publications are listed in Findley et al. (1996).

2. Many observations of mammals and other vertebrates recorded during 1805–1878 were reviewed and organized by Dorn (1986).

3. Russell (1921) for grizzlies, Whittlesey and Schullery (2011) for wolves, Lee (1916) for bighorn sheep near the Sweetwater River.

4. Krohn and Hoving (2010).

5. Jennings (1975), Parkman (2008).

6. Baird (1859).

7. Data are from online catalogues of the National Museum of Natural History and University of Kansas Museum of Natural History, accessed October 2011.

8. Figure reproduced from Cary (1917).

9. Robert Timm, KU, pers. comm.
10. Sterling (1989).
11. Cary (1911, 1917), Merriam (1898).
12. Murie (1940), Houston (1982), Craighead et al. (1995).
13. Franklin (2004).
14. Murie (1951).
15. Murie (1940) for Yellowstone coyotes, Murie (1944) for Alaskan wolves.
16. Rawson (2001), Franklin (2004).
17. Boxes no. 1–6, Murie Family Papers, Collection no. 11375, American Heritage Center, University of Wyoming, Laramie.
18. Long (1965).
19. This takes into account taxonomic synonomies recognized since 1965.
20. See Long (2008) for his mammal fauna of Wisconsin.
21. Clark et al. (1980), Web of Science, accessed 28 May 2013.
22. Clark and Stromberg (1987).
23. The original station was located along the Snake River 1.6 km (1 mi) downstream from the Jackson Lake Dam (Clarke 1955). The modern station is on the eastern shore of Jackson Lake, at the old AMK Ranch.
24. Altmann (1952).
25. Negus and Findley (1959).
26. Armitage (1961).
27. Houston (1967).
28. Calder (1969).

ZOOGEOGRAPHY

1. Peppers et al. (1996) for coyote; Anderson et al. (2004) for mountain lion.
2. See species account for black bear.
3. Epps et al. (2005).
4. Hall (1981).
5. Hafner and Smith (2010).
6. Merriam (1918, 9).
7. Rausch (1953).
8. Mayr (1942).
9. Paetkau et al. (1997).
10. Baker and Bradley (2006).
11. Moratelli (2014) discusses this controversy from the perspective of a mammalogist.
12. Sheets (1972), Clark, Campbell et al. (1984).
13. See species account for wolverine, Boyd and Pletscher (1999) for wolves.
14. Long et al. (2008).
15. Fitzpatrick et al. (2005).
16. McKelvey et al. (2008).
17. Frey (2009) for statistical detection of range expansions.
18. Hall and Kelson (1959), Long (1965).
19. Armstrong et al. (1986).
20. Grayson (2006).
21. See species accounts.
22. Knight et al. (2014).
23. Findley and Anderson (1956).

24. Also called the Great Divide Basin, which corresponds roughly to the northern Red Desert.

25. Among-site $F_{st} = 0.03$, Anderson et al. (2004).

26. Medicine Bow–Bighorn Mountains. $D_{LR} = 4.6$, Medicine Bow–Teton Mountains. $D_{LR} = 3.9$; David B. McDonald, University of Wyoming, pers. comm. The value for Canadian black bear over a distance of 3400 km (2000 mi) of boreal forest habitat was $D_{LR} = 5.3$ (Paetkau et al. 1997). The unforested basins of Wyoming are around one-tenth as permeable to black bears as is coniferous forest.

27. Klaus et al. (2001).

28. "Big sagebrush" is the common name of *Artemisia tridentata*. See the list of plant names for equivalencies.

29. Thimmayya and Buskirk (2012).

PREHISTORY

1. Kielan-Jaworowska et al. (2004).

2. McKenna and Bell (1997).

3. Martin (2006), Luo (2007).

4. Cope (1892).

5. Lillegraven and Ostresh (1988), Snoke (1993).

6. MacLeod (2014) reviewed findings that show a number of non-avian dinosaur lineages making their final appearance in the stratigraphic record in the Hell Creek Formation of Montana between 2.4 and 10 m below the K–Pg boundary.

7. Wilson (2014).

8. Springer et al. (2003).

9. Alroy (1998, 1999).

10. Paul (1992).

11. Four times for birds and mammals in the body size class of 10 g (0.4 oz.); 100 times for animals in the body size class of 1 kg (Silva et al. 1997).

12. Rohl et al. (2007).

13. Gingerich (2003, 2006).

14. Prothero and Heaton (1996), Zanazzi et al. (2007).

15. Wyoming Geological Survey data, based on Love and Christianson (1985).

16. Webb and Barnosky (1989).

17. Woodburne (2010). See Bacon et al. (2015) for mammalian fossil evidence and Montes et al. (2015) for geological evidence.

18. See respective species accounts.

19. Elias et al. (1996), Marincovich and Gladenkov (1999).

20. Anderson (1984).

21. A period of even greater total mammalian exchange, although not a higher rate, between North America and Eurasia occurred during early in the Eocene, when the Bering Land Bridge was open, and another dry route connected temperate North America, Ellesmere Island, Greenland, and northwestern Europe. The High Arctic at that time featured mixed conifer-broad-leaved rain forests, and such tropical vertebrates as alligators, turtles, and tapirs. Then, the mammal communities of temperate North America and Europe shared more than 50% of their genera—the highest degree of faunal sharing between the two continents in mammalian evolutionary history (McKenna 1975, Eberle and Greenwood 2012).

22. Webb (1985).

23. Webb and Barnosky (1989) synthesized data from various accounts, including Webb (1985).

24. Brugger (2006).

25. Gilbert and Martin (1984).

26. Lyons (2003), Williams and Jackson (2007).

27. Anderson (1968), Hager (1972), Kohn and McKay (2012).

28. Walker and Frison (1980). Other mammalian taxa found at the site included *Lepus* sp., Mexican ass (*Equus conversidens*), modern pronghorn, American camel (*Camelops* sp., with evidence of use as a tool), and bison (*Bos* sp.).

29. Koch and Barnosky (2006).

30. Fiedel (2009), Manley (2002).

31. Guthrie (1984).

32. Grayson (1984). This does not imply that humans killed every single member of these species. Cascading ecological effects involving habitat changes resulting from major declines in herbivore densities, along with demographic processes in the newly reduced populations, could have accounted for further population losses.

33. Haynes (1969), Rinella (2009).

34. Curry (2012).

35. Martin (1984).

36. Fiedel (2009).

VEGETATION AS HABITAT

1. Morrison (2006).

2. Maps 6.2–6.7 depict contiguous mapped patches of at least 250 pixels (5.63 km^2) of the indicated habitat type.

3. I relied heavily on Knight et al. (2014) in preparing this section.

4. Dorn (1986).

5. See species accounts for prairie dogs and badgers.

6. It covers around one-third of the state (Knight et al. 2014).

7. Theis and Lerdau (2003).

8. Knight et al. (2014).

9. Merrill et al. (1996).

10. Beuchat (1990).

11. See species accounts.

12. Douglas et al. (2005).

13. Buskirk et al. (1989).

14. Bentz et al. (2010).

15. Johnson et al. (2014).

16. The elevation of upper tree line is around 440 m (1450 ft) lower at the northern border of Wyoming than at the southern border.

17. Hultén (1968).

18. Streeter and Braun (1968) for martens. Weasels of unknown species commonly are observed in alpine talus fields statewide.

19. Inman et al. (2007) for the Greater Yellowstone Area.

20. Merrill and Boyce (1991).

21. Hoffman (1974).

22. Buechner (1960, 122).

CONSERVATION AND MANAGEMENT

1. Wyo. Stat. §23-1-103: "For the purpose of this act, all wildlife in Wyoming is the property of the state. . . There shall be no private ownership of live animals classified in this act as big or trophy game animals or of any wolf or wolf hybrid."

2. Kingham (1990).

3. Bureau of Land Management (2011), unpublished data.

4. Williams (2005).

5. Wild Free Roaming Horse and Burro Act (Public Law 92-195), amended in 1976 and 1978.

6. See relevant species account.

7. Black (2012).

8. Of Yellowstone National Park's area, 96% lies within Wyoming.

9. Blair (1987, 25).

10. Ward v. Race Horse (163 U.S 505 [1896]) was an important decision in the evolution of American wildlife law, the Supreme Court having found that nothing in the Indian treaty was specifically intended to supersede subsequent state law. The paradigm of state ownership of all wildlife on federal lands—areas with exclusive federal jurisdiction excepted—prevailed nationwide for the next 70 years (Bean 1983).

11. Blair (1987).

12. Bean (1983).

13. Crowe (1990) provides a detailed and colorful account of this period.

14. Leopold et al. (1963).

15. Executive Order No. 11643, signed February 8, 1972. The order was amended by several subsequent executive orders, then revoked by No. 12342 in 1982. The effect of the original order was largely maintained by other laws enacted by the time of the revocation order.

16. Alagona (2013) explained how the Endangered Species Act attained its current national significance, using case histories from California.

17. Van Manen et al. (2013).

18. Wyoming Game and Fish Department (2010a).

19. 16 US Code 2901–2911.

20. Outdoor Industry Association 2010 annual report, accessed online, April 18, 2013.

21. Map modified from Murie (1951) and Robbins et al. (1982).

22. Harvey (1994).

23. Rompola and Anderson (2002, 2004) showed that these species were most common in areas near cliffs; most cliffs in the area were submerged by the impoundment.

24. Honess and Frost (1942, 30).

25. Online searches of U.W. Coe Library, Web of Science, and Google Scholar conducted on August 13 2013. Google Scholar generated over five million hits for "Yellowstone wolf."

26. Wolf Management Committee (1991), McNamee (1997), Smith and Ferguson (2005).

27. Corbet (1978).

28. See references at species account for gray wolf.

29. Vucetich et al. (2005), Middleton, Morrison et al. (2013).

30. See ungulate species accounts.

31. Berger (2004).

32. Sawyer and LeBeau (2013).

33. Wyoming Department of Transportation unpublished data.

PART II

SPECIES ACCOUNTS

Morphological Keys

Accurate identification is more difficult for some taxa. In these cases, readers will find morphological keys at the beginning of the following discussions: shrews and moles (Soricomorpha); bats (Chiroptera); rodents (Rodentia); squirrels (Sciuridae); pocket mice and kangaroo rats (Heteromyidae); voles, mice, lemmings, and rats (Cricetidae); pocket gophers (Geomyidae); and pikas, rabbits, and hares (Lagomorpha). These keys are modified versions of ones that appear in the following sources: Hoffmeister (1986), Clark and Campbell (1987), and Armstrong et al. (2011). The key for bats is derived from one prepared by D. Keinath, Wyoming Natural Diversity Database.

Wyoming Distribution Maps

State-level distributions depicted with points are based on specimens or observations, and are supported by the online appendices (www.wyomammals.com). Distributions depicted by shading are based on models or approximations, explained in the introduction.

Notes and References

Numbered end notes appear last in each account. Note numbers that are repeated within the text of an account indicate the relevance of a previously cited note.

New World Opossums—Order Didelphimorphia

This relatively new order replaces the old Order Marsupialia, which is now recognized to contain several groups that themselves deserve the status of orders, having radiated for very long time spans. The didelphimorphs are those members of Infraclass Metatheria that radiated in the Americas, spreading to Antarctica as well, and are fairly underived (unchanged through time). The reproductive biology of the group differs markedly from that of eutherian mammals, with differing anatomy, a brief gestation period, and highly altricial neonates that nurse during an extended development. The group contains only one family.

NEW WORLD OPOSSUMS—FAMILY DIDELPHIDAE

This family has undergone recent major reorganization and now comprises 17 genera and around 90 living species restricted to the Americas. The group is morphologically generalized, having five digits per limb, a prehensile tail, and a walking gait with strong climbing abilities. The family occupies a wide range of mostly Neotropical habitats and includes *Chironectes*, a semiaquatic fish predator of Central and South America.

Virginia opossum, *Didelphis virginiana*

DESCRIPTION: A house-cat-sized marsupial with coarse grayish fur on the back, lighter face, pink nose (in the northern part of the distribution), and long, sparsely haired prehensile tail. There are five digits per foot, and the gait is typically a walk. The opposable medial digit of the hind limb has no claw. The dentary bones have an angular process reflected medially: characteristic of metatherians and distinct from all eutherians. Measurements (**Sex** [n] mean (mm, g), *range*): Pennsylvania, L = **F** [18] 753, *666–828*; **M** [10] 779, *698–883*; TL = **F** [18] 289, *216–319*; **M** [10] 291, *221–356*; HF = **F** [18] 64, *60–68*; **M** [10] 69, *63–75*; EFN = **F** [18] 55, *49–60*; **M** [10] 54, *44–60*; WT = **F** [18] 2400, *1820–2990*; **M** [10] 3400, *2190–4000*.[1] Dental formula: 5/4, 1/1, 3/3, 4/4 = 50.

STATUS: A non-protected nongame animal, which likely began to occur in Wyoming as a breeding population in the mid-twentieth century. It is a native in the sense of having arrived in the state via dispersal, likely facilitated by human-caused changes to the environment: land-use changes toward agriculture and climatic warming. It is now common in a very limited area of Wyoming. Unconfirmed reports of opossums from anomalous sites around the state suggest that they may be transported by vehicles.

NOMENCLATURE AND SYSTEMATICS: The scientific name has been stable for over a century. The species is the sole representative of a group that has diverged and radiated little over its 100-Ma history.[2] The lineage arose in North America, became extinct here during the

Locations from which the Virginia opossum has been identified in Wyoming.

Middle Miocene (while persisting in South America), and then recolonized North America from the south during the Great American Biotic Interchange.[3] It appeared at latitudes as far north as Wyoming during the late Pleistocene.[4] A sibling species (*D. marsupialis*) from Mexico to South America has been treated by some as conspecific.

DISTRIBUTION AND HABITAT ASSOCIATIONS: From its extent at the time of European contact (shown), the distribution in North America has expanded northward by >200 km (160 mi) in some places during the historical period, and the species was introduced to the Pacific Coast as early as 1915 and to sites in Colorado and New Mexico as a food source during the Great Depression.[5] The distribution in Kansas and Nebraska in the mid-twentieth century was restricted to the eastern halves of those states, but has since expanded to west the Wyoming boundary.[6] The known distribution in Wyoming in 1965 was a single record from northeastern Converse County; by 1987 the species was known to occur predictably in the riparian zone of the North Platte River just upriver from the state line.[7] Since 1987, the distribution has expanded upstream (northwesterly) another 70 km (43 mi), to as far upstream as Guernsey Reservoir in Platte County.[8] Habitats in Wyoming tend toward mesic woodland, riparian zones, and sites near human developments. Arid areas without water or vegetative cover are barriers to dispersal, and very cold environments without sheltering sites are not occupied.

ECOLOGY: The species has an unusual suite of characteristics among Wyoming mammals, being poorly adapted to cold or to highly arid environments. Although not studied in or near Wyoming, it clearly uses a narrow range of the habitats available in Wyoming, is a trophic generalist, and has a "fast" life history, with a short time to sexual maturity, large litters (up to 12), multiple litters per year in some locations, and brief life expectancy at birth. The diet includes insects, carrion, fruits, grains, and human foods. It does not hibernate, but may remain inactive for extended periods.[9]

CONSERVATION AND MANAGEMENT: The species is more often considered a pest than a conservation concern, and is seldom considered in conservation plans. Because of its sensitivity to low temperatures and its northward expansion over the historical period, the species is often considered to have expanded its geographic range in response to climatic warming. Such an ongoing change would likely involve further range expansion upstream on the North Platte River and possibly other inroads into the state on eastward-flowing drainages, for example, the Niobrara River and tributaries to the South Platte River.

NOTES AND REFERENCES

1. Blumenthal and Kirkland (1976).
2. McKenna and Bell (1997).
3. See chapter on prehistory.
4. Kurtén and Anderson (1980).
5. Grinnell (1915), McManus (1974).
6. Benedict et al. (2000), Gardner and Sunquist (2003).
7. Long (1965), Clark and Stromberg (1987).
8. Recent range extension along the North Platte River were based on specimens from Craig Acres, USDA Wildlife Services, Casper, Wyoming. This extension is approximately 40 km (25 mi) beyond that shown by Clark and Stromberg (1987).
9. Gardner and Sunquist (2003).

Shrews and Moles—Order Soricomorpha

Shrews and moles represent a taxon that was formerly grouped in the now-discarded Order Insectivora. The group now has been vastly revised based on improved understandings from molecular data. Two North American families are recognized.

KEY TO SPECIES OF SORICOMORPHA

1a. Tail length less than 20% of total length, forefeet flattened, enlarged; zygomatic arch complete; 36 teeth: . . . . ***Scalopus aquaticus***
1b. Tail length greater than 25% of total length, forefeet not flattened; zygomatic arch absent; fewer than 36 teeth: . . . . ***Sorex***, 2

2a. Length less than 130 mm (5.1 in); dorsal pelage brownish; no fringe of stiff hairs on hind feet: 3
2b. Length greater than 130 mm (5.1 in); dorsal pelage blackish gray; fringe of stiff hairs on hind feet: . . . . ***Sorex navigator***

3a. Only three unicuspids visible when viewed laterally: . . . . ***Sorex hoyi***
3b. More than three unicuspids visible when viewed laterally: 4

4a. Third unicuspid similar in size or larger than fourth: 5
4b. Third unicuspid obviously smaller than fourth: 8

5a. Whitish underparts, crowded unicuspid row, no medial tine on first upper incisor: . . . . ***Sorex merriami***
5b. Brownish underparts, uncrowded unicuspid row, medial tine present on first upper incisor

6a. Condylobasal length less than 14.2 mm, maxillary breadth less than 4.2 mm in northwestern mountains: . . . . ***Sorex preblei***
6b. Condylobasal length and maxillary breadth not as in 6a: 7

7a. Pigmentation on lower teeth continuous from one tooth to the next, from northeastern Wyoming: . . . . ***Sorex haydeni***
7b. Pigmentation on lower teeth discontinuous from one tooth to the next, being separated by unpigmented areas: . . . . ***Sorex cinereus***

8a. Having at least two of the following traits: tail no more than 40 mm long, hindfoot shorter than 11 mm, condylobasal length of skull less than 15.2 mm: . . . . **Sorex nanus**

8b. Having at least two of the following traits: tail longer than 40 mm, hindfoot longer than 11 mm, braincase convex when viewed laterally: 9

9a. Tail no longer than 41 mm, total length no more than 108 mm, medial tine on first upper incisor approximately at upper edge of pigmentation: . . . . **Sorex vagrans**

9b. Tail longer than 41 mm, total length greater than 108, medial tine on first upper incisor well below upper edge of pigmentation: . . . . **Sorex monticolus**

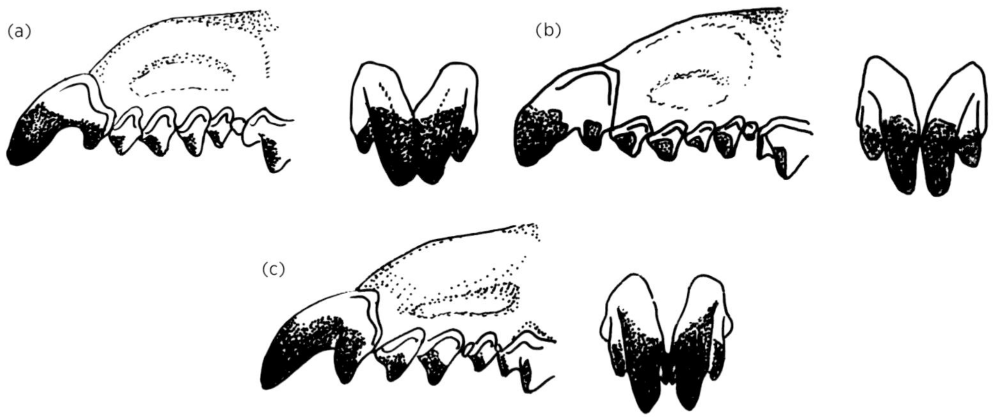

Lateral views of upper first incisors and unicuspid teeth, and frontal views of first upper incisors of three morphological subgroups of *Sorex* shrews (not congruent with putative subgenera) from Wyoming. (a) *S. cinereus, S. haydeni, S. merriami,* and *S. preblei,* showing the third unicuspid at least as large as the fourth, and the fifth unicuspid tiny (the upper first incisor has two cusps); and no medial tines visible on the first upper incisors. (b) *S. monticolus, S. nanus,* and *S. vagrans,* showing the third unicuspid smaller than the fourth, and the fifth tiny; and medial tines of the first upper incisors intermediate in length. (c) *S. hoyi,* showing only three unicuspid teeth visible laterally and medial tines on first upper unicuspids long, extending well below the upper limit of pigment. Note: *S. navigator* is not illustrated. Source: Drawings are modified from Diersing (1980).

SHREWS—FAMILY SORICIDAE

This family comprises a group that has remained unchanged in the face of newly reported molecular data. Three major subdivisions within the extant shrews are recognized: Crocidurini, Myosoricini, and Soricini; these are regarded as tribes.[1] All Wyoming shrews are soricine, of the genus *Sorex*: very small-bodied predators and scavengers that are active all year and tend to associate with moist habitats. The systematics of *Sorex* is among the most challenging and unstable of any North American mammalian genus. Major revisions have been made at the subgeneric level, and others can be expected; some of the species' taxonomies presented here should be considered tentative. Identification of live shrews in the field is most challenging: water shrews can be recognized moving underwater in creeks in montane and alpine areas, or in hand, but other shrews are similar in size, pelage, and shape, and are most reliably identified to species on the basis of number, size, and shape of incisors, unicuspid teeth, and tail length. Molecular markers for species identification are becoming available. All shrews known from Wyoming have the same dental formula (3/1, 1/1, 3/1, 3/3 = 32), but not all North American shrews do; the least shrew found in Nebraska near the Wyoming state line has 30 teeth instead. Shrews have extreme life history characteristics because of their small body sizes: high metabolic rates, poor tolerance of fasting, frequent bouts of foraging and resting, year-round activity, and brief lives. Like most shrews in North America, those in Wyoming have teeth that have chestnut pigmentation on their tips. At a continental level, shrews achieve their highest species richness in areas with dense vegetation and abundant rainfall: the Cascade and Appalachian Mountains have as many as eight species in a single area, whereas some sites in the Great Plains have only one.[2] It is surprising therefore that Wyoming has so much shrew diversity: nine species statewide and a five-species assemblage in a single shrub-steppe community in Lincoln County.[3] And a 10th species may be present but undetected: the least shrew (*Cryptotis parva*) has been collected less than 0.5 km (0.3 mi) from the state line along the North Platte River in westernmost Nebraska; it likely occurs undetected in Wyoming as well.[4]

NOTES AND REFERENCES

1. Wilson and Reeder (2005).
2. Berman et al. (2007).
3. This assemblage was described based on a seven-year effort involving 514,060 trap days at several sites in southern Lincoln County (Kirkland et al. 1997). From that project, 156 specimens were preserved in the Museum of Southwestern Biology (catalogue numbers MSB 139100–139255).
4. Geluso, White et al. (2004).

Masked shrew, *Sorex cinereus*

DESCRIPTION: A small to medium-sized shrew, with brown dorsal pelage and silvery undersides. Compared with the very similar Hayden's shrew, this shrew has more brown in its pelage, and its tail is longer.[1] The four unicuspid teeth generally decrease in size from anterior to posterior, viewed laterally. Measurements (**Sex** [n] mean (mm, g), *range*): various counties, L = **F** [18] 92, *84–104*; **M** [11] 97, *87–111*; TL = **F** [18] 39, *35–43*; **M** [11] 39, *31–46*; HF = **F** [18] 11, *9–12*; **M** [11] 11, *10–13*; EFN = **F** [18] 4, *2–6*; **M** [11] 4, *3–7*; WT = **F** [18] 3.1, *2.5–4.6*; **M** [11] 4.0, *2.8–5.9*.[2] Dental formula: 3/1, 1/1, 3/1, 3/3 = 32.

STATUS: This is a non-protected nongame animal by regulation.[3]

NOMENCLATURE AND SYSTEMATICS: The species was recognized in the late eighteenth century, and the scientific name has been stable for over a century. Of the seven subspecies

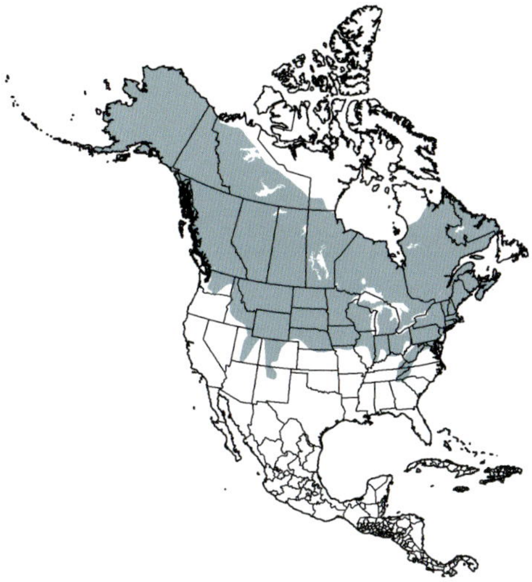

Distribution of the masked shrew in Wyoming.

recognized as of 2004, only *S. c. cinereus* occurs in Wyoming. The common names "cinereus shrew" and "common shrew" also are used.[1]

DISTRIBUTION AND HABITAT ASSOCIATIONS: This primarily boreal species occurs statewide in suitable habitats, although not in arid regions. Habitats are quite diverse, as suggested by the species' vast geographic range, but tend toward moist sites within grasslands, forests, woodlands, and alpine areas. Masked shrews were found in all but one of several habitats in the Medicine Bow Mountains and Laramie Basin, being absent only from shortgrass prairie.[4] Boggy areas and sites with mosses are preferred.

ECOLOGY: This—the most widespread shrew in North America—was the second-most abundant of six shrew species captured in montane and basin habitats in Albany County, and was by far the most abundant of five species captured in sagebrush steppe in Lincoln County.[5] The diet consists of a wide range of animal matter: insects, other arthropods, earthworms, mollusks, and small vertebrates. Carrion is readily consumed when available. Shrews are minor foods of small carnivores such as weasels, sometimes being killed and then left uneaten. Some are killed by raptors as well. Population densities have been estimated, but not near Wyoming; densities undergo intra-annual fluctuations of over an order of magnitude, mostly reflecting seasonal reproduction and overwinter deaths.

CONSERVATION AND MANAGEMENT: Masked shrews are common and not a perceived conservation concern; neither are they economic pests. The species was introduced to Newfoundland Island in the 1950s to control the larch sawfly; this attempt at biological control achieved some success.[3]

NOTES AND REFERENCES

1. Whitaker (2004).
2. DMNS specimens.
3. Commission Regulations Chapter 52, Section 6.
4. Brown (1967a).
5. Brown (1967a), Kirkland et al. (1997).

Hayden's shrew, *Sorex haydeni*

DESCRIPTION: A shrew smaller in overall dimensions and skull size than *S. cinereus*. The difference in body length is less than 1 mm (0.04 in) and that in skull length is slightly more than 1 mm; overlap between the two species in other measurements and pelage coloration is too great to be used for identification. Differential identification relative to masked shrew requires statistical treatment of multiple measurements, particularly total length, skull length, and length of the unicuspid tooth row. Molecular markers can distinguish between the two species as well.[1] Measurements (**Sex** [n] mean (mm, g), *range*): various counties, L = **F** [5] 94, *82–100*; **M** [7] 97, *92–100*; TL = **F** [5] 39, *37–43*; **M** [7] 38, *36–43*; HF = **F** [5] 12, *11–12*; **M** [7] 12, *11–12*.[2] Dental formula: 3/1, 1/1, 3/1, 3/3 = 32.

STATUS: This was a protected nongame animal and a state Species of Greatest Conservation Need as of 2012.[3]

NOMENCLATURE AND SYSTEMATICS: Formerly considered a subspecies of masked shrew, this taxon was raised to the level of species in the 1980s based on morphological traits, and later molecular analyses have corroborated this classification. Recent molecular studies have suggested that the closest phylogenetic ties are to an as-yet-unnamed lineage from the Southwest, and to *S. preblei*.[4] Subspecies have not been described. The common name "prairie shrew" also is used.

DISTRIBUTION AND HABITAT ASSOCIATIONS: The geographic range in the north-central states, Alberta and Saskatchewan, represents a hole in the geographic range of *S. cinereus*, with a zone of sympatry around its perimeter, including northeastern Wyoming.[5] Habitats tend to be drier sites with more grassland than is the case for the masked shrew.[6]

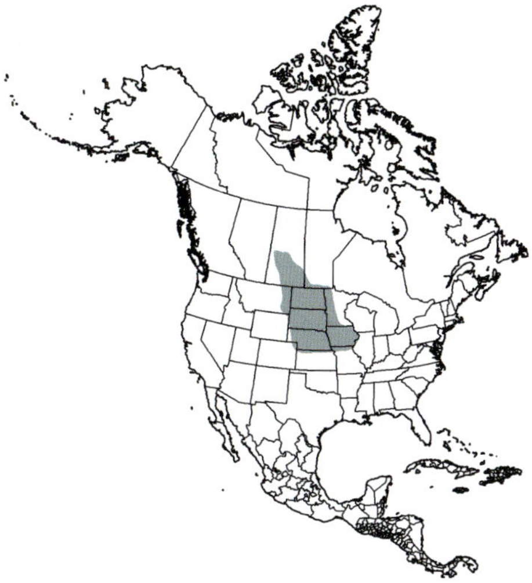

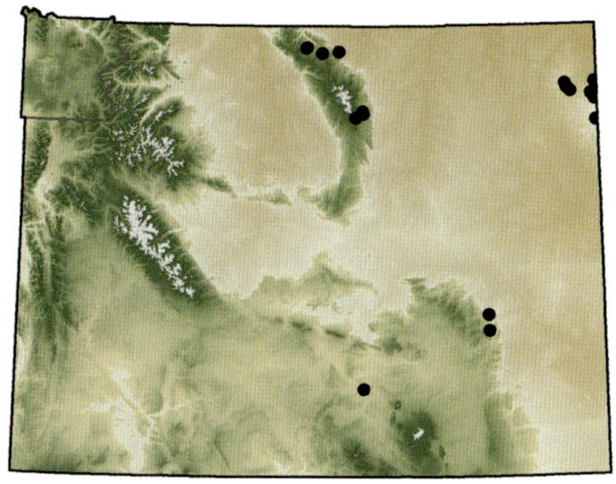

Locations from which Hayden's shrew has been identified in Wyoming.

ECOLOGY: This species is considered, mostly by nature of its geographic range, to be a closer associate of dry grassland vegetation than is the case for the masked shrew, which is more likely to be found in mesic forested habitats.[6] Having been recognized as a species so recently, its ecology is little known.

CONSERVATION AND MANAGEMENT: The species is of conservation concern because of a scarcity of information about its distribution, population status, and trends.

NOTES AND REFERENCES

1. van Zyll de Jong (1980), van Zyll de Jong and Kirkland (1989), Foresman and Jensen (1992).
2. UWMV specimens.
3. Commission Regulations Chapter 52, Section 11.
4. van Zyll de Jong and Kirkland (1989), Hope et al. (2012).
5. van Zyll de Jong and Kirkland (1989).
6. Foresman and Jensen (1992).

American pygmy shrew, *Sorex hoyi*

DESCRIPTION: This species contends for the title of smallest shrew in North America, with adult body weights as low as 2.2 g. Viewed laterally, the third unicuspid tooth is much smaller than the second.[1] Measurements (**Sex** [n] mean (mm, g), *range*): various counties, L = **B** [53] 86, *71–104*; TL = **B** [53] 31, *24–35*; HF = **B** [55] 10, *8–11.5*; WT = **B** [9] 3.6, *2.4–5.4*.[2] Dental formula: 3/1, 1/1, 3/1, 3/3 = 32.

STATUS: The species is a protected nongame animal and a state Species of Greatest Conservation Need as of 2012.[3]

NOMENCLATURE AND SYSTEMATICS: The species was originally described from Wisconsin in the mid-nineteenth century as "Hoy's pigmy shrew (*Microsorex hoyi*)."[1] It was moved to

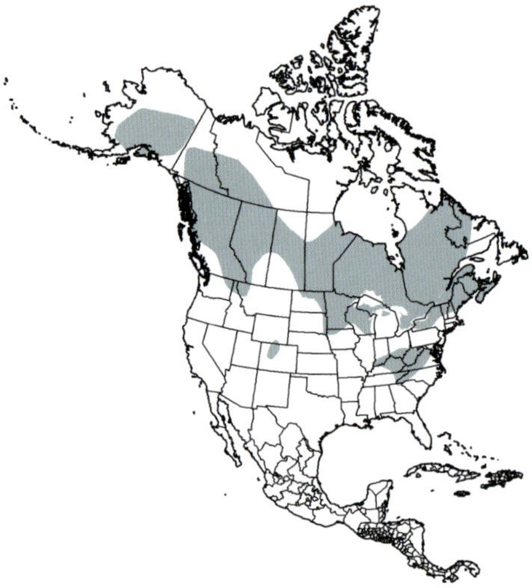

Distribution of the pygmy shrew in Wyoming.

genus *Sorex* in the 1980s, but remains in subgenus *Otisorex*.[4] Wyoming animals are of a disjunct subspecies, *S. h. montanus*.[1]

DISTRIBUTION AND HABITAT ASSOCIATIONS: The distribution in Wyoming and northern Colorado represents a disjunct part of a vast boreal and Appalachian range. This part of the geographic range was first detected near Cameron Pass in northern Colorado in 1961, and is now known to extend as far south as Gunnison County, Colorado, and as far north as the Snowy Range and Sierra Madres (Medicine Bow Mountains).[5] Habitats in the central Rocky Mountains are subalpine forests and meadows; most records are from elevations above 2900 m (9600 ft).

ECOLOGY: Almost nothing is known about the species, particularly in the central Rockies, except for the habitat associations characterized above and that insects have been detected in stomachs of captured animals.

CONSERVATION AND MANAGEMENT: The conservation concern regarding the species results from its highly restricted distribution in the state and narrow range of habitats occupied.[6]

NOTES AND REFERENCES

1. Long (1974).
2. van Zyll de Jong (1983).
3. Commission Regulations Chapter 52, Section 11.
4. *Otisorex* is synonymous with *Microsorex*. Other members of the subgenus are disputed. The distinctness of *S. hoyi* from more eastern and smaller *S. thompsoni* was confirmed by Stewart et al. (2003), although proposed much earlier. The complex history of phylogenetic understandings in this group is reviewed by Wilson and Reeder (2005).
5. Pettus and Lechleitner (1963), Armstrong et al. (2011).
6. Beauvais and McCumber (2006).

Merriam's shrew, *Sorex merriami*

DESCRIPTION: A medium-sized shrew with gray dorsal pelage, light underside, and bicolored tail. The dentition is heavy, by soricid standards, with molars as wide as long. The first three unicuspids are similar in size; the fifth much smaller than the first four.[1] Measurements (**Sex** [n] mean (mm, g), *range*): northern Arizona, L = **B** [11] 96, *89–101*; TL = **B** [11] 36, *33–42*; HF = **B** [11] 12, *11–13*.[2] Dental formula: 3/1, 1/1, 3/1, 3/3 = 32.

STATUS: This uncommon shrew is a protected nongame animal by regulation.[3]

NOMENCLATURE AND SYSTEMATICS: The species was first described, using its current scientific name, in 1890. It has undergone systematic revision in the area of Arizona, but not in or near Wyoming; no subspecies were recognized as of 2014.[4]

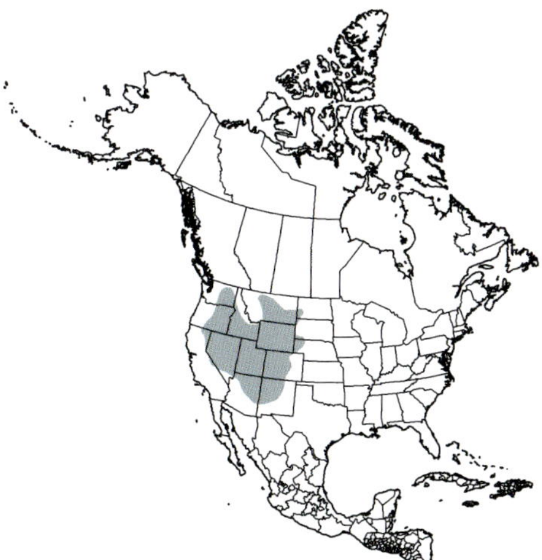

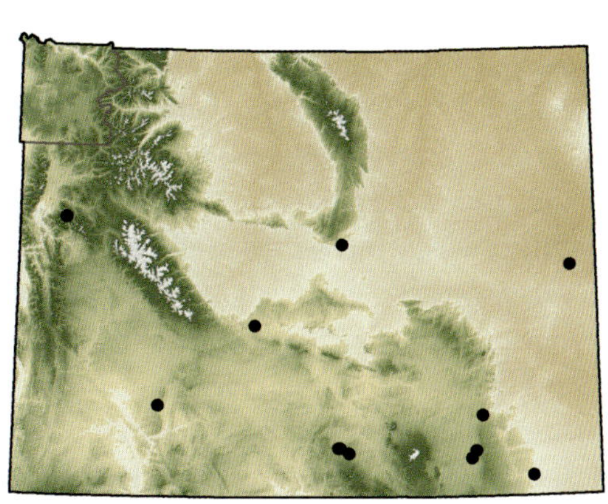

Locations from which Merriam's shrew has been identified in Wyoming.

DISTRIBUTION AND HABITAT ASSOCIATIONS: This western intermountain species potentially occurs statewide, but is so uncommon that it has been identified from only 11 localities. Habitats in and around Wyoming are variably reported as moist sites in subalpine forest, alpine meadows, grasslands, and mountain mahogany.[5] Specimens from Colorado have typically been collected at 1370–2920 m (4500–9600 ft) elevation.[6]

ECOLOGY: Only 106 specimens had been collected from throughout the species' known range by 1977, so it is not well known, aside from the habitats in which it is caught. It has been identified from the pellets of great horned owls.[7]

CONSERVATION AND MANAGEMENT: The conservation concern regarding this mammal results from its uncommonness and our lack of understanding of population status or habitat requirements.

NOTES AND REFERENCES

1. Armstrong and Jones (1971).
2. Hoffmeister (1986).
3. Commission Regulations Chapter 52, Section 11.
4. Diersing and Hoffmeister (1977), but see Hall (1981).
5. Long and Kerfoot (1963), Brown (1967a), Armstrong et al. (2011).
6. Armstrong et al. (2011).
7. Long and Kerfoot (1963).

Dusky shrew, *Sorex monticolus*

DESCRIPTION: A small to medium-sized shrew with a long tail. Compared to the very similar *S. vagrans*, this shrew has medial tines on its upper incisors that are large and positioned below the upper limit of the pigmentation on the upper incisors. The third unicuspid is smaller than the fourth. The toes are longer than for *S. vagrans*, and have more (5–6) paired callosities on hind digits 2–4 (*S. vagrans* has four or fewer). The latter character can only be used on fresh or liquid-preserved specimens.[1] Measurements (**Sex** [n] mean (mm, g), *range*): various Wyoming counties; Routt County, Colorado, L = **F** [19] 103, *87–112*; **M** [5] 108, *100–119*; TL = **F** [19] 43, *35–48*; **M** [5] 42, *41–49*; HF = **F** [19] 12, *11–14*; **M** [5] 11, *11–12*; WT = **F** [19] 4.2, *2.8–6.7*; **M** [5] 6.2, *3.8–10*.[2] Dental formula: 3/1, 1/1, 3/1, 3/3 = 32.

STATUS: A non-protected nongame animal by regulation.[3]

NOMENCLATURE AND SYSTEMATICS: The species was first described in the late nineteenth century and has carried the original scientific name without revision. This species is in the *S. vagrans* species group, and *S. monticolus* and *S. vagrans* were regarded by some biologists as conspecific before 1977, so literature prior to that date may refer to either species.[4] Around 18 subspecies were recognized as of 1996; these are mostly disjunct on coastal islands or mountain ranges. The subspecies in Wyoming extends to the Arctic Ocean: *S. m. obscurus*. The common name "montane shrew" also is used.

DISTRIBUTION AND HABITAT ASSOCIATIONS: This primarily Cordilleran species occurs in montane coniferous forests in all the major mountain ranges in the state; no records are

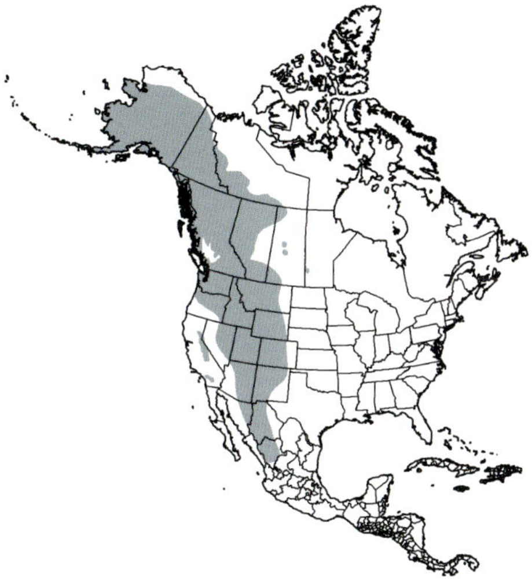

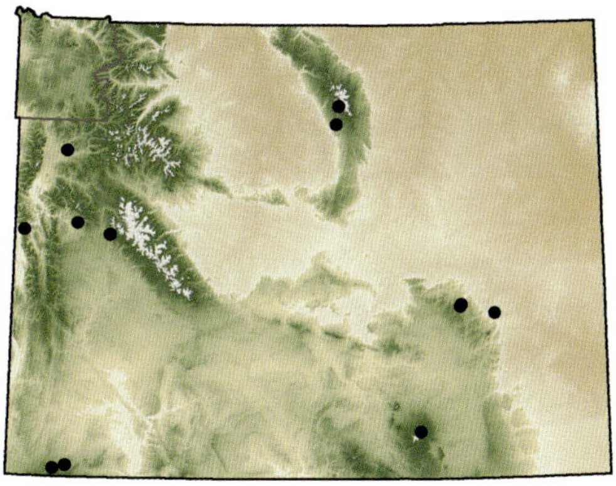

Locations from which the dusky shrew has been identified in Wyoming.

known for the Black Hills. The species was represented by Clark and Stromberg (1987) as occurring in the Green and Ferris Mountains and isolated, conifer-forested ridges in southern Sweetwater County, but there are no known specimens associated with these reports.[5] The species is variously reported from montane forested habitats, regenerating clear-cuts with herbaceous cover, pinyon–juniper woodlands, shrub-steppe dominated by low sagebrush, and big sagebrush–antelope bitterbrush.[6] The only feature that all these habitats share is abundant physical structure near the ground.[7]

ECOLOGY: The diet is typical of shrews, comprising mostly surface-dwelling invertebrates, conifer seeds, and some fungi. Little else is known about the ecology of the species.

CONSERVATION AND MANAGEMENT: The species is common and widely distributed; it is not generally considered a conservation concern.

NOTES AND REFERENCES

1. van Zyll de Jong (1982), Smith and Belk (1996).
2. DMNS, KU specimens.
3. Commission Regulations Chapter 52, Section 6.
4. Hennings and Hoffman (1977).
5. Clark and Stromberg (1987).
6. Smith and Belk (1996), Sutter et al. (1999).
7. Belk et al. (1990).

Dwarf shrew, *Sorex nanus*

DESCRIPTION: A small shrew with brown to buff dorsal pelage; the underside is smoke gray. The third and fifth unicuspid teeth are smaller than the fourth, as for *S. monticolus* and *S. vagrans*.[1] The tail is subtly bicolored. Measurements (**Sex** [n] mean (mm, g), *range*): various counties, Colorado, L = **B** [11] 84, *77–94*; TL = **B** [12] 41, *37–47*; HF = **B** [11] 11, *9–12*; WT = **B** [11] 2.1, *1.5–3*.[2] Dental formula: 3/1, 1/1, 3/1, 3/3 = 32.

STATUS: A protected nongame animal and state Species of Greatest Conservation Need as of 2012.[3]

NOMENCLATURE AND SYSTEMATICS: The species was first described in the late nineteenth century as a subspecies of *S. tenellus*, then raised to species level in 1928.[1] Subspecies have not been named, in spite of its disjunct distribution across the Rocky Mountains. The common name "Rocky Mountain dwarf shrew" also is used.

DISTRIBUTION AND HABITAT ASSOCIATIONS: This central and northern Rocky Mountain species was first detected in Wyoming in the Snowy Range of the Medicine Bow Mountains in 1947.[4] A capture in South Cascade Canyon of the Teton Mountains in

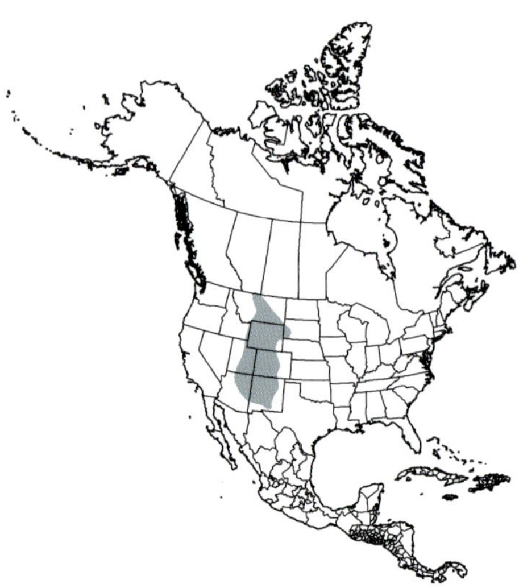

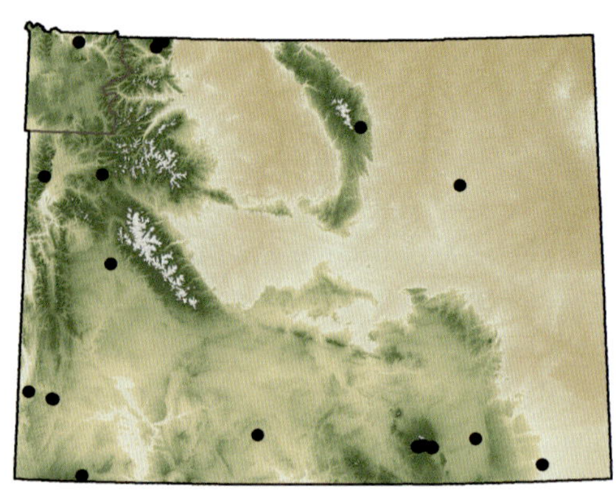

Locations from which the dwarf shrew has been identified in Wyoming.

1952 was the northernmost state record at that time.[5] Since then it has been found more extensively in the northwestern mountains of Wyoming.[6] Habitat in northern Colorado has been described as regenerating clear-cuts, and in southern Wyoming as a subalpine rockslides.[7]

ECOLOGY: The apparent rarity of this shrew is suggested by the fact that between its description in 1895 and 1966, only 18 specimens were reported across a geographic range that is now known to extend from southern New Mexico to the Canadian border.[1] Still, some biologists have been able to capture fairly large numbers in small areas; it has been common near Cameron Pass in northern Colorado.[7] Its rarity combined with the great difficulty of studying living animals through time limit our ecological knowledge of the species.

CONSERVATION AND MANAGEMENT: Conservation concern regarding the species results from our poor knowledge of its distribution, abundance, and life needs.

NOTES AND REFERENCES

1. Hoffman and Owen (1980).
2. UCM specimens.
3. Commission Regulations Chapter 52, Section 11.
4. Mickey (1948).
5. Durrant and Lee (1955).
6. Hoffman and Taber (1960).
7. Spencer and Pettus (1966), Brown (1967a).

Western water shrew, *Sorex navigator*

DESCRIPTION: A large shrew—the size of a mouse—with pelage that varies geographically and seasonally and stiff hairs along the lateral and medial margins of the hind feet (to a lesser extent, the front feet). The fourth unicuspid is larger than the third. In Wyoming, dorsal pelage is blackish with some white hairs; the underside is grayish. The tail is distinctly bicolored (dorsal–ventral) in most animals. In some animals from Wyoming the tip (5–10 mm) of the tail is white.[1] Measurements (**Sex** [n] mean (mm, g), *range*): Albany County, L = **F** [5] 143, *140–146*, **M** [5] 145, *130–155*; TL = **F** [5] 67, *65–74*; **M** [5] 71, *61–75*; HF = **F** [5] 20, *20–21*; **M** [5] 18, *15–19*.[2] Dental formula: 3/1, 1/1, 3/1, 3/3 = 32.

STATUS: This is a non-protected nongame animal by state regulation.[3]

NOMENCLATURE AND SYSTEMATICS: The species was first described in the early nineteenth century using *S. palustris*. Around nine subspecies were recognized in 1987, with only the nominal *S. p. navigator* found in Wyoming. Recently, molecular data have shown there to be four North American water shrew species, of which the Cordilleran and north coastal lineage now carries the specific epithet *S. navigator*.[4] Because the species is restricted to montane areas of the temperate western United States, biologists have tested for population genetic structure based on isolation by barrier, with the presumptive barrier being dry land. Isolation by land has not been confirmed; multiple lineages can share a single mountain

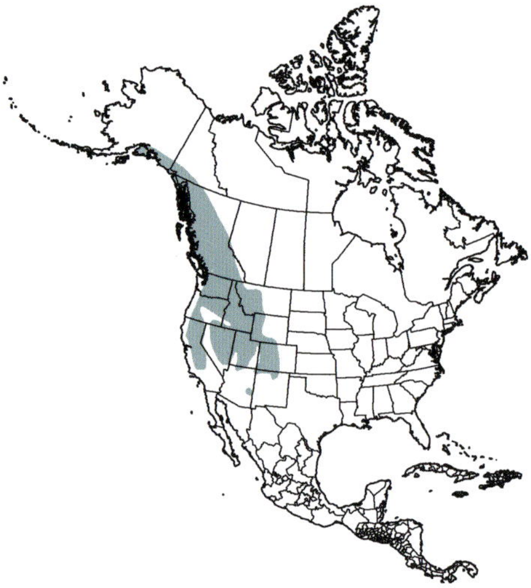

Locations from which the American water shrew has been identified in Wyoming.

range, a single lineage can occupy several isolated ranges, and other departures occur from the expected pattern.[5]

DISTRIBUTION AND HABITAT ASSOCIATIONS: The newly recognized species occupies the Cordillera as far north as the Aleutian Range, and coastal habitats.[6] The species occurs in montane areas statewide, including the isolated mountain ranges that parallel the Sweetwater River and in the Uinta Mountains. The Black Hills area is not occupied.

ECOLOGY: This is one of our best-known shrews, in part because its small body and semiaquatic environment have drawn the attention of physiologists.[7] Water shrews forage on aquatic and streamside animals: snails, slugs, stonefly nymphs, mayflies, caddis flies, spiders, fish, and carrion at night, using a combination of tactile sensation and olfaction.[8] Studying them in the Togwotee Pass area, William Calder found that water shrews entered the water frequently and dove below the surface for 30–45 seconds, moving along the bottom with a rapid walking motion, or swimming. They also enter water to evade danger.[9] Water shrews are preyed upon by a range of species, including garter snakes, weasels, and brown trout

CONSERVATION AND MANAGEMENT: The species is common within occupied areas of Wyoming and is not typically considered to be a conservation concern.

NOTES AND REFERENCES

1. Beneski and Stinson (1987) for general morphology. Brian Zinke (Nongame Program, Wyoming Game and Fish Department, Lander, pers. comm.) observed white tips on tails of around one-third of 73 water shrews caught in the Bighorn Mountains in 2014.
2. UWMV specimens.
3. Commission Regulations Chapter 52, Section 6.
4. Hope et al. (2014).
5. Himes and Kenagy (2010).
6. Continental distribution modified from NatureServe according to Hope et al. (2014).
7. Gusztak et al. (2005).
8. Catania (2013).
9. Calder (1969), Beneski and Stinson (1987).

Preble's shrew, *Sorex preblei*

DESCRIPTION: A small shrew, comparable in size to the pygmy and dwarf shrews. Dorsal pelage is gray and the underside is silver gray. The feet are buff colored and the tail bicolored. The third unicuspid is as large as or larger than the fourth. The foramen magnum is set low on the occiput. Relative to *S. merriami*, this species possesses a medial tine on the first upper incisor (see figure in Soricomorpha key) and a condylobasal length of less than 14.9 mm. Compared to *S. haydeni*, it exhibits a mandibular (lower) tooth row less than 4.2 mm long.[1] Measurements (**Sex** [n] mean (mm, g), *range*): Oregon, L = **F** [9] 83, *75–92*, **M** [15] 91, *75–99*; TL = **F** [9] 35, *32–37*; **M** [15] 36, *29–40*; HF = **F** [9] 11, *10–12*; **M** [15] 11, *10–11*; WT = **F** [9] 2.2, *2.1–2.3*; **M** [15] 2.4, *1.9–2.8*.[2] Dental formula: 3/1, 1/1, 3/1, 3/3 = 32.

STATUS: This is a protected nongame animal by state regulation and a state Species of Greatest Conservation Need as of 2012.[3]

NOMENCLATURE AND SYSTEMATICS: The species was first described on the basis of a single specimen from Oregon in 1922. Additional existing museum specimens were subsequently reassigned to this species. It has become somewhat better known in recent decades and is now considered to be within subgenus *Otisorex*.[4]

DISTRIBUTION AND HABITAT ASSOCIATIONS: This species of the northwestern United States was first noted in Wyoming in northernmost Yellowstone National Park, but a previously collected specimen from Uinta County was recognized by Tomasi and Hoffman in 1984 to be *S. preblei*.[5] The area of distribution in the southwest corner of the state is

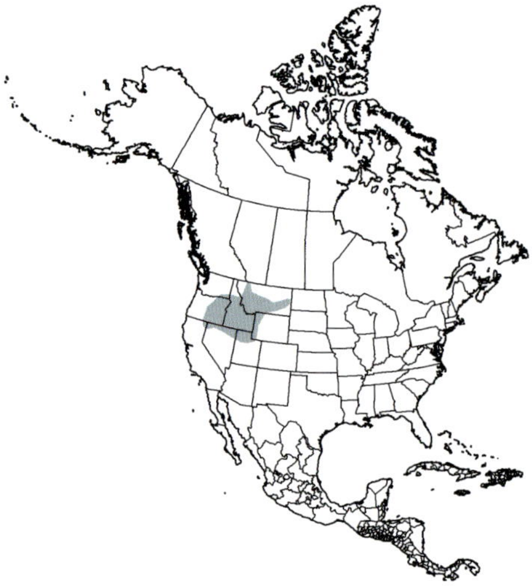

Locations from which Preble's shrew has been identified in Wyoming.

not well documented or understood. Habitats are not specifically known, but include moist sites with abundant vegetation near the ground. Dry sites within coniferous forests seem to be occupied as well.

ECOLOGY: This species is, if anything, less well known than the dwarf shrew. From the time of its description in 1922 until the next specimen was trapped in the wild and recognized as this species, 35 years had passed.[6] This clearly is one of North America's rarest mammals and most poorly understood with respect to ecology.

CONSERVATION AND MANAGEMENT: The species is of conservation concern because of its rarity, restricted distribution in the state, and our poor knowledge of its status.

NOTES AND REFERENCES

1. Cornely et al. (1992).
2. Verts and Carraway (1998).
3. Commission Regulations Chapter 52, Section 11.
4. Hoffman and Fisher (1978), Wilson and Reeder (2005).
5. Tomasi and Hoffman (1984). The specimen had been collected in 1948 and labeled as *S. cinereus* (KU Specimen 25178).
6. Hoffman and Fisher (1978).

Vagrant shrew, *Sorex vagrans*

DESCRIPTION: A small, moderately short-tailed shrew with, in the Rocky Mountains, grayish or brownish dorsal pelage. The underside is lighter, and the tail indistinctly bicolored in adults. Digits 2–4 on the hind limbs feature four or fewer paired callosities (cf. *S. monticolus*). The species is distinguished from other Wyoming *Sorex* by a condylobasal length generally 15.5–17.5 mm, and the third unicuspid noticeably smaller than

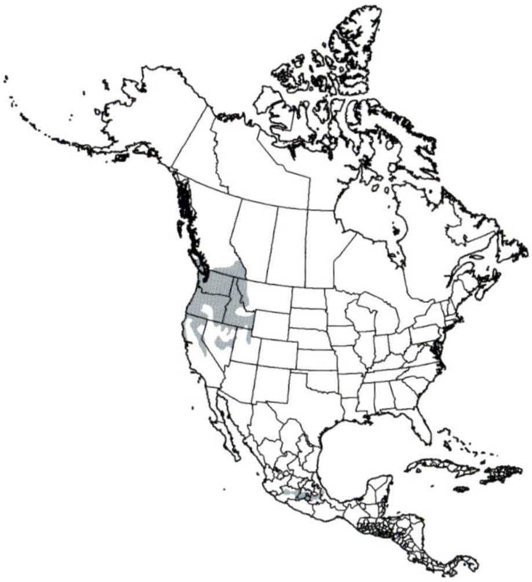

Distribution of the vagrant shrew in Wyoming.

the fourth, viewed laterally.[1] Measurements (**Sex** [n] mean (mm, g), *range*): Utah, L = **B** [6] 113, *102–119*; TL = **B** [6] 36, *34–39*; HF = **B** [6] 12, *12–14*.[2] Dental formula: 3/1, 1/1, 3/1, 3/3 = 32.

STATUS: A protected nongame animal by regulation and a state Species of Greatest Conservation Need as of 2012.[3]

NOMENCLATURE AND SYSTEMATICS: The vagrant or wandering shrew presents a challenging series of systematic issues, particularly as related to Wyoming. The species has long been placed in the *S. vagrans* complex, which comprises a series of lineages in a complex phylogeographical pattern. A number of these lineages were recognized before molecular tools were brought to bear on the systematic issues; molecular markers provided further clarity.[4]

DISTRIBUTION AND HABITAT ASSOCIATIONS: The geographic range in Wyoming includes two small areas in westernmost Lincoln County, one in the Star Valley and the other along the Bear River. This area is all that remains of the former extensive area depicted by Long (1965).[5] This range predated the incorporation of *S. vagrans obscurus* into the newly designated *S. monticolus*, thereby reducing the nominal geographic range of *S. vagrans* to parts of westernmost Wyoming. Habitats tend to be more open and grassy than is the case for *S. monticolus*; the importance of coarse woody debris on the forest floor is not clear.

ECOLOGY: Vagrant shrews in Oregon were found to eat—in decreasing order of importance—earthworms, spiders, crickets, lepidopteran larvae, slugs, snails, and beetles. Predators are highly varied and include owls and some mammals.[1]

CONSERVATION AND MANAGEMENT: The species is of some conservation concern as a result of its very restricted distribution in the state and the absence of knowledge of its population status and life needs.

NOTES AND REFERENCES

1. Gillihan and Foresman (2004).
2. Durrant (1952).
3. Commission Regulations Chapter 52, Section 11.
4. Hennings and Hoffman (1977), Demboski and Cook (2001).
5. Long (1965).

MOLES—FAMILY TALPIDAE

Moles are fossorial mammals that feed mostly on invertebrates, distributed across the Nearctic and Palearctic regions and most common in mild climates and fine soils. Wyoming is mostly too cold and its soils too rocky for moles, but one species does occupy part of the southeastern corner.

Eastern mole, *Scalopus aquaticus*

DESCRIPTION: A stocky small mammal with flattened forefeet, webbed toes, no pinnae, no visible eyes, and a short, lightly haired tail. The fur is dense, velvety, and gray or blackish in color. The feet are large, unfurred on the plantar surface, and the forefeet wider than long. Measurements (**Sex** [n] mean (mm, g), *range*): Nebraska, L = **F** [12] 145, *121–180*; **M** [18] 155, *136–169*; TL = **F** [12] 25, *20–35*; M [18] 25, *18–37*; HF = **F** [12] 20, *17–22*; **M** [18] 21, *19–22*; WT = **F** [12] 54.7, *31.3–82.0*; **M** [11] 64.7, *39.7–82*.[1] Dental formula: 3/2, 1/0, 3/3, 3/3 = 36.

STATUS: A non-protected nongame animal by regulation.[2]

NOMENCLATURE AND SYSTEMATICS: The species was first described from Pennsylvania in the mid-eighteenth century and has carried its current scientific name since 1905. The genus contains only this species, and 16 subspecies were recognized as of 1978, of which only the putative *S. a. caryi* occurs in Wyoming.[3]

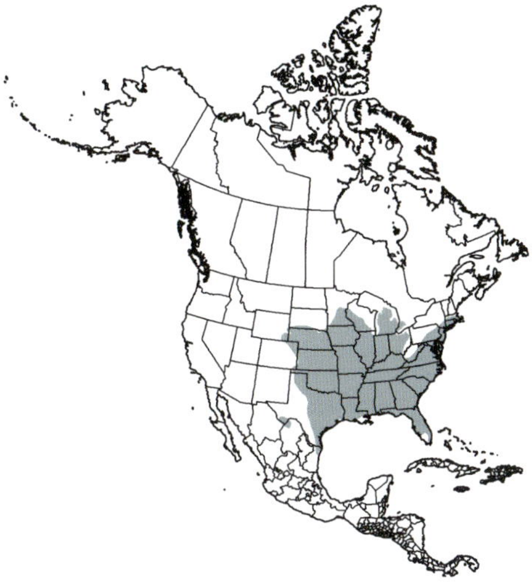

Locations from which the eastern mole has been identified in Wyoming.

DISTRIBUTION AND HABITAT ASSOCIATIONS: This eastern temperate species is limited to the southeastern part of the state, primarily in the floodplain of the North Platte River. It appears to be limited by the rocky soils beyond this narrow band.

ECOLOGY: This mole's diet is primarily invertebrates: earthworms, insects in various life stages, and some carrion and vegetation. The species seldom comes above ground, but below ground travels more widely than pocket gophers, using shallow tunnels for foraging and deeper tunnels for resting. The relatively wide movements by moles—facilitated by the sandy, loamy soils they prefer—mean that population genetic structure is weaker than is the case for fossorial rodents. Moles may serve important community or ecosystem functions by aerating soil and consuming soil invertebrates, but these processes have been little studied.[3]

CONSERVATION AND MANAGEMENT: This species—highly restricted in its Wyoming distribution—is neither a conservation concern nor an economic pest in the state. In the eastern United States, moles are in places a nuisance because their surface burrows disturb lawns and gardens. However, they eat little plant material and are not likely to pose economic hardships for agricultural producers in the small area of Wyoming where they occur.

NOTES AND REFERENCES

1. UNSM specimens.
2. Commission Regulations Chapter 52, Section 11.
3. Yates and Schmidly (1978).

Bats—Order Chiroptera

Bats are the only mammals that fly by flapping. They have highly modified forelimbs, including elongated metacarpals and phalanges, and have lost the distal forelimb bones in the most modified forms. Flight (patagial) membranes span the fore-digits and extend from the fifth digit to the trunk, and similar membranes (uropatagia or interfemoral membranes) extend between the hind limbs. Males possess an os penis, or baculum, and in most families are smaller-bodied than females. Bats are poorly known from the fossil record because their skeletons are so delicate, but the earliest fossils are from the Eocene Epoch, and several of the best preserved early specimens are from the Green River Formation of southwestern Wyoming.[1]

Bats represent the second most species-rich order of mammals in the world. They are widespread and abundant in tropical and subtropical regions, but also migrate seasonally to as far north as the Arctic. Bats were poorly understood until the twentieth century, partly because their nocturnal habits made them hard to observe, but modern research tools have revealed their natural histories, so that most North American species are better known than many small rodents. They are now regarded as most closely related to other lineages that had their earliest radiations in North America: carnivores, ungulates, and whales.[2] They are also now recognized to provide important ecosystem services, pollenating plants, dispersing fruit seeds, and feeding on insects at levels that can be economically important to agriculture.[3] Bats are the basis of locally important wildlife tourism industries in some areas.[4] Our understanding of the Wyoming bat community is immature; only 11 species were known from the state in 1965, 16 by 1987, and 18 in 2013.[5] For about one-third of the species described here, our only Wyoming-specific knowledge is that they occur in the state. Fortunately, federal and state agencies have funded, and the Wyoming Game and Fish Department and the Wyoming Natural Diversity Database have conducted efforts to inventory bat diversity since the early 1990s.

Bats need warm weather, access to surface water, and abundant insects, fleshy-fruited plants, or bat-pollinated plants for food; Wyoming lacks all of these features and therefore a diverse bat community. Bats from Wyoming can be generalized as entirely insectivorous, as mostly nonmigratory and as having low reproductive rates, typically one young per female per year or even less often. Whether a species is considered migratory depends on the distance of its movement, but hoary bats, eastern red bats, silver-haired bats, and the two species of free-tailed bats are among the most migratory of Wyoming bat species.[6] Most Wyoming bat species are known or suspected to overwinter in caves or abandoned mine adits. Many are sexually segregated, with males and nonreproductive females occupying different habitats than reproducing females during summer, or migrating at different times.[7] Some Wyoming bats, for example, pallid bats, are gleaners, sensing insects on the ground or on vegetation and landing to capture them, but most catch insects in flight. Bats can be further classified with regard to their roosting habitat: bats that roost in forests, for example, silver-haired bats, spend the day solitarily, on branches, in tree cavities, beneath peeling

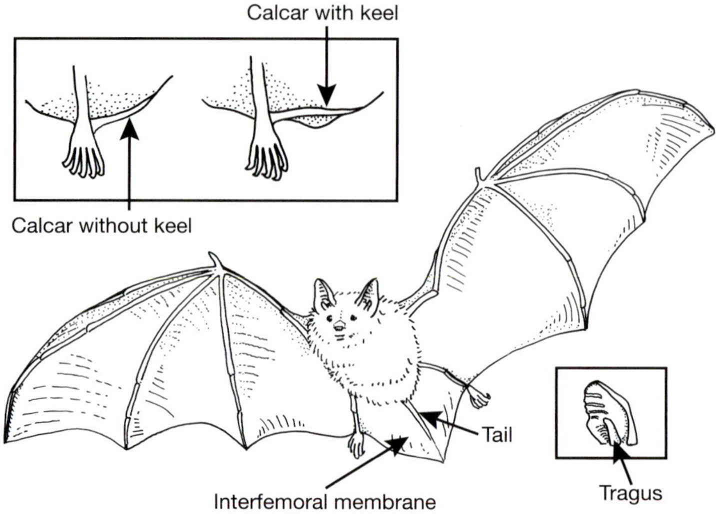

Important morphological characters used in identification of bats to species.

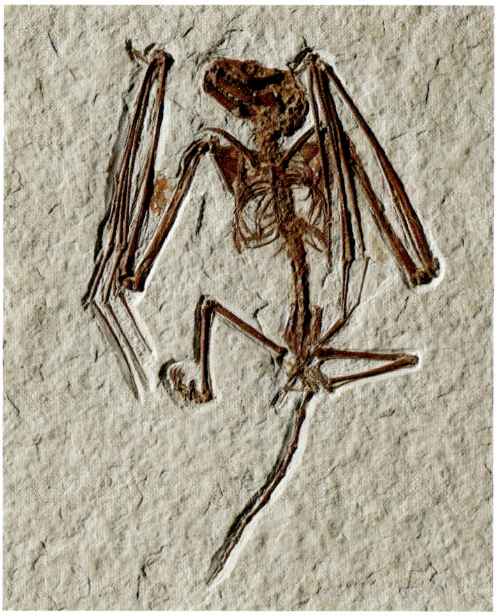
Icaronycteris index, one of the earliest complete fossil bat skeletons, from the Green River Formation of western Wyoming.

bark, or in rock crevices. Other species find caves, abandoned mine adits, and bridges that have suitable nooks and pockets where roosting can be solitary or communal.

Bats have some public health significance because they can transmit the rabies virus and other diseases to humans. During the period 2007–2010, an average of 10 cases per year of rabies in bats was reported from Wyoming.[8] This incidence is rare enough that people

TABLE 4 Number of bats, by species, collected in two different monitoring efforts: specimens submitted to the Wyoming State Veterinary Laboratory (WSVL) for rabies testing 1979–1992, and those caught and released in mist-netting surveys conducted by the Wyoming Game and Fish Department (WGFD) aimed at documenting forest bats during summers of 2009–2010. Bats submitted for rabies testing mostly appear near developed areas; those caught in mist nets were from forested habitats not associated with development.

Species	Rabies testing, WSVL, 1979–1992[a]	Mist nets, WGFD, 2008–2011[b]
Myotis californicus	9	0
M. ciliolabrum	139	14
M. evotis	29	50
M. lucifugus	365	449
M. septentrionalis	0	27
M. thysanodes	7	22
M. volans	106	153
M. yumanensis	0	5
Lasiurus borealis	0	5
L. cinereus	42	96
Lasionycteris noctivagans	235	281
Eptesicus fuscus	328	153
Euderma maculatum	0	2
Corynorhinus townsendii	2	0
Antrozous pallidus	15	0
Tadarida brasiliensis	3	0
Totals	1280	1257

a. Bogan and Cryan (2000).
b. Centers for Disease Control and Prevention (CDC) annual reports, 2007–2010 (e.g., Blanton et al. 2011). For the period 1981–1992, 113 of 1280 bats submitted to the Wyoming State Veterinary Laboratory tested positive for rabies, a mean rate of 9%. *Eptesicus fuscus* had the second-highest number of bats submitted, and the highest infection rate: 60% (Bogan and Cryan 2000). An average of 1708 cases per year of bat rabies were reported from the United States and Puerto Rico for the same period.

in Wyoming are at very low risk of contracting rabies via bats. Still, bat rabies is a concern because most of the human cases of rabies reported from the United States and Puerto Rico are thought to have been transmitted by bats.[9] The significance of rabies to the bats themselves is not fully understood, but major die-offs of bat colonies due to rabies seem to be rare. Histoplasmosis, a fungal disease of humans most common in the southern states, can be contracted where bats concentrate at roosts, but it is rare or nonexistent in Wyoming.

Survey efforts since 1990 have shown a range of bat species to be present more broadly than previously understood (Table 4).[10] Such an area is the Red Desert, where water sources and trees for roosting are limited. Earlier, bat surveys in the Red Desert had been few, because the general absence of surface water and roosting trees were assumed to be limiting to bat

The arid, unforested zone of Wyoming within which surface water and bat detections are uncommon. Exceptions are along the Sweetwater and Green Rivers, and in the forested mountains south of the Sweetwater River.

diversity. However, surveys conducted in the Red Desert in 2012 and 2013, represented in the accompanying maps, showed a surprising number of bat species occurring there, representing some small range extensions. Still, there appears to be a region of Wyoming where few bats occur, within which surface water, trees, and bat detections—even for habitat-generalist species such as the little brown myotis—are scarce. This zone extends from near the northeastern corner of the state diagonally to the southwest corner, and most bat detections within it have been near the Sweetwater River, the forested mountains just south of that river, and the Green River. Whether this region has been surveyed as intensively as some others in the state remains unclear. The bat fauna of Wyoming is sufficiently incompletely understood that several bat species now known only from adjacent states could easily cross into Wyoming intermittently: the western pipistrelle (*Parastrellus hesperus*) occurs along the Yampa River, less than 50 km (30 mi) from the Wyoming state line in Moffat County, Colorado.[11] Likewise, the western red bat (*Lasiurus blossevillii*) is found just south of the Uinta Mountains in Utah and the evening bat (*Nycticeius humeralis*) occurs in Hitchcock County, Nebraska, only 270 km (170 mi) from the southeast corner of Wyoming.[12]

Bat conservation represents an important branch of environmental biology, and growing evidence shows that bat populations have declined over large areas of North America and other continents, for some species by more than 90% in a few decades.[13] The mechanisms underlying these declines include the widespread use of agricultural insecticides, which are common in the insect foods of bats.[14] Further, roosting habitats of forest-dwelling bats have declined via intensive forest management, which reduces the availability of old trees and snags.[15] Disturbance and destruction of colonial bat roosts and hibernacula in caves and abandoned mine adits has long been a perceived threat, but major steps have been taken to reduce this disturbance by placing barriers to humans across entrances.[16] Wind turbines are known to cause bat deaths, although reliable estimates of the magnitude of this effect are scarce.[17] An emerging concern is the spread of white-nose syndrome (WNS) from eastern North America westward. This fungal disease, which grows on hibernating bats, is capable of killing nearly all the bats in a hibernaculum, has killed millions of North American bats in a few years, and has spread very rapidly to as far west as western Missouri.[18] Only three *Myotis* species that occur in Wyoming are particularly vulnerable to WNS because of their colonial roosting behaviors and only one, *Myotis lucifugus*, is common. This species' roosts and hibernacula in the state are relatively small, reducing somewhat the vulnerability of Wyoming populations to WNS. Whether and when this disease will reach Wyoming is not known, but it could have major effects on migratory bat populations in Wyoming if bats infected further south carry the fungus northward via migration.[19] All bat species in Wyoming are protected nongame animals, which may not be intentionally killed or collected without a permit.[20]

NOTES AND REFERENCES

1. Gunnell and Simmons (2005).
2. Springer et al. (2004).
3. Kunz et al. (2011).
4. Pennisi et al. (2004).
5. Long (1965), Clark and Stromberg (1987).
6. Adams (2003).
7. Cryan et al. (2000).
8. Centers for Disease Control and Prevention (CDC) annual reports, 2007–2010 (e.g., Blanton et al. 2011). For the period 1981–1992, 113 of 1280 bats submitted to the Wyoming State Veterinary Laboratory tested positive for rabies, a mean rate of 9%. *Eptesicus fuscus* had the second-highest number of bats submitted, and the highest infection rate: 60% (Bogan and Cryan 2000). An average of 1708 cases per year of bat rabies were reported from the United States and Puerto Rico for the same period.
9. Twenty of 29 cases during 2001–2011 (Blanton et al. 2011, Table 3). I could find no record of bat-transmitted human rabies in Wyoming.
10. Wyoming State Veterinary Laboratory data are from Bogan and Cryan (2000); Wyoming Game and Fish Department data are from Abel and Grenier (2012).
11. Bogan and Cryan (2000).
12. Adams (2003), Serbousek and Geluso (2009), Armstrong et al. (2011).
13. Geluso et al. (1976), Betke et al. (2008).
14. Clark (2001).
15. Miller et al. (2003).
16. Ducummon (1999).
17. Barclay et al. (2007), Arnett et al. (2008).
18. Frick et al. (2010). The fungus that causes the disease is *Pseudogymnoascus destructans* (Lorch et al. 2011).
19. Abel and Grenier (n.d.).
20. Commission Regulations Chapter 52, Section 11.

KEY TO SPECIES OF CHIROPTERA

1a. Tail projects beyond interfemoral membrane for one-fourth to half of its length; fibula well developed relative to tibia: . . . . **Molossidae, 2**

1b. Tail does not project beyond interfemoral membrane, or does so for less than one-sixth of its length; fibula small relative to tibia: . . . . **Vespertilionidae, 3**

2a. Length less than 110 mm (4.3 in); pinnae not united at bases and extend only slightly beyond nose when laid forward: . . . . ***Tadarida brasiliensis***
2b. Length greater than 120 mm (4.7 in), pinnae united at bases and extend well beyond nose when laid forward: . . . . ***Nyctinomops macrotis***

3a. Six teeth posterior to canines in each upper jaw; pelage brownish:. . . . . ***Myotis*, 4**
(molecular markers are available as of 2014 to identify all but one myotis bat species found in Wyoming)
3b. Fewer than six upper teeth posterior to canines; pelage color variable across species: 9

4a. Calcar distinctly keeled: 5
4b. Calcar without a distinct keel: 7

5a. Underside of wing furred from thorax to elbow or beyond; foot generally shorter than 8.5 mm (0.33 in):. . . . . ***Myotis volans***
5b. Underside of wing not furred as far as elbow; foot generally longer than 8.5 mm (0.33 in): 6

6a. Braincase rising abruptly from rostrum; third metacarpal as long as or longer than forearm; dorsal pelage dull: . . . .***Myotis californicus***
6b. Braincase rising gently from rostrum; third metacarpal not as long as forearm; dorsal pelage glossy: . . . .***Myotis ciliolabrum***

7a. Visible, but perhaps inconspicuous fringe of hairs on trailing edge of interfemoral membrane: 8
7b. No visible fringe of hairs on trailing edge of interfemoral membrane: 9

8a. Conspicuous fringe of hairs extending 1–1.5 mm (0.04–0.06 in) beyond trailing margin of interfemoral membrane; pinnae less than 20 mm (0.75 in) long; calcar with indistinct keel: . . . . ***Myotis thysanodes***
8b. Short, scattered, inconspicuous hairs on trailing margin of interfemoral membrane; adult pinnae longer than 20 mm (0.75 in): . . . . ***Myotis evotis***

9a. Skull generally longer than 14.2 mm (0.56 in): 10
9b. Skull generally shorter than 14.2 mm (0.56 in): . . . . ***Myotis yumanensis***

10a. Pinna long, extending 2 mm (0.1 in) beyond tip of nose when laid forward; tragus long and sharply pointed; pelage non-glossy: . . . . ***Myotis septentrionalis***
10b. Pinna shorter, not extending beyond the tip of the nose when laid forward; tragus shorter and less pointed; pelage glossy: . . . . ***Myotis lucifugus***

11a. One upper incisor on each side: 12
11b. Two upper incisors on each side: 14

12a. Ears longer than 25 mm (1 in), upper surface of interfemoral membrane mostly unfurred; color pale yellow to off-white; postcanine teeth 4/6: . . . .*Antrozous pallidus*
12b. Ears shorter than 19 mm (0.75 in), more than half of upper surface of interfemoral membrane furred: . . . . *Lasiurus*, 13

13a. Dorsal pelage brown with white tips on hairs; shoulders and wrists marked with small patches of white fur; pinnae with black rim: . . . . *Lasiurus cinereus*
13b. Dorsal pelage reddish brown (in females, yellowish) with some hairs white-tipped; white tufts of longer hairs on shoulders; pinnae without black rim:. . . . . . . *Lasiurus borealis*

14a. Dorsal pelage black with three large white spots; postcanine teeth 5/5: . . . . *Euderma maculatum*
14b. Dorsal pelage otherwise: 15

15a. Four upper postcanine teeth: . . . . *Eptesicus fuscus*
15b. Five upper postcanine teeth: 16

16a. Dorsal pelage black with light tips on hairs, giving a grizzled appearance; proximal half or so of dorsal surface of interfemoral membrane densely furred: . . . . *Lasionycteris noctivagans*
16b. Dorsal pelage and interfemoral membrane otherwise: 17

17a. Pinnae shorter than 16 mm (0.6 in); dorsal pelage distinctly banded, with bases and tips of hairs dark, middles paler: . . . . *Perimyotis subflavus*
17b. Pinnae longer than 30 mm (1.2 in), joined at bases; color brownish; postcanine teeth 5/6: . . . .*Corynorhinus townsendii*

EVENING BATS—FAMILY VESPERTILIONIDAE

Vesper bats occur in suitable habitats worldwide and make up most of the bats we see in the evening in temperate and northern North America. They are small to medium-sized, have pinnae that are highly variable in size, and lack elaborate outgrowths on the nose. They roost in a range of sites, most obviously caves or abandoned mine adits, but in trees, rock crevices, and human structures as well. The family name is based on the Latin "evening." Vesper bats mate in the fall, before entering hibernation, and females store sperm cells until they are ready to ovulate in the spring—a mechanism known as delayed fertilization. This family poses the most challenging problems of species identification in Wyoming; several vesper bat species look very similar in hand. The key identifying characteristics are the color and glossiness of the fur, the color and length of the ears, the presence of a keel on the calcar, the size and shape of the tragus, the size of the foot, the presence of fur on the ventral shoulders, and the presence of stiff hairs on the trailing margin of the interfemoral membrane. In addition, molecular tools are now available; seven of the eight myotid bats described here have species-specific genetic markers available for them.[1]

NOTES AND REFERENCES

1. Zinck et al. (2004).

California myotis, *Myotis californicus*

DESCRIPTION: This very small bat has light brown fur, a steeply sloping forehead, and a tail that does not extend beyond the interfemoral membrane. The tragus is sharply pointed and the calcar is keeled. The sexes are similar in size and coloration.[1] Measurements (**Sex** [n] mean (mm, g), *range*): eastern Oregon, L = **F** [18] 84, *75–92*; **M** [21] 83, *76–90*; TL = **F** [18] 30, *34–47*; **M** [21] 39, *31–46*; FA = **F** [21] 32.5, *29.8–34*; **M** [21] 32.7, *30–37*; EFN = **F** [18] 14, *10–15*; **M** [21] 14, *12–16*; WT = **F** [18] 5.0, *4.2–6.5*; **M** [21] 4.7, *2.5–9.0*.[2] Dental formula: 2/3, 1/1, 3/3, 3/3 = 38.

STATUS: The species is a protected nongame animal.[3] It is uncommon within its small geographic range within the state.

NOMENCLATURE AND SYSTEMATICS: The genus and species names have been stable for over a century. Of the four subspecies recognized in 1993, only *M. c. stephensi* is depicted as occurring in Wyoming.[4]

DISTRIBUTION AND HABITAT ASSOCIATIONS: This western species occurs along the Green River drainage as far upstream as the towns of Green River and Rock Springs. Reports from farther north in the state, including the Bighorn Basin and Yellowstone National Park, have so far been unconfirmed. This is a species associated with varied habitats, but in the interior of the continent tends to associate with water bodies, and forages near vegetation edges. The sexes hibernate together in caves and abandoned mines, but are segregated during summer, when daytime roosting tends to be in rock crevices. Hibernacula have been found in southwestern Utah, but not in Wyoming.[4]

ECOLOGY: The diet is a range of small flying insects, including flies, beetles, and true bugs. Breeding occurs in fall, and a single young is born in July. The species is thought to be long-lived, with some animals known to survive, and presumably reproduce, for longer than 15 years.[4]

CONSERVATION AND MANAGEMENT: The species is not a major conservation concern, in part because of its extensive geographic range outside Wyoming.

NOTES AND REFERENCES

1. van Zyll de Jong (1984), Simpson (1993).
2. Verts and Carraway (1998).
3. Commission Regulations Chapter 52, Section 11.
4. Bogan and Cryan (2000) described eight specimens (not mapped) from near Green River and Rock Springs. Adams (2003).

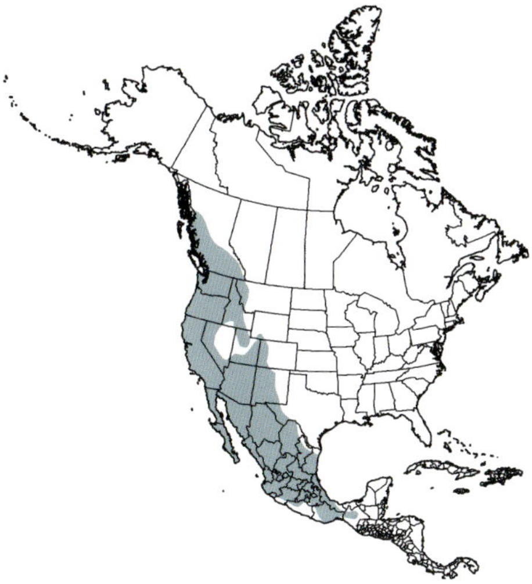

Locations from which the California myotis has been identified in Wyoming.

Western small-footed myotis, *Myotis ciliolabrum*

DESCRIPTION: A small bat similar in appearance to, but larger than, the California bat. Its fur is light to yellowish brown and not glossy, and it has black ears, wing membranes, and limbs, with a dark facial mask. The tail extends at least 4 mm (0.2 in) beyond the margin of the interfemoral membrane, and the calcar is keeled. The foot is small and lightly built.[1] Measurements (**Sex** [n] mean (mm, g), *range*): various counties, L = **F** [8] 79, *68–90*; **M** [9] 81, *67–88*; TL = **F** [8] 30, *25–41*; **M** [9] 34, *19–40*; HF = **F** [8] 7.6, *5–9*; **M** [9] 7.7, *6–9*; EFN = **F** [8] 12.75, *11–14*; **M** [9] 13.0, *12–15*; WT = **F** [8] 5.8, *3.9–9*; **M** [9] 3.8, *2.5–5*.[2] Dental formula: 2/3, 1/1, 3/3, 3/3 = 38.

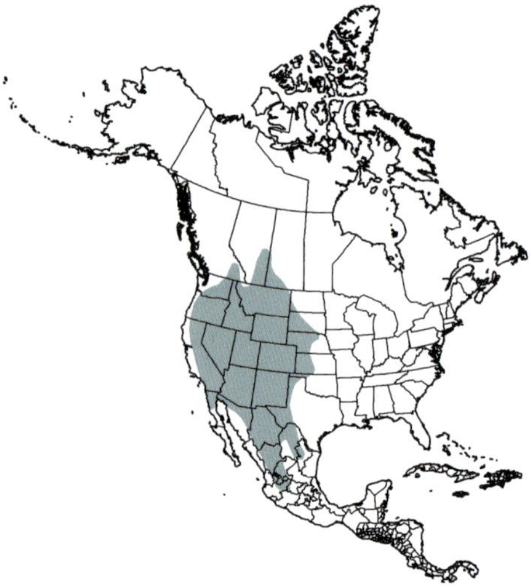

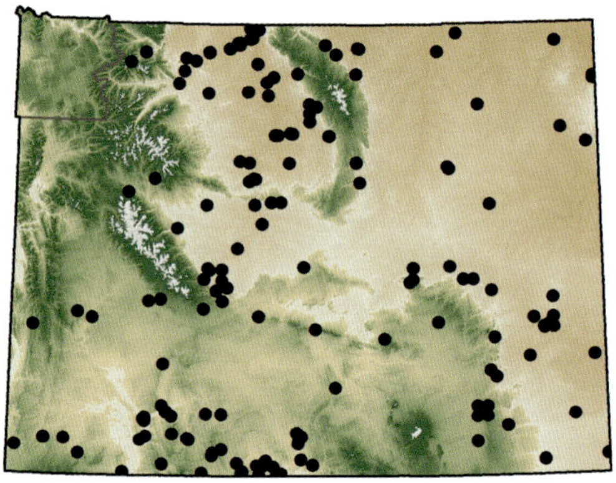

Locations from which the western small-footed myotis has been identified in Wyoming.

STATUS: This is a protected nongame animal and a state Species of Greatest Conservation Need as of 2012.[3] It is observed in suitable habitats statewide, but is not abundant.

NOMENCLATURE AND SYSTEMATICS: The taxonomy has undergone considerable revision, the species being referred to as *M. subulatus*, or as a subspecies of *M. leibii* in various accounts. The species name is Latin for "hairy-lipped." Two subspecies were recognized as of 2001; only *M. c. ciliolabrum* is mapped as occurring in Wyoming.[4]

DISTRIBUTION AND HABITAT ASSOCIATIONS: This western and intermountain species occurs in suitable habitats over most of the state, although it has not yet been detected in the northwestern mountains and has been scarce in the basins of the northeastern part. It is a habitat generalist, less likely to be limited to areas near water bodies than is the case for the California myotis. Females with young in Wyoming have not been reported, but several dozen wintering sites have been located, all in caves, mines, and tunnels in north-central and southwestern Wyoming.[5]

ECOLOGY: In spite of its broad distribution, the ecology of this species is poorly understood. The diet is similar to that of the California myotis: moths, flies, true bugs, and beetles, which it is more likely to catch near rocky bluffs than over open water.[2] It roosts in a wide range of site types: rock crevices, caves, tunnels, abandoned swallow nests, beneath loose tree bark, and in human structures. Adult females are suspected to produce a single offspring per year, although few maternity colonies have been examined.[6]

CONSERVATION AND MANAGEMENT: The concern regarding the species in Wyoming relates to the lack of reliable information on its distribution and habitats and its susceptibility to disturbance at caves, cliffs, and mines. As related to the Black Hills area, forest harvesting, spelunking, and rock climbing have all been suggested as possible negative factors, although scant information accompanies each concern.[7]

NOTES AND REFERENCES

1. Holloway and Barclay (2001).
2. KU specimens.
3. Commission Regulations Chapter 52, Section 11.
4. Holloway and Barclay (2001). *M. l. ciliolabrum* was the western subspecies of the small-footed myotis.
5. Bogan and Cryan (2000).
6. Geluso et al. (2013) described pregnant females and neonates from the Wildcat Hills in westernmost Nebraska, just east of Goshen County.
7. Schmidt (2003a).

Long-eared myotis, *Myotis evotis*

DESCRIPTION: A medium-sized myotis with glossy, variably brown fur. Ears, face, and flight membranes are blackish. The ears extend well (up to 7 mm [0.3 in]) past the tip of the nose when laid forward; this is the longest-eared North American myotis. The hairs along the posterior margin of the interfemoral membrane are small and inconspicuous, unlike in the fringed myotis.[1] Measurements (**Sex** [n] mean (mm, g), *range*): various counties, Montana and Colorado; L = **F** [10] 92, *83–105*; **M** [8] 93, *86–100*; TL = **F** [10] 44, *39–48*; **M** [8] 43, *41–45*; HF = **F** [10] 10, *7–11*; **M** [8] 10, *9–10*; EFN = **F** [10] 21, *21–23*; **M** [8] 21, *15–23*; WT = **F** [10] 7.0, *5.5–9.4*; **M** [8] 6.4, *5–7.5*.[2] Dental formula: 2/3, 1/1, 3/3, 3/3 = 38.

STATUS: A protected nongame animal, and a state Species of Greatest Conservation Need as of 2012.[3]

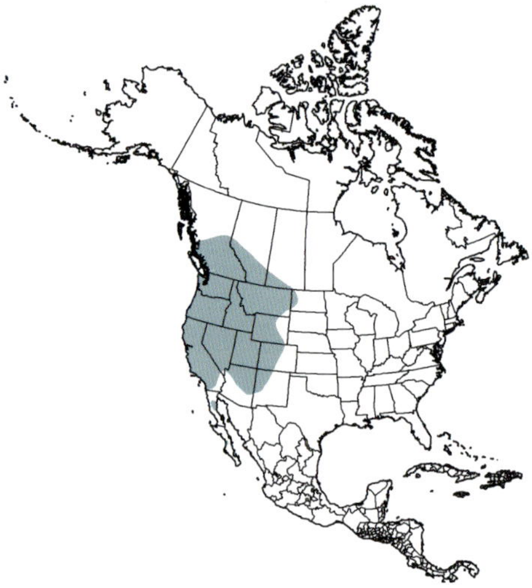

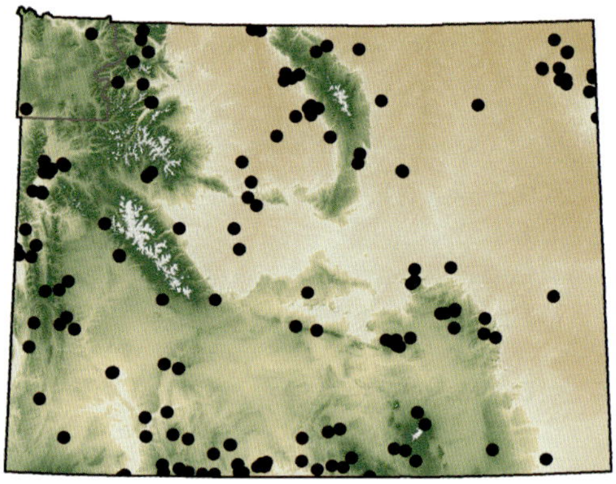

Locations from which the long-eared myotis has been identified in Wyoming.

NOMENCLATURE AND SYSTEMATICS: The genus and species names have been stable for over a century. Two subspecies are recognized; only *M. e. evotis* is generally regarded as occurring in the Rocky Mountains. However, a dissenting view holds that Wyoming animals belong to a disjunct subspecies—*M. e. chrysonotus*—the type specimen of which was collected from Sweetwater County in the 1890s.[4] The species was described from Late Pleistocene deposits in Little Box Elder Cave, Converse County.[5]

DISTRIBUTION AND HABITAT ASSOCIATIONS: This species of the Pacific Northwest and southwestern Canada was suspected of occurring statewide in 1965, and surveys since then have tended to confirm this view.[6] One winter hibernaculum has been identified in Wyoming.[7] Habitats tend to be coniferous forests and woodlands: ponderosa pine, lodgepole pine, and spruce–fir, but include deciduous riparian woodlands. Foraging tends to occur over water bodies, but also in openings, such as above small meadows and campgrounds.[1]

ECOLOGY: The diet is both flying and nonflying insects; this species can locate prey by biosonar as well as by passive listening and gleaning. Thus, prey include flying insects as well as moths perched on vegetation. The diet is dominated by moths, beetles, flies, and lacewings.[8] This species can live and forage at higher elevations than other bats; insects are less active at night high in the mountains, and this species can capture them on substrate. After foraging, the species uses nighttime roosts, 24 of which have been located in Wyoming—almost all in caves or abandoned mine adits. Daytime roosts have tended to be associated with tree boles; the only maternity roost found in Wyoming was in an abandoned building.[8] Adult females are suspected to produce one young per year.

CONSERVATION AND MANAGEMENT: This species is relatively uncommon in the state, in spite of being widespread. The reason for its species-of-concern status is the lack of reliable knowledge of its life history and ecology, and its potential vulnerability to wind energy development. It is also hypothesized to be vulnerable to timber management in coniferous forests, and the species could be affected by major changes in stand structural characteristics in response to extensive bark beetle outbreaks.

NOTES AND REFERENCES

1. Manning and Jones (1989).
2. UWMV, KU specimens.
3. Commission Regulations Chapter 52, Section 11.
4. Manning (1993).
5. Anderson (1968).
6. Long (1965), Clark and Stromberg (1987), Cerovski et al. (2004).
7. Priday and Luce (1999a), Bogan and Cryan (2000).
8. VonHof and Barclay (1997), Priday and Luce (1999a).

Little brown myotis, *Myotis lucifugus*

DESCRIPTION: A medium-sized myotis characterized by the absence of a keel on the calcar and the absence of hairs on the margin of the interfemoral membrane. The fur is variably brown and glossy, and the wing membranes are dark brown, rather than black. Long hairs extend beyond the tips of the toes. A separate morphological character is used to distinguish the species from each sympatric species of myotis in Wyoming.[1] Measurements (**Sex** [n] mean (mm, g), *range*): various counties; L = **F** [13] 90, *84–103*; **M** [6] 88, *83–93*; TL = **F** [13] 40, *31–44*; **M** [6] 33, *34–41*; HF = **F** [13] 11, *9–12*; **M** [6] 10, *10–12*; EFN = **F** [13] 15, *12–17*; **M** [6] 14, *12–15*; WT = **F** [13] 7.6, *5.7–9.5*; **M** [6] 7.3, *6.3–8.2*.[2] Dental formula: 2/3, 1/1, 3/3, 3/3 = 38.

STATUS: A protected nongame species, and a state Species of Greatest Conservation Need as of 2012.[3]

NOMENCLATURE AND SYSTEMATICS: The species has been recognized for nearly two centuries, although the current genus-species epithet has been in use only since around 1900;

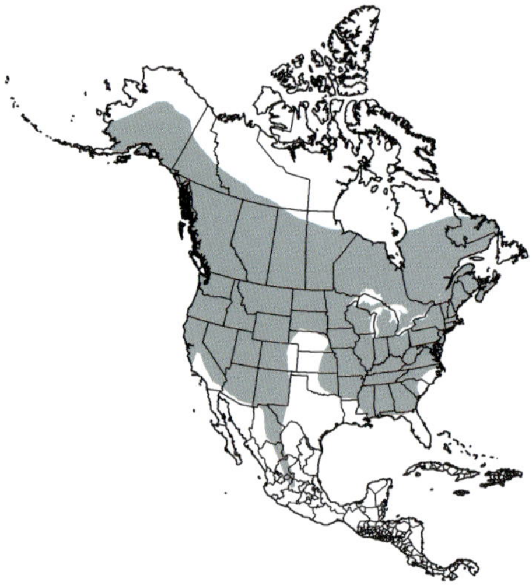

Locations from which the little brown myotis has been identified in Wyoming.

M. occultus is considered a synonym applied to the Southwest by some authors.[4] Six subspecies have been recognized traditionally, of which only the putative *M. l. carissima* is considered to occur in Wyoming.[1] Recently, however, genetic structure has been investigated across the putative boundary between *M. l. carissima* and *M. l. lucifugus* to the east. These nominal subspecies were found to interbreed widely and vary clinally, so the authors concluded that the subspecies distinction was not warranted.[5] A more recent study has considered whether females show high fidelity to natal colonies, a traditional notion regarding this species. That study showed considerable dispersal among females, which is consistent with a lack of population genetic structure across the geographic range.[6] Considering the lack of evidence of high breeding-site fidelity, the traditionally recognized subspecies should be considered tentative. The common name "little brown bat" also is used.

DISTRIBUTION AND HABITAT ASSOCIATIONS: This, the most widely distributed North American bat species, occurs statewide in Wyoming. It was recorded in 18 of 23 Wyoming counties by 1987, and in 27 of 28 latitude–longitude blocks by 2004.[7] Winter hibernacula have been described from various Wyoming sites, and maternal colonies have been reported from most regions of the state.[8] Habitats are varied, including coniferous forest, riparian zones, agricultural lands, and urban areas. It tends to avoid extensive hot, arid areas and typically forages near water or at vegetation boundaries. Daytime roosts in summer include virtually the full range of natural and human structures: caves, abandoned mine adits, bridges, old buildings, tree cavities, and rock crevices. Several winter hibernacula have been located in Wyoming, all in caves.[9]

ECOLOGY: This is arguably the best known North American evening bat in terms of ecological understandings. The diet spans the range of flying insects: caddisflies, mayflies, moths, midges, and beetles. Roosting occurs in colonies that can number into the hundreds, even in Wyoming, and roosting habitats are believed to be more limiting to the distribution than food resources.[10] Females give birth to one young per year. Of a sample of 155 animals from Wyoming, females were around twice as common as males.[11] A major factor in population ecology in eastern North America is WNS, which has been reported to kill 75–99% of bats of this species in various eastern hibernacula.[12]

CONSERVATION AND MANAGEMENT: This is one of the most widespread and numerically abundant bats in Wyoming, but is a conservation concern because of its susceptibility to devastating outbreaks of WNS in the eastern United States. This fungal disease has yet to reach Wyoming, but no viable treatment or management options are available, were it to arrive. General concerns regarding bat conservation—insecticide use and roosting habitats—also apply.

NOTES AND REFERENCES

1. Fenton and Barclay (1980).
2. KU specimens.
3. Commission Regulations Chapter 52, Section 11.
4. Hoffmeister (1986), for example.
5. Lausen et al. (2008).
6. Dixon (2011).
7. Hester and Grenier (2005).
8. Clark and Stromberg (1987), Cerovski et al. (2004), Hester and Grenier (2005).
9. Priday and Luce (1999a).
10. Adams (2003).
11. Bogan and Cryan (2000).
12. Dzal et al. (2011).

Northern long-eared myotis, *Myotis septentrionalis*

DESCRIPTION: A medium-sized temperate evening bat with non-glossy medium-brown fur. The fur and flight membranes are similar in darkness, and the free margin of the interfemoral membrane is bare or has scattered hairs. The tragus is long and sharply pointed. The tail is longer than for *M. lucifugus*.[1] Measurements (**Sex** [n] mean (mm, g), *range*): location not specified, L = **B** [49] 86, *80–96*; TL = **B** [49] 38, *29–46*; HF = **B** [47] 9, *8–11*; FA = **B** [14] 36, *35–405*; EFN = **B** [27] 16, *15–18*; WT = **B** [21] 7.4, *4.3–10.8*.[2] Dental formula: 2/3, 1/1, 3/3, 3/3 = 38.

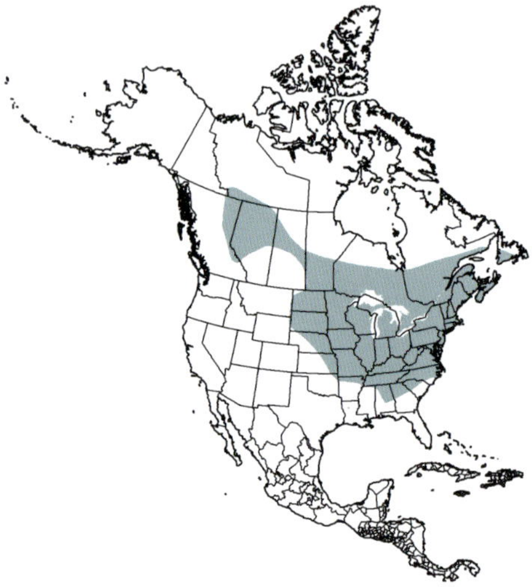

Locations from which the northern long-eared myotis has been identified in Wyoming.

STATUS: A protected nongame animal, and a state Species of Greatest Conservation Need as of 2012.[3] The species was listed as E.S.A. Threatened throughout its geographic range in 2015.[4]

NOMENCLATURE AND SYSTEMATICS: This species was considered one of two subspecies of *M. keenii* during 1928–1979; the corresponding common name was Keen's myotis. The two subspecies were raised to the level of species based on morphological characteristics, and the new common name (sometimes given as "northern long-eared bat") applied to this taxon.[5] No subspecies have been recognized.

DISTRIBUTION AND HABITAT ASSOCIATIONS: This mostly northeastern species occurs only in and around the Black Hills, including the Bear Lodge Mountains at intermediate elevations (1200–1950 m [4000–6500 ft]).[6] Habitats are mostly coniferous forests in Wyoming, but are mostly deciduous stands farther east.

ECOLOGY: The diet consists of flying insects and those gleaned from substrates. Foraging typically occurs under a forest canopy, above small water bodies, or above roads or openings. The species migrates less than other members of the genus; some populations overwinter as far north as Canada. Hibernacula are typically caves and abandoned mine adits; summer roosts tend to be in trees, although human structures also are used. In the Black Hills of South Dakota, roosting was observed in recently harvested stands, as well as in mature forest.[7] Roosting is solitary or in small groups. As for various other vesper bats, females are more likely to remain near where they are born, whereas males disperse farther.[8] A single young is produced by each adult female yearly, and longevity is high, with wild animals over 18 years old recorded.[1]

CONSERVATION AND MANAGEMENT: The species has been common in mist-net surveys in the Black Hills of South Dakota, but rarer in northeastern Wyoming. Conservation concern across the species range is based on marked population declines in the face of WNS.[4] On Prince Edward Island, its occurrence responded to various landscape-scale attributes, suggesting sensitivity to forest cutting.[9] Its listing as E.S.A. Threatened in 2015 was in response to observed population declines, which are expected to expand westward.[4]

NOTES AND REFERENCES

1. Barbour and Davis (1969, see *M. keenii* account); van Zyll de Jong (1985); Caceres and Barclay (2000).
2. van Zyll de Jong (1985).
3. Commission Regulations Chapter 52, Section 11; United States.
4. US Fish and Wildlife Service (2015).
5. van Zyll de Jong (1979).
6. Hester and Grenier (2005).
7. Cryan et al. (2001).
8. Arnold (2007).
9. Henderson et al. (2008).

Fringed myotis, *Myotis thysanodes*

DESCRIPTION: A myotis with long pinnae and a conspicuous fringe of stiff hairs along the posterior margin of the interfemoral membrane, more conspicuous than in the long-eared myotis (only these two myotid bats have such a fringe). The dorsal pelage is yellowish brown to olive and the ventral pelage is lighter. The ears and flight membranes are very dark. A well-developed calcar is indistinctly keeled.[1] Measurements (**Sex** [n] mean (mm, g), *range*): Oregon, L = **F** [3] 85, *81–92*; **M** [9] 87, *80–93*; TL = **F** [3] 38, *35–43*; **M** [9] 38, *33–41*;

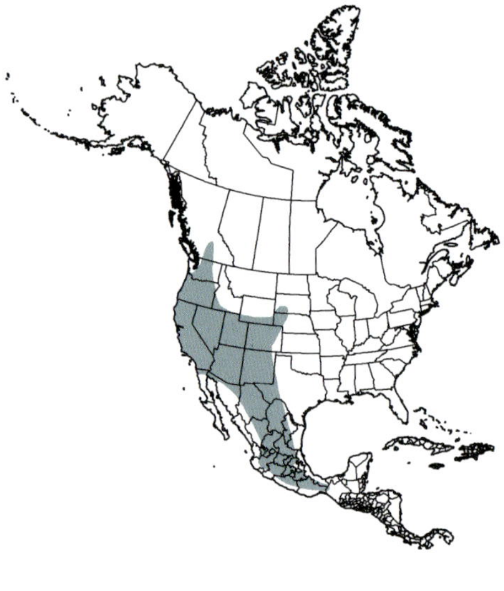

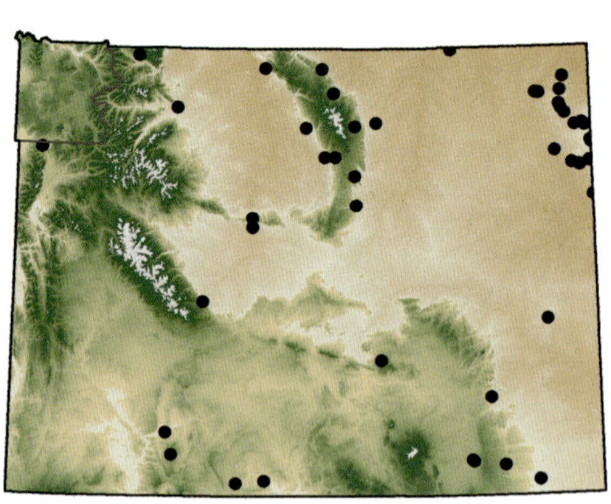

Locations from which the fringed myotis has been identified in Wyoming.

FA = **F** [3] 38.8, *36.5–41.9*; **M** [9] 38.8, *36.5–41.9*; EFN = **F** [3] 19, *19–19*; **M** [9] 17, *13–20*; WT = **F** [3] 7.4, *7.4–7.4*, **M** [9] 6.2, *4.7–7.0*.[2] Dental formula: 2/3, 1/1, 3/3, 3/3 = 38.

STATUS: A protected nongame animal and a state Species of Greatest Conservation Need as of 2012.[3] It is uncommon within Wyoming; only 20 specimens from the state were known as of 2000.[4]

NOMENCLATURE AND SYSTEMATICS: The genus, species, and common names have been stable since around the turn of the twentieth century. Three putative subspecies were recognized as of 1980, two of which—*M. t. thysanodes* and *M. t. pahasapensis*—occur in Wyoming, but most locations for the species in Wyoming fall outside the mapped ranges of any of the three subspecies.[5] So, assignment of Wyoming animals to a subspecies is problematic. The species was noted from deposits of Late Pleistocene age in Little Box Elder Cave, Converse County.[6]

DISTRIBUTION AND HABITAT ASSOCIATIONS: This mostly western US species is widespread in central Wyoming and eastern Wyoming.[7] Habitats are varied, including coniferous forest, juniper woodlands, and shrub-steppe.

ECOLOGY: The diet in northern Colorado was observed to be similar to that of other temperate vesper bats: moths, beetles, bees, and lacewings.[8] The species may migrate short distances, although movements by Wyoming animals are not understood well. A number of night roosts have been found in Goshen County, and two hibernacula have been located, both near the eastern border of the state. Summer roosting sites include rock crevices, tree cavities, caves, mine adits, and human structures.[9] A single young is born per female per year.

CONSERVATION AND MANAGEMENT: The Western Bat Working Group considers this species to be a particular concern in the Pacific States and in parts of the Rocky Mountains—most of Wyoming included. The concern is centered on several factors, particularly the rarity of the species and possible disturbance at maternity colonies and hibernacula. Also, the removal of large-diameter cavity-forming trees tends to decrease roosting habitat, and modification of forest structure around roosting structures is seen as detrimental.[10] It is plausible that outbreaks of bark beetles will negatively affect roosting habitat for this species into the future.

NOTES AND REFERENCES

1. O'Farrell and Studier (1980), van Zyll de Jong (1985).
2. Verts and Carraway (1998).
3. Commission Regulations Chapter 52, Section 11.
4. Bogan and Cryan (2000).
5. See Boyce (1980), O'Farrell and Studier (1980) for interpretations.
6. Anderson (1968).
7. Keinath (2004).
8. Adams (2003).
9. O'Farrell and Studier (1980), Priday and Luce (1999a), Lacki and Baker (2007).
10. Western Bat Working Group (1998), Keinath (2004).

Long-legged myotis, *Myotis volans*

DESCRIPTION: This is a medium-sized myotis with medium to dark brown fur. The ears and flight membranes are black, and the undersides of the wing have long, dense fur extending from the ventral torso laterally to the elbow. It can also be distinguished in the field from other myotid bats by its keeled calcar and short, rounded pinna, which almost covers the nostrils when laid forward.[1] Measurements (**Sex** [n] mean (mm, g), *range*): various counties; L = **F** [7] 95, *92–98*; **M** [6] 94, *90–97*; TL = **F** [7] 43, *34–50*; **M** [6] 42, *39–45*; HF = **F** [7] 9.1, *8–10.5*; **M** [6] 9.8, *9–11*; EFN = **F** [7] 13.0, *12–14*; **M** [6] 13.1, *11–14.5*; WT = **F** [7] 8.2, *5.1–9.5*; **M** [6] 5.7, *4.3–7.7*.[2] Dental formula: 2/3, 1/1, 3/3, 3/3 = 38.

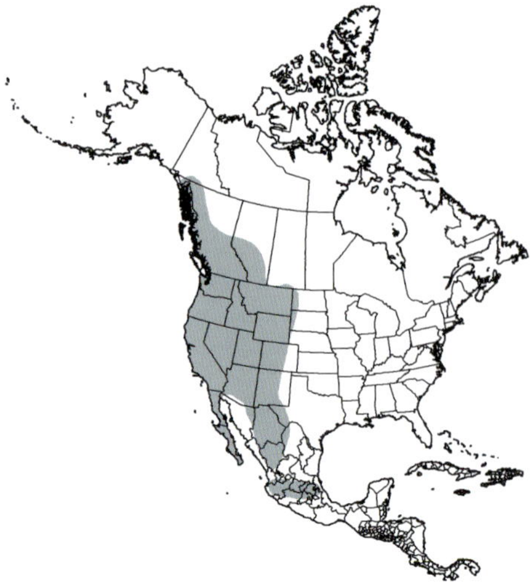

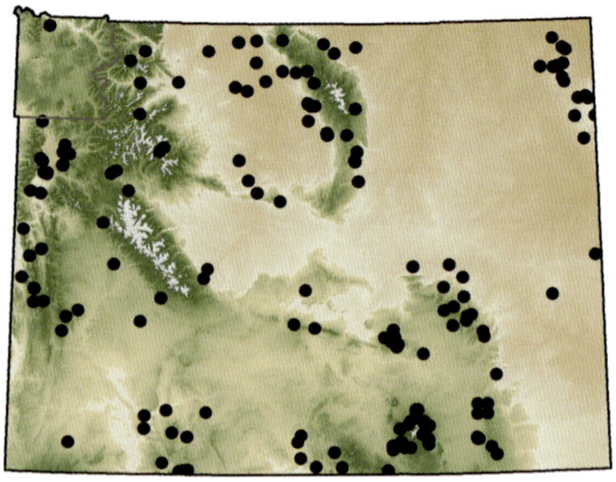

Locations from which the long-legged myotis has been identified in Wyoming.

STATUS: A protected nongame animal and a state Species of Greatest Conservation Need as of 2012.[3] It is common in Wyoming.

NOMENCLATURE AND SYSTEMATICS: The species was first recognized in the mid-nineteenth century, and the current genus-species epithet first used around 1914.[1] Of the four subspecies recognized in 1984, only the putative *M. v. interior* occurs in Wyoming. Some taxonomists consider the Baja California subspecies to warrant species-level recognition; if this were to occur, Wyoming populations would require renaming.[4] A specimen from the Late Pleistocene was described from Little Box Elder Cave, in Converse County.[5]

DISTRIBUTION AND HABITAT ASSOCIATIONS: This western species was recognized to be widely distributed over Wyoming by Long (1965), and was documented in all 25 latitude–longitude blocks by 2004, with an apparent gap in the northeastern basins.[6] A number of night roosts have been located in Wyoming, most of them in caves and abandoned mine adits.[7] The species mostly occupies forested habitats in Wyoming, including juniper-dominated woodlands, ponderosa pine, higher-elevation meadows, and lower-elevation shrub-steppe.

ECOLOGY: The species tends to roost in tree cavities, snags, old buildings, and rock crevices.[8] The diet in Idaho was observed to be dominated by moths and beetles.[9] The species is thought to migrate short distances, but the distances moved by Wyoming animals are not known. Small numbers of hibernators have been described from the Black Hills of South Dakota.[4] A single young is produced per female per year.

CONSERVATION AND MANAGEMENT: The species is of conservation concern in spite of being one of the most numerically abundant and widespread bats in Wyoming. The concerns involve general issues of disturbance at roosting sites, alteration of habitat, and possible cascading effects resulting from bark-beetle-caused tree death.[10]

NOTES AND REFERENCES

1. Warner and Czaplewski (1984).
2. UWMV, KU specimens.
3. Commission Regulations Chapter 52, Section 11.
4. Bogan and Cryan (2000).
5. Anderson (1968).
6. Long (1965), Cerovski et al. (2004).
7. Priday and Luce (1999a).
8. Adams (2003).
9. Johnson et al. (2007).
10. Schmidt (2003b).

Yuma myotis, *Myotis yumanensis*

DESCRIPTION: This small-bodied member of the genus is similar in size and appearance to the little brown myotis, with dorsal fur pale or tawny and non-glossy. The underside is paler than the dorsum, and the pinnae are paler than for the little brown bat. The calcar is unkeeled, but features a lobe; the tail barely extends beyond the interfemoral membrane. Although field morphological characters are not reliable in separating this species from the

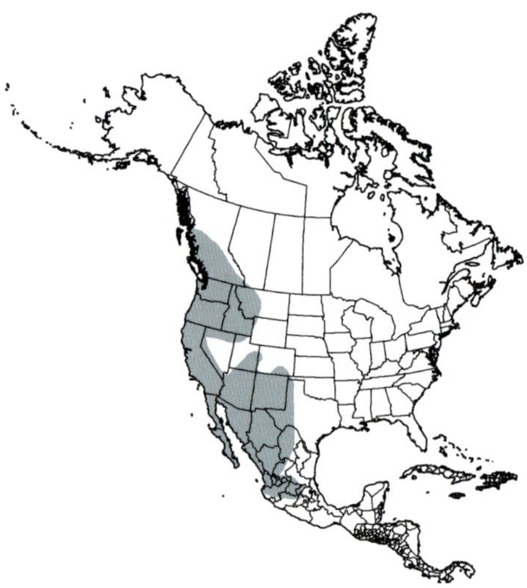

Locations from which the *Yuma myotis* has been identified in Wyoming.

much more common and widespread little brown myotis, morphology in combination with frequencies of echolocation vocalizations can be diagnostic.[1] Measurements (**Sex** [n] mean (mm, g), *range*): eastern Oregon, L = **F** [30] 85, *77–90*; **M** [16] 83, *77–89*; TL = **F** [30] 34, *30–38*; **M** [16] 33, *30–38*; FA = **F** [30] 34.7, *31.7–36.7*; **M** [16] 34.1, *32.6–35.1*; EFN = **F** [30] 14, *10–15*; **M** [16] 14, *12–16*; WT = **F** [30] 6.0, *3.8–8.3*; **M** [16] 5.3, *4.0–9.0*.[2] Dental formula: 2/3, 1/1, 3/3, 3/3 = 38.

STATUS: This is a protected nongame species.[3] It is rare in the state.

NOMENCLATURE AND SYSTEMATICS: The species has been recognized since the 1860s, and the current scientific name used since around the turn of the twentieth century. Six subspecies were recognized as of 1980; *M. y. yumanensis* is the nominal species likely to occur in southern Wyoming.[4]

DISTRIBUTION AND HABITAT ASSOCIATIONS: The distribution of this western species in Wyoming is difficult to infer, because of the dearth of museum specimens; most records are for animals captured in mist nets and released. The species was first reported from Wyoming in 1982, but that specimen was later attributed to *M. lucifugus*.[5] I rely here on three specimens in the Museum of Southwestern Biology, identified to *M. yumanensis* by Michael Bogan and Ernest Valdez.[6] However, additional specimens,[7] as well as several animals examined and released, are consistent with the two foci of distribution suggested by the three specimens mapped. One is in Lincoln County along the western border with Utah and the other is in northern Big Horn County. These foci are presumed extensions of the east-projecting peninsula in Idaho and Montana. In Colorado the species occurs in riparian woodlands, semidesert shrublands, and in pinyon–juniper.[8]

ECOLOGY: The species forages on flying insects, hunting mostly over open water. Day roosts in summer are in rock crevices or abandoned swallow nests.[9]

CONSERVATION AND MANAGEMENT: The species is fairly well known from the Pacific states, but not well-enough known from Wyoming to construct a case that it occurs here, let alone is a conservation concern. It is at the periphery of its geographic range here.

NOTES AND REFERENCES

1. van Zyll de Jong (1985), Adams (2003). Rodhouse et al. (2008) found pelage sheen to be the most reliable morphological character in discriminating from *M. lucifugus* (96% accurate), but relied on molecular markers (Zinck et al. 2004) for definitive identification.
2. Verts and Carraway (1998).
3. Commission Regulations Chapter 52, Section 11.
4. Hall (1981).
5. See Clark and Stromberg (1987). The 1981 specimen reported by Clark and Stromberg (1987) was refuted by Bogan and Cryan (2000).
6. They examined five putative specimens from Wyoming on January 22, 2015, and concluded that three specimens (114,478, 114,479, and 125,182) were *Myotis yumanensis* (mapped); two other specimens from Sheridan County (110,471, 110,936) could be identified to genus Myotis, but not to species.
7. For example, 8814, 8817, and 8818 in the Sam Noble Oklahoma Museum of Natural History.
8. Armstrong et al. (1994).
9. Adams (2003).

Eastern red bat, *Lasiurus borealis*

DESCRIPTION: This distinctive bat is medium-sized, with dense brick-red or orange fur on the dorsum of males, marked by white flecks. Females are yellowish, and the underside of both sexes is paler. Each shoulder has a white tuft. The interfemoral membrane is well furred on the dorsal surface, and the tail extends slightly beyond it.[1] This species cannot be confused with any other Wyoming bat. Measurements (**Sex** [n] mean (mm, g), *range*):

location not specified, L = **B** [21] 104, *87–120*; TL = **B** [19] 46, *36–65*; HF = **B** [17] 8, *7–10*; FA = **B** [8] 40.0, *36–42*; EFN = **B** [5] 11.4, *10–12*; WT = **B** [4] 12.5, *10–17.4*.[2] Dental formula: 1/3, 1/1, 2/2, 3/3 = 32.

STATUS: A protected nongame animal and a state Species of Greatest Conservation Need as of 2012.[3]

NOMENCLATURE AND SYSTEMATICS: The genus and species names have undergone recurring revision. Before the 1990s, this species and its western sibling taxon were considered conspecific: the red bat, *Lasiurus* (= *Nycteris*) *borealis*.[4] Subsequent molecular studies showed the eastern and western forms to be separate species (western form: western red bat, *L. blossevillii*).[5] No subspecies of *L. borealis* are recognized.

DISTRIBUTION AND HABITAT ASSOCIATIONS: This mostly eastern species occurs in the eastern half of Wyoming, with several animals captured around the Medicine Bow Mountains.[6] It has been rare here, and years can pass between detections in mist-net surveys. The species is closely associated with forested habitats, particularly deciduous bottomland forest.[7]

ECOLOGY: This rapid flier forages by capturing insects in flight. Insect species eaten include virtually the entire range of flying species of suitable size. Roost sites have been found in tree boles, foliage, and ground vegetation, including vines and woody debris, but not caves or mines.[8] The species is generally regarded as migratory, but specific seasonal movement distances are not known. The species is preyed upon by a wide range of opportunistic mammals and birds.[9]

CONSERVATION AND MANAGEMENT: Data from Wyoming are too sparse to detect population trends, but the species is of concern in relation to wind development. Eastern red bats declined by 52–85% over a period of several decades in southern Michigan, a trend reflected in submissions of animals for rabies testing. The hypothesized cause of the declines was wind energy development.[8] Wind turbine construction would be a concern in Wyoming if the species distribution overlapped with areas of high surface winds, which is not the case at this time.

NOTES AND REFERENCES

1. Shump and Shump (1982a).
2. van Zyll de Jong (1985).
3. Commission Regulations Chapter 52, Section 11.

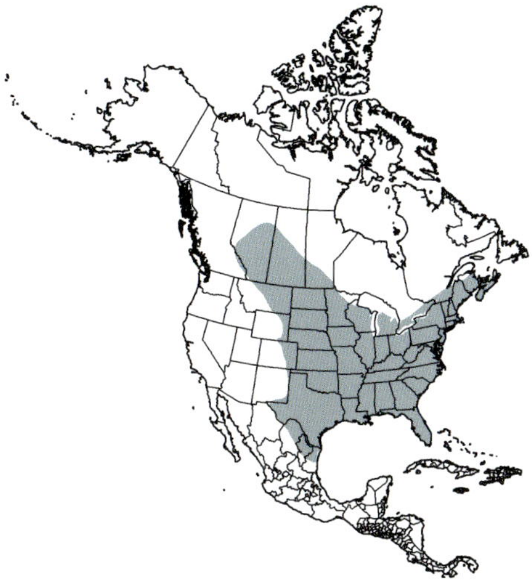

Locations from which the eastern red bat has been identified in Wyoming.

4. Hall (1981), Shump and Shump (1982a).
5. Morales and Bickham (1995).
6. R. Luce, cited by Bogan and Cryan (2000).
7. Shump and Shump (1982a).
8. Mager and Nelson (2001).
9. Winhold et al. (2008).

Hoary bat, *Lasiurus cinereus*

DESCRIPTION: A large bat for Wyoming, distinctive in appearance, with long, dense, brownish or grayish fur, tipped with silver gray. The dorsal fur shows strong banding along the lengths of the hairs. Shoulders and wrists feature white patches of fur. The interfemoral membrane is furred, the short ears black-rimmed, and the tragus short and wide. Females are around 4% larger than males.[1] Measurements (**Sex** [n] mean (mm, g), *SD*): Oregon; L = **F** [3] 136, *133–139*; **M** [17] 133, *120–143*; TL = **F** [3] 59, *53–65*; **M** [17] 57, *48–64*; FA = **F** [3] 54.9, *54.1–54.5*; **M** [17] 52.7, *41.3–54.0*; EFN = **F** [3] 17.0, *17–17*; **M** [17] 16, *12–20*; WT = **F** [3] 25.4, *25.4–25.4*; **M** [17] 24.1, *19.3–28.4*.[2] Dental formula: 1/3, 1/1, 2/2, 3/3 = 32.

STATUS: This is a protected nongame animal, fairly uncommon over much of the state.[3] The perceived uncommonness is partly because the species seldom visits developed areas or roosts in human structures, where it might be encountered. But it is frequently detected in surveys, especially acoustic surveys, away from human development. It has been common in the intensive surveys conducted in the northwestern mountains in recent years.

NOMENCLATURE AND SYSTEMATICS: The scientific name has been stable since the 1860s. The genus name means hairy-tailed; the species name refers to the silver-gray color. Of the three subspecies recognized, only *L. c. cinereus* occurs in North America.[1]

DISTRIBUTION AND HABITAT ASSOCIATIONS: This is among the most widely distributed Western Hemisphere bats, occurring from northern Canada to southern Argentina and Chile,

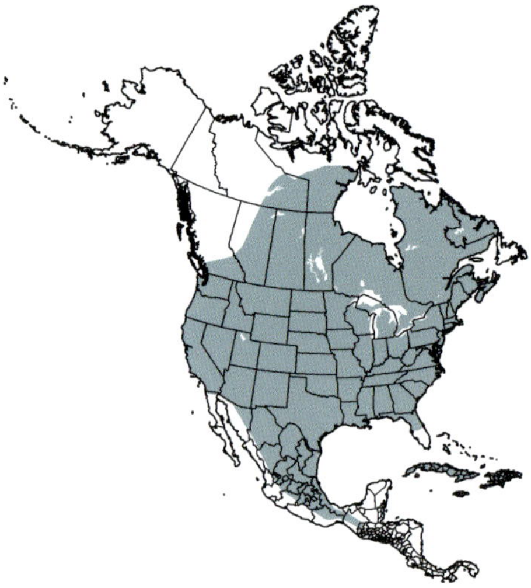

Distribution of the hoary bat in Wyoming.

and includes a resident subspecies in Hawaii—the only native land mammal in that island state. The migratory North American subspecies moves from wintering areas in southern California, the Gulf Coast, and central Mexico in the spring to most of the western and plains states and provinces during July and August.[4] The species was known only from Campbell County in 1965, but was suspected to occur statewide even then. By 1987, it had been found in four more counties, and extensive surveys more recently have extended the known Wyoming distribution to nearly every county. The species tends to fly at greater heights above the ground and forage at higher elevations than other bats.

ECOLOGY: This is a solitary, migratory, insectivorous, foliage-roosting species. The diet during autumn migration is dominated by moths and true bugs.[5] The species has a particularly wide range of temperature tolerances and has been observed to roost in exposed sites in a torporous state in near-freezing temperatures. The young are cared for in exposed sites as well. Roosting sites in southeastern Wyoming have been observed in tall lodgepole pine trees or tall shrubs.[6] The species has a relatively high reproductive rate by bat standards—most females give birth to two young each spring. The sexes migrate separately, with females leaving southern California and Mexico earlier, and traveling farther north and east, than males. During migration through New Mexico, both sexes were found to feed mostly (93%) on moths, likely caught in flight.[7] Hoary bats migrate at greater heights above the ground than other bat species, making them difficult to capture in mist nets and perhaps more vulnerable to injury from wind turbines.[8]

CONSERVATION AND MANAGEMENT: The species is considered a concern in the state because of the lack of knowledge of its distribution and life history and because it is believed to be vulnerable to a range of human-caused effects.[9] This species is frequently considered in planning and siting for wind turbine facilities and is believed to suffer some of the highest losses from wind-turbine-related causes of any North American bat species.[10]

NOTES AND REFERENCES

1. Shump and Shump (1982b).
2. Verts and Carraway (1998).
3. Commission Regulations Chapter 52, Section 11.
4. Cryan (2003).
5. Reimer et al. (2010).
6. Gruver (2002).
7. Valdez and Cryan (2009).
8. Barclay et al. (2007).
9. Hester and Grenier (2005).
10. Arnett et al. (2008).

Silver-haired bat, *Lasionycteris noctivagans*

DESCRIPTION: This distinctive medium-sized bat has dorsal pelage that is long, glossy, and dark brown or black, with silvery-white frosting. The ventral fur is slightly paler, and the interfemoral membrane is furred on the dorsal surface. The flight membranes and short ears are black or nearly so.[1] Measurements (**Sex** [n] mean (mm, g), *range*): various counties, L = **F** [11] 100, *91–105*; **M** [6] 99, *95–104*; TL = **F** [11] 38, *33–43*; **M** [6] 36, *31–43*; HF = **F** [11] 9, *7–11*; **M** [6] 9, *8–11*; FA = **F** [11] 41, *39–42*; **M** [6] 42, *40–43*; EFN = **F** [11] 12, *9–14*; **M** [6] 13, *11–15*; WT = **F** [11] 11.9, *6.9–15.3*; **M** [6] 10.7, *9.3–11.8*.[2] Dental formula: 2/3, 1/1, 2/3, 3/3 = 36.

STATUS: The species is a protected nongame animal.[3]

NOMENCLATURE AND SYSTEMATICS: The species was first described in the mid-nineteenth century, and the current scientific name has been in use since then. This is a monotypic genus, and the species differs in several morphological respects from other North American vesper bats. No subspecies were recognized as of the early 1980s.[4] The species is known from Late Pleistocene deposits of Little Box Elder and Bell Caves.[5]

DISTRIBUTION AND HABITAT ASSOCIATIONS: This widely distributed North American species occurs across Wyoming and adjacent states. Habitats are varied, but tend toward forest, particularly riparian areas in late-successional coniferous forest. However, the species has been captured far from trees in the Red Desert. It roosts solitarily in snags; animals work their way under loose bark, into cracks and into bole cavities.[6] This bat hibernates away from Wyoming in winter.[7] Only one record from Wyoming (in Cheyenne) in winter has been reported.[8]

ECOLOGY: This is one of the better-known temperate North American bats in terms of its ecology. The diet tends to include smaller flying prey than is the case for hoary bats. During summer, moths, flies, true bugs, aphids, and wasps are eaten. During autumn migration in southern Alberta, the species was observed to consume mostly moths and flies.[9] Females commonly give birth to two young per year, compared to one per year for most Wyoming bat species. Bats of this species caught in mist nets in Wyoming have tended to be mostly males. One 2010 collecting effort (n = 110) in northeastern Wyoming showed a sex ratio of nine males per female.[10]

CONSERVATION AND MANAGEMENT: This bat's conservation status is puzzling. On the one hand, it is one of our commoner bats across the continent and has a higher reproductive rate than most species. On the other hand, it is often found dead at wind turbine sites and is thought to be at risk because of forestry practices that limit the availability of old standing trees and snags.[11] Better understanding of its conservation status awaits additional ecological data.

NOTES AND REFERENCES

1. Kunz (1982).
2. DMNS specimens.
3. Commission Regulations Chapter 52, Section 11.
4. Hall (1981), Kunz (1982).
5. Anderson (1968), Kurtén and Anderson (1980).
6. Mattson et al. (1996), Campbell et al. (1996).
7. Geluso, Huebschman et al. (2004).

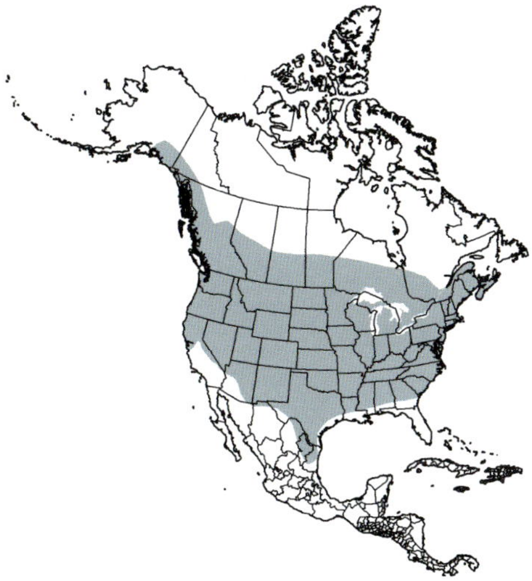

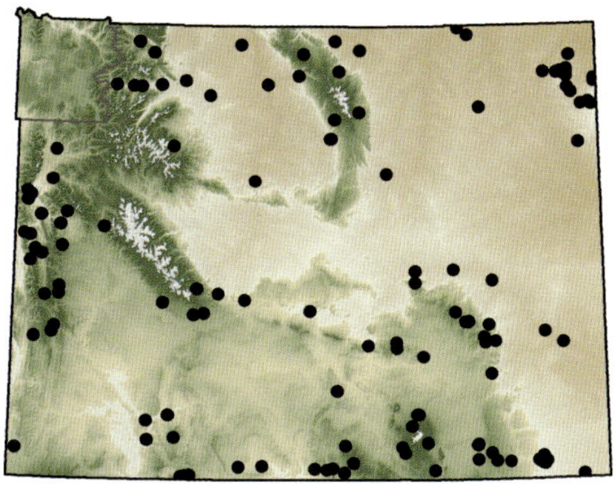

Locations from which the silver-haired bat has been identified in Wyoming.

8. Bogan and Cryan (2000).
9. Reimer et al. (2010).
10. Cudworth et al. (2011).
11. Campbell et al. (1996), Reimer et al. (2010).

American perimyotis, *Perimyotis subflavus*

DESCRIPTION: This small bat has distinctively banded fur—dark at the base, lighter yellowish brown in the middle of the hair shaft, and dark again at the tip. The ventrum is paler. The proximal one-third of the interfemoral membrane is furred, and the leading edge of the wing is paler than the other wing membranes.[1] The species is a weak and erratic flier, readily affected by wind. Measurements (**Sex** [n] mean (mm, g), *range*): location not specified, L = **B** [28] 85, *74–98*; TL = **B** [28] 38, *30–46*; HF = **B** [21] 9, *8–11*; FA = **B** [12] 34, *32–36*; EFN = **B** [16] 13, *11–15*; WT = **B** [7] 6.9, *6.0–7.9*.[2] Dental formula: 2/3, 1/1, 2/2, 3/3 = 34.

STATUS: This is a protected nongame animal.[3] It has been found at one site in the state.

NOMENCLATURE AND SYSTEMATICS: The species has been recognized since the early nineteenth century, but the genus name has undergone recurring revision. Mid-twentieth-century morphologists asserted and recent molecular analyses have confirmed that the eastern and western forms, rather than being the New World members of the diverse Old World genus *Pipistrellus*, are in fact the respective monotypic representatives of two new genera, *Perimyotis* and *Parastrellus*, respectively.[4] Of the four subspecies recognized as of 1981, only *P. s. subflavus* is likely to be in or near Wyoming.[1] The common name "tricolored bat" also is used.

DISTRIBUTION AND HABITAT ASSOCIATIONS: This mostly eastern species has undergone considerable westward expansion during the past three decades. In 1981, it occurred only east of South Dakota and barely entered the southeastern corner of Nebraska. Since then, it has been recorded in western South Dakota, across Nebraska, and in northern New Mexico and

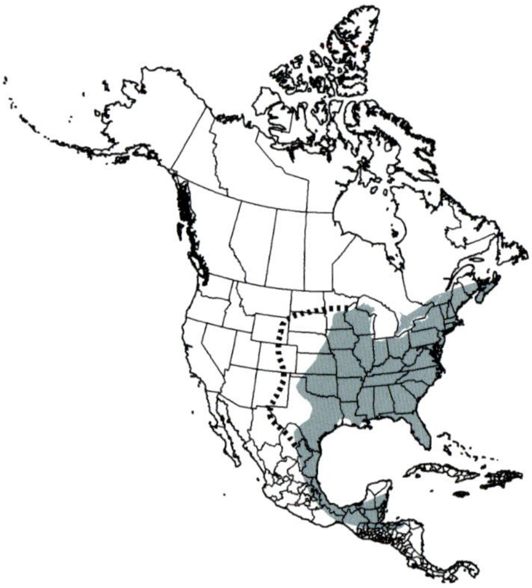

The single location from which the American perimyotis has been identified in Wyoming.

Colorado (continental distribution map shows earlier geographic range with recent expansion westward indicated by dotted line). The first record from Wyoming (three animals) was in a cave hibernaculum in northern Goshen County in 1996. Thus, the species seems to have expanded its geographic range westward along a broad front, from Texas to Minnesota.[5] No mechanism underlying this apparent range expansion has been proposed.

ECOLOGY: The ecology of the species is well known from the eastern states, but not from the western plains states, where it is an apparently new arrival. The diet tends to be small flying insects: beetles, aphids, flies, and wasps. This bat hibernates in caves or mines, typically roosting singly rather than in clusters. Maternity roosts have tended to be in barns or other human structures. In the Mississippi Valley, the species roosts in summer in tree canopies and human structures. Most pregnant females give birth to two (less commonly one) young. Some marked animals in the wild have lived to over 14 years.[6]

CONSERVATION AND MANAGEMENT: The species is not considered a conservation concern in Wyoming because it is an apparently recent arrival, and is widespread and common farther east.

NOTES AND REFERENCES

1. Fujita and Kunz (1984).
2. van Zyll de Jong (1985).
3. Commission Regulations Chapter 52, Section 11.
4. Hoofer et al. (2006).
5. Bogan and Cryan (2000). Of the three specimens found, one was collected (DMNS 8510). The mapped locality is estimated from Geluso et al. (2005).
6. Fujita and Kunz (1984), Perry and Thill (2007).

Big brown bat, *Eptesicus fuscus*

DESCRIPTION: This bat is larger than other Wyoming species with which it could possibly be confused. It is heavily built, and the fur is glossy, pale to dark brown on the dorsal side, depending on subspecies. The tail extends beyond the interfemoral membrane by about 3 mm (0.1 in), and the calcar is keeled. The forearm is longer and the tragus shorter and less pointed than for sympatric myotid bats.[1] Measurements (**Sex** [n] mean (mm, g), *range*): various counties, L = **F** [11] 109, *100–133*; **M** [7] 104, *85–112*; TL = **F** [11] 45, *37–50*; **M** [7] 44, *32–49*; HF = **F** [11] 10, *8–13*; **M** [7] 10, *8–12*; FA = **F** [11] 47, *41–51*; **M** [7] 44, *35–48*; EFN = **F** [11] 14, *11–18*; **M** [7] 16, *14–18*; WT = **F** [11] 14.3, *9–20*; **M** [7] 11.4, *6–15*.[2] Dental formula: 2/3, 1/1, 1/2, 3/3 = 32.

STATUS: This is a protected nongame animal, and a state Species of Greatest Conservation Need as of 2012.[3]

NOMENCLATURE AND SYSTEMATICS: The species has been recognized for over 200 years, and the current genus-species epithet used since the turn of the twentieth century. One of the most widely distributed North American bats, it comprises 10 subspecies, some disjunct on the Caribbean Islands. The subspecies occurring throughout the Rocky Mountains has been regarded as *E. f. pallidus*, although recent studies show that the zone of contact with *E. f. fuscus* passes through easternmost Wyoming.[1]

DISTRIBUTION AND HABITAT ASSOCIATIONS: The species occurs across most of Wyoming and the central Rockies. This is one of the most common and widely distributed Wyoming and North American bats. Habitats are perhaps more diverse than for any other Wyoming vesper bat: areas near and away from water bodies, forests, and other vegetation types,

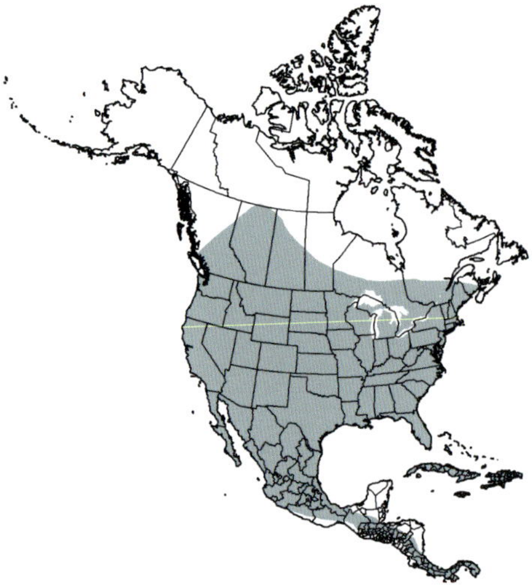

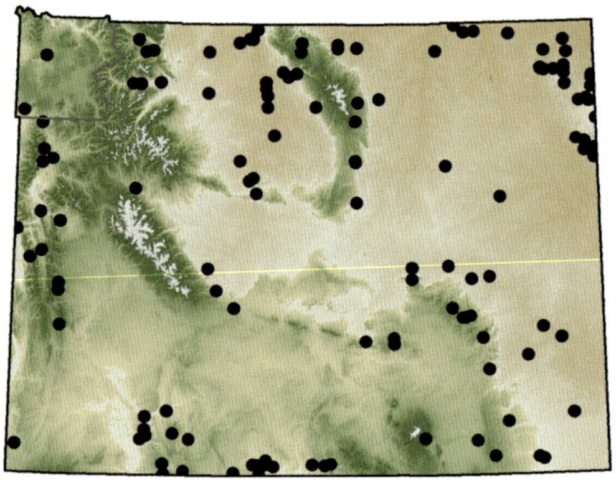

Locations from which the big brown bat has been identified in Wyoming.

eastern plains to mountains over 3050 m (10,000 ft) elevation, and urban and wildland settings.[4] Still, detections in the diagonal area between the northeast and southwest corners of the state are uncommon. Hibernation has been documented in southeastern and north-central Wyoming, as has at least one maternity colony.[5] This is the most common bat in cities along the Colorado Front Range in summer, where it roosts in buildings, typically of older construction.[6]

ECOLOGY: Mean litter size is near one in the West, but closer to two in the southeastern United States and Cuba.[7] This species tends to forage somewhat higher above the ground than most other Wyoming bats and the diet is a wide range of insects captured in flight. Daytime roosts in summer are highly varied and include tree boles, human structures, caves, and abandoned mine adits. Mean annual adult survival in the Fort Collins, Colorado area has been found to be 0.79 in a population that was increasing in size. Longevities in the wild of over 19 years have been reported.[8]

CONSERVATION AND MANAGEMENT: The species is considered a conservation concern in Wyoming primarily because of the potential impacts of wind turbine installations. The species is a small public health concern because of the high incidence of rabies relative to other Wyoming bat species, and the species' proximity to humans in urban and suburban environments.

NOTES AND REFERENCES

1. Kurta and Baker (1990), Hoffman and Genoways (2008).
2. DMNS specimens.
3. Commission Regulations Chapter 52, Section 11.
4. Agosta (2002).
5. Bogan and Cryan (2000), Hester and Grenier (2005).
6. O'Shea, Neubaum et al. (2011).
7. O'Shea et al. (2010).
8. O'Shea, Ellison et al. (2011).

Spotted bat, *Euderma maculatum*

DESCRIPTION: This is a medium-sized to large vesper bat of striking appearance. The dorsal fur is black with a large white spot on each shoulder and the rump. The fur on the underside is white-tipped. The very large ears and flight membranes are pink. Further aiding identification in the field are biosonar vocalizations, which are audible to trained humans as high-pitched "tics."[1] Measurements (**Sex** [n] mean (mm, g), *range*): location not specified, L = **F** [3] 111, *107–115*; **M** [5] 110, *107–114*; TL = **F** [3] 49, *47–50*; **M** [5] 47, *46–48*;

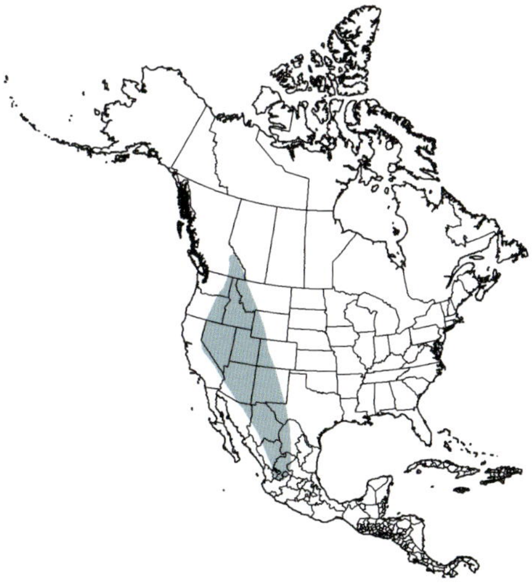

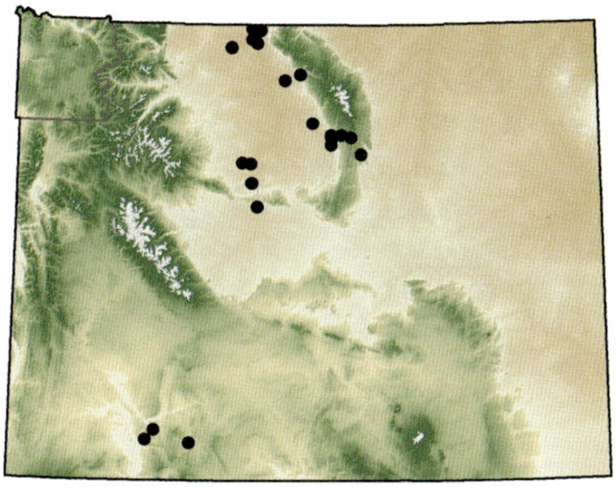

Locations from which the spotted bat has been identified in Wyoming.

HF = **F** [4] 11, *10–12*; **M** = [2] 10, *9–10*; FA = **F** [4] 50, *48–51*; **M** [6] 49, *48–50*; EFN = **F** [2] 42, *37–47*; **M** [5] 41, *39–43*.[2] Dental formula: 2/3, 1/1, 2/2, 3/3 = 34.

STATUS: A protected nongame animal, and state Species of Greatest Conservation Need as of 2012.[3]

NOMENCLATURE AND SYSTEMATICS: The species was first described in the 1890s and the current genus-species epithet has been used since then. No subspecies have been described.[1] The common name "pinto bat" also is used.

DISTRIBUTION AND HABITAT ASSOCIATIONS: The distribution in Wyoming of this intermountain species is uncertain; only two records from Wyoming were known before 1990.[4] The largest concentrations of records are from the lower Green River and the Bighorn Basin. The species is sufficiently rare in mist-net samples that knowledge about its distribution accumulates slowly.[5] Habitats are varied across the geographic range and include desert, shrub-steppe, and coniferous forest; roosting is in cliff faces. Most foraging occurs around water bodies, but the species also is sometimes thought of as an arid-lands specialist.[4]

ECOLOGY: The species is poorly known because of its general rarity. The diet in northern Arizona was recently found to be dominated by moths of the families Noctuidae, Lasiocampidae, and Geometridae; 88–100% of the diet was made up of these taxa.[6] Foraging tends to be at greater heights above the ground than is the case for other bat species; one study estimated a mean foraging height of 10 m (33 ft) in canyon country of northwestern Colorado.[7] Few summer roosts have been described; rock crevices in cliff faces several miles from permanent water were the most common ones in northern Arizona.[8]

CONSERVATION AND MANAGEMENT: The species is a major conservation concern because of its perceived rarity. Bob Luce summarized evidence for the species north of the Mexican border and found 50 records before 1981 and only 60 during 1981–2003.[9] Thus, the species' reputation as one of North America's rarest mammals may be deserved. The specific mechanisms that may negatively affect the species are loss or alteration of wet meadows by livestock or water diversion, collecting activities of scientists, and the use of pesticides.[10]

NOTES AND REFERENCES

1. Watkins (1977).
2. van Zyll de Jong (1985).
3. Commission Regulations Chapter 52, Section 11.
4. Priday and Luce (1999b).
5. Bogan and Cryan (2000).
6. Painter et al. (2009).
7. Navo et al. (1992).
8. Chambers et al. (2011).
9. Luce (2004).
10. Luce and Keinath (2007).

Townsend's big-eared bat, *Corynorhinus townsendii*

DESCRIPTION: A medium-sized bat with gray or brown dorsal fur and very long, wide pinnae that merge across the middorsal line. The wings, ears, and interfemoral membrane are gray or light brown. The face features a prominent sebaceous gland on either side of the snout; this species is easily distinguished in the field from other Wyoming bats.[1] Measurements (**Sex** [n] mean (mm, g), *range*): Platte and Weston Counties; South Dakota; L = **F**

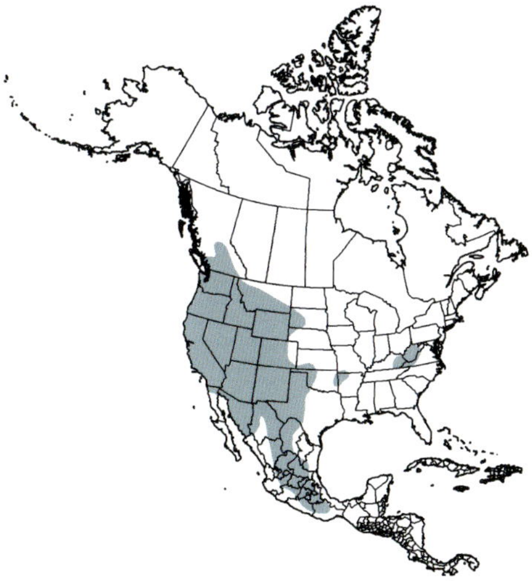

Locations from which Townsend's big-eared bat has been identified in Wyoming.

[5] 100, *90–108*; **M** [5] 94, *86–103*; TL = **F** [5] 48, *44–55*; **M** [5] 45, *40–52*; HF = **F** [5] 11, *10–12*; **M** [5] 10, *9–11*; EFN = **F** [5] 35, *32–37*; **M** [5] 34, *32–35*; WT = **F** [5] 9.6 , *8.5–11.1*; **M** [5] 7.9, *6.5–9.5*.[2] Dental formula: 2/3, 1/1, 2/3, 3/3 = 36.

STATUS: A protected nongame species, and a state Species of Greatest Conservation Need as of 2012.[3]

NOMENCLATURE AND SYSTEMATICS: The species has been recognized since the early nineteenth century, the genus name *Plecotus* being applied until the early 1990s, when this and two other *Plecotus* species were placed into a new genus: *Corynorhinus*.[4] The species epithet refers to Charles H. Townsend, an early twentieth-century naturalist who traversed the Rockies. Of the four putative subspecies recognized in 1981, two of them disjunct in the Midwest and East, only *C. t. pallescens* occurs in Wyoming and adjacent states. The common names "western long-nosed bat" and "ram-eared bat" are also encountered, the latter term referring to the species' tendency to curl its pinnae into rolls—resembling horns—while resting.

DISTRIBUTION AND HABITAT ASSOCIATIONS: This species of the western United States and central Mexico is widespread in Wyoming, with records in the north-central and southeastern parts of the state. It is a year-round resident, although some small-scale seasonal movements may occur. Habitats are quite varied, and include shrub-steppe, juniper woodland, and coniferous forest.[5] Water bodies are important for drinking, and the availability of caves and abandoned mine adits for roosting appears to limit the distribution of this species, although other kinds of roosting structures are occasionally used.[6]

ECOLOGY: The diet is mostly small moths, lacewings, beetles, flies, and wasps, primarily taken in flight. Whether this bat gleans prey from substrate is uncertain. The species is known to be relatively sedentary—moving short distances seasonally—and is constrained to areas with suitable roosting sites. In northern Utah the species was found to day-roost in caves and abandoned mine adits almost exclusively; highway bridges were not used. A single young is born per female. Hibernation occurs singly or in small groups in the Rocky Mountains—one hibernaculum in Wyoming had 49.[7]

CONSERVATION AND MANAGEMENT: The species is a conservation concern because of the perceived importance of mine closures and disturbances in mines and caves by recreationalists and miners.[8] Reductions of forest cover and loss of wetlands are likewise thought to negatively affect the species.[6]

NOTES AND REFERENCES

1. Kunz and Martin (1982).
2. KU specimens.
3. Commission Regulations Chapter 52, Section 11.
4. Frost and Timm (1992), Tumlison and Douglas (1992).
5. Hester and Grenier (2005).
6. Gruver and Keinath (2006).
7. Kunz and Martin (1982), Priday and Luce (1996), Sherwin et al. (2000).
8. Western Bat Working Group (2005).

Pallid bat, *Antrozous pallidus*

DESCRIPTION: A large, large-eared bat with pale yellow pelage tipped with darker shades. The fur on the undersides is lighter. The pinnas do not join across the middorsal line, as they do for the Townsend's big-eared bat. The eyes are large and the face features wart-like sebaceous glands on the sides of the rostrum.[1] Measurements (**Sex** [n] mean (mm, g), *range*): L = **F** [22] 119, *107–130*; **M** [15] 116, *111–124*; TL = **F** [22] 43, *37–49*; **M** [15] 44, *35–44*; HF = **F** [21] 14, *12–16*; **M** [15] 13, *11–15*; FA = **F** [13] 58, *54–60*; **M** [10] 56, *53–58*; EFN = **F** [22] 30, *26–33*; **M** [14] 29, *27–31*; WT = **F** [11] 14.3, *9–20*; **M** [7] 11.4, *6–15*.[2] Dental formula: 1/2, 1/1, 1/2, 3/3 = 28.

STATUS: A protected nongame animal and a state Species of Greatest Conservation Need as of 2012.[3]

NOMENCLATURE AND SYSTEMATICS: The species has been recognized since the mid-nineteenth century, and the current genus-species epithet has been in use since the 1930s. Either six or no subspecies were recognized in accounts before 1990. However, more recent molecular analyses have shown three major phylogeographic clusters: one in California, one centered on western Texas and New Mexico, and the third extending from the southern tip of Baja California to the southern Wyoming. The distributions of these clades are not concordant with traditional subspecies ranges.[4] The genus name is the Latinization of "cave animal."

DISTRIBUTION AND HABITAT ASSOCIATIONS: The distribution of this western species in and near Wyoming is curious—animals along the Green River are a logical extension of the range in Colorado, but the species is also documented in Goshen and Fremont Counties and in the Pryor Mountains, just north of Big Horn County.[5] Hibernacula in Wyoming are not known, and the species may migrate short distances in winter. Habitats are mostly low-elevation desert and shrub-steppe, including riparian woodlands.[6] In Wyoming they have been observed in badlands near Worland and in Shirley Basin.[5] The species is uncommon in or absent from montane or subalpine coniferous forests in Wyoming, unlike in the Pacific Northwest.[7]

ECOLOGY: The species is a gleaning insectivore, capturing mostly large insects on the ground or vegetation surfaces. Some prey may also be captured in flight near the ground. This is one of the most intensively studied bat models for passive hearing, by which pallid

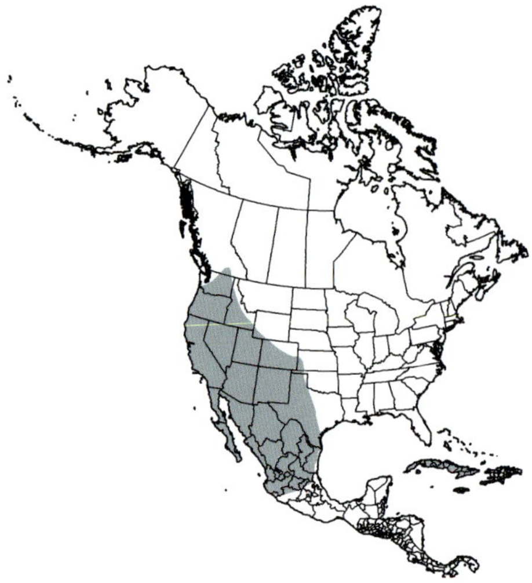

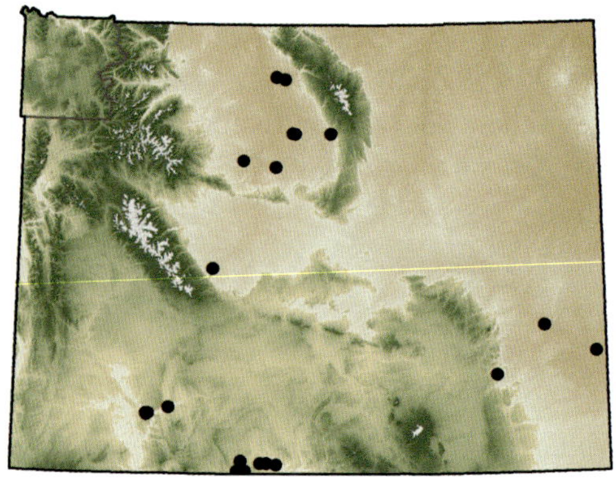

Locations from which the pallid bat has been identified in Wyoming.

bats detect and localize sounds made by moving insects, while at the same time using biosonar to navigate through terrain.[8]

CONSERVATION AND MANAGEMENT: The species is a conservation concern because of the lack of strong knowledge of its distribution, abundance, and limiting factors in the state. The suite of factors thought to negatively affect other temperate bats are likewise presumed important to this species.

NOTES AND REFERENCES

1. Hall (1981), Hermanson and O'Shea (1983).
2. Orr (1954).
3. Commission Regulations Chapter 52, Section 11.
4. Hall (1981), Hermanson and O'Shea (1983), Weyandt and Van Den Bussche (2007), Lack et al. (2010).
5. Robert J. Luce, Sierra Vista, AZ, pers. comm.
6. Baker et al. (2008).
7. Bogan and Cryan (2000).
8. Fuzessery et al. (1993), Johnston and Fenton (2001).

FREE-TAILED BATS—FAMILY MOLOSSIDAE

This family occurs on all temperate continents, but does not extend as far north or south of the equator as do vesper bats. The family is characterized by large, forward-leaning ears, a simple nose with no outgrowths, and a tail that extends well beyond the interfemoral membrane. Free-tailed bats are swift fliers—the falcons of the bat world—which catch insects on the wing.

Brazilian free-tailed bat, *Tadarida brasiliensis*

DESCRIPTION: A medium-sized bat with about half the tail extending beyond the interfemoral membrane. The dorsum is uniformly medium brown, although somewhat variable across individuals; the upper lip features vertical wrinkles. The rounded ears are angled forward, but not joined along the dorsal midline, as is the case for other members of the genus. The ears and flight membranes are darker than the dorsal fur, with short, sparse pelage.[1] Measurements (**Sex** [n] mean (mm, g), *range*): Louisiana, L = **F** [46] 95, *90–100*; **M** [23] 102, *90–109*; TL = **F** [46] 33, *29–42*; **M** [23] 35, *29–44*; HF = **F** [46] 8, *7–12*; **M** [23] 8, *7–14*;

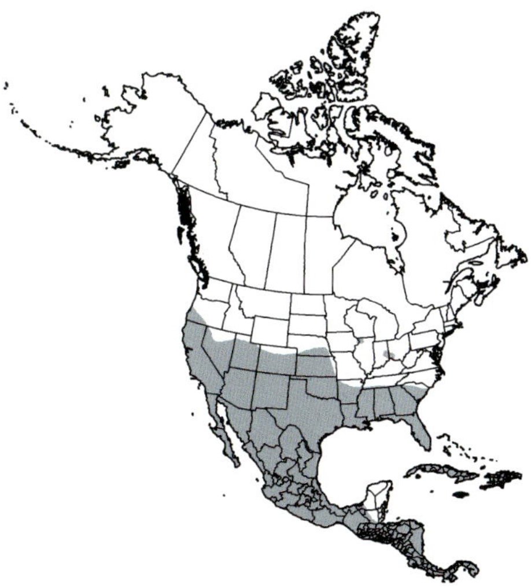

Locations from which the Brazilian free-tailed bat has been identified in Wyoming.

FA = **F** [46] 43, *38–46*; **M** [23] 42, *40–45*; EFN = **F** [46] 16, *14–19*; **M** [23] 15, *10–20*.[1] Dental formula: 1/3, 1/1, 2/2, 3/3 =32.

STATUS: This is a protected nongame animal.[2]

NOMENCLATURE AND SYSTEMATICS: The species was recognized in the early nineteenth century, although the current genus-species epithet has been the predominant one only since 1920. Several subspecies have been identified from North and Central America; the nominal subspecies in and near Wyoming is *T. b. mexicana*.[3] The common name "Mexican free-tailed bat" is also used, especially in North America.

DISTRIBUTION AND HABITAT ASSOCIATIONS: One of the most widely distributed New World mammals, this species' northern distributional limit is just south of Wyoming, and records from this state are considered extralimital by some.[4] It was not known from the state prior to 1981.[5] Populations in more tropical parts of the geographic range are nonmigratory, but those in the states south of Wyoming migrate seasonally to various parts of Mexico; some bats winter as far north as Carlsbad Cavern, New Mexico.[6] Habitats in Mexico and the Southwestern United States tend to be arid areas.

ECOLOGY: This is one of the better-known New World bat species, as a result of its relative abundance in the southern states, where the species roosts in concentrations that attract scientific and popular attention. Summer roosting colonies can number tens or hundreds of thousands, and tourism based on evening emergences from caves or bridges can be locally important.[7] Nightly trips from roosts to foraging areas can be as far as tens of miles, and the bats fly at several thousand feet above the ground. Foraging occurs only a few tens of feet above the ground, and prey is mostly moths and beetles, all taken in flight.[8]

CONSERVATION AND MANAGEMENT: The species has been detected too recently in the state and is too widespread and abundant farther south to be a concern in Wyoming. While clearly abundant in roosting colonies, the species may have undergone declines in the Southwest. The uncertainty lies in estimating numbers of bats emerging from caves that in the 1950s were thought to hold tens of millions of roosting bats per cave. Those same caves now are known to collectively hold a fraction of that number, but whether this represents population declines or more accurate estimation of bat numbers is not clear.[9] The species has long been thought to be vulnerable to exposure to insecticides, in part because of its foraging in agricultural areas.[10]

NOTES AND REFERENCES

1. Wilkins (1989).
2. Commission Regulations Chapter 52, Section 11.
3. Hall (1981), Wilkins (1989).
4. Genoways et al. (2000).
5. Stromberg (1982).
6. Geluso (2008).
7. Pennisi et al. (2004).
8. Davis et al. (1962), Matthews et al. (2010).
9. Betke et al. (2008).
10. Thies et al. (1996).

Big free-tailed bat, *Nyctinomops macrotis*

DESCRIPTION: A large bat with about half the length of its tail extending beyond the interfemoral membrane. The fur is glossy, reddish brown to dark brown; the base of each hair is light-colored. The large forward-angled ears are joined along the middorsal line, unlike in most *Tadarida*. Each upper lip is deeply grooved with vertical wrinkles.[1] Measurements (**Sex** [n] mean (mm, g), *range*): Arizona, L = **F** [11] 132, *120–139*; TL = **F** [11] 50, *40–57*; HF = **F** [11] 9, *7–11*; FA = **F** [11] 60, *58–63*; EFN = **F** [11] 27, *25–32*.[1] Dental formula: 1/2, 1/1, 2/2, 3/3 = 30.

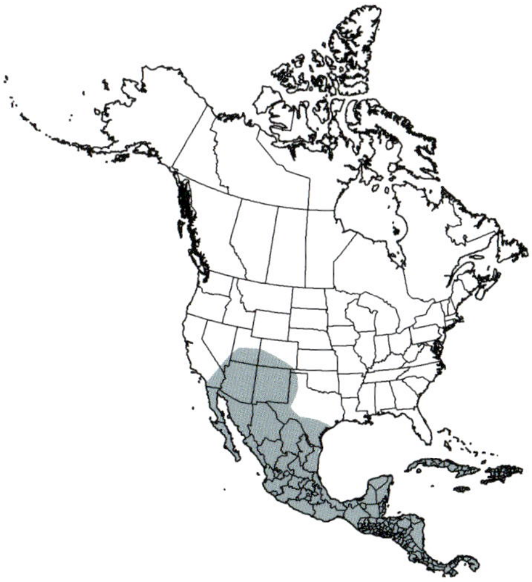

The single location from which the big free-tailed bat has been identified in Wyoming.

STATUS: A protected nongame animal, detected only once in the state.[2]

NOMENCLATURE AND SYSTEMATICS: The species was first described from Cuba in the mid-nineteenth century; the current genus-species epithet has been in use since the early twentieth century. Some taxonomists retain the earlier-applied genus *Tadarida*. No subspecies have been generally recognized.[1]

DISTRIBUTION AND HABITAT ASSOCIATIONS: This broadly distributed New World species reaches its northernmost extent in western North America, having been detected as far north as British Columbia. The single record from Wyoming comes from Teton County.[3] Specimens have been collected from various parts of Colorado, including as far north as Routt County, about 50 km (30 mi) south of Sweetwater County.[4] Habitats in the southwestern United States tend to be rocky, arid lands. The species is seasonally migratory over most of its North American geographic range.

ECOLOGY: The species is a rapid flier and straight-line pursuer of flying insects, which has been found to prey mostly on large moths in Big Bend National Park, Texas. Roosting tends to be in rock crevices high on cliff faces, but buildings and tree cavities also are used. Females produce a single young per year. Little else is known about the species in temperate North America.[5]

CONSERVATION AND MANAGEMENT: The species is not a conservation concern in Wyoming, being considered outside its normal geographic range in the state.

NOTES AND REFERENCES

1. Milner et al. (1990).
2. Commission Regulations Chapter 52, Section 11
3. Bogan and Cryan (2000).
4. Navo and Gore (2001).
5. Easterla and Whitaker (1972), Hasenyager (1980).

Rodents—Order Rodentia

Rodents make up the most diverse mammalian order—an ecologically important and highly cosmopolitan group. Rodent fossils and DNA have drawn differing pictures of the time of rodent divergence from the nearest relative—the lagomorphs; recent molecular estimates have placed the divergence at around 62 Ma in what is now Eurasia.[1] The major lineages of rodents diverged shortly after that.[2] Rodents are distinct from other mammals in having two upper and two lower chisel-like, continuously growing incisors. They have no canine teeth, and feature a toothless gap between incisors and cheek teeth. The cheek teeth differ among rodent families from open-rooted, with enamel ridges on the occlusal surface, to closed-rooted, with enamel covering the entire occlusal surface. Most rodents weigh 10–200 g and are herbivorous, but insectivorous forms, vertebrate predators, and omnivores occur. Most species, with the exception of most squirrels, are nocturnal.

The major suborders of rodents have undergone almost continuous revision and debate for the last century, being distinguished by their morphologies of jaw musculature, soft tissue attributes, and more recently their DNA. Traditional rodent families have persisted at a different level, however; taxonomic families recognized 75 years ago bear a close resemblance to the suborders accepted today: Sciuromorpha (squirrels), Castorimorpha (beavers, kangaroo rats, pocket mice, and pocket gophers), Myomorpha (jumping mice, mice, rats, and voles), and Hystricomorpha (porcupines).[3]

Rodents have important ecological functions: participating in food webs and creating burrows that are used by other vertebrates. They are arguably the largest vertebrate consumers of plant production in Wyoming, with aboveground and fossorial species consuming leaves, seeds, roots, and flowers of many plant types. They disperse some seeds and prey on others, and are the primary and obligatory prey for many or most small and medium-sized predators in Wyoming, avian and mammalian. Some species manipulate vegetation aside from their herbivory: prairie dogs clip woody vegetation and beavers alter riparian community composition.

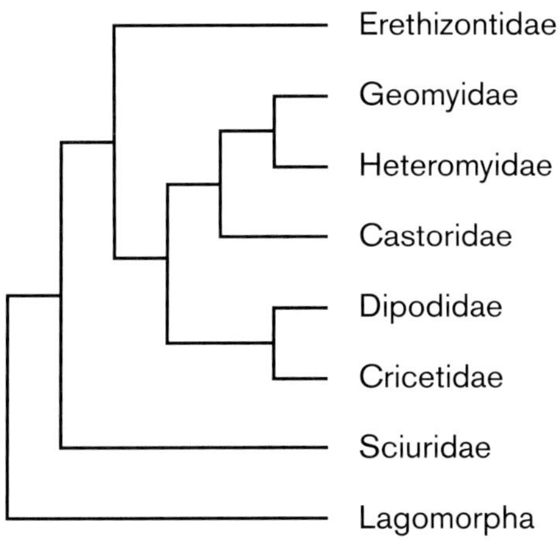

Molecular phylogeny of families of Rodentia that occur in Wyoming, with Lagomorpha as an out-group, based on protein-coding genes and a maximum-likelihood model. Branch lengths are not informative.

Rodents have considerable economic and health significance as well. Beavers and muskrats have much less value for their fur than in the nineteenth and early twentieth centuries, when they were a major geopolitical force, but still are trapped today. A modest amount of squirrel hunting occurs as well. Some species are economic pests in various agricultural settings, and can be nuisances in gardens and lawns. Rodents carry several zoonotic diseases, including the plague bacterium and hantaviruses. Some mice could carry the bacterium that causes Lyme disease as well, although human cases are very rare in Wyoming. Lastly, people enjoy eastern fox and gray squirrels, and have translocated them to various towns to enrich city parks and neighborhoods.

NOTES AND REFERENCES

1. Springer et al. (2004), Blanga-Kanfi et al. (2009), Wu et al. (2012).
2. Blanga-Kanfi et al. (2009).
3. Wilson and Reeder (2005). Family Castoridae is traditional, but the inclusion of kangaroo rats, pocket mice, and pocket gophers with beavers in the suborder is recent.

KEY TO FAMILIES OF RODENTIA

1a. Infraorbital canal absent or small with no part of masseter muscle passing through it to the rostrum: 2
1b. Infraorbital canal clearly visible. Part of masseter muscle passes through it to originate on lateral rostrum: 5

2a. Postorbital processes present, external fur-lined cheek pouches absent: . . . . **Sciuridae**
2b. Postorbital processes absent, either external fur-lined cheek pouches present or tail large and flattened: 3

3a. Tail wide and flattened, cheek pouches absent: . . . . **Castoridae**
3b. Tail narrow and cylindrical or tapered, cheek pouches present: 4

4a. Distal forelimbs flattened for burrowing, auditory bullae not inflated, pinnae barely visible above fur: . . . . **Geomyidae**
4b. Distal forelimbs not flattened for burrowing, auditory bullae inflated, pinnae easily visible above fur: . . . . **Heteromyidae**

5a. Infraorbital canal moderately enlarged, with branch of masseter muscle originating just forward of the canal on the lateral maxilla, no dorsal guard hairs modified as quills: 6
5b. Infraorbital canal highly enlarged, with part of masseter muscle originating well forward of canal, some dorsal guard hairs modified as quills: . . . . **Erethizontidae**

6a. Tail length less than one-fourth of body length: . . . . **Cricetidae**
6b. Tail length more than one-fourth of body length: . . . . **Dipodidae**

SQUIRRELS—FAMILY SCIURIDAE

This important family is nearly cosmopolitan—native to all continents except Antarctica—and contains a number of major subgroups (or tribes), of which three occur in Wyoming. Most members of the family are active in daytime and herbivorous–omnivorous. Tribe Sciurini comprises the tree squirrels—arboreal species that tend to be diurnal, territorial, solitary, and nonhibernators. Four members of this group occur in Wyoming, two of them recently arrived from neighboring states. The Pteromyini includes flying squirrels of the Old and New Worlds, of which only a single nocturnal species occurs in Wyoming. And the Marmotini is composed of ground squirrels and chipmunks. Ground squirrels climb poorly, and many are semi-fossorial hibernators. Classification of the ground squirrels at the level of genus has undergone considerable recent revision.[1] Chipmunks are better climbers than ground squirrels, readily ascending angled logs and steep rocks, and hibernate seasonally.[2] Squirrels are some of the most visible and audible mammals in Wyoming because many are diurnally active, highly visible, and audible vocalizers, animating forest and steppe landscapes with their barks, whistles, and chatters.

The three Holarctic genera of chipmunks show a curious biogeographic pattern: a single species (*Eutamias sibiricus*) occupies a vast area of northeastern Asia, whereas 23 species (*Tamias* spp.) have radiated in western North America and differ in external morphology, bacular shape, chromosome number, and ecology. Some of these species hybridize and introgress where sympatric.[3] Western chipmunks are best distinguished (morphologically)

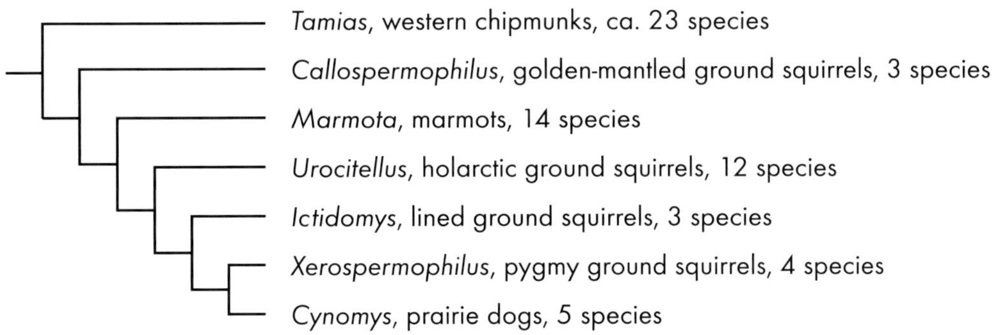

Molecular phylogeny of ground-dwelling squirrel genera that occur in Wyoming based on complete sequences of the cytochrome *b* gene. Branch lengths are not informative. Note: Numbers of species are those worldwide.

by their bacula, which provide somewhat of a lock-key basis for reproductive isolation between species. Bacular shape is quite consistent within, but variable among western chipmunk species. Bacular examination is challenging for nonexperts to use for species identification, however, because it requires dissecting the penis, removing soft tissue from the baculum, and comparing it to reference collections or illustrations under magnification. The biogeographic pattern of these 23 species is complex, and additional species not detectable through morphology are suspected to occur. Finally, a single species (the eastern chipmunk, *Tamias striatus*) occupies eastern North America.[4]

NOTES AND REFERENCES

1. Based on Helgen et al. (2009, fig. 2a).
2. Wilson and Reeder (2005). Richardson and Vander Wall (2007) have shown that yellow-pine chipmunks are unable to climb aspen trees, affecting their activity as canopy nest predators.
3. Sullivan et al. (2014).
4. Banbury and Spicer (2007).

KEY TO GENERA AND SPECIES OF GROUND-DWELLING SCIURIDAE

1a. A dark stripe, bordered by two light stripes, passes through each eye: . . . . **Neotamias, 7**
1b. Facial stripes lacking or not as above: 2

2a. A single light stripe, bordered by two shorter dark stripes, passes from neck to base of tail: . . . . **Callospermophilus lateralis**
2b. Markings on trunk otherwise: 3

3a. Length of adults generally greater than 400 mm: . . . . **Marmota flaviventris**
3b. Length of adults generally less than 400 mm: 4

4a. Trunk fairly uniformly tan or buff, with faint stippling: 5
4b. Trunk marked with multiple stripes or spots: 6

5a. Length less than 320 mm: . . . . **Urocitellus** (identification to species is a task for a specialist)
5b. Length greater than 340 mm: . . . . **Cynomys, 8**

6a. Alternating dark brown and light brown stripes, solid or dotted, run from neck to rump: . . . . **Ictidomys tridecemlineatus**
6b. Light spots in indistinct rows extend from neck to rump, no stripes: . . . . **Xerospermophilus spilosoma**

7a. Dorsal stripes faint: . . . . ***Neotamias dorsalis***
7b. Dorsal stripes distinct:. . . . . ***Neotamias* spp.**
(identification to species is a task for a specialist)

8a. Tip of tail mostly white: . . . . ***Cynomys leucurus***
8b. Tip of tail mostly black: . . . . ***Cynomys ludovicianus***

Yellow-pine chipmunk, *Tamias amoenus*

DESCRIPTION: A medium-sized chipmunk, with facial stripes and dorsal pelage similar to that for other Wyoming species, but the underside is more buff than white. The rump and lateral thighs are gray.

Measurements (**Sex** [n] mean (mm, g), *range*): Lincoln and Park Counties, L = **F** [9] 208, *187–215*; **M** [8] 201, *183–210*; TL = **F** [9] 88, *55–106*; **M** [8] 86, *78–92*; HF = **F** [9] 33, *30–35*; **M** [8] 32, *30–34*; EFN = **F** [9] 20, *16–22*; **M** [8] 20, *15–20*; WT = F [9] 57.1, *41.3–69.7*; M [8] 44.5, *32.1–52.5*.[1] Dental formula: 1/1, 0/0, 2/1, 3/3 = 22.

STATUS: A nongame species that may not be intentionally taken and a state Species of Greatest Conservation Need as of 2012.[2] Its biological status is uncertain within occupied habitats, because of infrequent surveys.

NOMENCLATURE AND SYSTEMATICS: The genus name has undergone recurring revision and is still seen in some sources as *Neotamias*. The specific epithet has long been stable, and this species is more or less distinct from other western chipmunk complexes. Phylogeographic structure within the species is complicated and subspecies described from morphology are not well supported by molecular data.[3] The nominal subspecies found in Wyoming is *T. a. luteiventris*, which extends north through the central and Canadian Rockies to around 53° N. The species is prone to introgression in Montana and Idaho, primarily with the red-tailed chipmunk, *T. ruficaudis*.[4]

DISTRIBUTION AND HABITAT ASSOCIATIONS: This Pacific Northwestern species occurs in the northwest corner of Wyoming, apparently not including the Wind River Mountains.[3] This species illustrates the challenge of updating the geographic range in the absence of substantial trap-capturing activities over the past few decades. Only a few specimens have been collected since 1980, but these are mostly outside the geographic range known in 1965. It is implausible that the range has shifted in so short a time as much as these data suggest. The geographic range shown here is modified from that of Long (1965), adding occurrence records on the north slope of the Uinta Mountains. Habitats are lodgepole pine and adjacent aspen forests and shrublands. These shrubs include snowberry, mountain mahogany, antelope bitterbrush, and currant.[5]

ECOLOGY: The species is diurnally active during the warmest months; during winter the species disappears above ground and presumably enters torpor in nests. The diet includes seeds, fruits, flowers, insects, bird eggs, plant leaves, and invertebrates. Pine seeds and other plant parts are collected and cached below ground, sometimes in large quantities. Although hibernation is well documented, the volume of stored foods below ground raises the

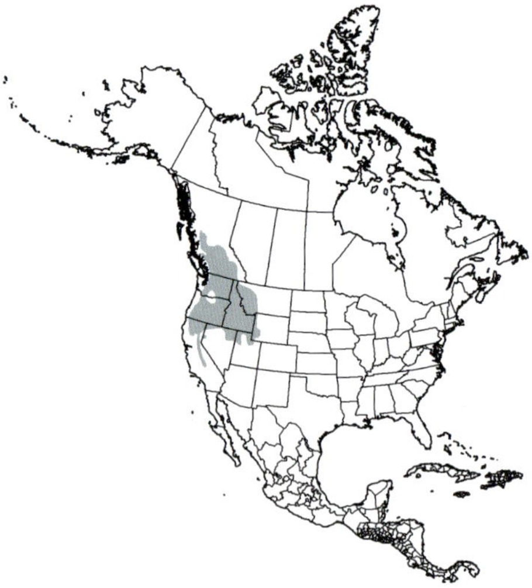

Locations from which the yellow-pine chipmunk has been identified in Wyoming.

possibility of the animals spending more time below ground in arousal than is the case for most hibernators.[6]

CONSERVATION AND MANAGEMENT: The species likely is common within the documented range and perhaps outside it. It is not generally considered a conservation issue.

NOTES AND REFERENCES

1. KU specimens.
2. Commission Regulations Chapter 52, Section 11.
3. Demboski and Sullivan (2003).
4. Good et al. (2008), Sullivan et al. (2014).
5. Sutton (1992).
6. Kenagy (1981), Sutton (1992).

Cliff chipmunk, *Tamias dorsalis*

DESCRIPTION: This medium-sized chipmunk lacks the bright white longitudinal stripes down the dorsum of other Wyoming species; the stripes are indistinct and not visible at a distance. The facial stripes are apparent, but their contrast is lower than that for other Wyoming species. Vocalizations are varied and not useful for species identification.[1] Measurements (**Sex** [n] mean (mm, g), *range*): Sweetwater County, Wyoming, Moffat County, Colorado, L = **F** [6] 210, *198–215*; **M** [6] 201, *191–213*; TL = **F** [6] 93, *90–97*;

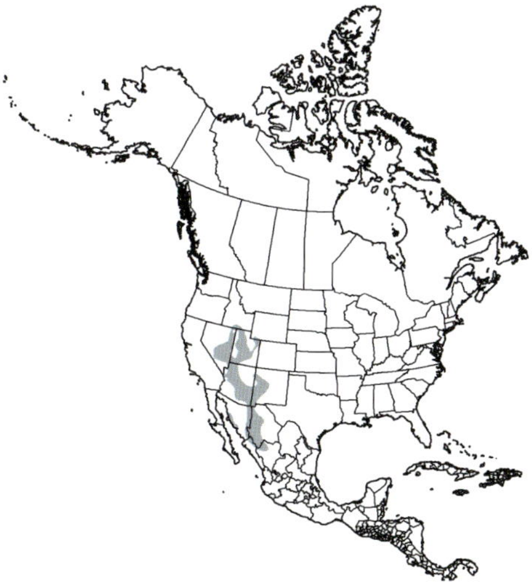

Locations from which the cliff chipmunk has been identified in Wyoming.

M [6] 88, *81–98*; HF = **F** [6] 34, *31–39*; **M** [6] 32, *30–33*; EFN = **F** [6] 19, *17–21*; **M** [6] 18, *17–19*; WT = **F** [6] 53.3, *47.8–55.7*; **M** [6] 47.6, *41.8–54.3*.[2] Dental formula: 1/1, 0/0, 2/1, 3/3 = 22.

STATUS: This is a nongame mammal that may not be intentionally taken and a state Species of Greatest Conservation Need as of 2012.[3] The species' distribution and status are uncertain across the limited area occupied in Wyoming.

NOMENCLATURE AND SYSTEMATICS: The genus name has undergone considerable revision (see account for *T. amoenus*), but the specific epithet has been stable for many decades. This species is placed close to *T. umbrinus* in recent chipmunk phylogenies. Six subspecies are recognized, about half of them representing isolated populations. Wyoming animals represent the northern extension of an isolated pocket of *T. d. utahensis* that resides near the Utah–Colorado–Wyoming corner.[1]

DISTRIBUTION AND HABITAT ASSOCIATIONS: Within Wyoming, this mammal of the Sonoran Desert and Colorado Plateau occurs in southern Sweetwater County, on east and west sides of Flaming Gorge Reservoir.[4] Habitats generally are rocky, dry areas, including in woodlands and open shrublands. It was common in a small area east of Flaming Gorge Reservoir in 1998–1999, where it was caught at 13 of 14 sites sampled.[5] There, the species was found in juniper-dominated sites that were rocky or cliffy with less grass cover, less forb cover, and more rock cover than sites generally available.[4]

ECOLOGY: The species is a diurnally active, generalist forager. As for other chipmunks, foods include tree seeds (particularly juniper and conifer), other plant parts, and invertebrates, as available. Caching of plant parts, particularly of seeds, is common. Predators include the usual suite of vertebrates that kill rodents of this size: weasels, American badgers, foxes, coyotes, raptors, and rattlesnakes.[1]

CONSERVATION AND MANAGEMENT: The species has a very small geographic range within the state. Its distribution was reduced to an unknown extent by the filling of Flaming Gorge Reservoir in the 1960s and early 1970s.[6]

NOTES AND REFERENCES

1. Hart (1992).
2. KU specimens.
3. Commission Regulations Chapter 52, Section 11.
4. Rompola and Anderson (2004).
5. Rompola (2000, Appendix III).
6. Long (1965) mentions five specimens collected 6 km (4 mi) northeast of the former Linwood, Utah (now inundated) at 1758 m (5800 ft) elevation by Merritt Cary in 1911 (USNM catalogue numbers 177470; 177497–177500); the current elevation of Flaming Gorge Reservoir fluctuates around 1830 m (6040 ft) elevation. So, the site where Cary found this species, which has a very restricted distribution within Wyoming, is now below the surface of Flaming Gorge Reservoir.

Least chipmunk, *Tamias minimus*

DESCRIPTION: Small for western chipmunks, the species has alternating dark and light striping on the back and face, prominent pinnae, orange-tinged flanks, and whitish undersides. As for other chipmunks, the tail is carried erect while running. Movement is quick, jerky, and bounding. Vocalizations are chirps, but not useful for species identification.[1] Measurements (**Sex**, month [n] mean (mm, g), *range*): Albany County, L = **F** [37] 204, *184–219*; **M** [41] 198, *183–210*; TL = **F** [38] 89, *75–98*; **M** [41] 86, *73–95*; HF = **F** [41] 31, *27–33*; **M** [47] 31.4, *30–32*; WT = **F** [7, each a mean of several measurements within a year] 52.5, *45.2–57.2*; **M** [7, each a mean of several measurements within a year] 44.2, *39–48.5*.[2] Dental formula: 1/1, 0/0, 2/1, 3/3 = 22.

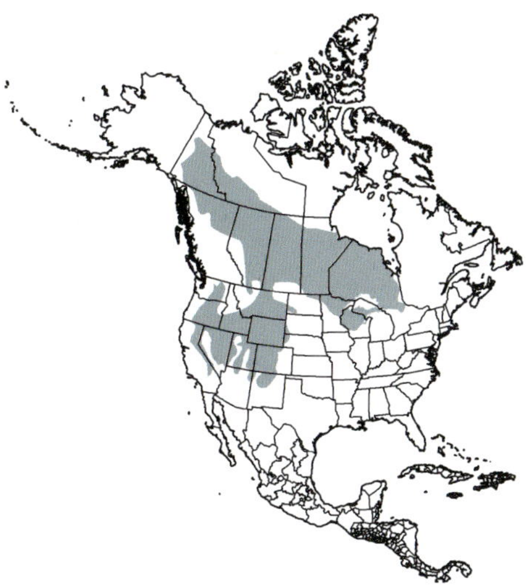

Distribution of the least chipmunk in Wyoming.

STATUS: A non-protected nongame species, common in occupied habitats statewide.[3]

NOMENCLATURE AND SYSTEMATICS: Of the 18 subspecies recognized, six (*T. m. confinis, T. m. consobrinus, T. m. minimus, T. m. operarius, T. m. pallidus*, and *T. m. silvaticus*) occur in Wyoming, with only the first of those limited to a discrete physiographic unit, the Bighorn Mountains.[1]

DISTRIBUTION AND HABITAT ASSOCIATIONS: This boreomontane species occurs statewide in suitable habitats: shrub-steppe, alpine meadow, rockslides, and other open or disturbed areas. Thus, it is the only chipmunk found in eastern Wyoming north of the Medicine Bow Mountains. It occupies shrub-steppe to a greater degree than the yellow-pine chipmunk in northwestern Wyoming, the latter tolerating forest conditions better and behaving aggressively toward the smaller least chipmunk. Near Jackson Hole, least chipmunks occur in a wider range of cover types than the two other chipmunk species present.[4]

ECOLOGY: This is a small, diurnal squirrel that climbs logs and angled tree boles, and a generalized feeder. Summer diets in northern Colorado alpine zones included seeds (49–55%), arthropods (7–23%), leaves (8–10%), fruits and flowers (9–22%), and fungi (5–6%), depending on year.[5] The species hibernates, but may also remain active below ground, eating stored food. Population densities in a Colorado alpine zone ranged from 11 to 22/ha (2800–5600/mi^2) with season and year. In Utah, densities were lower, 0.3–6/ha (77–1500/mi^2), depending on habitat and season.[6] Predators include the full range of mid-sized mammals and raptors: weasels, foxes, coyotes, and small hawks.

CONSERVATION AND MANAGEMENT: The species is common to abundant and is seldom hunted. It is not typically considered a conservation issue, nor is it often considered in conservation planning.

NOTES AND REFERENCES

1. Verts and Carraway (1980).
2. Skryja (1970, Tables 11 and 12).
3. Commission Regulations Chapter 52, Section 6.
4. Negus and Findley (1959).
5. Vaughan (1974).
6. Vaughan (1974), Andersen et al. (1980).

Uinta chipmunk, *Tamias umbrinus*

DESCRIPTION: This species is large for a chipmunk, with the most lateral of the dark dorsal stripes indistinct or missing.[1] Measurements (**Sex** [n] mean (mm, g), *range*): Fremont, Sublette, and Teton Counties, L = **F** [9] 225, *192–235*; **M** [6] 216, *205–228*; TL = **F** [9] 98, *57–111*; **M** [6] 96, *86–115*; HF = **F** [9] 32, *31–34*; **M** [6] 32, *30–34*; EFN = **F** [9] 18, *12–21*; **M** [6] 19, *17–20*; WT = F [9] 69.5, *62.8–76.4*; M [6] 59.7, *41–92*.[2] Dental formula: 1/1, 0/0, 2/1, 3/3 = 22.

STATUS: A nongame mammal that may not be taken intentionally and a state Species of Greatest Conservation Need as of 2012.[2] It is common in occupied habitats.

NOMENCLATURE AND SYSTEMATICS: The genus name has undergone revision, but the species name has been stable for over a century. The species is closely related to *T. dorsalis* and other species in the *T. quadrivittatus* group.[3] Seven subspecies are recognized, three of which occur in Wyoming in mountain ranges: *T. u. fremonti* (northwestern mountains), *T. u. umbrinus* (Uinta Mountains), and *T. u. montanus* (Medicine Bow Mountains).

DISTRIBUTION AND HABITAT ASSOCIATIONS: This mammal of the Intermountain West occurs in mountains of the northwestern and southern parts of the state. The southern Wyoming distribution comprises the Medicine Bow Mountains and northern slope of the Uinta Mountains. Habitats are mainly coniferous forests, including ponderosa pine, lodgepole pine (including admixtures with juniper and

Distribution of the Uinta chipmunk in Wyoming.

Douglas-fir), and Engelmann spruce–subalpine fir, but the species extends into woodlands below and alpine zones above. In the Colorado Front Range, the lower elevational limit is around 2130 m (7030 ft), where it is parapatric with the Colorado chipmunk (*N. quadrivittatus*).[4] Rocky slopes are considered preferred sites.[1]

ECOLOGY: The diet is typical of western chipmunks: seeds, other plant parts, and invertebrates. Population biology has been studied little, but one coarse estimate of annual survival was <0.5.[1] The species is assumed to hibernate most of the time when not active above ground, but could be active below ground, feeding on cached food.[5]

CONSERVATION AND MANAGEMENT: The species' classification as a Species of Greatest Conservation Need is due to a lack of reliable knowledge of abundance or ecology.

NOTES AND REFERENCES

1. Bergstrom and Hoffman (1991).
2. KU specimens.
3. Commission Regulations Chapter 52, Section 11.
4. Piaggio and Spicer (2000).
5. Armstrong et al. (2011).

Golden-mantled ground squirrel, *Callospermophilus lateralis*

DESCRIPTION: This small ground squirrel is larger and stockier in appearance than a chipmunk, with alternating stripes (black-white-black) on each side of the dorsum. The face lacks striping, but light bands above and below the eyes resemble those of red squirrels. The tail is shorter and less well furred than in chipmunks. The underside is buff or off-white.[1] Vocalizations are simple clicks and not useful in identification. Golden-mantled ground squirrels are not as fast or agile as chipmunks. The species can climb trees, but is more often seen climbing angled logs or rocks to survey the area. Measurements (**Sex** [n] mean (mm, g), *range*): Sweetwater and Albany Counties, L = **F** [6] 260, *220–283*; **M** [6] 258, *243–275*; TL = **F** [6] 75, *58–98*; **M** [6] 85, *53–97*; HF = **F** [6] 40 *36–44*; **M** [6] 40, *36–43*; EFN = **F** [6] 16, *11–23*; **M** [6] 17, *10–22*; WT = **F** [6] 169.8, *125–219*; **M** [6] 178.2, *155.7–207*.[2] Dental formula: 1/1, 0/0, 2/1, 3/3 = 22.

STATUS: A non-protected nongame animal, common in occupied habitats statewide.[3]

NOMENCLATURE AND SYSTEMATICS: The species was one of three closely related ones in subgenus *Callospermophilus*, which was recently raised to the level of genus. The species name has been stable for over a century, and of the 13 subspecies recognized as of 1993, four occur in Wyoming (*C. l. cinerascens*, *C. l. castanurus*, *C. l. lateralis*, and *C. l. wortmani*), only the last of them mostly occurring within Wyoming.[1]

DISTRIBUTION AND HABITAT ASSOCIATIONS: This dweller of western mountains occurs from forest margins at the lower elevational limit of trees to above 3600 m (11,900 ft). The distributions of named subspecies in Wyoming do not follow recognizable physiographic boundaries, with *C. l. cinerascens* occurring in northwestern mountains north of around 44° N. latitude, *C. l. castanurus* south of that line to the Red Desert, and a third subspecies, *C. l. lateralis*, in the southern Wind River, Uinta, Medicine Bow, and Laramie

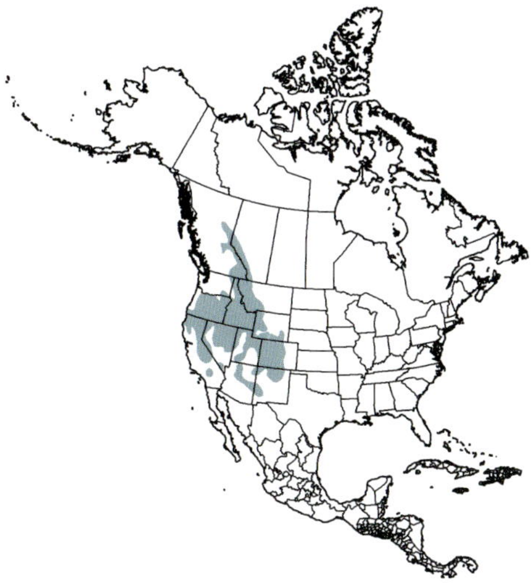

Distribution of the golden-mantled ground squirrel in Wyoming.

Mountains. The fourth subspecies, *C. l. wortmani,* has been depicted east of the Green River in Sweetwater County.[4] The systematic patterns implied by these subspecies boundaries are difficult to reconcile with geography; therefore, this subspecific classification should be considered tentative. Habitat is open forest or woodland dominated by various pine species, Douglas-fir or aspen. The species avoids dense spruce–fir habitats, but ranges into meadows and shrublands some distance from trees.

ECOLOGY: The species is diurnal, hibernating omnivore, and digs its own burrows, typically associated with rocks, tree stumps, or other protective structures. The diet is varied, including fungi, conifer seeds, and acorns.[5] Soft mast is important, and fruits of rose, serviceberry, western thimbleberry, currant, and herbaceous species are consumed. Insects are eaten during summer, as are small bird eggs, neonatal mice, and carrion. Human foods are highly sought. The species enters torpor from late August to November, depending on elevation and latitude. Emergence is in March or April.[1] Population biology is not known from Wyoming. The species is important prey for various mid-sized mammals and raptors. It was the second most frequently taken prey—after red squirrels—by northern goshawks in the Medicine Bow Mountains.[6]

CONSERVATION AND MANAGEMENT: This is not generally considered a species of conservation concern, although the complex pattern of subspecific biogeography admits the possibility of some endemic subspecific forms that are rare, isolated, in decline, or all of these.

NOTES AND REFERENCES

1. Bartels and Thompson (1993).
2. UWMV specimens.
3. Commission Regulations Chapter 52, Section 6.
4. Long (1965).
5. Cork and Kenagy (1989), Bartels and Thompson (1993).
6. Squires (2000).

Yellow-bellied marmot, *Marmota flaviventris*

DESCRIPTION: A "giant" ground squirrel with tan or buff fur, dorsal guard hairs banded or tipped with light color, tending to tan or orange on the underside. The tail is well furred with black and orange hairs, and the pinnae are visible but short. The population in the Teton Mountains has a high incidence of melanism; around 23% of animals are nearly black with lighter facial markings.[1] Several vocalizations are used, all distinctive, communicating different levels and sources of alarm. Visually, marmots are easily identified at a distance with experience, but are challenging for novice observers, being confused with badgers, wolverines, fishers, porcupines, and beavers. The behavior and microhabitat of the species can be strong clues to identity: marmots tend to perch atop large rocks and often stand on their hind legs to scan for predators or peer at intruders. Measurements (**Sex** [n] mean (mm, g), *range*): various counties, L = **F** [30] 504, *300–630*; **M** [5] 504, *376–714*; TL = **F** [30] 139, *90–186*; **M** [5] 136, *108–180*; HF = **F** [30] 70, *53–81*; **M** [5] 70, *61–75*; EFN = **F** [30] 26, *20–39*; **M** [5] 21, *17–23*; WT = **M** [5] 1560, *671–2079*.[2] Dental formula: 1/1, 0/0, 2/1, 3/3 = 22.

STATUS: A non-protected nongame mammal that is common within its geographic range.[3]

NOMENCLATURE AND SYSTEMATICS: The genus and species names have been stable for decades; the specific name means "yellow-bellied." Of the 11 recognized subspecies, three occur in Wyoming: *M. f. luteola* in the Laramie and Medicine Bow Mountains, *M. f. nosophora* in the Uinta, Bighorn, and northwestern mountains, and *M. f. dacota*, the only geographically disjunct subspecies, in the Black Hills and Bear Lodge Mountains.[4]

DISTRIBUTION AND HABITAT ASSOCIATIONS: This species of temperate western mountains occurs in montane and alpine habitats statewide. Typical habitat is alpine

Yellow-bellied marmot with typical pelage.

Yellow-bellied marmot with the melanistic pelage common in the Teton Mountains.

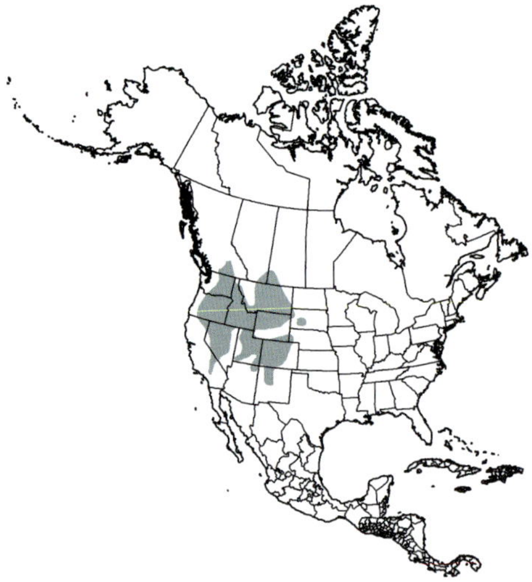

Distribution of the yellow-bellied marmot in Wyoming.

tundra or woodland edge with rock outcroppings or talus. Such features protect marmots from excavation by large digging carnivores: bears, badgers, and wolverines. Areas above and below the forest zone are used, as are forest edges, but dense forest and the interiors of woodlands are avoided. In the Great Basin, most of the marmot burrows examined in one study were within the elevation range 2100–3000 m (6900–9900 ft).[5]

ECOLOGY: Yellow-bellied marmots are diurnal, semi-fossorial, seasonally hibernating herbivores. Their annual pattern of activity includes emergence in late April or May, adults emerging before juveniles and males before females. Cessation of aboveground activity occurs from August to September, in roughly the same demographic order. The continuity of hibernation depends on age, with adults able to withstand prolonged torporous fasts, while juveniles must arouse to feed on vegetation when weather conditions and plant phenology allow.[6] Marmots are generalist feeders, eating plants based on abundance, phenology, and defensive compounds. Most foods are grasses, forbs, or seeds.[4] Burrows are widely distributed across the home range and vary in function from predator escape to hibernation. Hibernation is believed to be somewhat social, as for other colonial marmots, in that adult males synchronize torpor with adult females, and mothers with offspring. But thermal energetic benefits of co-hibernation do not appear to be a benefit of such synchrony.[7] The main cause of death during the active season appears to be predation; in southern Colorado predators caused 96% of summer deaths. Predators include various mammals and large raptors of montane or alpine settings, especially coyotes and badgers. In Colorado, martens caused as many deaths as black bears.[8] Yellow-bellied marmots in Montana were found to be the third most common prey taken by golden eagles.[9]

CONSERVATION AND MANAGEMENT: The species traditionally has not been a conservation concern. However, climate warming is believed to be affecting dates of emergence from hibernation at lower elevations, with animals in Gunnison County, Colorado, emerging 10–16 days earlier in 2003 than they had in early 1990s.[7] In the same area over the period

1976–1999, the first sighting of a marmot in spring advanced by 23 days.[10] On the other hand, the changing climate may feature more precipitation and longer-duration snowpack at higher elevations, so that emergence times in those areas could be delayed.[11] Clearly, the effect of a changing climate on species such as the yellow-bellied marmot is a multifaceted issue, involving possibly opposing forces at various sites within the geographic range.

NOTES AND REFERENCES

1. Armitage (1961). This local anomaly in pelage color was first noted by Olaus Murie in the 1930s (Murie 1934).
2. UWMV specimens for males, KU specimens for females.
3. Commission Regulations Chapter 52, Section 6.
4. Frase and Hoffman (1980).
5. Floyd (2004).
6. Armitage (1991).
7. Blumstein et al. (2004).
8. Armitage (1961), Van Vuren (2001).
9. McGahan (1967).
10. Inouye et al. (2000). This estimate is based on a regression of the date of first sighting of a marmot in the area on year. The corresponding change in temperature at Crested Butte, Colorado, for the period studied was +1.4°C (2.5°F), showing the apparently high sensitivity of yellow-bellied marmots to temperature.
11. van Vuren and Armitage (1991).

Uinta ground squirrel, *Urocitellus armatus*

DESCRIPTION: A medium-sized ground squirrel slightly larger than the Wyoming ground squirrel. Dorsal fur is tan, and the tail is buff-colored, with black on dorsal and ventral surfaces, and pink along the margins.[1] Melanistic animals are rare, but have been reported from near Lyman, Wyoming.[2] Measurements (**Sex** [n] mean (mm), *range*): Sublette County, L = **F** [8] 270, *256–284*; **M** [11] 282, *257–308*; TL = **F** [8] 66, *61–72*; **M** [11] 68, *62–74*; HF = **F** [8] 42, *40–43*; **M** [11] 42, *40–45*.[3] Dental formula: 1/1, 0/0, 2/1, 3/3 = 22.

STATUS: A non-protected nongame mammal, common or abundant within the occupied range.[4]

NOMENCLATURE AND SYSTEMATICS: The species name has been stable for over a century, but the genus name has undergone recurring revision through the same period.[5] Subspecies have not been designated, although some Idaho populations are isolated on mountain ranges. The zone of contact or area of sympatry with the Wyoming ground squirrel is uncertain with respect to its location, and regarding the extent and depth of any introgression.

DISTRIBUTION AND HABITAT ASSOCIATIONS: This species has a small geographic range. In Wyoming, it occurs in a north–south band along the western border of the state, around 130 km (80 mi) in east–west dimension. Marked genetic structure within the species' range has been linked to volcanic and glacial events during the Pleistocene. In particular, the species is low in genetic variability, consistent with a major volcanic eruption

Distribution of the Uinta ground squirrel in Wyoming. South of the Wind River Mountains, the zone of parapatry or introgression with the Wyoming ground squirrel is uncertain.

at the Lava Creek Caldera in the Yellowstone region 640,000 years ago. This eruption would have deposited 1–10 m (3–30 ft) of ash on the ground, near the middle of the species' restricted range.[6] Subsequent glaciations isolated intermontane basins from each other, further structuring the remaining genetic variability.[7] Habitats are open areas, particularly shrub-steppe dominated by sagebrush and rabbitbrush.

ECOLOGY: The species is a diurnally active, semi-fossorial, seasonally hibernating herbivore. Diets are a range of plants, particularly grasses, forbs, and seeds. In northern Utah, spring foods were 90% grass leaves, shifting to forb leaves later in summer. *Artemisia* leaves represented around 10% of foods throughout the active season.[8] In Jackson Hole, males were first observed above ground around April 9, 1987–1989; females emerged four days later.[9]

In northern Utah, population density was found to reach 28/ha (7200/mi^2), not including juveniles, during a year of high densities. Annual survival rates in northern Utah varied with sex and age, averaging 0.27 for males and 0.37 for females, 0.29 for juveniles and 0.42 for yearlings and adults. Density-dependent survival was apparent.[10] Predators include the full range of sympatric mid-sized carnivores and raptors.[1]

CONSERVATION AND MANAGEMENT: The species is common or abundant within the Wyoming portion of its range and not ordinarily considered a conservation concern.

NOTES AND REFERENCES

1. Eshelman and Sonnemann (2000).
2. Long (1963).
3. Long (1965).
4. Commission Regulations Chapter 52, Section 6.
5. See Wyoming ground squirrel for genus-level taxonomy.
6. Perkins and Nash (2002).
7. van Tuinen et al. (2008).
8. Walker (1968).
9. Rieger (1996).
10. Slade and Balph (1974).

Wyoming ground squirrel, *Urocitellus elegans*

DESCRIPTION: A medium-sized ground squirrel slightly smaller than the Uinta ground squirrel. The dorsum is tan with dark tones, creating the impression of tan speckling on a darker background. The underside is tan, and the top of the head uniform gray-tan. The tip of the medium-length tail is dark. The pinnae are visible, but not large. Measurements (**Sex** [n] mean (mm, g), *range*): Albany and Sweetwater Counties, L = **F** [6] 247, *200–270*; **M** [5] 273, *241–300*; TL = **F** [6] 62, *51–69*; **M** [5] 70, *68–76*; HF = **F** [6] 40, *38–44*; **M** [5] 42, *39–45*; EFN = **F** [6] 14, *9–16*; **M** [5] 13, *11–16*; WT = **F** [6] 213.6, *108–289*; **M** [5] 325.8, *194–453.6*.[1] Dental formula: 1/1, 0/0, 2/1, 3/3 = 22.

STATUS: This is a non-protected nongame animal, common to abundant within its geographic range.[2] It is considered a "designated pest" by the Wyoming Department of Agriculture, which provides for local reduction of densities.[3]

NOMENCLATURE AND SYSTEMATICS: The genus name has undergone frequent revision and reversal, including from *Citellus* back to *Spermophilus* in the early 1980s, and to *Urocitellus* recently.[4] Prior to the mid-1970s this taxon was considered a subspecies of the Richardson's ground squirrel (*S. richardsonii elegans*); it was combined with two other disjunct subspecies (see continental map) and raised to the level of species based on chromosome number and molecular evidence.[5] Most recently, phylogenetic relationships among ground squirrels (including prairie dogs and marmots) have been reexamined and some new generic names assigned, or reassigned.[6] Three subspecies of Wyoming ground squirrel are recognized, each a disjunct population: only *S. e. elegans* occupies southern Wyoming and adjacent parts of Utah and Colorado. The identity of animals in the zone of sympatry with Uinta ground

Distribution of the Wyoming ground squirrel in Wyoming. The zone of parapatry or introgression with the Uinta ground squirrel is uncertain.

squirrels—in southwestern Wyoming—has been problematic for decades and apparent hybrids have been described.[7] But introgression has not been documented using molecular markers, leaving an uncertain zone of contact. Several morphological differences have been described between these two nominal species, including subtle differences in pelage color, size, and the shape of the baculum.[8]

DISTRIBUTION AND HABITAT ASSOCIATIONS: The species is patchily distributed across the width of Wyoming in the south, tapering northward to an apex in the central Bighorn Basin. The eastern boundary of the distribution follows the western flanks of the Bighorn and Laramie Mountains. Although the species occurs just above the upper elevational limit of trees, its distribution in alpine areas is spotty, and the highest alpine areas have been suspected of being geographic barriers.[9] Habitats are highly varied, but mostly shrublands, grasslands, and meadows. Rocky slopes and disturbed sites can be important microhabitats.

ECOLOGY: Wyoming ground squirrels are diurnally active, semi-fossorial, and territorial. Hibernation is obligatory; males enter torpor before females in latest July or early August, and adults enter torpor before juveniles. Males also emerge in spring two weeks before females, typically during March. The diet is generalized and includes a variety of vascular plants, dominated by forbs rather than grasses, and animal matter.[10] Foods are not stored for the winter. Densities of resident animals vary by season and have been reported for the Laramie Basin to be 0.2/ha (51/mi^2) in March and 1.2/ha (307/mi^2) in June.[11] Including transients, a much higher density of 44/ha (>11,000/mi^2) was reported for the same area.[12] Annual survival rate in the Laramie Basin was around 50%. Causes of death are not well documented, but predation likely is important, with a wide range of mammalian and avian predators contributing.

CONSERVATION AND MANAGEMENT: The species is common or abundant in occupied habitats and is not often considered a species of conservation concern. It is controlled in a few areas—but not as often as the black-tailed prairie dog—because of conflicts with agriculture.

NOTES AND REFERENCES

1. UWMV specimens.
2. Commission Regulations Chapter 52, Section 6.
3. The authority is provided by W.S. 11-5-102-a-xii.
4. Zegers (1984). Genus *Spermophilus*, as previously used, is now believed to have been paraphyletic, and the new name *Urocitellus* has been proposed for a group that includes this species (Helgen et al. 2009).
5. Nadler (1974); *S. richardsonii richardsonii* from Montana 2N = 36; *S. r. elegans* from southern Wyoming 2N = 34. Harrison et al. (2003) for molecular evidence.
6. Harrison et al. (2003), Helgen et al. (2009).
7. Hall (1943).
8. Zegers (1984).
9. Long (1965).
10. Over 65% of volume, in Albany County during every two-week period from April 1 to August 15, 1963 (House 1964).
11. House (1964).
12. Clark (1970).

Spotted ground squirrel, *Xerospermophilus spilosoma*

DESCRIPTION: A small, slender ground squirrel with dull dorsal pelage and whitish dorsal spots, arranged in irregular lines. The underside is paler. Measurements (**Sex** [n] mean (mm, g), *range*): Nebraska, L = **F** [5] 205, *193–221*; **M** [12] 215, *162–250*; TL = **F** [5] 60, *55–63*; **M** [12] 64, *45–80*; HF = **F** [5] 32, *31–34*; **M** [12] 33, *26–38*; EFN = **F** [5] 8, *5–12*; **M** [11] 8, *4–13*; WT = **F** [5] 99.0, *84.2–127.6*; **M** [10] 116.4, *101.5–133.5*.[1] Dental formula: 1/1, 0/0, 2/1, 3/3 = 22.

STATUS: This is a protected nongame animal that may not be taken intentionally and a state Species of Greatest Conservation Need as of 2012.[2] It is uncommon or rare within the Wyoming portion of the geographic range.

NOMENCLATURE AND SYSTEMATICS: The genus name underwent revision from *Spermophilus* to *Xerospermophilus* during the early 2000s. The new classification places this species with three other Mexican or southwestern US species generically called pygmy ground squirrels.[3] The number of recognized subspecies has fluctuated over the last 40 years; *X. s. obsoletus* is the nominal subspecies in Wyoming.[4]

DISTRIBUTION AND HABITAT ASSOCIATIONS: The Wyoming distribution of this southern plains and Mexican desert species includes the three counties in the southeastern corner. The species tends to associate with sandy or loamy sand soils and moderately tall vegetation. In southern Goshen County, the primary vegetation associations in which the species was found were yucca–grass (*Aristida, Bouteloua, Hesperostipa*), big sagebrush–grass (*Bouteloua, Calamovilfa, Sporobolus*), and sand dune (*Salsola, Andropogon*).[5]

ECOLOGY: This is a solitary, hibernating, burrowing herbivore. The diet is thought to include less animal material than that of the thirteen-lined ground squirrel: mostly seeds and green plant parts.

CONSERVATION AND MANAGEMENT: The species is too poorly known to characterize its conservation status well. It is considered a concern because of its very restricted distribution within Wyoming, its rarity within that range, and our lack of reliable information about factors limiting populations.

NOTES AND REFERENCES

1. UNSM specimens.
2. Commission Regulations Chapter 52, Section 6.
3. Helgen et al. (2009).
4. Streubel and Fitzgerald (1978a), Hall (1981), Helgen et al. (2009).
5. Maxwell and Brown (1968).

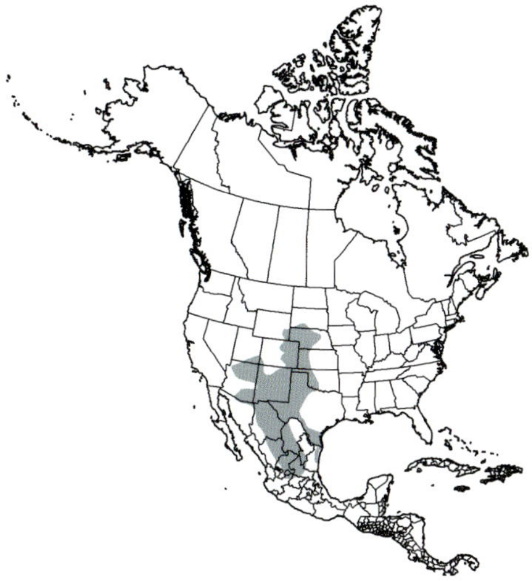

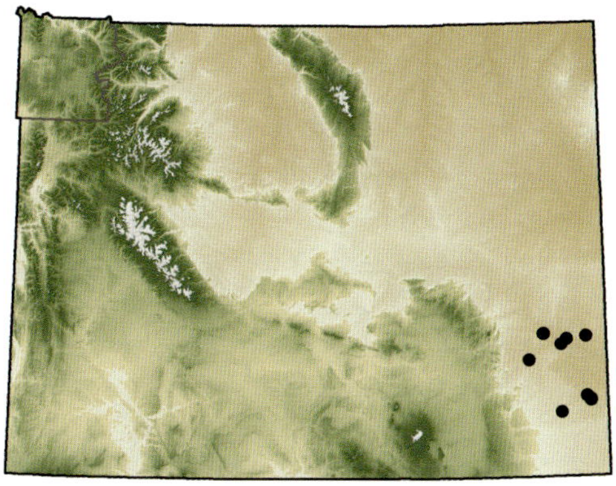

Locations from which the spotted ground squirrel has been identified in Wyoming.

Thirteen-lined ground squirrel, *Ictidomys tridecemlineatus*

DESCRIPTION: A slender, medium-sized ground squirrel with 13 alternating solid and dotted light stripes on a dark background. The pinnae are small, and the tail moderately long and well furred. Measurements (**Sex** [n] mean (mm, g), *range*): Albany, Goshen, and Sweetwater Counties, L = **F** [8] 180, *177–226*; **M** [6] 209, *193–227*; TL = **F** [8] 62, *30–76*; **M** [6] 62, *58–65*; HF = **F** [8] 30, *27–34*; **M** [6] 30, *27–33*; EFN = **F** [8] 7, *4–10*; **M** [6] 8, *4–10*; WT = **F** [8] 89, *52–118*; **M** [6] 88.3, *64.5–110*.[1] Dental formula: 1/1, 0/0, 2/1, 3/3 = 22.

STATUS: A non-protected nongame animal.[2] It is common but discontinuous within its geographic range.

NOMENCLATURE AND SYSTEMATICS: Formerly placed within *Spermophilus* and believed to be a sibling species with the spotted ground squirrel, this species was recently reclassified to a genus of three species of "lined ground squirrels" of southwestern North America. Other members are the Mexican ground squirrel (*I. mexicanus*) and Rio Grande ground squirrel (*I. parvidens*).[3] The number of described subspecies has varied across accounts, with Hall (1981) showing a non-disjunct subspecies (*I. tridecemlineatus olivaceus*) in the Black Hills and Streubel and Fitzgerald (1978b) depicting a non-isolated one in the northwestern mountains near Union Pass.[4] The subspecies found in Wyoming, according to Hall (1981) and Long (1965), comprise *I. t. alleni, I. t. pallidus, I. t. parvus,* and *I. t. olivaceous*. Other authors depict only the first three of these.[5] The subspecies boundaries presented by these two reports are not concordant with each other or with any apparent geographical discontinuities. The putative *I. t. alleni* has been depicted as endemic to Wyoming, distributed in montane habitats from the crest of the Bighorn Mountains to the western slope of the Wind River Mountains.[6]

DISTRIBUTION AND HABITAT ASSOCIATIONS: This broadly distributed mid-continent species occurs over the eastern 3/4 of Wyoming in a highly discontinuous pattern,

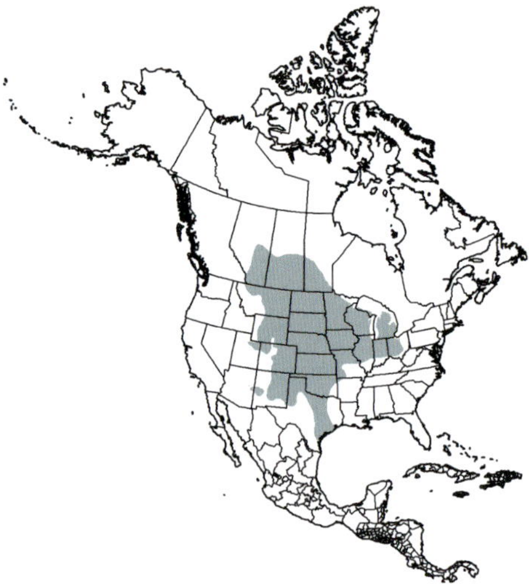

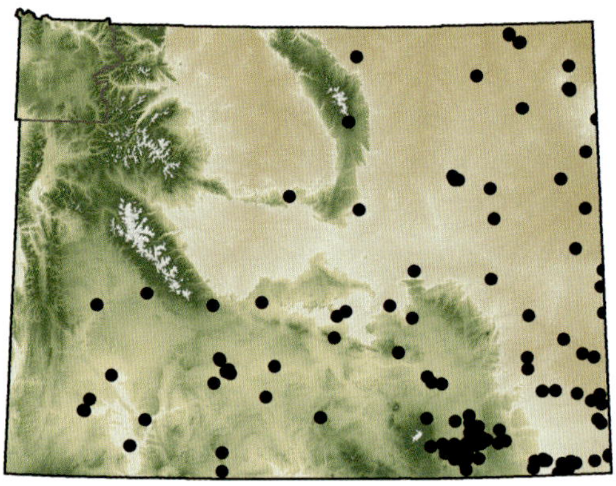

Locations from which the 15-lined ground squirrel has been identified in Wyoming.

although this must be considered a dated and conjectural decpiction of distribution.[7] Although Long (1965) and Hall (1981) showed the entire Bighorn Basin as being occupied, none of the points mapped here fall within that area. Hall conjectured that the species had been reduced in distribution in that basin by agricultural pesticides.[8] Typical habitats are grasslands with short vegetation below the elevational limit of coniferous forest. In southern Goshen County, the species was found primarily in areas with loamy sand soils, and grass mixtures (*Bouteloua–Hesperostipa* and *Calmovilfa–Bouteloua*) typically less than 25 cm (10 in) tall.[9] In the Laramie Basin, shortgrass was found to be preferred by a factor of two over greasewood.[10]

ECOLOGY: This is a solitary, burrowing omnivore that hibernates during winter; diets can include almost 50% animal matter, and insect abundance has been found to be a predictor of the presence of the species in northern Colorado.[11] This species is regarded as one of the most important nest predators of shortgrass prairie birds in this region.[12] It is exposed to plague during episodic epizootics. Population-level effects are not known, but few animals have been found to carry antibodies, suggesting that mortality rates resulting from infection may be high.[13]

CONSERVATION AND MANAGEMENT: The species is sufficiently common and widely distributed that it is not generally considered a conservation concern. However, agricultural pesticide use has arisen as a potential policy issue in the Rocky Mountain states, with some questioning how pesticide use has affected the distributions of some subspecies.

NOTES AND REFERENCES

1. UWMV specimens.
2. Commission Regulations Chapter 52, Section 6.
3. Harrison et al. (2003), Helgen et al. (2009).
4. Streubel and Fitzgerald (1978b), Hall (1981).
5. Howell (1938), Streubel and Fitzgerald (1979b).
6. This distribution was estimated based on six specimens from four localities. See Howell (1938), Long (1965), Streubel and Fitzgerald (1979b), Hall (1981).
7. Of the 213 specimens that contribute to the locations depicted, 118 were collected before 1955, and very few since 1990.
8. Long (1965, 579).
9. Maxwell and Brown (1968).
10. Clark (1971a).
11. Higgins and Stapp (1997).
12. Grant et al. (2006), Stanley (2010).
13. Stapp et al. (2008).

White-tailed prairie dog, *Cynomys leucurus*

DESCRIPTION: A medium-bodied, colonial ground squirrel with a distinctive bark-like vocalization. Pelage is tan or buff over most of the body, with a short white- or light-tipped tail and a black band above and below the eye. The species is sexually dimorphic for size (adult males are 20–35% heavier, depending on season), but not body form or pelage color. In adults standing erect, the thorax and head appear slender, but the abdomen plump. Colonies in shrub-dominated steppe are less conspicuous from a distance than those of black-tailed prairie dogs. This could be because burrow entrances are not rimmed as they are for black-tailed prairie dogs, population densities are lower, or because white-tailed prairie dogs modify vegetation less by clipping than do black-tailed prairie dogs.[1] Measurements: (**Sex** [n] mean (mm, g), *SD*) Larimer County, Colorado, L= **F** [41] 353, *13.2*; **M** [28] 371, *11.3*; TL = **F** [41] 54, *5.2*; **M** [28] 55, *5.6*; WT = **F** [41] 925, *147*; **M** [28] 1139, *256*.[2] Dental formula: 1/1, 0/0, 2/1, 3/3 = 22.

Distribution of the white-tailed prairie dog in Wyoming.

STATUS: An unprotected nongame species.[3] It is common but patchily distributed in suitable habitats. Classified as a "designated pest" by the Wyoming Department of Agriculture, it is subject to local control actions.[4]

NOMENCLATURE AND SYSTEMATICS: The genus and species names have been stable for decades; no subspecies have gained wide recognition. The genus comprises five species, three of which (*C. leucurus, C. gunnisoni, C. utahensis*) occupy higher-elevation regions farther west than *C. ludovicianus*. These three high-elevation species are placed in the subgenus *Leucocrossuromys*, and the white-tailed prairie dog has the northernmost distribution of them, extending just over the state line into Montana. A 2009 reclassification of Holarctic ground squirrels confirmed that prairie dogs were a monophyletic group, with its sibling

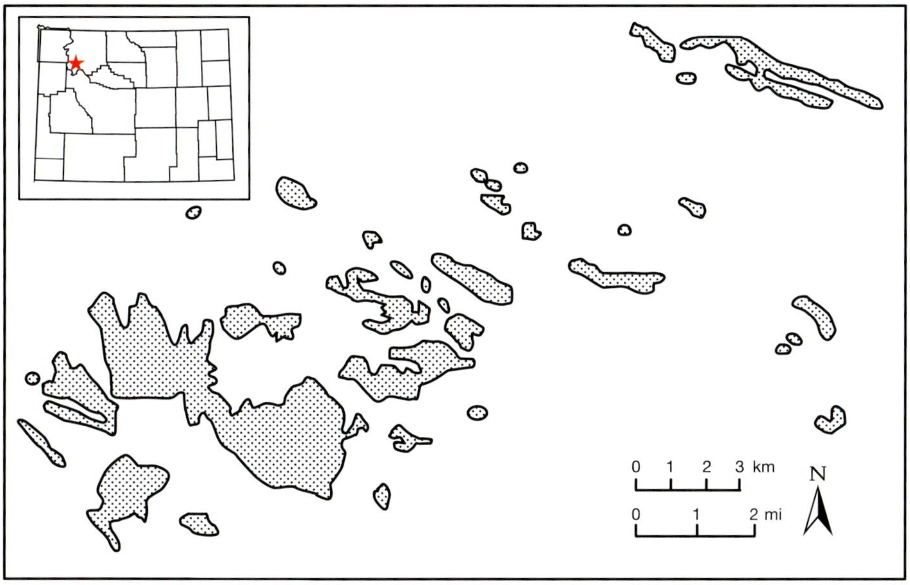

Map of a white-tailed prairie dog colony complex near Meeteetse, Wyoming, showing the tendency for colonies to align with soil or vegetative features, which in this case run from northwest to southeast.

lineage the pygmy ground squirrels.[5] Hybridization with other prairie dog species has not been reported.

DISTRIBUTION AND HABITAT ASSOCIATIONS: This is a high plains shrub-steppe species that occurs today (shown) over much of the high-elevation shrub-steppe of Wyoming, Colorado, and eastern Utah. Estimates of presettlement colony area are difficult to generate and tend to be species-wide in scope, not specific to Wyoming. Most accounts suggest a smaller distribution today than 120 years ago. The 2006 colony area in Wyoming was estimated at 185,988 ha (718 mi^2), which represents 55% of the area of the entire species range.[6] The species occurs in shrub-steppe below the lower elevational limit of trees over a triangle-shaped region with the northern corner where the Clarks Fork of the Yellowstone River crosses the state line into Montana, one in southeastern Albany County, and the southwestern one in southern Uinta County. The geographic range overlaps slightly with that of black-tailed prairie dogs at the zone of contact between the Bighorn and Laramie Mountains. Those two mountain ranges are barriers between the ranges of the two species in the northern and southern parts of the state. Colonies are distributed across the landscape in a pattern that reflects soil and vegetative features.[7]

ECOLOGY: This is a diurnal, colonial, burrowing herbivore that superficially resembles black-tailed prairie dogs. But their ecology differs in important ways. They are less socially structured than black-tailed prairie dogs, although both species have territorial home ranges occupied by adult females that mate with a single male that limits access by other males.

Colonies are associated with level or rolling topography, tending to avoid steep slopes. White-tailed prairie dogs hibernate in winter, plugging their burrows with dirt that, when frozen, becomes a physical barrier to burrowing predators. Hibernation is obligate, deep, and prolonged, with 70–80% of animals in torpor at any one time in midwinter. Emergence from burrows begins in late February, depending on location. Adult males disappear from aboveground areas by late summer, presumably entering torpor then.[8] Diets vary seasonally: shrub leaves in early spring, then forbs as green-up progresses. In late summer, diets tend to be grass leaves and seed heads. Strong selection for particular plant species at a site has not been reported.[9] The species is shorter-lived than the black-tailed prairie dog, producing their first litters at one year of age; litters are larger than those for black-tailed prairie dogs. Survival rates are highly variable by age and year; juveniles survived to their second year at rates of 9–23%, while adults (>1 year old) had annual survival rates of 23–70%.[10] The lowest of these values may have reflected the effects of plague.[11] Population densities within colonies vary by year and season. In northern Colorado, densities were 3.5/ha (910/mi^2) in the spring, but 8.4/ha (2200/mi^2) after juveniles emerged from burrows. On-colony densities are typically lower than those for black-tailed prairie dogs.[12] White-tailed prairie dogs are eaten by many mammalian and avian predators: coyotes, badgers, black-footed ferrets, red foxes, golden eagles, and various hawks. They are vulnerable to flooding in low-lying areas. Unknown numbers are killed by recreational shooters, but perhaps the single greatest factor in their abundance is plague, an introduced disease to which they are susceptible, and which reaches epizootic proportions episodically.[13]

Prairie dogs have important ecological roles aside from their herbivory and importance as prey. Their underground burrow systems are credited with providing physical structure in an environment that is otherwise physically simple. The burrows are important to various animal species: burrowing owls, snakes, mice, and insects. White-tailed prairie dogs clip shrubs from the vicinity of their burrow entrances, although less so than black-tailed prairie dogs, and these open sites are preferred by mountain plovers. Some consider prairie dogs generally to be keystone species because of their addition of physical structure (burrows) in a habitat that otherwise has little soil turnover and vegetation dynamics.[14]

CONSERVATION AND MANAGEMENT: White-tailed prairie dogs are not controlled for conflicts with agriculture to the extent that black-tailed prairie dogs are. This is primarily because the former do not attain the high local densities of the latter—at least in shrub-dominated steppe—and white-tailed prairie dogs are less likely to occur near crop agriculture. The primary human manipulations are shooting for recreation and localized control by poisoning. The species is critical to the recovery of the black-footed ferret in Wyoming.

NOTES AND REFERENCES

1. Hoogland (2003).
2. Tileston and Lechleitner (1966).

3. Commission Regulations Chapter 52, Section 6.
4. The authority for this designation is provided by Wyoming Statutes 11-5-102-a-xii.
5. Helgen et al. (2009).
6. Pauli, Stephens et al. (2006).
7. Map is modified from Forrest et al. (1988).
8. Harlow and Menkens (1986).
9. Clark et al. (1971).
10. Menkens and Anderson (1989).
11. Forrest et al. (1988).
12. Hoogland (1995).
13. Ubico et al. (1988).
14. Whicker and Detling (1988) and Miller et al. (2000), but see Stapp (1998) and Kotliar et al. (1999).

Black-tailed prairie dog, *Cynomys ludovicianus*

DESCRIPTION: A colonial ground squirrel slightly smaller bodied than the white-tailed prairie dog. Its tail is longer and black-tipped, and it has no dark band over the eye. The dorsum is tan with a reddish cast; the underside is cream-colored. Males are slightly smaller than females in late April, but about 24% larger by late June. Vocalizations are more varied than for white-tailed prairie dogs. Colonies tend to be conspicuous, with high densities of entrances, most of them surrounded by rimmed mounds. Vegetation is clipped short or nearly absent near burrows.[1] Measurements: (**Sex** [n] mean (mm, g) *SD*, Larimer County, Colorado, L = **F** [32] 368, *19.1*; **M** [21] 375, *15.4*; TL = **F** [32] 76, *10.0*; **M** [21] 78, *8.0*; WT = **F** [32] 864, *174*; **M** [21] 776, *152*.[2] Dental formula: 1/1, 0/0, 2/1, 3/3 = 22.

STATUS: This is a non-protected nongame species.[3] It is common but patchily distributed within the geographic range, and classified a designated pest by the Wyoming Department of Agriculture.[4] Petitioned during the 1990s for listing under E.S.A., the species was found not warranted for listing in 2009.[5]

NOMENCLATURE AND SYSTEMATICS: The genus and species names have been stable for decades. Together with the very similar species or subspecies *C. l. mexicanus* (Mexican prairie dog), the species is placed in subgenus *Cynomys*. Two subspecies were recognized by 1996; only *C. l. ludovicianus* occurs north of New Mexico.[6]

DISTRIBUTION AND HABITAT ASSOCIATIONS: This Great Plains species occurred from Mexico to Canada when Europeans explored the area (shown); today it occurs in grassland and steppe habitats up to the lower elevational limit of trees (<2133 m, [<7000 ft]) in eastern Wyoming. The distribution of this species and that of the white-tailed prairie dog are roughly parapatric along a line connecting the southern end of the Bighorn Mountains and the northern end of the Laramie Mountains.

ECOLOGY: This is a diurnal, territorial, colonial, facultative hibernator. Although superficially resembling the white-tailed prairie dog, the two species' life histories differ markedly. First, reproduction in this species occurs at two years and litters are smaller than for white-tailed prairie dogs. Black-tailed prairie dog colonies have generally higher population

Juvenile black-tailed prairie dogs.

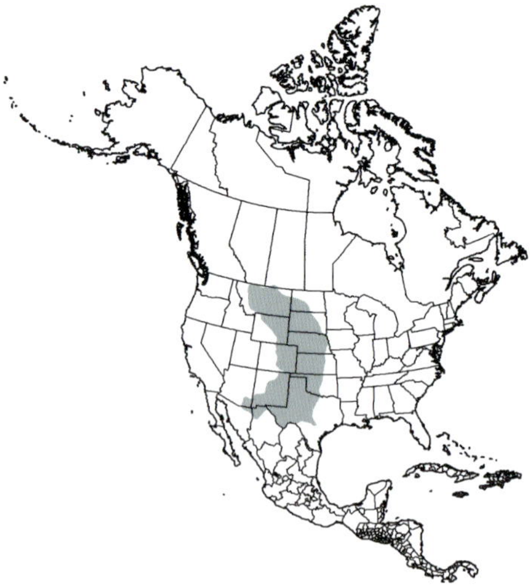

Distribution of the black-tailed prairie dog in Wyoming. The southern part of the geographic range in Wyoming is highly discontinuous.

densities within colonies, >10/ha (>2590/mi^2), not including young of the year.[7] Vegetation within colonies is markedly lower than outside, typically <20 cm (<8 in) tall. This results from the preference by this prairie dog for areas with low vegetation, routine foraging on plants, and the instinctive clipping of shrubs >20 cm (>8 in) tall, a presumed adaptive behavior to improve predator detection around burrow entrances. Diets in summer are dominated by grasses (*Pascopyrun*, *Bouteloua*, *Bromus*), scarlet globemallow, and rabbitbrush. In winter, the species eats prickly pear cactus and thistle, but avoids sagebrush, dyssodia, and horseweed.[8] The species forms more complex social structures than does the white-tailed

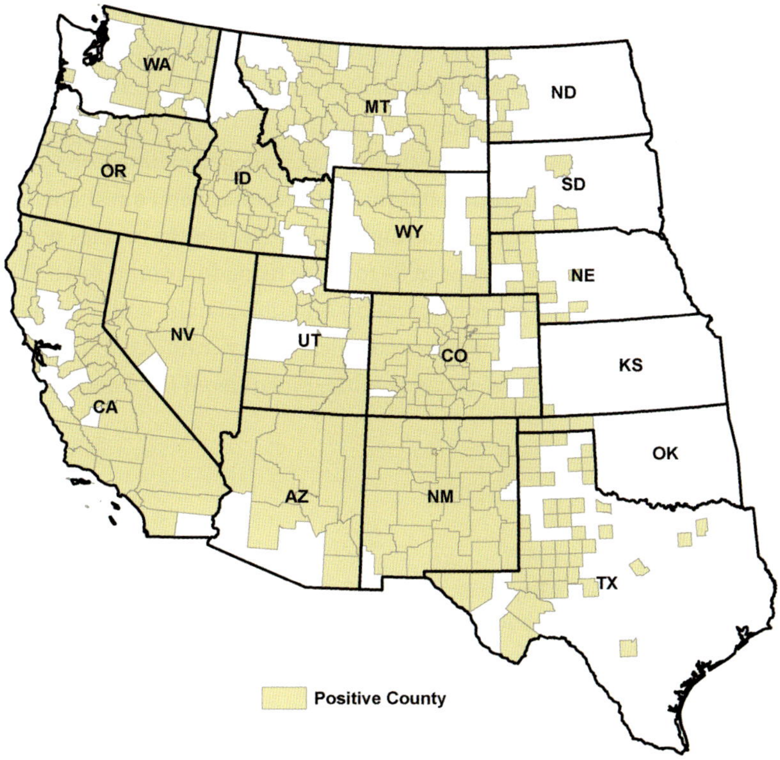

Distribution of plague in counties of the western United States during 1970–2011, based on animal carcasses, positive serology, fleas, and tissues. Human cases are not represented.

prairie dog; colonies are organized around female-based units called "coteries" that hold one or a few adult males, several adult females, and—in late summer—up to several dozen juveniles and yearlings.

Survival of juveniles for the first year of life has been estimated at 54% for females and 47% for males. The most common cause of neonatal deaths is infanticide by related but nonmaternal females. Predation losses are high and are caused by coyotes, bobcats, badgers, black-footed ferrets, golden eagles, and other raptors. Females live considerably longer in the wild—up to eight years in South Dakota—than males, which tend not to live beyond five years.[9] The species is highly vulnerable to epizootics of plague and colony-wide mortality of 95% over a few weeks has been reported.[10] The species has evolved resistance to plague in places where it has been repeatedly exposed, but populations in eastern Wyoming have been exposed only a few times, accounting for the high mortality rates.[11] The distribution of plague across eastern Wyoming has been discontinuous, with some areas affected multiple times and others apparently never having had reported plague outbreaks.[12] Recovery from such die-offs can require years or decades. The species has been the focus of some academic debate regarding its functions in ecosystems, being credited by some with making keystone contributions, and others challenging that view.[13] The species has been demonstrated to facilitate occupancy and nest survival by mountain plovers.[14] It also affects nutrient fluxes strongly, and bison and pronghorns are attracted to colony areas because of vegetation changes caused by prairie dogs.[15]

CONSERVATION AND MANAGEMENT: The species has been alleged to have lost >95% of its geographic range since presettlement times, but that estimate has been questioned.[16] Extensive poisoning during the twentieth century was highly effective, but the species-wide area occupied by colonies has increased by a factor of 8–10 since its nadir around 1960. The primary management activities are shooting for recreation on public lands, with local poisoning on private lands. Black-tailed prairie dogs are ecologically very influential, possibly keystone community members, and key to the recovery of the endangered black-footed ferret in eastern Wyoming.

NOTES AND REFERENCES

1. Hoogland (1996, 2003).
2. Tileston and Lechleitner (1966).
3. Commission Regulations Chapter 52, Section 6.
4. The authority for this designation is provided by Wyoming Statutes 11-5-102-a-xii. The statute requires county weed and pest districts to undertake programs of control for this species, but the level of effort varies by county.
5. US Fish and Wildlife Service (2009).
6. Hoogland (1996), based on Hall (1981).
7. Hoogland (1995), Slobodchikoff et al. (2009).
8. Summers and Lindner (1978).
9. Hoogland (1995).
10. Pauli, Buskirk et al. (2006).
11. Rocke et al. (2012).
12. Data used by permission of Centers for Disease Control and Prevention, Fort Collins, Colorado.
13. See discussion for white-tailed prairie dogs.
14. Augustine and Skagen (2014).
15. Fahnestock and Detling (2002).
16. Knowles et al. (2002), Virchow and Hygnstrom (2002).

Abert's squirrel, *Sciurus aberti*

DESCRIPTION: A large tree squirrel, 3–4 times the weight of a red squirrel. Pelage is variable but typically medium to dark gray with some red along the dorsal midline, and white undersides. The long, curved, well-furred tail is highly mobile, and dark ear tufts are pronounced. Most animals from Wyoming and northernmost Colorado are variably melanistic—brown to nearly black over the entire body.[1] They are active on the ground and in trees during the day and are common near developed areas and at bird feeders.[2] Measurements (**Sex** [n] mean (mm, g), *range*): Boulder, Douglas, and Jefferson Counties, Colorado, L = **F** [10] 516, *473–562*; **M** [7] 491, *450–560*, TL = **F** [10] 249, *193–298*; **M** [7] 228, *205–280*; HF

Abert's squirrel, showing the melanistic pelage typical of the northern Colorado Front Range and southern Wyoming.

Locations from which Abert's squirrel has been identified in Wyoming.

= **F** [10] 66, *55–70*; **M** [7] 68, *65–71*; EFN = **F** [10] 45, *31–66*; **M** [7] 47, *35–70*; WT = **F** [10] 617, *508–752*; **M** [7] 601, *531–648*.[3] Dental formula: 1/1, 0/0, 2/1, 3/3 = 22.

STATUS: A protected nongame mammal.[4] It is uncommon to common within the very limited Wyoming part of the range.

NOMENCLATURE AND SYSTEMATICS: The scientific terminology is stable; "tassel-eared squirrel" is also used as the common name. Six subspecies have been recognized, mostly isolated from each other by low-lying non-forested areas; the subspecies in the Front Range, including Wyoming, is *S. a. ferreus*.[5] John James Abert was a US soldier who participated in nineteenth-century expeditions to map the American West; John James Audubon named the species in his honor.

DISTRIBUTION AND HABITAT ASSOCIATIONS: Abert's squirrels are found in ponderosa pine forests from Mexico and Arizona northward across the Colorado Front Range to the Wyoming border. The species extends about 1.6 km (1 mi) north of the state line near the old Union Pacific railroad stop of Harriman, and in the Boulder Ridge area. Near Harriman, residents report it a common visitor at bird feeders.[6] The distribution in the Laramie Mountains represents the northern limit of contiguous ponderosa pine in the region, from the perspective of Abert's squirrels. Although large areas of ponderosa pine occur farther north in the Laramie Mountains, a break in contiguous stands (shortgrass prairie extends for several miles) has prevented this species from expanding northward to the extensive ponderosa pine forests in the Pole Mountain area and northward.

ECOLOGY: This territorial, diurnally active species has an association with ponderosa pine that is obligatory; Abert's squirrel populations cannot persist in association with other coniferous tree species. This resource specialization has important implications for the ecology of the squirrels and of ponderosa pine.[7] The squirrels feed on various plant parts seasonally—cambium of small stems, terminal buds, and seeds—but they do not cache cones as do red squirrels. Hypogeous fruiting bodies of fungi are also consumed, as is carrion. The species is active all year.

CONSERVATION AND MANAGEMENT: The very restricted distribution (only a few square miles) in Wyoming puts the occurrence of the species in the state at some risk. A mountain pine beetle outbreak was apparent in the Harriman area in 2011, and a severe insect outbreak, severe fire, or combination of factors could affect a substantial part of the species' occupied habitat in Wyoming.[8]

NOTES AND REFERENCES

1. Hancock and Nash (1982). Farentinos (1972) reported 57% and Ramey and Nash (1976) reported 68% of animals in Front Range of northern Colorado were of the black color phase.
2. Allred (2010).
3. DMNS specimens.
4. Commission Regulation Chapter 52, Section 11.
5. Hoffmeister and Diersing (1978).
6. The first specimen from Wyoming was collected by L. N. Brown in 1963 (USNM 356286).
7. Snyder (1993).
8. Keith (2003).

Eastern gray squirrel, *Sciurus carolinensis*

DESCRIPTION: A medium-sized tree squirrel, around 20% smaller than the fox squirrel, with gray fur, and a pale or buff, rather than orange or rust, underside. The head and flanks may be washed with pale reddish brown. The large, bushy tail has dark and white guard hairs, and variably reddish tan underfur.[1] Measurements (**Sex** [n] mean (mm, g), *range*): Minnesota, WT = **F** [14] 724, *703–793*; **M** [13] 728, *507–853*.[2] Dental formula: 1/1, 0/0, 2/1, 3/3 = 22.

STATUS: A small game animal by Wyoming Statute.[3] This is a naturalized nonnative species.

NOMENCLATURE AND SYSTEMATICS: The genus and species names have been stable for over 200 years; which subspecies was introduced to Wyoming is not known, but might be inferred from molecular markers.

DISTRIBUTION AND HABITAT ASSOCIATIONS: This east-temperate and southeastern species occurred east of 96° longitude before European settlement (shown); it has since expanded westward and now occurs in the towns of Sheridan, Dayton, Ranchester, Buffalo, and Gillette. It may have a limited distribution in nearby rural riparian areas with mature trees, but seems to be commensal on humans and likely does not occur

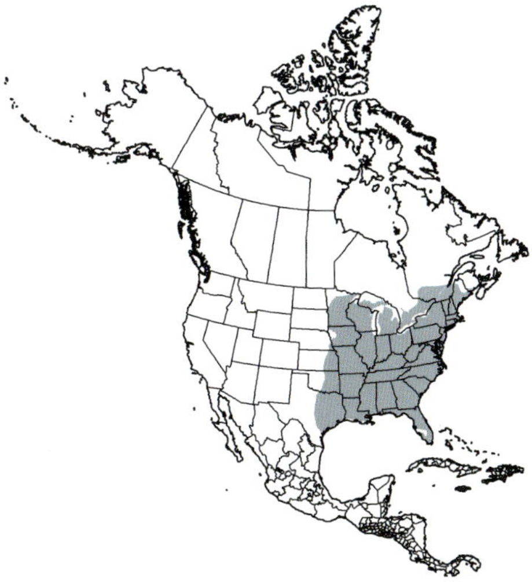

Locations from which the eastern gray squirrel has been identified in Wyoming.

away from human foods. The species has been introduced, mostly likely from cities in southern Montana, to which the species was introduced from farther east.[4] It was not mentioned in Clark and Stromberg (1987).[5]

ECOLOGY: In native habitats, the species feeds primarily on tree nuts, buds, and flowers of various eastern hardwoods and shrubs. In Wyoming towns, the diet likely is largely human foods: bird seed at feeders, food provided by homeowners, and refuse. Nesting cavities are important in winter and for reproduction. Competitive interactions with fox squirrels are common in the Midwest and East, where local co-occurrence is common. The larger fox squirrels displace them in some interactions, but gray squirrels can outnumber fox squirrels in forest interiors, and under some food resource conditions.[6]

CONSERVATION AND MANAGEMENT: The species is not a conservation concern, as a result of its introduced status. Although it is classified as a small game animal, very few are taken by hunters because the species scarcely occurs outside city limits, within which hunting is illegal.

NOTES AND REFERENCES

1. Koprowski (1994a).
2. Thoma and Marshall (1960).
3. Wyoming Statutes 23-1-101-a-11.
4. Foresman (2012).
5. Clark and Stromberg (1987).
6. Sexton (1990), Derge and Yahner (2000).

Eastern fox squirrel, *Sciurus niger*

DESCRIPTION: A medium-sized tree squirrel with a long, well-furred, animated tail. The fur is distinctive: the dorsum is grizzled gray guard hairs with a dull or light orange undercoat. The underside and sides of the head are orange- or rust-colored. The sexes are similar. Vocalizations include barks similar to those of the red squirrel, but no red-squirrel-like chatter. Measurements (**Sex** [n] mean (mm, g), *range*): various counties, L = **M** [6] 486, *425–550*; TL = **M** [6] 236, *208–283*; HF = **M** [6] 61, *41–68*; EFN = **M** [6] 30, *26–33*; WT = **M** [6] 558, *396–721*.[1] Dental formula: 1/1, 0/0, 1/1, 3/3 = 20.

STATUS: This small game animal by Wyoming Statute[2] is common or abundant in many Wyoming towns and cities, where it has been introduced.

NOMENCLATURE AND SYSTEMATICS: The genus and species names have been stable for decades. Of the 10 subspecies recognized in 1994, only the nominal *S. n. rufiventer* occurs in Wyoming.[3]

DISTRIBUTION AND HABITAT ASSOCIATIONS: The distribution of this eastern species in South Dakota and Nebraska at the time of European settlement (shown) can only be inferred, because of weak documentation in the early historical record, and early translocations in those states. By 1908, its distribution was described as "eastern Nebraska, but practically confined to the timber bordering the Missouri River and its tributaries, west to about the 98th meridian."[4] The species has expanded westward across Nebraska and North Dakota, but how much of that

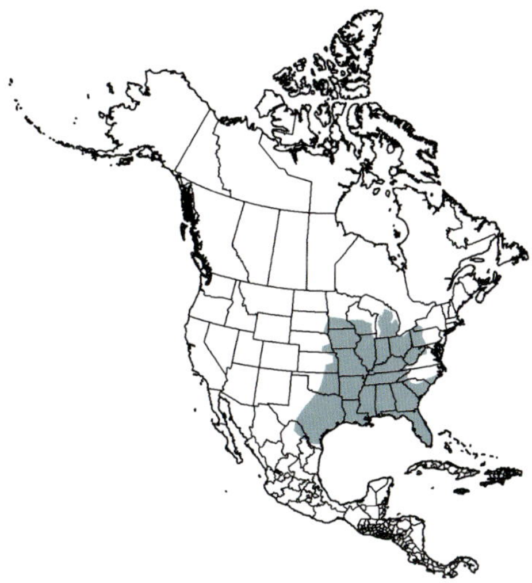

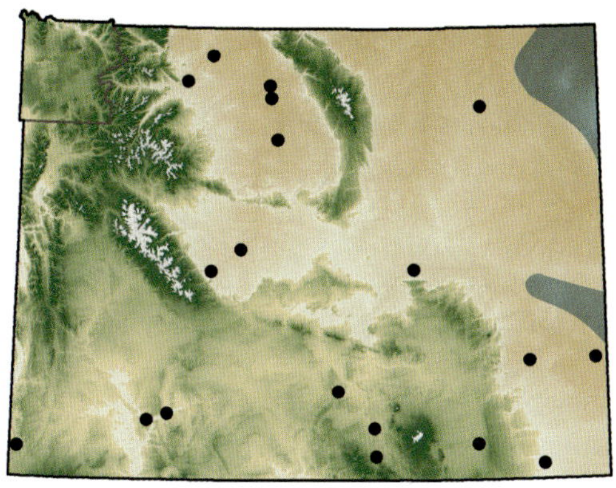

Locations from which the eastern fox squirrel has been identified in Wyoming. Points represent towns where the species occurs, shaded areas are nonurban occupied habitat.

expansion is due to dispersal, and how much to human translocation, is uncertain. It is clear that the species did not occur in Wyoming in 1850.[5] It occupies riparian gallery forests in the Black Hills and Bear Lodge Mountains, and has been extensively introduced to city parks and treed neighborhoods. Some towns, including Kaycee and Moorcroft, seem never to have had eastern fox squirrels, according to local residents. As of 2013, the species was reported to be present in Laramie, Cheyenne, Wheatland, Torrington, Casper, Sinclair, Rawlins, Saratoga, Riverside-Encampment, Green River, Rock Springs, Evanston, Lander, Riverton, Cody, Powell, and Gillette.[6] (Gillette has eastern gray squirrels as well, occurring in separate parts of the town from the fox squirrels.) It has been introduced recently to Basin and Greybull, and has spread along the Bighorn River as far north as the Yellowtail Wildlife

Habitat Area. Eastern fox squirrels have been in Riverside-Encampment for less than a decade. Farther east, habitat is deciduous and mixed riparian forests, especially those with open understories, and urban or residential areas with trees. In the northeast corner of the state, riparian galleries holding bur oak and paper birch are important habitats.[7]

ECOLOGY: The species is a generalist herbivore, but attains its highest densities farther east in stands that produce hard mast: acorns, pine nuts, and seeds of other trees and shrubs. This squirrel eats hypogeous fruiting bodies of fungi as well. In wild settings, buds and flowers of such species as hawthorn can be important when hard mast is not available. In urban settings, bird seed, pet foods, and refuse are important.[2] Fox squirrels can coexist with other tree squirrel species to some extent; they have been replaced through time by gray squirrels in parts of the Midwest for unclear reasons, and they dominate the much smaller red squirrel where syntopic.[8]

CONSERVATION AND MANAGEMENT: The species is an important game species farther east, but is seldom hunted in Wyoming outside of the Black Hills. The species is not commonly regarded as one of conservation concern and may be a nuisance in some areas, leading to translocations to areas away from towns.

NOTES AND REFERENCES

1. UWMV specimens.
2. Wyoming Statutes 23-1-101-a-11.
3. Koprowski (1994b).
4. Swenk (1908).
5. Hibbard (1956), Jones (1964).
6. Personal communications from WGFD biologists Heather O'Brien (Casper and Gillette), Tom Easterly (Basin and Greybull), Greg Hiatt (Sinclair and Rawlins), Will Schultz (Saratoga and Riverside-Encampment), Martin Hicks (Cheyenne, Wheatland, and Torrington), Mark Zones (Green River, Rock Springs, Evanston), Greg Anderson (Lander and Riverton), Bart Kroger (Worland), Tim Wooley (Cody, Powell).
7. Turner (1974).
8. Sexton (1990).

Red squirrel, *Tamiasciurus hudsonicus*

DESCRIPTION: This small tree squirrel with generally reddish dorsal pelage exhibits significant variation among Wyoming subspecies. The upper parts are grayish yellow washed with reddish olive in *T. h. baileyi*, darker and more reddish brown in *T. h. ventorum*, and tawny olive or orange-cinnamon in *T. h. dakotaensis*. The dorsum is dark brown in *T. h. fremonti* and can appear almost black in poor light. True melanistic—nearly black—individuals are encountered in Jackson Hole. All subspecies have light buff to white undersides, dark lateral lines, and white bands above and below the eyes.[1] Vocalizations are distinctive sharp barks or chatters, audible from over 100 m (330 ft). Field sign includes conifer cone bracts, either scattered or consolidated into piles called middens. Mushrooms stored in tree branches, and leaf and grass nests in trees of various species are also distinctive. Measurements (**Sex** [n] mean (mm, g), *range*): Albany, Teton, and Natrona Counties, L = **F** [8] 314, *268–351*; **M** [9] 326, *310–342*; TL = **F** [8] 118, *90–134*; **M** [9] 120, *65–131*; HF = **F** [8] 44, *20–56*; **M** [9] 45, *25–60*; EFN = **F** [8] 19, *6–23*; **M** [9] 24, *10–32*; WT = **F** [8] 197, *132–230*; **M** [9] 214, *152–252*.[2] Dental formula: 1/1, 0/0, 1 or 2/1, 3/3 = 20–22.

STATUS: A small game animal, although not taken in large numbers by hunters in Wyoming.[3] It is common or abundant in occupied habitats.

NOMENCLATURE AND SYSTEMATICS: The genus and species names have been stable for decades. Around 25 subspecies have been recognized, four of them in Wyoming (*T. h. ventorum* in the northwestern mountains; *T. h. baileyi* in the Bighorn and Laramie Mountains and mountains flanking the lower Sweetwater River; *T. h. fremonti* in the Medicine Bow and Uinta Mountains, and *T. h. dakotensis* in the Black Hills and Bear Lodge Mountains). *T. h. dakotensis* shows the most pronounced enlargement of skull and dental features of any Rocky Mountain subspecies, which reflects the diet of almost exclusively ponderosa pine seeds encased in large, tough cones.[4] The common name for members of *Tamiasciurus* is "pine squirrel," a term at times used for the red squirrel specifically.

DISTRIBUTION AND HABITAT ASSOCIATIONS: Among the most widely distributed of boreomontane New World species, it occurs across forested mountain ranges of the state. Isolated populations are found in many small mountain ranges: Shirley Mountains, Bear Lodge Mountains, Chalk Mountain, and Hart Mountain. Molecular studies have shown that the Green River and Wyoming Basin form a major divide for this species, but the current subspecies boundaries are not concordant with molecular evidence. The latter separates red squirrels in the Black Hills and Bighorn, Laramie, and Medicine Bow Mountains from those farther west, suggesting that the Bighorn Mountains and Black Hills populations have shared genetic material during the Holocene.[5] This is plausible, given the presence of remnant coniferous woodlands and riparian galleries that connected these mountain ranges. The Laramie and Medicine Bow Mountain populations are closely related genetically, and have nearly contiguous coniferous forest habitat, except for the discontinuity in the distribution of ponderosa pine near Harriman.[6] Red squirrels are found in

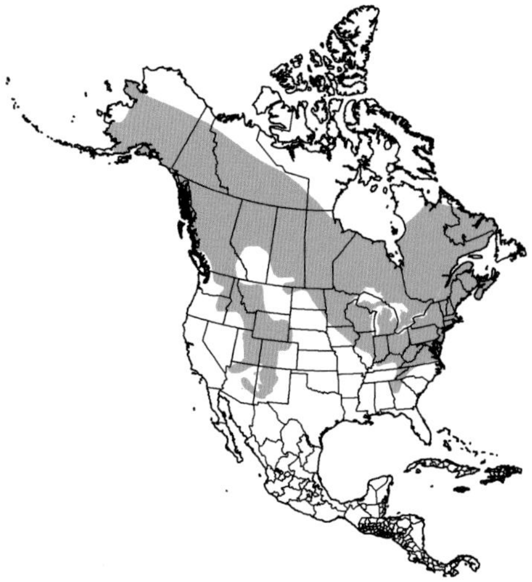

Distribution of the red squirrel in Wyoming; the species occurs in coniferous forests of sufficient extent statewide.

most coniferous forest habitats of Wyoming to the upper elevational limit of trees. The species also occurs in riparian gallery forests dominated by deciduous species; such corridors may allow episodic movements among forested mountain ranges. Mature stands of spruce–fir, lodgepole pine, and whitebark pine are important habitats in Wyoming; Douglas-fir may form admixtures with the former species. Red squirrels are found in ponderosa pine-dominated forests of the Laramie Mountains and Black Hills, but mostly in moist north-facing areas where fungi are more common and where canopy is continuous. Red squirrels are rare in dry, open-canopy stands of ponderosa pine in Wyoming. They are nearly absent from pure stands of ponderosa pine in northern Colorado because larger Abert's squirrels competitively exclude them.[7]

ECOLOGY: This is a diurnal, territorial conifer-seed consumer, which also eats fungi and animal material (e.g., eggs and nestlings) seasonally. Red squirrels collect green cones by cutting them from branches, removing some seeds while green, but gathering most cut cones into caches that are typically incorporated into middens of cone bracts for use over winter. Nest sites are critical for surviving winter and have been described for the Pole Mountain area of the Laramie Mountains: all were in lodgepole pine trees, typically 4–8 m (13–26 ft) above the ground, adjacent to the trunk, and in the interior of a stand. One or multiple cone caches/middens were typically located within 30 m (100 ft) of the nest tree.[8] In the coldest weather, squirrels remain in belowground nests associated with middens. Red squirrels fulfill several important ecological functions, aside from being important prey for Pacific martens, Canada lynx, raptors, and other predators.[9] First, they are important seed predators, influencing the reproductive biology of food trees, and the evolution of cone morphology in various conifer species.[10] Red squirrels also bury some cones in the duff layer, but are not thought to facilitate seed dispersal or germination in this way. Second, they are important predators of nestling birds and eggs, structuring breeding bird communities across their range.[11] Third, they effectively compete with crossbills for lodgepole pine seeds, limiting crossbill densities where sympatric. Fourth, their consolidation of whitebark pine cones into caches is critical in providing access for grizzly bears and likely for Pacific martens to whitebark pine seeds. Whitebark pine seeds gathered from red squirrel caches are the almost exclusive food of grizzlies during some seasons, and martens have been reported to use the same resource in the same way.[12]

CONSERVATION AND MANAGEMENT: The species has not been one of concern until the early twenty-first century outbreak of forest insects in coniferous forests of temperate and boreal North America. Beginning around 1996, a major outbreak of beetles (mountain pine beetle, *Dendroctonus ponderosae*; spruce bark beetle, *D. rufipennis*) caused massive and extensive death of conifer trees of species and cone-bearing ages important to red squirrels in Wyoming. The species primarily affected by the mountain pine beetle are ponderosa pine, lodgepole pine, and limber pine; spruce bark beetles kill Engelmann spruce. Large trees are the most important for cone and seed production for red squirrels and are the most likely to be infested; nearly 100% of the trees >30 cm (>12 in) diameter at breast height on a site have been killed.[13] The long-term implications of these massive beetle outbreaks, which extend to British Columbia and Alaska, for red squirrel populations are not understood, but red squirrels could be affected at the population and biogeographical levels.

NOTES AND REFERENCES

1. Steele (1998).
2. UWMV specimens.
3. Wyoming Statutes 23-1-101-a-11.
4. Lindsay (1987).
5. Wilson et al. (2005).
6. See discussion of this discontinuity for Abert's squirrel.
7. Ferner (1974).
8. Rothwell (1979).
9. Squires (2000) reported that red squirrels were the most commonly taken prey of northern goshawks, being found in 50% of regurgitated pellets.
10. Benkman (1995), Benkman et al. (2010), Talluto and Benkman (2013).
11. Siepielski (2006) regarded the red squirrel as a key nest predator across coniferous and mixed forests within its geographic range.
12. Mattson and Reinhart (1997).
13. Johnson et al. (2014).

Northern flying squirrel, *Glaucomys sabrinus*

DESCRIPTION: A small, nocturnal tree squirrel with large, dark eyes, beige-gray dorsal fur, and white undersides. It weighs about one-half as much as the red squirrel in Wyoming. The tail is horizontally flattened, and a gliding membrane—the flight patagium—connects fore and hind legs and the torso. Typically observed only in artificial light at night, northern flying squirrels are attracted to bird feeders in occupied areas.[1] Measurements (**Sex** [n] mean (mm, g), *range*): Weston, Fremont, Teton, and Park Counties,

L = **F** [10] 314, *282–338*; **M** [6] 310, *255–335*; TL = **F** [10] 142, *114–153*; **M** [6] 146, *128–156*; HF = **F** [10] 40, *35–44*; **M** [6] 39, *31–42*; EFN = **F** [10] 26, *16–30*; **M** [6] 24, *20–27*; WT = **F** [10] 154.6, *45.1–208.5*; **M** [6] 131.1, *72–152*.[2] Dental formula: 1/1, 0/0, 2/1, 3/3 = 22.

STATUS: A protected nongame animal and a state Species of Greatest Conservation Need as of 2012.[3] The biological status of the species is difficult to assess because detection methods are species-specific, although it is occasionally caught in traps set for other rodents.

NOMENCLATURE AND SYSTEMATICS: Flying squirrels across the Holarctic evolved as a single lineage, tribe Pteromyini. The genus name is stable; the two North American species are broadly separated along latitudinal lines, but sympatric in the mixed hardwood forest of the Great Lakes, Appalachian, and Laurentian regions. Of the 25 or so putative subspecies of northern flying squirrel recognized as of 1984, only one occurs in Wyoming: *G. s. bangsi*.[1] *G. s. lucifugus* occurs in the Uinta Mountains of Utah, close to the Wyoming border. Flying squirrels in the Black Hills represent the faunal relics of a Pleistocene refuge, show molecular evidence of isolation, and have been proposed to warrant subspecies-level distinction. They exhibit closer phylogenetic ties to populations west of the Black Hills than to those farther east.[4]

DISTRIBUTION AND HABITAT ASSOCIATIONS: The species is more boreal than montane in its faunal associations; it is disjunct and anomalous in its Wyoming distribution, occurring in some small mountain ranges, but not in other, larger ones. It occupies the northwestern mountains and Black Hills (including Bear Lodge Mountains), but is absent from the Medicine Bow, Laramie, and Bighorn Mountains.[5] The species occurs in the Uinta Mountains, and may have been detected on those mountains' northern slope within Wyoming, but that is not confirmed.[6] The regional geographic pattern is curious, because the species occupies smaller habitat islands in southern Utah, the Alexander Archipelago of Alaska, and the Appalachians than in some Wyoming mountain ranges from which it is absent. So, simple persistence since the Late Pleistocene glacial retreat does not seem to account for the distributional pattern. Across the species' range, typical habitat is coniferous, mixed, and deciduous forests. In Wyoming, habitats are dominated by coniferous species: lodgepole pine in the northwestern mountains, and ponderosa pine and white spruce in the

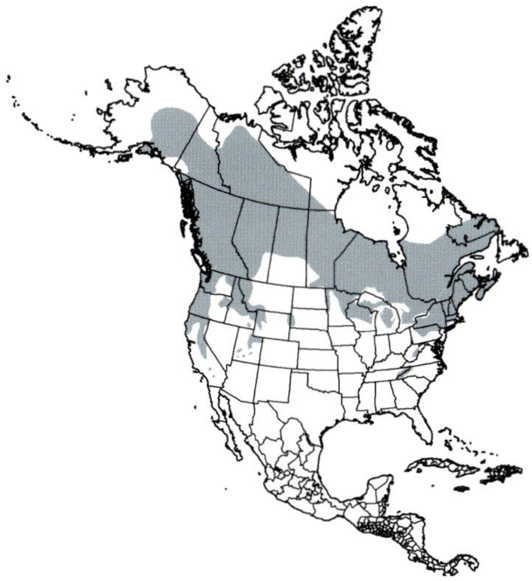

Distribution of the northern flying squirrel in Wyoming. The species has been reported from the Uinta Mountains in Wyoming, but no specimens are known.

Black Hills. Within those types, northern flying squirrels associate with late-successional stands, or younger stands nearby. The key features of such stands are moist conditions, tall trees and well-developed canopy, and abundant decaying wood. These features are consistent with the fungal diet component: northern flying squirrels occupy sites where they can forage for fruiting bodies of fungi, some of which are commensal with mature conifers. An association with riparian sites has been noted in dry regions, suggesting a need for access to water or moist-site trees near water. Structural stage is important because of the need for tall trees from which to glide, and because the squirrel occupies tree cavities typically resulting from interior decay and the actions of cavity-forming woodpeckers. Flying squirrels perceive recent clear-cuts as barriers to movement. At a Black Hills study site less than 10 km (6 mi) outside Wyoming, red squirrels preferred mature ponderosa pine forests, whereas flying squirrels preferred aspen-paper birch forests interspersed with ponderosa pine.[7]

ECOLOGY: Unlike most tree squirrels, this species does not rely on hard mast—seeds or nuts—although it occasionally eats them. Instead, its dietary staples are fungi, lichens, plant buds, and soft mast. It is regarded as an obligate of the hypogeous fruiting bodies of fungi in the Pacific Northwest, excavating these truffle-like bodies from beneath the duff layer and locating them by olfaction. The species stores fungi as do red squirrels, and remains active almost all winter, although capable of shallow torpor and reduced activity during severe weather.[1] Population attributes are not well known, although density estimates are available for various forest types, mostly in the Pacific Northwest. Densities in stands identified as old growth are in the range of 0.5–3.3/ha (130–840/mi^2), with most estimates around 1–2/ha (256–512/mi^2).[1] Several predators are recognized as potentially important: owls, northern goshawks, martens, weasels, and fishers. The question of possible competition with red squirrels has been considered, but the two species' differing resource needs and opposite diel activity patterns are thought to minimize possible competition. In addition to participating in food webs, the species' mycophagous diet makes it an important disperser of fungal spores.[8]

CONSERVATION AND MANAGEMENT: The species is regarded as a species of conservation concern in various jurisdictions, in spite of the lack of any E.S.A. review. South Dakota considered it to be "imperiled" as of 2012.[9] Conservation concerns center on three mechanisms: population isolation and its possible genetic consequences, clear-cutting and forest fragmentation, and climate change. Because many populations in the southern part of the range occur as islands, possible genetic consequences have been predicted and in some cases demonstrated. Clear-cutting has clear negative effects on the species, because flying squirrels avoid clear-cuts or stands lacking tall trees from which to launch their glides. Young stands also lack trees with which mycorrhizal fungi—important foods—can associate. And climate change is thought to be the mechanism underlying a number of predicted and observed changes not favorable to northern flying squirrels. Among them are the massive outbreaks of the mountain pine beetle and spruce bark beetle, which kill most large conifers on a site. Such massive die-offs likely reduce northern flying squirrel abundances, including in areas where populations are small and isolated. Isolated populations thus reduced must be considered at risk of extinction in coming decades.[10]

NOTES AND REFERENCES

1. Wells-Gosling and Heaney (1984).
2. UWMV, KU and DMNS specimens.
3. Commission Regulations Chapter 52, Section 11.
4. Kiesow (2008, fig. 5.2).
5. Long (1965), Kiesow (2008). Howell (1918, 38) described two specimens from Middle Hay Creek in the Bear Lodge Mountains, and Alyssa Kiesow (pers. comm.) collected 16 from the Bear Lodge Mountains in 2006.
6. Durrant (1952) for Utah distribution. It was shown as occurring in an "isolated population" in the Uinta Mountains in Clark and Stromberg (1987), but that point is not associated with a specimen or other data. The species was first described from Colorado by Holmes and Neubaum (2013).
7. Kiesow (2008).
8. Carey (2000).
9. Kiesow et al. (2012).
10. Smith (2007), Weigl (2007).

VOLES, MICE, LEMMINGS, AND RATS—FAMILY CRICETIDAE

This New World family shares an unsettled taxonomic relationship with family Muridae. These two taxa have been combined, separated, and subordinated to each other in various configurations for over a century. During the latter half of that period, subordinating the members of this group, as subfamily Cricetinae, to family Muridae has been the prevailing taxonomic paradigm. Here, I employ the classification of Wilson and Reeder, which places Cricetidae separately and equal to Muridae.[1] Among the family members are typical mouse- or vole-shaped animals that spend the day in burrows and forage on the ground at night. The group also includes North American forms that live entirely in trees, some fossorial species, and the semiaquatic common muskrat. The Cricetidae comprises 17 Wyoming representatives, including the most numerically abundant mammal species in the state.

NOTES AND REFERENCES

1. Wilson and Reeder (2005).

KEY TO GENERA AND SPECIES OF CRICETIDAE (MODIFIED FROM HOFFMEISTER 1986, ARMSTRONG ET AL. 2011).

1a. Ears prominent, sparsely haired, tail typically as long as body: 10
1b. Ears variable in length and hairiness; tail variable, lightly furred: 2

2a. Length greater than 300 mm, tail sparsely haired, scaly, tail compressed laterally: . . . . ***Ondatra zibethica***
2b. Length less than 300 mm, tail haired or furred, with scales not prominent, tail more round in cross section:. 3

3a. Cheek teeth with multiple roots, crown enamel-covered, enamel wearing away to show enamel edges with age: 4
3b. Cheek teeth with single, ever-growing root, occlusal surface showing crescents or loops of enamel surrounding dentine throughout life: 5

4a. Dorsal trunk pelage strongly reddish, enamel loops extending roughly evenly to lingual and buccal sides: . . . . ***Myodes gapperi***
4b. Dorsal trunk pelage grayish brown, enamel loops extending farther to lingual than buccal side, giving tooth row a saw-like appearance: . . . . ***Phenacomys intermedius***

5a. Tail short, about the length of the hindfoot:. . . . . ***Lemmiscus curtatus***
5b. Tail clearly longer than hindfoot: 6

6a. Tail one-third or more of total length: . . . . ***Microtus longicaudus***
6b. Tail less than one-third of total length: 7

7a. Second (of three) upper cheek tooth with five dentine surfaces surrounded by enamel, the posteriormost of them round or oblong, rather than crescentric: . . . . ***Microtus pennsylvanicus***
7b. Second (of three) upper cheek tooth with four dentine surfaces surrounded by enamel: 8

Lower cheek tooth row of *Phenacomys intermedius* (anterior at top, lingual side to left, buccal side to right), showing asymmetrical enamel extending much farther to the lingual side than the buccal side.

Second upper cheek tooth (nnn molar) of *Microtus pennsylvanicus* (anterior at top, lingual side to right), showing five enamel loops, of which the posteriormost is small and rounded.

8a. Ventral trunk buff-colored: . . . . ***Microtus ochrogaster***
8b. Ventral trunk gray or off-white:. 9

9a. Adult length less than 185 mm (7.3 in): . . . . ***Microtus montanus***
9b. Adult length greater than 190 mm (7.5 in): . . . . ***Microtus richardsoni***
10a. Each upper incisor with a groove on its anterior surface; body length typically less than 80 mm (3.1 in); tail length about the same as or greater than body length: . . . . ***Reithrodontomys*** (identification to species is a task for a specialist)

10b. Upper incisors lack groove on anterior surface; body length typically greater than 80 mm (3.1 in); if less, tail obviously shorter than body: 11

11a. Tail length less than 60% of that of body; coronoid process of dentary projects above top of articular process of dentary:. . . . . ***Onychomys leucogaster***
11b. Tail length greater than 60% of that of body; coronoid process of dentary does not project above the top of the articular process of dentary: . . . . ***Peromyscus*****, 12**

12a. Tail fairly well haired, with tuft near tip: . . . . ***Peromyscus crinitus***
12b. Tail lightly haired, lacking tuft near tip:. 13

13a. Tail length less than that of body; without obvious tuft near the tip: 14
13b. Tail length greater than that of body; with tuft near the tip; ear noticeably longer than hindfoot: . . . . **Peromyscus truei**

14a. Tail bicolored with dorsal stripe narrow, distinct; no noticeable dark stripe along dorsal midline: . . . . **Peromyscus maniculatus**
14b. Tail with broad dorsal stripe; tail not sharply bicolored; noticeable dark stripe along dorsal midline: . . . . **Peromyscus leucopus**

Canyon mouse, *Peromyscus crinitus*

DESCRIPTION: A small to medium-sized member of a diverse North American genus, this mouse is characterized by dorsal pelage that is long, and silky (hairs are fine and easily bent), with a yellowish cast. The pinnae are about as long as the hindfeet (small relative to those of *P. truei*), and the slender bicolored tail is longer than the body by 5–43%, with a more pronounced tuft of hairs near the tip than seen in *P. truei*.[1] Measurements (**Sex** [n] mean (mm, g), *range*): Utah and Arizona; L = **F** [73] 170, *118–190*; **M** [96] 171, *154–187*; TL = **F** [73] 90, *67–101*; **M** [96] 89, *76–100*; HF = **F** [73] 21, *19–22*; **M** [95] 21, *18–23*; EFN = **F** [53] 21, *18–26*; **M** [85] 21, *18–23*; WT = **F** [42] 15.8, *8.9–22.5*, **M** [64] 16.2, *10–20.5*.[2] Dental formula: 1/1, 0/0, 0/0, 3/3 = 16.

STATUS: A protected nongame animal and a state Species of Greatest Conservation Need as of 2012.[3]

NOMENCLATURE AND SYSTEMATICS: The species was first recognized and the current scientific name first applied during the 1890s.[4] Of the eight subspecies recognized as of 1981, only *P. c. doutti* is depicted as occurring in Wyoming.[5] The common name "canyon deer mouse" is also used.

DISTRIBUTION AND HABITAT ASSOCIATIONS: The Wyoming distribution of this Colorado Plateau and intermountain basin species is very restricted, including areas along Flaming Gorge east and west of the reservoir. The area occupied is small and disjunct, in rocky canyon sites. East of Flaming Gorge, it was trapped in two of seven sites with cliffs nearby.[6] Vegetation associations are diverse and include occurrences in conifer woodlands and shrub-dominated semiarid types on sites with cliffs nearby.

ECOLOGY: As for other members of the genus, the diet is diverse, dominated by seeds, insects, and green plant material at various seasons. Population densities have not been estimated in habitats like those in southern Wyoming, but elsewhere they have varied widely within a single study.[1]

CONSERVATION AND MANAGEMENT: As is the case for cliff chipmunks, the species' geographic range in the state likely was drastically reduced by the inundation of Flaming Gorge Reservoir in the late 1950s. The canyon mouse was trapped by Merritt Cary in an area just north of the state line that was subsequently inundated, and much of the cliffy riverine

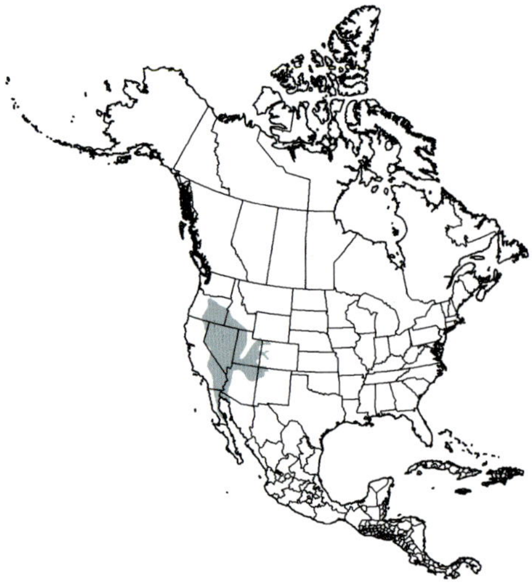

Locations from which the canyon mouse has been identified in Wyoming.

habitat likely occupied by canyon mice was lost. It is a conservation concern because of the limited and disjunct habitats that the species now occupies in the state.[7]

NOTES AND REFERENCES

1. Hall and Hoffmeister (1942), Hoffmeister (1986), Johnson and Armstrong (1987).
2. UCM specimens.
3. Commission Regulations Chapter 52, Section 11.
4. Wilson and Reeder (2005).
5. Hall (1981), Johnson and Armstrong (1987).
6. Goin (1944), Rompola and Anderson (2004).
7. Two specimens collected by Merritt Cary in 1911 (USNM 177481, 177484) were from 1768 m (5834 ft) elevation. The surface of Flaming Gorge Reservoir now fluctuates around 1840 m (6040 ft) elevation. The elevation reported by Long (1965, 627) for these specimens—1151 m (3800 ft)—is a typographical error.

White-footed mouse, *Peromyscus leucopus*

DESCRIPTION: A small to medium-sized member of the genus with brownish to grayish dorsal pelage, a very light underside, medium-sized pinnae, and white fur on the dorsum of each foot. The tail is bicolored, but not as strongly as in some members of the genus, for example, the deer mouse. The species is morphologically similar to the deer mouse, but ear length <17 mm indicates white-footed mouse in nearly all cases.[1] Measurements (**Sex** [n] mean (mm, g), *range*): Nebraska, L = **F** [7] 175, *154–188*; **M** [8] 151 *145–188*; TL = **F** [7] 77, *70–87*; **M** [8] 78, *66–89*; HF = **F** [7] 22, *21–23*; **M** [8] 21, *14–23*; EFN = **F** [7] 17, *15–19*; **M** [8] 15, *15–17*; WT = **F** [7] 28.3, *21.7–33.5*; **M** [8] 25.5, *21.8–29.2*.[2] Dental formula: 1/1, 0/0, 0/0, 3/3 = 16.

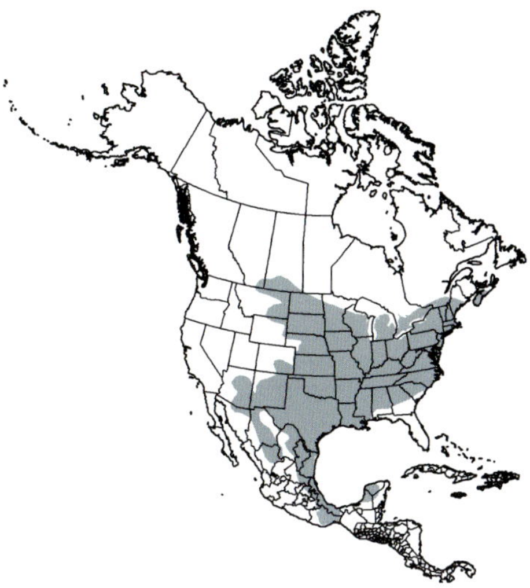

Distribution of the white-footed mouse in Wyoming.

STATUS: A protected nongame species by state regulation.[3]

NOMENCLATURE AND SYSTEMATICS: The species was first recognized in the early nineteenth century, and the current scientific name was first applied in the 1890s. Of the 17 subspecies recognized as of 1985, only *P. l. aridulus* occurs in Wyoming.[1]

DISTRIBUTION AND HABITAT ASSOCIATIONS: The Wyoming distribution of this eastern species includes riverine areas along the Little Missouri River in Crook County. Clark and Stromberg (1987) mapped the species in northern Sheridan County and along the Niobrara River in Niobrara County, but no specimens or other data are associated with these points.[4] More recent records are not known, and the species could undergo considerable distributional shifts without that being detected. Habitats in this region tend to be brushy areas or deciduous woodlands, commonly in riparian settings. The species tends to avoid grassy areas that lack shrubs or riparian trees.

ECOLOGY: White-footed mice are fairly omnivorous, eating green plant material, seeds, fruits, and insects, as reported in various studies. Although this species has been intensively studied in the eastern part of its geographic range, it is not well known from states near Wyoming. The species is commonly infected with *Borrelia burgdorferi*, the bacterium that causes Lyme disease in humans, but infections in captive mice seem not to affect their physical vigor. Infection rates in and near Wyoming are not known.[5]

CONSERVATION AND MANAGEMENT: The species is of conservation concern in Wyoming because of a lack of information about its distribution and ecology. It is also of some public health interest because it is considered an important reservoir of zoonotic disease agents: *Borrelia*, hantavirus, and *Giardia lamblia*.[6]

NOTES AND REFERENCES

1. Lackey et al. (1985) for general characteristics. Ear length <0.67 in (17 mm) successfully classified 98% of animals of *P. leucopus* and *P. maniculatus* in Wisconsin (Stephens et al. 2014); definitive identification in that study was from restriction fragment length polymorphisms in *cytochrome b* regions of mitochondrial DNA (Meece et al. 2005).
2. UWMV specimens.
3. Commission Regulations Chapter 52, Section 6.
4. Clark and Stromberg (1987).
5. Schwanz et al. (2011).
6. Ubico et al. (1996), Ziegler et al. (2009), Luong et al. (2011).

Deer mouse, *Peromyscus maniculatus*

DESCRIPTION: A medium-sized member of the genus that is variable in dorsal pelage color across its wide geographic range; the ventral trunk and dorsal feet are white.[1] The tail tends to be more sharply bicolored and the pinna larger than for the very similar white-footed mouse (see that account) in some geographic areas, but these criteria have not been evaluated in northeastern Wyoming. Measurements (**Sex** [n] mean [mm, g], *SD*): Sweetwater County, L = **F** [10] 146, *17.4*; **M** [6] 146, *14.2*; TL = **F** [10] 63, *8.6*; **M** [6] 63, *7.7*; HF = **F** [10] 19, *1.1*; **M** [6] 20, *1.4*; EFN = **F** [10] 17, *2.1*; **M** [6] 15, *1.9*; WT = **F** [10] 17.7, *4.3*; **M** [6] 17.2, *1.8*; Albany County, L = **F** [19] 141, *14.0*; **M** [50] 141, *15.6*; TL = **F** [19] 59, *10.4*; **M** [50] 59, *9.2*; HF = **F** [19] 19, *1.6*; **M** [50] 19, *1.5*; EFN = **F** [19] 14, *2.4*; **M** [50] 15, *2.0*; WT = **F** [19] 16.8, *3.2*; **M** [50] 18.9, *4.4*.[2] Dental formula: 1/1, 0/0, 0/0, 3/3 = 16.

STATUS: A non-protected nongame species by regulation.[3]

NOMENCLATURE AND SYSTEMATICS: The species was first recognized in the mid-nineteenth century, and a large number of synonyms have been used for various intraspecific variants. This is arguably the most thoroughly studied North American mammal species with respect to taxonomy; three major revisions of the genus have been completed.[4] Over 65 subspecies

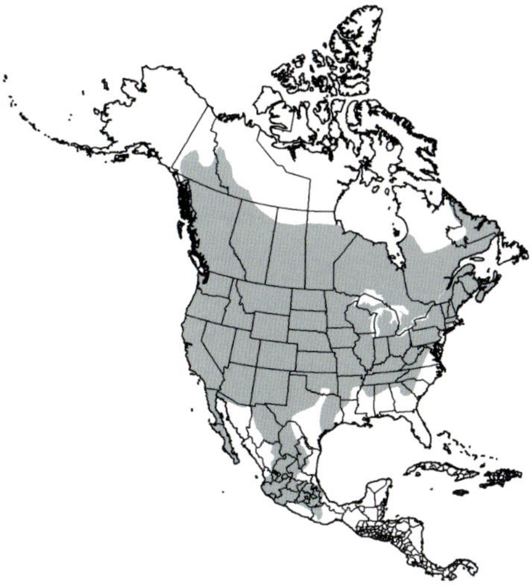

Distribution of the deer mouse in Wyoming.

were recognized across the extensive geographic range as of 1981; the status of many of these is poorly understood, having been named on the basis of morphology in the late nineteenth century.[5] A recent molecular study showed that what is now considered "the deer mouse" may comprise multiple species that diverged during Pleistocene isolation events. These periods of isolation and divergence may also account for the very different hantavirus lineages found in deer mouse populations across temperate North America.[6] The common name "North American deer mouse" is also used.

DISTRIBUTION AND HABITAT ASSOCIATIONS: This widely distributed North American rodent occurs statewide in almost all habitats. Deer mice tend to be more associated with grassy habitats than the brushy habitats preferred by white-footed mice. In Rocky Mountain forests, deer mice are found in a wide range of stand types, and tend to be favored by disturbance. So, in stands dominated by lodgepole pine or Engelmann spruce–subalpine fir following clear-cutting or fire, deer mice tend to surpass southern red-backed voles as the dominant small rodents.[7]

ECOLOGY: This was perhaps the first mammalian species to receive focused ecological study in Wyoming when the Murie brothers in 1929 investigated habitat associations and homing behavior in Teton County.[8] Deer mice are dietary generalists; in one study from Grand Teton National Park they were reported to eat mostly arthropods, with some green plant and seed material. A more geographically extensive study in Wyoming and Colorado found conifer and other plant seeds the dominant foods.[9] In western Montana, gall flies have been found to make up a high proportion of the diet year-round, and this may be an important food in northwestern Wyoming as well.[10] Deer mice are commonly the most abundant small mammals in trap samples in shrub-steppe and plains habitats across lower-elevation Wyoming.[11] Hantavirus and the Lyme disease bacterium are believed to not cause serious symptoms in them.[12]

CONSERVATION AND MANAGEMENT: This is one of the most abundant and widespread mammals in Wyoming, not considered to be a conservation concern. Deer mice are of considerable public health concern because they serve as important reservoirs of diseases transmissible to humans; hantavirus pulmonary syndrome occurs in Wyoming and neighboring states, and deer mice are known to be important reservoirs in this region.[13]

NOTES AND REFERENCES

1. King (1968). See Stephens et al. (2014) for ear length and body weight as criteria in Wisconsin; molecular markers for species are available for the genus.
2. UWMV specimens.
3. Commission Regulations Chapter 52, Section 6.
4. These were in 1909, 1968, and 1989, reviewed by Wilson and Reeder (2005).
5. Hall (1981).
6. Dragoo et al. (2006).
7. Campbell and Clark (1980).
8. Murie and Murie (1931).
9. Williams (1959), Clark (1975).
10. Reviewed by Pearson et al. (2000). The larvae of *Urophora affinis* and *U. quadrifasciata* are introduced biological control agents for introduced spotted knapweed (*Centaurea maculosa*), and attain high biomass densities compared with their native habitats in Eurasia. The diets of deer mice in western Montana can be dominated by *Urophora* larvae.
11. Clark (1977).
12. Klein and Calisher (2007), Previtali et al. (2010).
13. Centers for Disease Control and Prevention data. Fourteen cumulative cases of hantavirus pulmonary syndrome had been documented to be contracted in Wyoming as of 2013 (http://www.cdc.gov/hantavirus/surveillance/state-of-exposure.html, accessed August 26, 2013). This is a fairly high per capita rate relative to surrounding states.

Pinyon mouse, *Peromyscus truei*

DESCRIPTION: A medium-sized member of the genus with pinnae larger in absolute size than those of other Wyoming *Peromyscus*. Dorsal pelage is geographically variable—yellowish brown to brownish black. The distinctly bicolored tail is slightly shorter than the body in and near Wyoming, with a slight tuft of longer hairs near the tip. The dorsum of the hindfoot is white or light gray. The species is most similar morphologically to *P. leucopus*, but the Wyoming habitats of the two species are quite distinct, and the geographic ranges separated by over 200 mi (320 km).[1] Measurements (**Sex** [n] mean

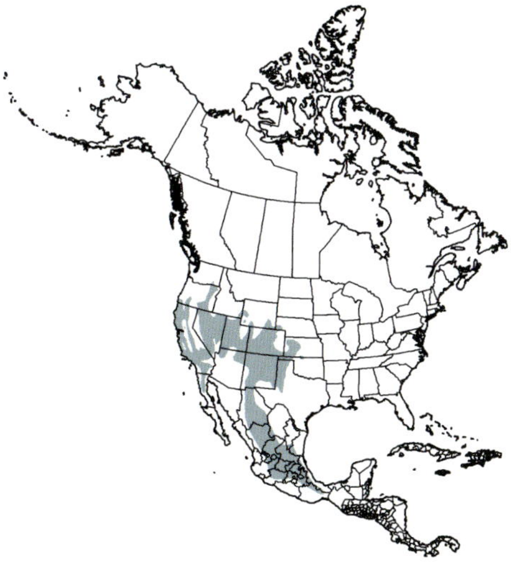

Locations from which the pinyon mouse has been identified in Wyoming.

(mm, g), *range*): Utah, L = **F** [59] 181, *152–215*; **M** [96] 177, *140–203*; TL = **F** [59] 90, *60–114*; **M** [96] 88, *66–105*; HF = **F** [9] 23, *21–25*; **M** [96] 23, *19–25*; EFN = **F** [59] 26, *20–30*; **M** [94] 25, *17–29*; WT = **F** [57] 22.7, *12.5–38.9*; **M** [90] 20.7, *11.8–30.6*.[2] Dental formula: 1/1, 0/0, 0/0, 3/3 = 16.

STATUS: A protected nongame species and a state Species of Greatest Conservation Need as of 2012.[3]

NOMENCLATURE AND SYSTEMATICS: The species was first recognized in the 1880s and the current scientific name applied in the 1890s. Of the 15 subspecies recognized as of 1981, only the nominal *P. t. truei* occurs in Wyoming.[1]

DISTRIBUTION AND HABITAT ASSOCIATIONS: This species of western and Mexican deserts and woodlands occurs along the southern edge of the state near the Green River. Early records of occurrence in Wyoming are difficult to interpret. Rompola and Anderson (2004, 91) commented that "the earliest published account of the pinyon mouse in Wyoming appears to be 1942 (Hoffmeister 1951)."[4] However, Merritt Cary trapped four specimens in 1911 in the same area along the Green River where he also trapped canyon mice and cliff chipmunks. The site where he trapped is beneath the waters of Flaming Gorge Reservoir today.[5] Long (1965) referred only to the 1911 specimens, and not to any 1942 ones. So, it is not clear what Rompola and Anderson meant; however, it is clear that they caught pinyon mice on the east side of Flaming Gorge in 1998.[4] The species is mostly associated with pinyon pine and junipers throughout its geographic range. It has also been detected in ponderosa pine forest, among bristlecone pines, and in riparian cottonwoods surrounded by sagebrush. East of Flaming Gorge, the species was found at sites with more canopy cover and farther from rock outcrops than sites generally available.[4]

ECOLOGY: The ecology of the species is poorly known. Diets are assumed to be dominated by seeds of pinyon pine and junipers, although no data support those assumptions.

CONSERVATION AND MANAGEMENT: The species is a conservation concern because the distribution was restricted to the Flaming Gorge area when Europeans arrived and likely was reduced by the inundation of the canyon. The species is known to carry hantavirus, but is not thought to be a major public health threat because it is much less likely than the deer mouse to occupy human structures, where humans commonly are infected.[6]

NOTES AND REFERENCES

1. Hoffmeister (1981).
2. UCM specimens.
3. Commission Regulations Chapter 52, Section 11.
4. Rompola and Anderson (2004, 91), referring to Hoffmeister (1951).
5. These specimens are USNM 177485–177488, of which 177486 does not appear on the 2012 online catalogue of the USNM.
6. Armstrong et al. (2011).

Western harvest mouse, *Reithrodontomys megalotis*

DESCRIPTION: A small, long-tailed mouse with medial grooves on the anterior surfaces of the upper incisors, prominent pinnae, and a lightly haired tail. Dorsal pelage is gray buff, with some dark guard hairs; the underside is very pale, and the tail is bicolored, dark on the dorsum.[1] Measurements (**Sex** [n] mean (mm, g), *range*): Albany, Campbell, Johnson, and Washakie Counties, L = **F** [8] 137, *124–150*; **M** [8] 133, *122–148*; TL = **F** [8] 64, *61–71*; **M** [8] 62, *58–68*; HF = **F** [8] 17, *15–17*; **M** [8] 17, *16–18*; EFN = **F** [8] 13, *12–15*; **M** [8] 13, *11–14*; WT = **F** [8] 12.1, *7–17*; **M** [8] 11.3, *9.5–13*.[2] Dental formula: 1/1, 0/0, 0/0, 3/3 = 16.

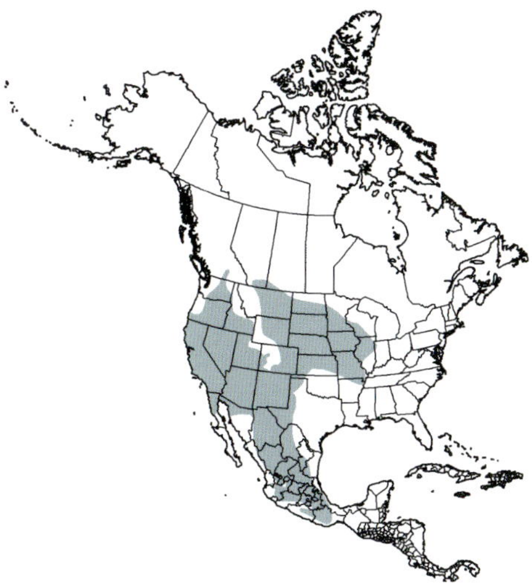

Distribution of the western harvest mouse in Wyoming.

STATUS: This is a non-protected nongame species, by state regulation.[3]

NOMENCLATURE AND SYSTEMATICS: The species was recognized in the mid-nineteenth century and has carried its current scientific name since 1893. The genus—very closely related to *Peromyscus*—includes two subgenera; both Wyoming species are in subgenus *Reithrodontomys*. Of the 16 subspecies recognized as of 1982, only the nominal *R. m. dychei* occurs in Wyoming.[1]

DISTRIBUTION AND HABITAT ASSOCIATIONS: This broadly distributed species occurs over roughly the eastern two-thirds of Wyoming, below the lower elevational limit of coniferous forest. Habitats include a wide range of non-forest types: grassy areas, recovering disturbed sites, fence lines, meadows, arid lands, marshes, and low-elevation woodlands.[1]

ECOLOGY: The species is primarily granivorous, but also eats green plant material and insect larvae. In southern Oregon, western harvest mice were found to have eaten hypogeous fruiting bodies of two fungi—*Cazia flexiascus* and *Tuber candidum/T. quercicola*—and were inferred to be potentially important spore dispersal agents for these species.[4] In tallgrass prairie of Kansas, the species was found to be quite generalized in relation to recent fire history and topography.[5] A 12-year study of western harvest mice in southern Arizona showed that the only discernable characteristic of populations was variation: seasonal and interannual variation in population sizes and sexual differences in movement rates and behaviors.[6]

CONSERVATION AND MANAGEMENT: The species is common, abundant in many areas, and very widely distributed outside Wyoming. It is not generally considered to be a conservation concern.

NOTES AND REFERENCES

1. Webster and Jones (1982).
2. KU specimens.
3. Commission Regulations Chapter 52, Section 6.
4. Frank et al. (2006).
5. Kaufman et al. (1988).
6. Skupski (1995).

Plains harvest mouse, *Reithrodontomys montanus*

DESCRIPTION: This species is very similar to the western harvest mouse in body size, pelage color, ear size, and the anterior groove on the upper incisors.[1] The only characters that fairly reliably distinguish between the two species are skull measurements and enamel patterns on the molars. In one case, researchers closely examined 37 specimens in the University of North Dakota Museum that had been labeled as *R. megalotis*. Fifteen of the skulls were found to have dimensions more consistent with *R. montanus*, and eight skulls had dimensions that indicated they could not be assigned to either species.[2] Clearly, identification of harvest mice to species on the Great Plains is a task for a specialists using reference collections. Measurements (**Sex** [n] mean (mm, g), *range*): various counties in Wyoming and

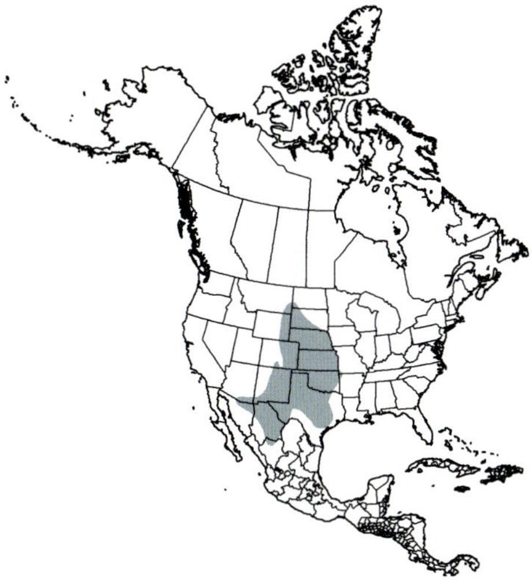

Locations from which the plains harvest mouse has been identified in Wyoming.

Nebraska, L = **F** [5] 118 *102–130*; **M** [8] 120, *111–129*; TL = **F** [5] 52, *43–61*; **M** [8] 51, *48–54*; HF = **F** [5] 16, *14–20*; **M** [8] 16, *15–17*; EFN = **F** [5] 13, *12–15*; **M** [8] 12, *11–14*; WT = **F** [5] 9.5, *7.3–11.5*; **M** [8] 9.85, *8–11.6*.[3] Dental formula: 1/1, 0/0, 0/0, 3/3 = 16.

STATUS: This was a state Species of Greatest Conservation Need as of 2012.[4]

NOMENCLATURE AND SYSTEMATICS: The species was recognized in the mid-nineteenth century and was given the current scientific name then. Of the three subspecies recognized as of 1986, only the nominal *R. m. albescens* occurs in Wyoming and adjacent states.[1]

DISTRIBUTION AND HABITAT ASSOCIATIONS: This Great Plains species is believed to occur through the eastern one-fourth of the state, with the western limit roughly in the Harriman area near the Colorado state line and in northwestern Crook County in the north. Habitats are generally regarded as more narrowly cast than is the case for the western harvest mouse: grass is more emphasized in descriptions of habitat for plains harvest mice. In Goshen County, plains harvest mice were taken at sites with more grass and less bare ground, and western harvest mice were taken in sites with taller, shrubbier vegetation.[5]

ECOLOGY: The ecology of the species is poorly known. It eats seeds of buffalograss and switch grass, and various insects. The life history is almost completely unknown.[6]

CONSERVATION AND MANAGEMENT: The species is a conservation concern because of poor information about distribution, abundance, and life needs.

NOTES AND REFERENCES

1. Wilkins (1986).
2. Seabloom and Shaffer (2005).
3. KU specimens.
4. Commission Regulations Chapter 52, Section 11.
5. Maxwell and Brown (1968).
6. Brown (1946).

Northern grasshopper mouse, *Onychomys leucogaster*

DESCRIPTION: A mouse with a robust torso. The pinnae are prominent, and the tail short and thick at its base, tapering sharply toward the tip. The pelage is distinctly bicolored, variably pale brown, gray, or pinkish on the dorsum, lighter on the flanks and underside.[1] Measurements (**Sex** [n] mean [mm, g], *SD*): Albany and Sweetwater Counties, L = **F** [8] 130, *6.8*; **M** [20] 135, *22.1*; TL = **F** [8] 34, *1.8*; **M** [20] 33, *9.3*; HF = **F** [8] 18, *3.0*; **M** [20] 19, *1.2*; EFN = **F** [8] 16, *1.5*; **M** [20] 16, *4.6*; WT = **F** [8] 28.4, *5.2*; **M** [20] 28.2, *8.3*.[2] Dental formula: 1/1, 0/0, 0/0, 3/3 = 16.

STATUS: A non-protected nongame animal by regulation.[3]

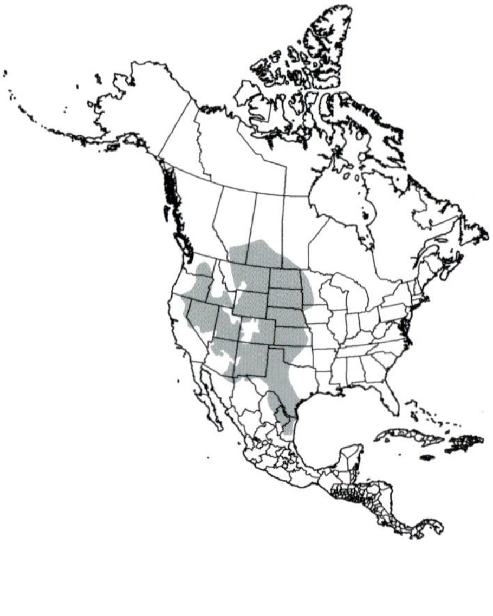

Distribution of the northern grasshopper mouse in Wyoming.

NOMENCLATURE AND SYSTEMATICS: The species was first recognized in the mid-nineteenth century, when the currently used scientific name was applied. This genus, which comprises two Nearctic species, is closely related to *Peromyscus* and its sibling genera.[4] Of the 13 subspecies recognized as of 1981, three are represented as occurring in Wyoming: *O. l. missouriensis* in the northeastern corner of the state, *O. l. brevicaudus* along the western border, and *O. l. arcticeps* over the remainder of the state except the northwestern mountains.[5] However, the four subspecies on the Great Plains have been found to be indistinguishable on the basis of morphology, and these subspecies—including the two in eastern Wyoming—have been proposed to represent a single one: *O. l. arcticeps*.[6] The specific epithet "*leucogaster*" is the Latinized "white-bellied."

DISTRIBUTION AND HABITAT ASSOCIATIONS: This species of the western Great Plains and intermountain basins occurs statewide below the lower elevational limit of coniferous forest. Habitats are quite varied and include the full range of arid and semiarid shrub-steppe and Great Plains types: shortgrass, mixed grass, sagebrush, and dunes. Grasshopper mice associate with various kinds of disturbed habitats: moderately overgrazed grasslands, pocket gopher mounds, and black-tailed prairie dog colonies have all been found to have higher densities than sites nearby.[7]

ECOLOGY: This mouse is remarkably predaceous and has the life history traits of a predator. Various studies have shown the diet to consist of 75–90% animal matter: grasshoppers, Mormon crickets, sand crickets, beetles, mantids, pupae of various insect orders, and other small mammals. The small mammals killed and eaten by grasshopper mice include deer mice and voles.[8] As for other predators, population densities are low and home ranges large for the mouse's body size. The species is considered to be a likely reservoir for *Yersinia pestis*—the bacterium that causes plague—in the wild. Considerable recent research has focused on how, and the extent to which, grasshopper mice serve as hosts for *Yersinia*-carrying fleas and sustain plague at low levels of incidence in rodent communities. Current understandings are that 11–23% of grasshopper mice carry antibodies to plague during epizootics among syntopic black-tailed prairie dogs, but that few or no grasshopper mice have antibodies two years after such outbreaks. This suggests that a substantial part of the population can survive infection with plague. Grasshopper mice may facilitate epizootics among prairie dogs by transporting infected fleas among prairie dog towns more effectively than do prairie dogs.[9]

CONSERVATION AND MANAGEMENT: The species is sufficiently common and widespread that it is not a conservation concern. Although it may serve an important ecological role in sustaining plague in rodent communities in Wyoming, no management actions have been directed at the species, perhaps because it is so common and widespread.

NOTES AND REFERENCES

1. McCarty (1978).
2. UWMV specimens.

3. Commission Regulations Chapter 52, Section 6.
4. Reviewed by Wilson and Reeder (2005).
5. Hall (1981).
6. Engstrom and Choate (1979).
7. Stapp (1999), Armstrong et al. (2011).
8. Bailey and Sperry (1929), Egoscue (1960), Flake (1973).
9. Thiagarajan et al. (2008); Stapp et al. (2008, 2009); Salkeld et al. (2010).

Bushy-tailed woodrat, *Neotoma cinerea*

DESCRIPTION: A rat-sized rodent with large, finely furred pinnae, a well-furred tail, furred soles on the hindfeet, and exceptionally long vibrissae—up to 100 mm (4 in). It is larger bodied than most species in the genus. The dorsal pelage is highly variable in color, from dull gray or buff to dark brown; the tail is sharply bicolored—dark above and light on the underside.[1] Measurements (**Sex** [n] mean (mm, g), *range*): Albany, Carbon, and Teton Counties, L = **F** [5] 334, *288–362*; **M** [8] 327, *240–413*; TL = **F** [5] 147, *136–162*; **M** [8] 146, *97–185*; HF = **F** [5] 40, *38–42*; **M** [8] 39, *30–45*; EFN = **F** [5] 32, *29–36*; **M** [8] 28, *22–38*; WT = **F** [5] 212.8, *122–261*; **M** [8] 308.5, *240–360*.[2] Dental formula: 1/1, 0/0, 0/0, 3/3 = 16.

STATUS: A non-protected nongame animal by regulation.[3]

NOMENCLATURE AND SYSTEMATICS: The species was first recognized in 1815, and the current scientific name was first applied in the 1820s. This species, one of 21 in this Nearctic genus, is taxonomically stable, in contrast to some other taxa in the genus that have been revised repeatedly. Of the 13 subspecies recognized as of 1997, four are depicted as occurring in Wyoming in a complex geographic pattern that is not concordant with geographic boundaries: *N. c. cinerea*, *N. c. orolestes*, *N. c. cinnamomea*, and *N. c. rupicola*.[4]

DISTRIBUTION AND HABITAT ASSOCIATIONS: This Rocky Mountain and Pacific Northwest species occurs nearly statewide, including in high-elevation talus and desert shrubs. Across the Intermountain West, the common habitat feature selected by this species is proximity to rocks or cliffs; in Wyoming and adjacent states, it is almost invariably associated with cliffs or large rocks with fissures or holes that provide protection for their nests.

ECOLOGY: The species is well studied because of its high importance in reconstructing past environments and biotas. Where bushy-tailed woodrats occur, they construct nests deep within rocky structures that are protected from predators and weather; these nests are seldom observed or described.[4] At the entrances to their living places, they also assemble objects that are variously interpreted as protecting nest entrances, or as attracting females. These objects include sticks, fecal droppings of other vertebrates, segments of cacti, bones, and miscellaneous objects of human origin. The behavior of carrying objects to their nest entrances gives rise to the nickname "packrat." The third kind of

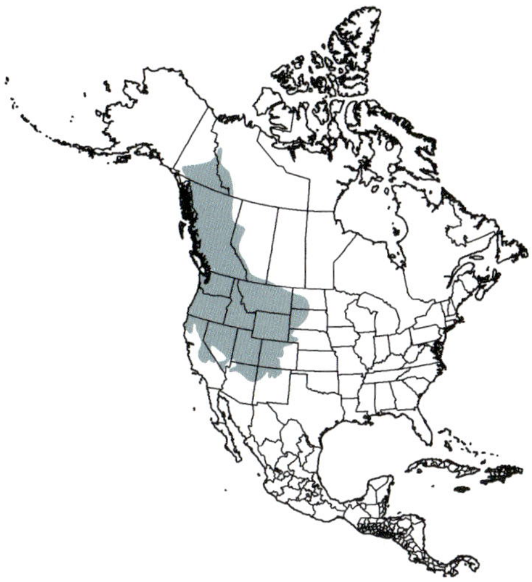

Distribution of the bushy-tailed woodrat in Wyoming.

structure created by this species is the latrine, where feces and plant fragments are cemented together by dried urine. Old latrines in places protected from weather can be visibly layered through time, can be dated via radiocarbon methods, and can represent tens of thousands of years of accumulation. Using them, paleoecologists and paleoclimatologists can reconstruct the biotas and climates through the Holocene and Late Pleistocene—a powerful investigative tool.[5] Bushy-tailed woodrats eat a wide range of plant species, and this fairly nonselective diet adds to the utility of middens as a record of the vegetation near a midden. Foods include leaves and fruits of many forbs and woody plants, including conifers and

A bushy-tailed woodrat midden of big sagebrush wood, ungulate long bones, and other found objects, near Medicine Bow. Woodrats are able to carry objects larger than themselves.

quaking aspen.[6] The species is important as prey for many mid-sized avian and mammalian predators.

CONSERVATION AND MANAGEMENT: Bushy-tailed woodrats are common, widespread, and generally considered neither a species of conservation concern nor an economic pest. They are capable of serving as reservoirs for *Borrelia burgdorferi*, the bacterium causing Lyme disease, but rarely do so in Wyoming.[7] The species has not been implicated as a hantavirus reservoir.

NOTES AND REFERENCES

1. Hoffmeister (1986). The Mexican woodrat approaches the Wyoming state line in northern Larimer County, Colorado, but has not been recorded in Wyoming. It differs in that the lateral hairs on its tail are shorter than those of the bushy-tailed woodrat, which can be over 19 mm (0.75 in) long.

2. UWMV specimens.

3. Commission Regulations Chapter 52, Section 6.

4. Finley (1990), Planz et al. (1996), Smith (1997), Edwards and Bradley (2001), Wilson and Reeder (2005).

5. Betancourt et al. (1990).

6. Vaughan (1990), Lyford et al. (2004).

7. See Gordus and Theis (1993) for a case from California.

Southern red-backed vole, *Myodes gapperi*

DESCRIPTION: A medium-sized vole with reddish, rusty, or chestnut dorsal pelage. It has a light underside and a bicolored tail. Males and females are similar in size. The tail is proportionally shorter than for most mice, yet longer than for some voles. Measurements (**Sex** [n] mean [mm, g], *SD*): Albany and Park Counties, L = **F** [7] 139, *7.1*; **M** [16] 131, *15.7*; TL = **F** [7] 41, *4.2*; **M** [16] 36, *8.1*; HF = **F** [7] 17, *0.7*; **M** [16] 17, *2.4*; EFN = **F** [7] 16, *1.5*; **M** [16] 14, *2.6*; WT = **F** [7] 27.2, *2.5*; **M** [16] 20.6, *3.8*.[1] Dental formula: 1/1, 0/0, 0/0, 3/3 = 16.

STATUS: A non-protected nongame species common or abundant in forested habitats.[2]

NOMENCLATURE AND SYSTEMATICS: The genus name has undergone recurring revision over decades; most changes have reflected new interpretations of the historical precedence of various generic epithets (e.g., *Evotomys, Clethrionomys*), rather than new understandings of phylogeny.[3] The specific epithet has been stable, although the species has in a few cases been considered conspecific with its sibling taxon—the northern red-backed vole—with which it is parapatric in northwestern British Columbia. Hybridization or introgression between the two species has not been reported. Of the 29 subspecies recognized as of 1981, three are depicted as occurring in Wyoming: *M. g. galei* in the Bighorn Mountains, Medicine Bows, Laramie Mountains, and Uintas; *M. g. idahoensis* in the Tetons and Wyoming Range; and *M. g. brevicaudis* in the Black Hills and Bear Lodge Mountains. The putative subspecies boundary (*M. g. galei–M. g. idahoensis*) is difficult to reconcile with any natural features.[4] The common name "Gapper's red-backed vole" is also used; Anthony Gapper was an English physician and naturalist who described some mammals of eastern Canada.

DISTRIBUTION AND HABITAT ASSOCIATIONS: This boreomontane species occurs in forested mountain ranges statewide. Even some small isolated ranges with coniferous forest are occupied: Hart Mountain, Shirley Mountains, Ferris Mountains, and Green Mountain. Habitats in the central Rockies are coniferous and mixed forests, especially those on moist sites and with abundant physical structure near the ground. The species also occurs in mid-seral forests, woodlands with abundant herbaceous growth, and meadows near coniferous forests. Stand age and decay stage of coarse woody debris have been found

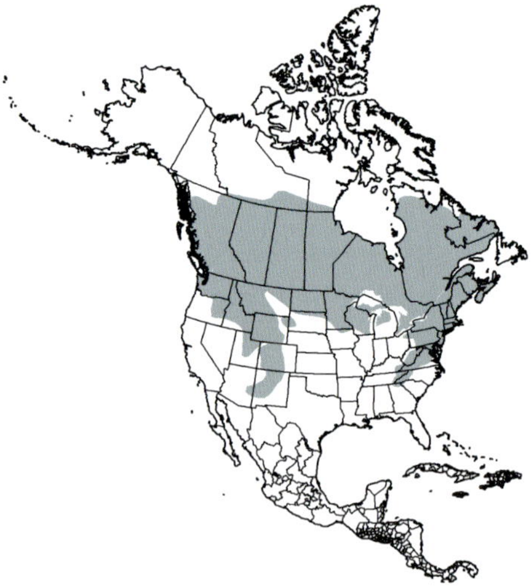

Distribution of the southern red-backed vole in Wyoming.

to be positively related to relative abundance of the species on lodgepole-pine-dominated sites in Wyoming.[5]

ECOLOGY: The diverse diet includes leaves, seeds, berries, arthropods, and fungi. The importance of hypogeous fruiting bodies of fungi that are mutualists of old conifer trees was recognized in the 1970s. This mutualistic association, by which voles eat fungal fruiting bodies and disperse the spores in their feces, has been investigated in Wyoming and helps explain the association of southern red-backed voles with old stands of conifers.[6] The species is closely associated with large volumes of well-decayed coarse woody debris, showing that volume and decay stage of woody debris after timber harvesting affect how this species responds.[7] This is an important prey species for several forest vertebrates: martens, short-tailed weasels, and boreal owls.

CONSERVATION AND MANAGEMENT: The species has been a topic of discussion in relation to forest management for decades. Typically, clear-cutting causes a decline in southern red-backed voles and an increase in deer mice. The latter species is less important as a prey species for forest-associated predators. The effect of selective cutting and patch cutting on southern red-backed voles is less pronounced than for clear-cutting, however. The centers of patch cuts in the Medicine Bow Mountains and selectively cut stands in Jackson Hole showed little change in abundance of the species. It is not known how long it takes for regenerating lodgepole pine stands to produce high densities of the species.[8]

NOTES AND REFERENCES

1. UWMV specimens.
2. Commission Regulations Chapter 52, Section 6.
3. Carleton et al. (2014) provided the most comprehensive review of the precedence of genus names for this taxon.
4. Hall (1981), Merritt (1981).
5. Nordyke and Buskirk (1991).
6. Merritt (1981), Dubay (2000), Dubay et al. (2008).
7. Fauteaux et al. (2012).
8. Campbell and Clark (1980), Hayward et al. (1999), Keinath and Hayward (2003).

Western heather vole, *Phenacomys intermedius*

DESCRIPTION: A rare small, short-tailed rodent with rounded snout, similar in appearance to *Microtus* voles. Dorsal pelage is grayish or brownish, and the underside whitish or gray. The best morphological character to distinguish from *Microtus* voles is the enamel pattern on the cheek teeth: the enamel lophs are much deeper on the lingual than the buccal side of the tooth in this genus (see figure in Cricetidae key).[1] C. Hart Merriam referred to the genus as "deceiver voles."[2] Measurements (**Sex** [n] mean (mm, g), *range*): Park, Albany, Carbon, and Lincoln Counties, L = **F** [12] 133, *106–168*; **M** [6] 132, *106–153*; TL = **F** [12] 34, *22–46*; **M** [6] 30, *25–38*; HF = **F** [12] 20, *15–26*; **M** [6] 18, *13–24*; EFN = **F** [12] 13, *9–16*; **M** [6] 13, *10–17*; WT = **F** [12] 25.9, *12.8–48.5*; **M** [6] 29, *18.2–43.6*.[3] Dental formula: 1/1, 0/0, 0/0, 3/3 = 16.

STATUS: A non-protected nongame species by regulation.[4]

NOMENCLATURE AND SYSTEMATICS: The species was originally described in 1889, when the currently used scientific name was applied. Of the nine subspecies recognized as of 1988, only the nominal *P. i. intermedius* occurs in Wyoming. Populations east of the Rocky Mountains are now considered a separate species, *P. ungava*.[5]

DISTRIBUTION AND HABITAT ASSOCIATIONS: This species of the Rocky Mountains, Sierra Nevada, and Coast Ranges occurs in Wyoming's northwestern mountains, Uinta Mountains, and Medicine Bow Mountains (both Snowy Range and Sierra Madres). Clark and Stromberg (1987) depicted the species as occurring in the isolated ranges south

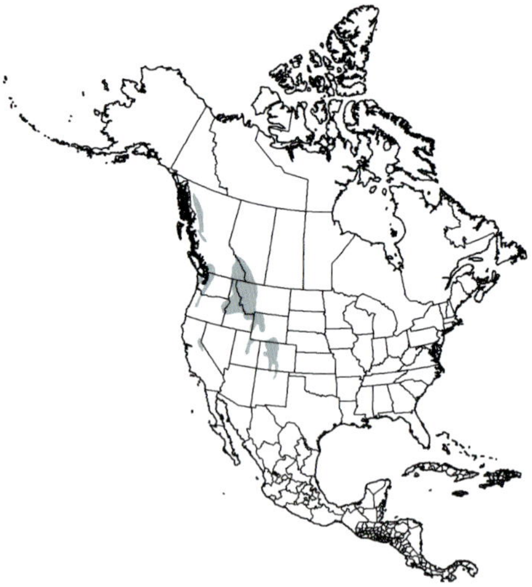

Distribution of the western heather vole in Wyoming.

of the Sweetwater River, but no specimens from these mountains are known.[6] Habitats are coniferous forests, tending toward mesic or dry types, in some cases near water. The species occurs in timberline krummholz conifers, alpine tundra, and woodland or streamside areas with ericaceous low shrubs, birch, or other understory shrubs. It was found in the Beartooth Mountains of Park County to be most common in krummholz conifers.[7]

ECOLOGY: More so than mice, heather voles are herbivorous, eating bark, leaves, seeds, and fruits of a wide range of woody, typically boreal taxa: willows, ericaceous low shrubs, dwarf birch, and soapberry. In captive feeding trials alongside other vole species, it was found to not prefer conifer seeds.[8] Food caches are common near burrow entrances, and cached food may be eaten all winter; the species is not known to hibernate. It is commonly eaten by boreal and alpine vertebrates, including martens, weasels, and owls. Population densities have not been estimated in or near Wyoming, but the species is regarded as less abundant than sympatric voles.

CONSERVATION AND MANAGEMENT: The species is sufficiently common that it is not generally considered a conservation concern.

NOTES AND REFERENCES

1. Banfield (1974).
2. Merriam (1889), McAllister and Hoffman (1988).
3. UWMV, DMNS, MSB specimens.
4. Commission Regulations Chapter 52, Section 6.
5. McAllister and Hoffman (1988).
6. Clark and Stromberg (1987).
7. Pattie and Verbeek (1967).
8. McAllister and Hoffman (1988), Lobo et al. (2009).

Long-tailed vole, *Microtus longicaudus*

DESCRIPTION: A medium-sized vole with a long tail relative to head and body. The tail is bicolored, as for many other mice and voles; tail length and color are not reliable field characters. Dorsal pelage is dull gray to brownish gray and the underside is dull gray to whitish.[1] Measurements (**Sex** [n] mean [mm, g], *SD*): various counties, L = **F** [11] 154, *21.2*; **M** [21] 147, *16.7*; TL = **F** [11] 52, *13.5*; **M** [21] 42, *9.7*; HF = **F** [11] 18, *2.4*; **M** [21] 20, *2.7*; EFN = **F** [11] 12, *3.4*; **M** [21] 12, *3.1*; WT = **F** [11] 29.3, *10.1*; **M** [21] 31.2, *8.8*.[2] Dental formula: 1/1, 0/0, 0/0, 3/3 = 16.

STATUS: This is a non-protected nongame mammal, by regulation.[3]

NOMENCLATURE AND SYSTEMATICS: The species was first recognized in the late nineteenth century, and the species epithet has been stable since then. The systematics of this genus are among the most complex of any mammalian taxon in Wyoming, and the position of *M. longicaudis* among vole-like rodents has undergone repeated revision.[1] Of the 15 or so subspecies that were recognized as of the 1980s, only *M. l. longicaudis* occurs in Wyoming. But more recent molecular studies have shown strong population structuring within the species, reflecting Pleistocene isolation and postglacial range recovery. The most strongly differentiated lineage is that which occupies the southern Rocky Mountains from Arizona and New Mexico north through Colorado to southern Wyoming. This lineage represents an isolated population southeast of the Wyoming Basin and Green River, which shows evidence of greater isolation than other subspecific lineages. The morphological subspecies identified by early twentieth-century systematists bear little resemblance to the lineages depicted via molecular methods.[4]

DISTRIBUTION AND HABITAT ASSOCIATIONS: This species of the Cordillera and Pacific mountain ranges occurs in higher-elevation areas nearly statewide, with the exception of a narrow

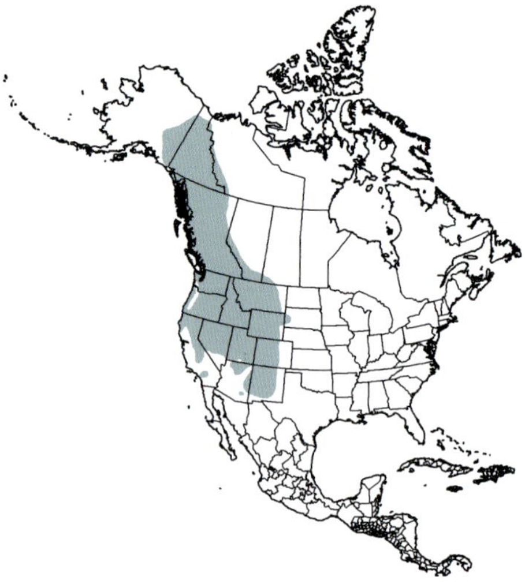

Distribution of the long-tailed vole in Wyoming.

north–south band in eastern Laramie and Goshen Counties. Habitats are fairly diverse, and in Wyoming include mesic and structurally complex forest habitats, alder–willow communities, aspens, and subalpine herbaceous meadows. Proximity to water has been found to be a positive predictor of presence, but this likely has to do with vegetation rather than the need for access to free water. Populations are thought to be positively affected by early postfire successional stages.[5]

ECOLOGY: Diets are geographically and seasonally variable, limited to plant and fungal material: green plant parts, seeds, fungi, and bark of shrubs. Populations have not been studied in environments similar to Wyoming's, but are known to fluctuate with season, year, and disturbance factors. Many small and mid-sized bird and mammal predators are known or suspected of eating long-tailed voles, but few detailed studies have been undertaken.[1] In one well-documented case, marten populations in southeast Alaska were shown to fluctuate in synchrony with those of long-tailed voles—the only vole species present—in an island environment.[6]

CONSERVATION AND MANAGEMENT: The species is common across a wide range of habitats and is not generally considered a conservation concern.

NOTES AND REFERENCES

1. Smolen and Keller (1987).
2. UWMV specimens.
3. Commission Regulations Chapter 52, Section 6.
4. Hall (1981), Smolen and Keller (1987), Conroy and Cook (2000), Spaeth et al. (2009).
5. Clark (1973), Armstrong et al. (2011), Sullivan and Sullivan (2012).
6. Flynn and Schumacher (2009).

Montane vole, *Microtus montanus*

DESCRIPTION: A medium-sized vole with brown dorsal fur washed with gray or buff. The underside is off-white to light gray and the tail bicolored as for many small rodents. The sexes are similar in size. Measurements (**Sex** [n] mean [mm, g], *SD*): Albany and Sweetwater Counties, L = **F** [9] 147, *18.0*; **M** [15] 137, *33.0*; TL = **F** [9] 44, *4.7*; **M** [15] 43, *7.9*; HF = **F** [9] 18, *2.6*; **M** [15] 19, *2.0*; EFN = **F** [9] 12, *3.5*; **M** [15] 12, *2.1*; WT = **F** [9] 30.9, *8.2*; **M** [15] 28.0, *10.0*.[1] Dental formula: 1/1, 0/0, 0/0, 3/3 = 16.

STATUS: This is a non-protected nongame animal, by regulation.[2]

NOMENCLATURE AND SYSTEMATICS: The species was first described in the mid-nineteenth century and has carried its current scientific name since 1898. Of the approximately 15 subspecies recognized as of 2003, three are depicted as occurring in Wyoming: *M. m. codiensis* in the Absaroka Mountains, *M. m. nanus* on the Yellowstone Plateau south and east to the southeastern corner of the state, and *M. m. zygomaticus* in the Bighorn Mountains.[3] The Absaroka and Bighorn Mountain subspecies are represented as being isolated by a narrow (<16 km, <10 mi) band of unoccupied habitat in the Bighorn Basin, but such a narrow unoccupied corridor is difficult to reconcile with the broad habitat tolerances of the species elsewhere in the state.[4]

DISTRIBUTION AND HABITAT ASSOCIATIONS: Montane voles occur in most mountains of the western states and in suitable habitats in the southwestern two-thirds of Wyoming. Habitats are herbaceous meadows near water in montane areas. The species tends not to occur above the elevational limit of trees and occurs only at low densities in shrub-steppe.

ECOLOGY: Like other voles, the species is a strict herbivore. In northern Colorado subalpine meadows, the diet was dominated by only a few species of forbs and grasses: small

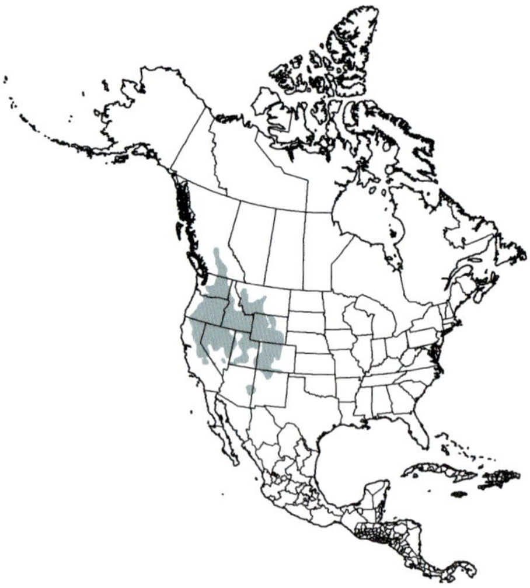

Distribution of the montane vole in Wyoming.

evening primrose, American milkvetch, pale agoseris, tiny trumpet, and arnica. Seeds are not an important food.[5] Population densities and dynamics are so variable geographically and temporally that few patterns are apparent.[6] Montane voles were the most abundant and widely distributed across habitats among four species of vole in Grand Teton National Park.[7]

CONSERVATION AND MANAGEMENT: The species is common and widespread and is not considered a conservation concern. An economic pest in some orchard agricultural settings because of damage to bark, the species is not known to cause such conflicts in Wyoming.

NOTES AND REFERENCES

1. UWMV specimens.
2. Commission Regulations Chapter 52, Section 6.
3. Sera and Early (2003).
4. Hall (1981), Sera and Early (2003).
5. Clark (1973), Vaughan (1974).
6. Vaughan (1974), Pinter (1986, 1988), Negus et al. (1986).
7. Clark (1973).

Prairie vole, *Microtus ochrogaster*

DESCRIPTION: Larger than most Wyoming voles, this species has grayish brown dorsal fur, the guard hairs giving it a grizzled appearance. The underside is gray to yellowish. Little or no sexual dimorphism of size has been noted.[1] The species is so morphologically similar to the meadow vole that field identification of live specimens is not reliable, but a molecular marker to distinguish between these species is now available.[2] Measurements (**Sex** [n] mean (mm, g), *range*): Converse, Crook, Fremont, Campbell, and Big Horn Counties, L = **F** [10] 155, *140–172*; **M** [13] 153, *128–181*; TL = **F** [10] 38, *34–46*; **M** [13] 38, *28–45*; HF = **F** [10] 20, *17–22*; **M** [13] 20, *16–22*; EFN = **F** [10] 14, *10–16*; **M** [13] 14, *11–17*; WT = **F** [10] 48.8, *32–59.5*; **M** [13] 44.8, *31.9–58.5*.[3] Dental formula: 1/1, 0/0, 0/0, 3/3 = 16.

STATUS: This is a non-protected nongame animal under state regulations.[4]

NOMENCLATURE AND SYSTEMATICS: This species was first recognized in the mid-nineteenth century, and the currently used scientific name has been stable since around the turn of the twentieth century. Of the approximately seven subspecies recognized as of 1990, two are depicted as occurring in Wyoming: *M. o. similis* in the southeast corner of the state, and

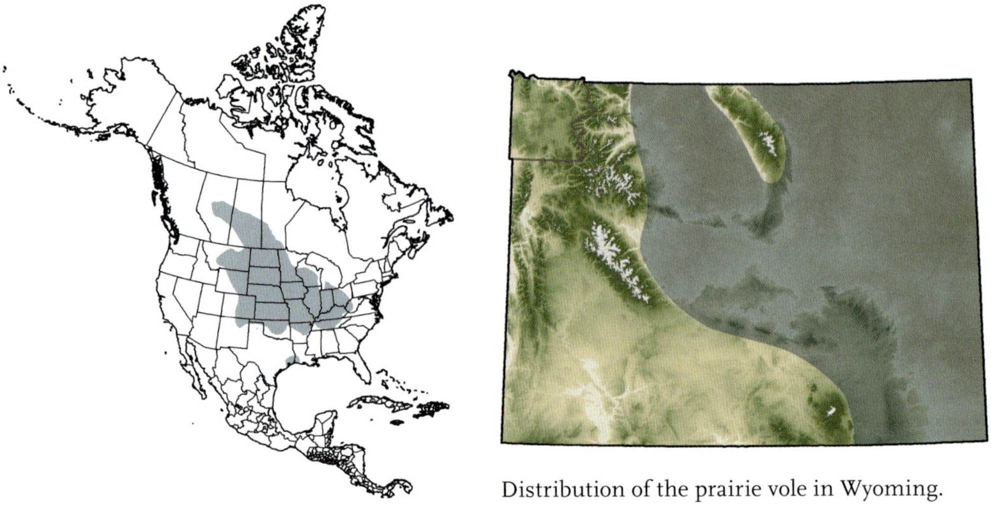

Distribution of the prairie vole in Wyoming.

M. o. haydeni in the central and northeastern parts. The specific epithet is the Latinized "ochre-bellied."[5]

DISTRIBUTION AND HABITAT ASSOCIATIONS: This Great Plains species occurs in the northeastern half of the state below the elevational limit of coniferous forest. It has undergone considerable range expansion into western Kansas and southeastern Colorado in recent decades, and undetected expansions in Wyoming may have occurred as well.[6] Habitats over most of the range have been reported as mesic grassland, but in southeastern Wyoming the species was associated with sagebrush-grass communities.[7]

ECOLOGY: The diet is mostly plant material: leaves, roots, and other plant parts. In South Dakota, the species was found to have a diverse diet, which at times included a substantial arthropod component.[8] Prairie voles have been intensively studied, including in captivity, because they exhibit much higher levels of sociality than other North American vole species. This sociality includes social and reproductive monogamy; the pair bonds are strong enough that when one member of a pair dies, the surviving one may remain unbonded for some time. Further, in winter multiple pairs of adults may nest together below ground, with a presumed thermal energetic benefit. Studies of the species in captivity have dealt with neural and hormonal correlates of monogamy.[9]

CONSERVATION AND MANAGEMENT: The species is not a conservation concern in Wyoming. Farther east, it is an economic pest in places because of damage to orchard trees.

NOTES AND REFERENCES

1. Mumford and Whitaker (1982), Stalling (1990).
2. Henterly et al. (2011).
3. KU specimens.
4. Commission Regulations Chapter 52, Section 6.
5. Stalling (1990).

6. Reed and Choate (1988), McCaffrey et al. (2003).
7. Maxwell and Brown (1968).
8. Agnew et al. (1988).
9. Getz et al. (1981), Lim and Young (2004).

Meadow vole, *Microtus pennsylvanicus*

DESCRIPTION: A medium-sized vole with grayish brown dorsal pelage. The underside is silver gray, and the tail bicolored. The species cannot be reliably distinguished from other voles, particularly the prairie vole, in the field.[1] Measurements (**Sex** [n] mean [mm, g], *SD*): various counties, L = **M** [5] 145, *17.9*; TL = **M** [5] 35, *5.4*; HF = **M** [5] 19, *1.6*; EFN = **M** [5] 10, *2.7*; WT = **M** [5] 29.4, *6.3*.[2] Dental formula: 1/1, 0/0, 0/0, 3/3 = 16.

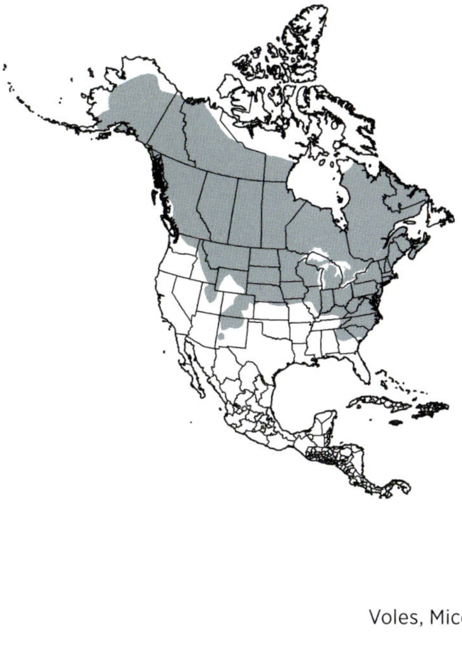

Distribution of the meadow vole in Wyoming.

STATUS: This is a non-protected nongame animal, by state regulation.[3] It is one of our most common voles.

NOMENCLATURE AND SYSTEMATICS: First recognized in 1815, this taxon is one of the most thoroughly studied voles, with respect to its systematics. As of the 1980s, 24–28 putative subspecies were recognized, around one-third of them on coastal islands; two occur in Wyoming: *M. p. pullatus* in the western part corner and *M. p. insperatus* in the northeastern one-fourth. The geographic ranges of two additional subspecies—*M. p. pennsylvanicus* and *M. p. uligocola*—abut the southeast corner of the state and may enter it.[1]

DISTRIBUTION AND HABITAT ASSOCIATIONS: This broadly distributed temperate and boreal species occupies the northwestern mountains except for the Wind River Mountains southeast of Union Pass, and the northeast corner of the state. The distributional limits in this latter area are unclear. Habitats are herbaceous meadows, riparian shrubs and woodlands, and forest edges. Where it is sympatric with other vole species, the meadow vole tends to occupy moister sites. Adolph Murie, working in Jackson Hole in 1933, was surprised to find the species living—and raising young—in old bulrushes a few inches above a marsh with water 1 m (3 ft) deep.[4]

ECOLOGY: This species is a strict herbivore, eating a range of grasses and forbs. It is believed to be competitively dominant to other voles and mice—southern red-backed voles, deer mice, and white-footed mice—in moist grassy habitats. Population biology, including cyclic population fluctuations, has been studied intensively in the Midwest and East. When abundant, meadow voles are capable of reducing standing crop in grasslands by 30–72%, comparable to the effects attributed to ungulates.[5] The species is eaten by a wide range of medium-sized mammalian and avian predators, and in Boulder County, Colorado, was found to be the most frequently eaten prey, among 26 mammalian prey species, by red-tailed hawks.[6]

CONSERVATION AND MANAGEMENT: The species is widespread and locally abundant; it is not generally considered a conservation concern.

NOTES AND REFERENCES

1. Reich (1981).
2. UWMV specimens.
3. Commission Regulations Chapter 52, Section 6.
4. Murie (1960).
5. Howe et al. (2006).
6. Blumstein (1989).

American water vole, *Microtus richardsoni*

DESCRIPTION: Except for the muskrat, this is the largest member of the subfamily that includes voles in the contiguous United States. The dorsal pelage is like that of other voles: grayish or reddish brown, but has longer fur than that of other voles.[1] Measurements (**Sex** [n] mean (mm, g), *range*): Fremont, Teton, and Uinta Counties, L = **F** [10] 222, *202–239*; **M** [8] 214, *185–262*; TL = **F** [10] 72, *66–86*; **M** [8] 65, *58–75*; HF = **F** [10] 26,

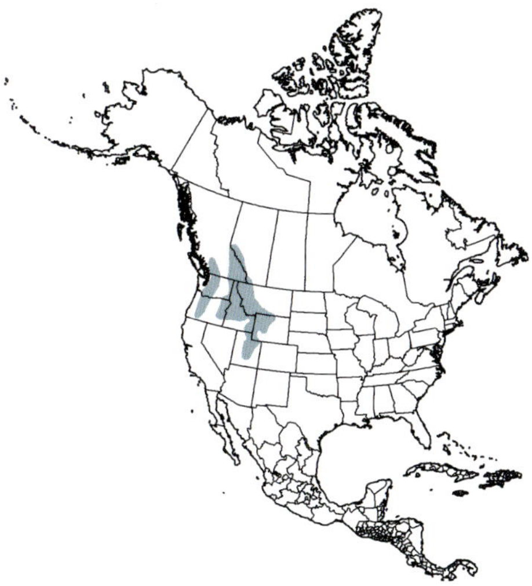

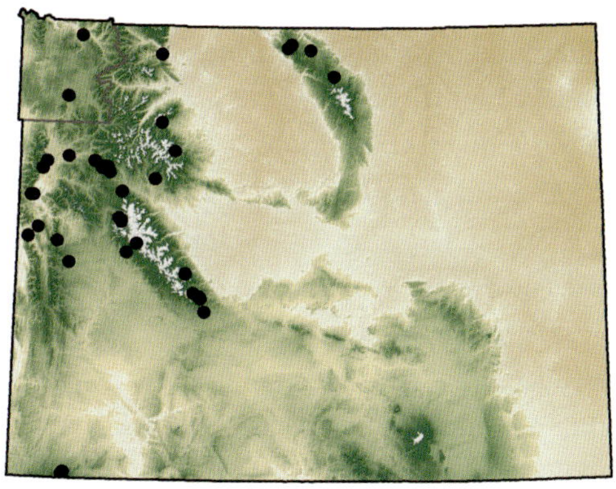

Locations from which the American water vole has been identified in Wyoming.

24–32; **M** [8] 25, *16.5–28*; EFN = **F** [10] 16, *14–17*; **M** [8] 15, *9.5–20*; WT = **F** [10] 83.2, *60.8–108.2*; **M** [8] 79.8, *56.2–112.0*.[2] Dental formula: 1/1, 0/0, 0/0, 3/3 = 16.

STATUS: A protected nongame animal and state Species of Greatest Conservation Need as of 2012.[3]

NOMENCLATURE AND SYSTEMATICS: The species was first recognized and the current species name applied in the mid-nineteenth century. The phylogenetic placement of the species has undergone revision, with the species earlier placed with Old World water voles in genus (or subgenus) *Arvicola*. Recent molecular studies have placed this species with North American *Microtus*, showing a sibling–species relationship either to *M. xanthognathus* or

M. pinetorum.[4] Of the four subspecies recognized in the 1980s, only *M. r. macropus* is shown as occurring in Wyoming.

DISTRIBUTION AND HABITAT ASSOCIATIONS: This species of the Rocky Mountains and Cascades occurs in subalpine and alpine areas of the Bighorn Mountains, the northwestern mountains, and the Uintas. Habitats in Wyoming are meadows or small herbaceous patches close to running water, near or above the upper limit of trees. Vegetation cover is typically >75%. These water bodies tend to be clear and spring- or snow-fed, with moderate slopes and gravelly bottoms. In the Bighorn Mountains, no water voles were caught below 2440 m (8000 ft) elevation, and abundance was greater on sites ungrazed by livestock than on livestock-grazed areas.[5]

ECOLOGY: This remarkable vole exhibits the unusual behavior of traveling by way of waterways—during daily movements as well as long-distance dispersal. It is also able to disperse up to several kilometers across land. It is rare and seldom observed outside a trap; in a six-year study in the Oregon Cascades, water voles made up 1% of small mammal captures.[6] Water voles have been variably common in the Bighorn and Beartooth Mountains.[7] Low abundances in the Bighorns in the 1990s was suggested to have been due to high livestock grazing there.[8] Working in the Beartooths, Marion Klaus noted greater abundance and wider local distribution during a particularly wet summer.[9] Population connectivity among sites within the Beartooth Mountains was high, but animals in Togwotee Pass were genetically distinct from those in the Beartooths.[10] The American water vole eats leaves and stems of forbs and grass-like plants, with willows, seeds, and insects appearing occasionally in the diet. Its predators are not well known, but likely include subalpine and alpine small carnivores and owls, weasels, and martens in particular.

CONSERVATION AND MANAGEMENT: This species is a conservation concern at the state level because of its restricted geographic range, specialized habitats, and possible sensitivity to grazing-induced changes to streamside vegetation. It was identified as a sensitive species for USDA Forest Service Region 2 in 1994. The isolated population on the Bighorn Mountains is of particular concern because of its isolation and restricted distribution.[11]

NOTES AND REFERENCES

1. Ludwig (1984).
2. KU specimens.
3. Commission Regulations Chapter 52, Section 11.
4. Wilson and Reeder (2005).
5. Ludwig (1984), Klaus (2003).
6. Hooven (1973).
7. Klaus and Beauvais (2004).
8. Klaus (2003).
9. Marion Klaus, Park City, Utah, pers. comm.
10. Klaus et al. (2001).
11. Klaus (2003), Klaus and Beauvais (2004).

Sagebrush vole, *Lemmiscus curtatus*

DESCRIPTION: This vole is small with a distinctly short tail, indistinctly bicolored, about the same length as the hindfoot. The dorsal pelage is gray or buff in color, dense, and soft; the underside is lighter.[1] Measurements (**Sex** [n] mean (mm, g), *range*): various counties, L = **M** [5] 108, *101–113*; TL = **M** [5] 17, *13–20*; HF = **M** [5] 15, *12–17*; EFN = **M** [5] 8, *7–9*.[2] Dental formula: 1/1, 0/0, 0/0, 3/3 = 16.

STATUS: This is a non-protected nongame animal by state regulation.[3]

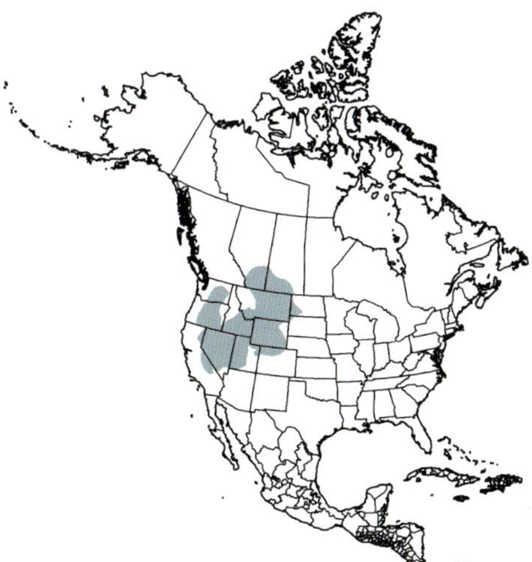

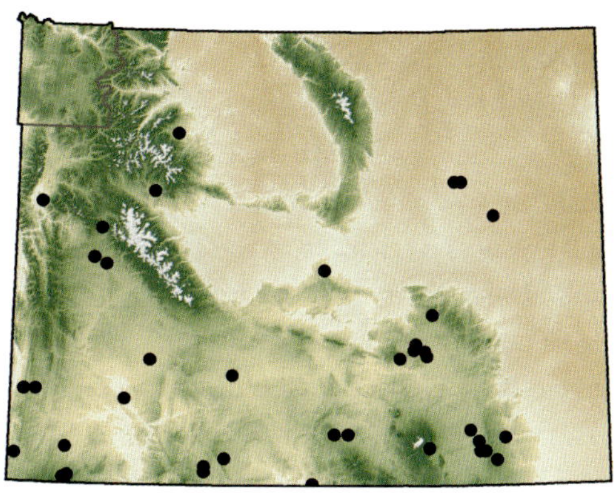

Locations from which the sagebrush vole has been identified in Wyoming.

NOMENCLATURE AND SYSTEMATICS: The species was first recognized in the mid-nineteenth century, and the species name has been stable since then. *Lemmiscus* was earlier considered a subgenus of the Holarctic *Lagurus*, but the sagebrush vole now is considered more closely related to the New World tribe Microtini, which includes *Microtus*. Of the six subspecies recognized as of the 1980s, only the nominal *L. c. levidensis* is shown as occurring in Wyoming.[4]

DISTRIBUTION AND HABITAT ASSOCIATIONS: This species' geographic range roughly corresponds with that of big sagebrush. It occurs at elevations below coniferous forest statewide, except along the eastern border of the state. Habitats are sagebrush- or rabbitbrush-dominated admixtures with bunchgrasses, particularly crested wheatgrass.[1]

ECOLOGY: This poorly understood species is almost entirely herbivorous and eats a wide range of herbaceous and shrubby plants: leaves, nodes, flowers, and young culms. The most common food item in southeastern Idaho in June was Indian paintbrush and in August was lupine.[5] In a study of effects of mechanical disturbance on small mammals in a Wyoming big sagebrush steppe in Albany County, this species was too uncommon to evaluate over the three-year study.[6]

CONSERVATION AND MANAGEMENT: The species is sometimes mentioned as a concern with regard to treatment or conversion of sagebrush habitats.

NOTES AND REFERENCES

1. Carroll and Genoways (1980).
2. UWMV specimens.
3. Commission Regulations Chapter 52, Section 6.
4. Carroll and Genoways (1980), Hall (1981).
5. Mullican and Keller (1986).
6. Borchgrevink et al. (2010).

Common muskrat, *Ondatra zibethicus*

DESCRIPTION: This is the largest-bodied of the microtine rodents, having proportionally short pinnae and a naked, scaled, vertically flattened tail. The hindfeet show slight webbing, and the toes are edged with stiff hairs, increasing their surface area. Dorsal pelage, generally observed wet in the field, is silvery tan to dark brown when dry. The underside fur is lighter, and the dorsal surfaces of the feet are nearly black. A range-wide analysis of body size concluded that areas with the most—and most seasonally consistent—precipitation produced the largest muskrats; Wyoming has small-bodied ones.[1] Distinctive field sign of muskrats includes houses made of emergent herbaceous vegetation cut and stacked, and "push-ups" visible on lake ice in winter and spring: piles of discarded vegetation that muskrats bring to the ice surface and eat. Measurements (**Sex** [n] mean [mm, g], *SD*): Goshen County, L = **F** [14] 499, *64.5*; **M** [25] 486, *72.3*; TL = **F** [14] 192, *28.6*; **M** [25] 187, *30.9*; HF = **F** [14] 72, *6.6*; **M** [25] 71, *6.0*; EFN = **F** [14] 19, *2.6*; **M** [25] 19, *1.6*; WT = **F** [14] 825, *274*; **M** [25] 824, *279*.[2] Dental formula: 1/1, 0/0, 0/0, 3/3 = 16.

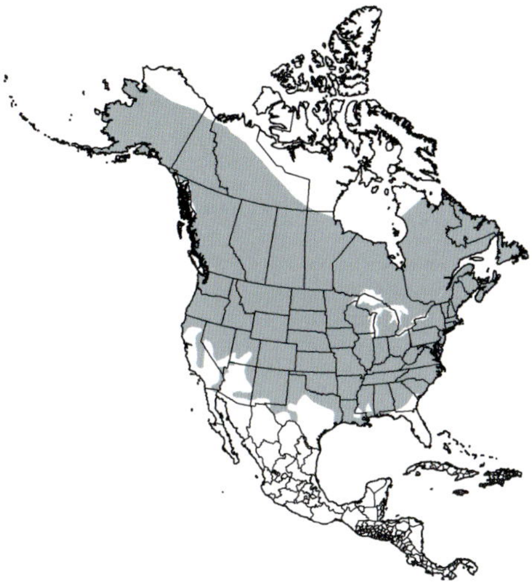

Distribution of the common muskrat in Wyoming.

STATUS: This is a furbearing animal by Wyoming Statute.[3]

NOMENCLATURE AND SYSTEMATICS: The species was first described from eastern Canada in the mid-eighteenth century. Of the approximately 16 subspecies recognized in the 1980s, two occur in Wyoming: *O. z. osoyoosensis* and *O. z. cinnamominus*, with the subspecies boundary running north–south roughly mid-state.[4] The concordance of the geographical subspecies with molecular evidence has recently been tested, and broad phylogeographic clustering was found, with populations from Newfoundland Island, other Canadian maritime provinces, and more western Canadian provinces showing genetic distinctness. So, molecular evidence supports an east–west cline in muskrat population genetic structure.[5]

DISTRIBUTION AND HABITAT ASSOCIATIONS: The species, widespread across temperate and boreal North America, has been introduced—or escaped from fur farms—into many sites within and beyond its range limits. It occurs statewide in suitable habitats; its distribution in the Red Desert is restricted to permanent water sources that are interconnected and do not freeze solid in winter. Habitats are bodies of fresh water, especially cattail and bulrush marshes, slow-moving streams, and shallow lakes seasonally.

ECOLOGY: Muskrats are semiaquatic omnivores that mostly eat submerged and emergent aquatic plants; invertebrates are eaten as well: crayfish, bivalve mollusks, and insect larvae. Muskrats are abundant in some marsh environments and can be important prey for American minks, Northern raccoons, and raptors.[6] Muskrats tolerate human development and agriculture well, and the species is common or abundant in human modified landscapes, including farm ponds and riverine areas within cities.[7]

CONSERVATION AND MANAGEMENT: The species is managed as a furbearer, with open seasons in Wyoming in the autumn and winter.[8] However, muskrat harvests by trappers have declined in eastern North America by 75% since 1986, likely reflecting lower prices and reduced interest in trapping.[9] The species is regarded as a pest in areas where its burrows damage earthen dams. Muskrats on the Ohio and Mississippi Rivers have been shown to selectively eat introduced zebra mussels over native bivalves, and may have importance in controlling the abundance of this introduced aquatic pest.[10]

NOTES AND REFERENCES

1. Boyce (1978).
2. UWMV specimens.
3. Wyoming Statutes 23-1-101-a-iii.
4. Willner et al. (1980).
5. Laurence et al. (2011).
6. Willner et al. (1980), Owen et al. (2011).
7. Cotner and Schooley (2011).
8. The open season during 2011–2012 was October 1 to April 30, with no limit on take (Wyoming Game and Fish Commission 2011).
9. Roberts and Crimmins (2010a).
10. Sietman et al. (2003).

JUMPING MICE AND JERBOAS—FAMILY DIPODIDAE

This Northern Hemisphere group has been shown, using molecular markers, to be monophyletic. Its organization below the level of family is uncertain, however, in spite of intensive study over decades. Jumping mice are tentatively in subfamily Zapodinae, comprising three genera, only one of which occurs in Wyoming.[1] *Zapus* mice are long-tailed, small-bodied mice that lack cheek pouches, use walking or hopping locomotion, and feed mostly on insects and seeds. Their pelage features a distinct yellowish-orange lateral band, lighter than the dorsum.

NOTES AND REFERENCES

1. Wilson and Reeder (2005).

Meadow jumping mouse, *Zapus hudsonius*

DESCRIPTION: A small mouse with a very long tail, and hind feet that are much longer than forefeet. Dorsal pelage is yellowish brown, with a dark middorsal stripe and paler sides. The underside is pale with yellowish tint. The upper incisors are strongly grooved on the anterior surfaces. Distinguishing between the two jumping mouse species in Wyoming requires repeated measurements of a cleaned skull by different people, followed by the use of statistical software. Alternatively, molecular markers can be used with tissue samples.[1] Measurements (**Sex** [n] mean (mm, g), *range*): Albany, Crook, Laramie, and Teton Counties, L = **F** [9] 224, *197–290*; **M** [5] 201, *187–223*; TL = **F** [9] 133, *119–146*; **M** [5] 125, *120–138*; HF = **F** [9] 30, *29–31*; **M** [5] 29, *26–32*; EFN = **F** [9] 13, *10–17*; **M** [5] 12.0, *8–15*; WT = **F** [9] 22.2, *14–31.2*; **M** [5] 16.2, *13.4–22.4*.[2] Dental formula: 1/1, 0/0, 1/0, 3/3 = 18.

STATUS: This is a protected nongame animal by state regulation.[3] The subspecies in Wyoming has a complex history of protection under E.S.A., having been originally listed as threatened in 1995, then threatened throughout the subspecies range in 1998. Critical habitat was designated in 2003, the subspecies was proposed for delisting in 2005, and the part of the subspecies range in Wyoming was delisted (but the protections in Colorado were retained) in 2007, but protection for the Wyoming portion of the geographic range was restored by court order requested by the US Fish and Wildlife Service in 2011.[4]

NOMENCLATURE AND SYSTEMATICS: The species was first recognized in the late eighteenth century based on specimens from Hudson Bay. Of the 11 or so subspecies recognized as of 1981, those in Wyoming—*Z. h. preblei* in particular—have been the most problematic with regard to phylogeny and distribution. Recent studies have confirmed that *Z. h. preblei* is a valid subspecies.[5] Most significantly, Malaney and Cook (2013) showed that *Z. h. preblei* represents a single lineage that extends from the Front Range area to northwestern Alaska, and reflects rapid Holocene expansion of the geographic range northward.[6]

DISTRIBUTION AND HABITAT ASSOCIATIONS: This is a boreomontane meadow species. The geographic range in Wyoming has become better resolved with the completion of targeted surveys and a recent review of all specimens and records.[7] The Front Range and eastern foothills, represented in Wyoming by the Laramie Mountains, is occupied by what is generally considered *Z. h. preblei*. The Black Hills and Bear Lodge Mountains were earlier believed to hold *Z. h. campestris*, but that subspecies may now be subsumed by *Z. h. preblei*. A newly located distributional patch located along the east face of the Bighorn Mountains may be contiguous with those in the Bear Lodge Mountains by way of populations extending across southern Montana.[8] Habitats of *Z. h. preblei* in Wyoming and Colorado are riparian meadows and wetlands with abundant grasses, sedges and rushes, willows, narrowleaf cottonwood, and thinleaf alder.[9]

ECOLOGY: This is a hibernating omnivore, inactive during the cold months. The diet in the eastern United States includes insects, seeds, fungi, and other plant material. Limited information on seasonal body condition suggests that animals increase body weight markedly just before entering hibernation. The overwinter mortality rate is suspected to be high.[10]

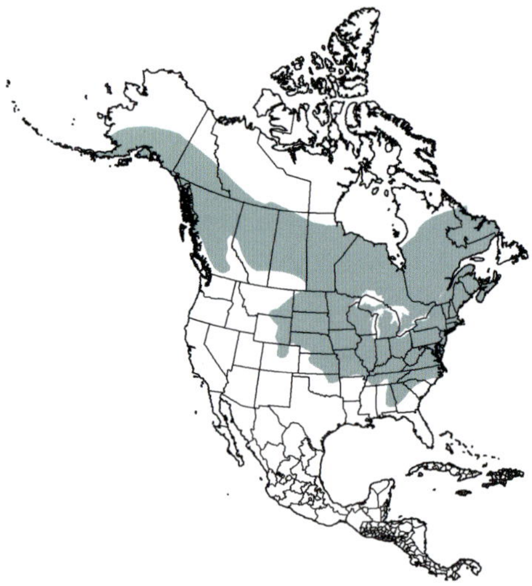

Distribution of the meadow jumping mouse in Wyoming.

CONSERVATION AND MANAGEMENT: As an E.S.A.–protected taxon, the Preble's subspecies of the meadow jumping mouse has received intense attention from scientists, government agencies, and the media over the last two decades. Critical habitat protections have constrained habitat disturbance and development in riparian zones along the Front Range, and these constraints have led to tensions between landowners, public lands users, and regulators. With the return of federal protections for the subspecies in Wyoming, these issues will continue to be raised, but the recent finding that the lineage extends all the way to Alaska, rather than being geographically restricted to Wyoming and neighboring states, reduces the apparent rarity and endemicity of the taxon.[6]

NOTES AND REFERENCES

1. Conner and Shenk (2003), King, Eackles et al. (2006).
2. UWMV, DMNS specimens.
3. Commission Regulations Chapter 52, Section 11.
4. US Fish and Wildlife Service (2008).
5. See reviews of the controversy in King, Switzer et al. (2006).
6. Malaney and Cook (2013).
7. Conducted by WYNDD (Bowe and Beauvais 2012).
8. Foresman (2012).
9. Trainor et al. (2007).
10. Whitaker (1972).

Western jumping mouse, *Zapus princeps*

DESCRIPTION: This mouse is similar in appearance to the meadow jumping mouse, although slightly smaller in size. Standard field measurements overlap strongly between species. Methods of distinguishing this species from the meadow jumping mouse from skull morphology were described for Colorado.[1] Measurements (**Sex** [n] mean (mm, g), *range*): June, Teton County; L = **F** [29] 230, *226–239*; **M** [25] 225, *216–238*; TL = **F** [29] 134, *130–146*; **M** [25] 130, *122–147*; HF = **F** [29] 26.2, *25–27*; **M** [25] 26.2, *25–27*; EFN = **F** [29] 31.1, *30–33*; **M** [25] 30.9, *29–32*; WT = **F** [29] 25.8, *18–30*; **M** [25] 23.3, *19–37*.[2] Dental formula: 1/1, 0/0, 1/0, 3/3 = 18.

STATUS: This is a protected nongame animal by regulation.[3]

NOMENCLATURE AND SYSTEMATICS: The species was first described in the 1890s, using the currently applied scientific name. Recent molecular studies have shown that two Sierra Nevadan lineages are quite divergent from other western jumping mice and may warrant elevation to species level.[4] Of the 11 or so subspecies recognized as of 2004, three occur in Wyoming: *Z. p. idahoensis* and *Z. p. utahensis* in the western half of the state and *Z. p. princeps* in the eastern half.

DISTRIBUTION AND HABITAT ASSOCIATIONS: The species occurs across the western states to as far north as Yukon Territory. Current understandings specific to Wyoming are based on a review conducted by Bowe and Beauvais (2012).[5] Specimens have been collected from the northern and southern parts of the Laramie Mountains, and here I depict those two populations as contiguous. *Zapus* of unknown species have been captured—but not collected—from the Shirley and Ferris Mountains; those animals likely were western jumping mice. Where it is parapatric with meadow jumping mice in southeastern Wyoming, this species seems to occur at higher elevations. In the Medicine Bow Mountains, the western jumping mouse formerly was common or abundant in willow–alder, quaking aspen forests, and subalpine meadows. It was completely absent from sagebrush, mountain mahogany, and alpine tundra communities and was rare in lodgepole pine and Engelmann spruce–subalpine fir forests. It was closely associated with water.[6] A comparable study in an area of Teton County lacking coniferous forest found a close association with quaking aspens.[7]

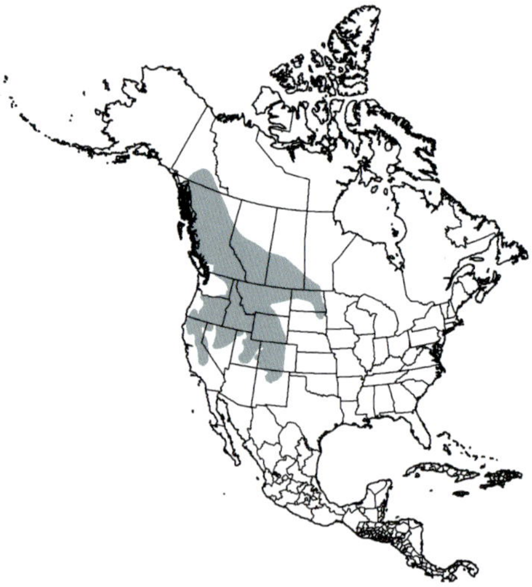

Distribution of the western jumping mouse in Wyoming.

ECOLOGY: This species is a hibernating omnivore, in the Medicine Bow Mountains spending 10½–11½ months in torpor, depending on elevation.[8] Diets are geographically variable, but dominated by seeds, invertebrates, fungi, fruits, and some green vegetation. Mice from Grand Teton National Park lacked a strong invertebrate component to the diet.[7] Population densities in the Laramie Mountains during 1965–1969 were relatively constant: 2.3–4.5/ha (590–1170/mi^2). Around one-half of juvenile mice were estimated to have perished over their first winter; the corresponding value for adults was 17%.[2]

CONSERVATION AND MANAGEMENT: The species is common or abundant in suitable habitats and not generally regarded as a conservation concern.

NOTES AND REFERENCES

1. Conner and Shenk (2003), Hart et al. (2004).
2. Brown (1970).
3. Commission Regulations Chapter 52, Section 11.
4. Malaney et al. (2013).
5. Bowe and Beauvais (2012).
6. Brown (1967b).
7. Clark (1971b).
8. Brown (1967c).

BEAVERS—FAMILY CASTORIDAE

This family—most closely related to the lineage that holds pocket mice and pocket gophers—includes a single genus and two species, both in the Northern Hemisphere, one each in the New and Old Worlds. The two beaver species differ in chromosome number and cranial features.

American beaver, *Castor canadensis*

DESCRIPTION: A large rodent, medium brown in color, with lighter pelage on the head and a horizontally flattened, scaly tail. The feet are prominently webbed. No other mammals slap their tails on the water surface in alarm before diving. In spite of their distinctiveness, beavers are commonly confused with other partially submerged semiaquatic species, particularly the muskrat, northern river otter, and American mink. The field sign of beavers is distinctive and abundant: cut stems of riparian trees and shrubs, maintenance activities on dams, lodges and caches, and muddy trails near ponds. Some of these signs of activity persist for years after beavers have left an area, and cut stems can float downstream for miles. As a result, it can be difficult to determine whether beavers are active at some sites. Measurements (**Sex** [n] mean (mm, kg), *SD*): Albany and Laramie Counties, >2 years, L = **B** [30] 1076, *45*; TL = **B** [30] 281, *15*; HF = **B** [26] 177, *6.9*; WT = **B** [31] 16.6, *2.5*.[1] Dental formula: 1/1, 0/0, 1/1, 3/3 = 20.

STATUS: A furbearing animal by regulation.[2] It is trapped for its fur outside of protected areas; special regulations give landowners the right to control beavers causing damage to crops, trees, or property.

NOMENCLATURE AND SYSTEMATICS: Around 24 subspecies of *C. canadensis* were recognized as of 1981; two putatively occur in Wyoming.[3] The geographic rationales for these subspecific designations are difficult to appreciate; in the Rocky Mountains some named subspecies putatively occur in headwater streams, while different ones occupy downstream stretches a few miles away. A transect drawn from northern Nevada east to southeastern Wyoming crosses the geographic ranges of no fewer than seven named subspecies, which is difficult to reconcile with geography or hydrography. Also worth noting is the reintroduction

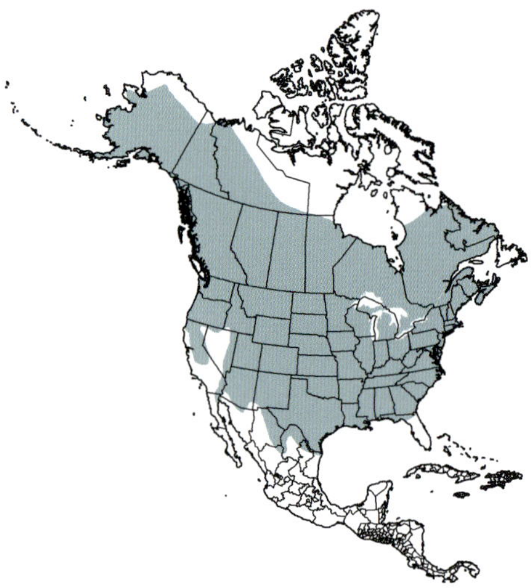

Distribution of the American beaver in Wyoming. The species occurs discontinuously in arid parts of the state.

of beavers to much of their former temperate North American range, including Wyoming, in the early to mid-twentieth century.[4] This must have had strong homogenizing effects on genetic structure that are not reflected in the traditional taxonomy below the species level. As a result, some recent taxonomic reviews recognize no subspecies.[5]

DISTRIBUTION AND HABITAT ASSOCIATIONS: In presettlement times, the species occurred across North America from northern Mexico to the Arctic Ocean (shown), except for very arid areas of the Southwest and the Florida peninsula (shown). Trapping and land conversion drastically reduced densities across the temperate United States—including Wyoming—but it is unclear whether populations were biologically extinct or merely restricted to the most remote areas. Reintroductions have restored the species in many states, but areas of intensive agricultural or water management continue to exclude them.[6] The distribution in Wyoming includes some small streams, as well as the largest Wyoming rivers.

ECOLOGY: Few North American rodents are understood in greater ecological depth than the beaver, which has been studied extensively in relation to population, behavioral, and community ecology. Beavers are well known as occupying headwater streams, where they construct dams, raise water levels, build caches of woody stems near lodges, and access these caches beneath the ice during winter. However, they also overwinter in rivers that are too large to be impounded and never freeze. In such sites, they build tunnels with entrances near or below water line. At the end of tunnels that can be over 10 m long, a chamber provides subterranean space for adults, yearlings, and juveniles. The beaver has a life history consistent with its large body size: sexual maturity occurs typically at two years, litters are small for a rodent, young typically remain with the monogamous parents for two years, and longevity is high in the absence of human-caused mortality. Beaver colonies typically are made up of extended family groups numbering 4–8, and population densities are variously reported as number of colonies per linear distance of stream or numbers of beavers per unit of pond area. The diet is seasonally variable, dominated by herbaceous vegetation during summer and the cambium of woody plants in winter. Among tree species in Wyoming, members of the genus *Populus*—aspens and cottonwoods—are preferred, but willow is eaten in large quantities as well.[7] The effects of beavers on communities are arguably the farthest reaching of any single North American mammal, particularly when evaluated on the basis of the numerical abundance or biomass of beavers. These effects include raising water tables near impoundments, which in turn affects growth of riparian woody plants. Meadows are increased in area and nutrient cycling strongly influenced.[8] Water quality is strongly affected by impoundment, and settled particulate material produces sediments, adding to floodplain soils.[9] In addition, dams block migrations of fish, limiting spawning movements and colonization. In some cases these barriers protect native species in headwaters from introduced species farther downstream.[10] Beavers can affect woody vegetation—particularly quaking aspen—profoundly and lead to elimination of large aspen boles and beaver abandonment of river reaches for decades at a time. This is less true of willow-dominated riparian habitats, which can support beavers more sustainably. The interactive effects among wolves, elk, and beavers have been debated and investigated in the case of Yellowstone's Northern Range. In this region, beavers were common in the 1920s, but became rare or extinct in the decades following the extinction of the wolf and the large increase in numbers of wintering elk. Following the reintroduction of wolves in 1995, beavers began to appear on the Northern Range, aided by translocations to adjacent areas in the Gallatin National Forest.[11] The return of beavers to the northern range, combined with declining elk densities, has led to some recovery of willows, which were rare for decades. Current understandings are that the reduction of browsing pressure by elk, combined with limited colonization by beavers, can lead to raised water levels and favorable conditions for willow recovery.[12] So, the interactions among members are somewhat conditional: although wolves eat beavers and beavers eat willows, the reintroduction of wolves has led only indirectly and partially to increased willows and beavers on the Northern Range.

CONSERVATION AND MANAGEMENT: The beaver may compare with the American bison in the degree of its depopulation in the temperate United States during the nineteenth century. However, beavers have made strong recoveries in many areas, in places aided by reintroductions and low trapping activities in recent decades. Beaver reintroductions continued in some areas of Wyoming into the 1980s, and beavers continue to be used to restore degraded watersheds. This was arguably the most important economic asset of the area that later became Wyoming in the early nineteenth century, but the value of beaver fur used in manufacturing hats collapsed by 1840. The direct economic value of the species today is shaped by modest fur values and local nuisance costs to ranchers and land agencies. Beavers commonly flood hay meadows and block road culverts, and they are considered nuisance animals in those settings. The ecological services are harder to evaluate economically, but are increasingly recognized as important.

NOTES AND REFERENCES

1. Osborn (1953). Weights given are for skinned carcasses; live animals were 18% heavier, on average.
2. Wyoming Statutes 23-1-101-a-iii.
3. Hall (1981).
4. Osborn (1953), Hill (1982).
5. Wilson and Reeder (2005).
6. Baker and Hill (2003).
7. Müller-Schwarze (2011).
8. Naiman et al. (1994).
9. Maret et al. (1987).
10. Collen and Gibson (2001).
11. Tyers (2008).
12. Warren (1926), Wolf et al. (2007), Bilyeu et al. (2008).

POCKET MICE AND KANGAROO RATS—FAMILY HETEROMYIDAE

This family comprises six genera of kangaroo rats and mice, pocket mice, and spiny mice, with pocket gophers the closest living relatives. It is a New World lineage with most of its representatives in the arid West.[1] The temperate members of the group are characterized by the presence of external fur-lined cheek pockets that can be everted and their contents emptied, and enlarged hindfeet. The group occupies arid shrub-steppe to desert habitats, and the diet is mostly seeds, other plant parts, and invertebrates.

NOTES AND REFERENCES

1. Alexander and Riddle (2005), Hafner et al. (2007).

KEY TO GENERA AND SPECIES OF HETEROMYIDAE

1a. Hindfeet greatly enlarged, length greater than 35 mm (1.4 in), face with distinct markings: . . . . *Dipodomys ordii*
1b. Hindfeet smaller, length less than 30 mm (1.2 in), face unmarked: 2

2a. Pelage coarse; hindfoot length typically greater than 25 mm (1 in), plantar feet unfurred:. . . . . *Chaetodipus hispidus*
2b. Pelage silky or smooth; hindfoot length typically less than 25 mm (1 in), plantar feet sparsely haired:. . . . . *Perognathus*
(identification to species is a task for a specialist)

Ord's kangaroo rat, *Dipodomys ordii*

DESCRIPTION: This is a medium-sized kangaroo rat with enlarged hind legs and feet, which moves by mostly quadrupedal running. The dorsal pelage of animals from Wyoming is orange-tan, with white dorsal feet, facial spots, and undersides. Each hindquarter features a white line extending from the anterior mid-leg to the base of the tail. The tail is strongly bicolored with a dark middorsal line and a slight fur tuft on the end. The hindfeet each have five toes versus four for all of the other kangaroo rats sympatric with Ord's. The auditory bullae are greatly inflated.[1] Measurements (**Sex** [n] mean (mm, g), *range*): Sweetwater, Goshen, Campbell, and Albany Counties, L = **F** [5] 258, *242–266*; **M** [10] 251, *237–280*; TL = **F** [5] 146, *131–152*; **M** [10] 139, *103–155*; HF = **F** [5] 41, *40–42*; **M** [10] 41, *40–43*; EFN = **F** [5] 13, *13–14*; **M** [10] 13, *11–19*; WT = **F** [5] 57.0, *40–68*; **M** [10] 62.0, *38–82.5*.[2] Dental formula: 1/1, 0/0, 1/1, 3/3 = 20.

STATUS: This is a non-protected nongame animal by regulation.[3]

NOMENCLATURE AND SYSTEMATICS: The species was recognized in the mid-eighteenth century, and the scientific name originally assigned is still used. Dozens of subspecies have been named over the large geographic range, of which three putatively occur in Wyoming: *D. o. terrosus* in the Bighorn Basin and northeastern corner, *D. o. luteolus* in central and eastern Wyoming, and *D. o. priscus* in the Green River Basin and Red Desert.[1] George Ord was the president of the Academy of Natural Sciences of Philadelphia at the time the species was described.

DISTRIBUTION AND HABITAT ASSOCIATIONS: This is a species of intermountain basins and the western Great Plains. The Wyoming distribution includes almost all of the eastern three-fourths of the state below the lower elevational limit of conifers. Habitats are highly varied over the extensive geographic range, including pinyon–juniper, big sagebrush, saltbush, and greasewood types. The species is regarded as especially common on sand dunes of southern Wyoming, and in yucca–sagebrush–shortgrass admixtures. One common trait of sites occupied is sandy, fine-textured soils.[4]

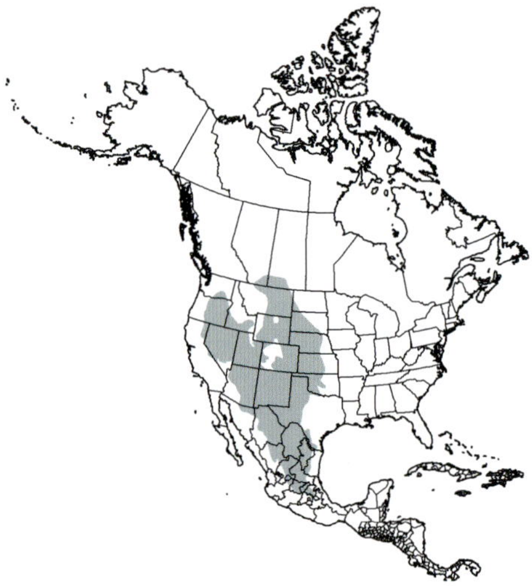

Distribution of Ord's kangaroo rat in Wyoming.

ECOLOGY: As for pocket mice, the diet is mostly known from examination of external cheek pockets, which in Colorado contained 74% seeds, 13% forbs, 5% grasses, and a few insect parts. Grama is often mentioned as a commonly eaten grass seed, and cryptantha, rice grass, and Russian thistle have been shown to be preferred over other seed species.[5] In the Sand Hills of Nebraska, the species was found to scatter-hoard in summer, but to larder-hoard in winter, when animals occupied a single burrow containing its larders of food, mostly seeds.[6]

CONSERVATION AND MANAGEMENT: The species is common in Wyoming and widely distributed across arid parts of the western United States. It is not generally considered to be a conservation concern.

NOTES AND REFERENCES

1. Hoffmeister (1986), Garrison and Best (1990).
2. UWMV, DMNS specimens.
3. Commission Regulations Chapter 52, Section 6.
4. Maxwell and Brown (1968), Garrison and Best (1990).
5. Flake (1973), Henderson (1990).
6. White and Geluso (2012).

Olive-backed pocket mouse, *Perognathus fasciatus*

DESCRIPTION: A medium-sized mouse, which in southwestern Wyoming has dorsal pelage ranging from dark with olive undertones to buff. The underside is pale or white. The tail—nearly as long as the head and body—is moderately bicolored, dark on the dorsum. The anterior surface of each upper incisor is grooved.[1] Measurements (**Sex** [n] mean (mm, g), *range*): Sweetwater County, L = **F** [9] 126, *115–142*; **M** [9] 131, *121–137*; TL = **F** [9] 59, *53–72*; **M** [9] 62, *58–70*; HF = **F** [9] 18, *17–20*; **M** [9] 18, *16–20*; EFN = **F** [9] 7, *5–8*; **M** [9] 7, *5.5–8*; WT = **F** [9] 10.2, *7.7–12.5*; **M** [9] 10.8, *10.1–11.8*.[2] Dental formula: 1/1, 0/0, 1/1, 3/3 = 20.

STATUS: A non-protected nongame animal, but a state Species of Greatest Conservation Need as of 2012.[3]

NOMENCLATURE AND SYSTEMATICS: The species was described in the early nineteenth century, with the current scientific name applied then. Of the two subspecies recognized as of 1988, both occur in Wyoming: *P. f. fasciatus* over the eastern two-thirds of the state and *P. f.*

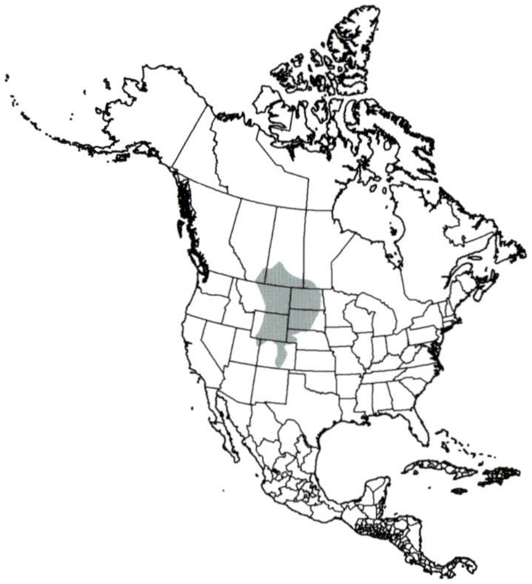

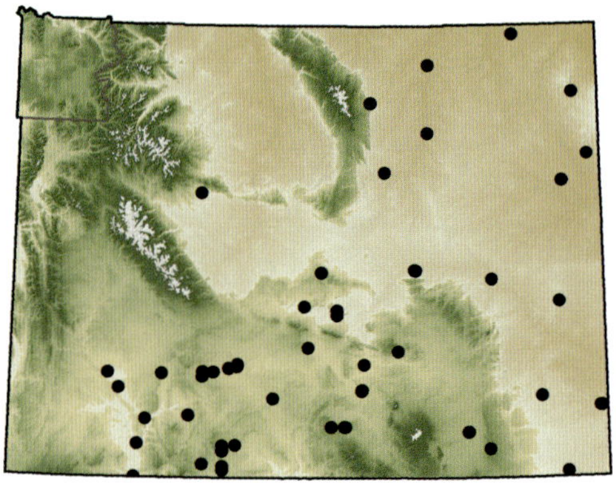

Locations from which the olive-backed pocket mouse has been identified in Wyoming.

callistus in a small area centered where Utah, Colorado, and Wyoming meet.[1] A more complex subspecific taxonomy from 1981 placed all five subspecies within or near Wyoming's borders.[4] The species name is the Latinization of "banded," referring to the lateral line along the flank.

DISTRIBUTION AND HABITAT ASSOCIATIONS: This northern Great Plains species occurs patchily over the eastern two-thirds of the state. The number of locations is not sufficient to depict a geographic range. Habitats are not well understood, but tend to be thinly vegetated dry grasslands on sandy or clayey soils.[1] The species was too rare in a study of rodent habitats in the Laramie Mountains to draw inferences about habitat association—yielding less than 0.9 animals per 1000 trap nights.[5]

ECOLOGY: The species is too poorly known to characterize ecologically, but is suspected of being a seed predator of various grasses, forbs, and weedy shrubs. These animals dig inconspicuous burrows at the bases of shrubs and yuccas and likely are preyed upon by the smallest nocturnal predators: small foxes, weasels, and small owls.

CONSERVATION AND MANAGEMENT: The conservation concern regarding this species relates to its apparent rarity and our poor understanding of its distribution and limiting factors.

NOTES AND REFERENCES

1. Manning and Jones (1988).
2. KU specimens.
3. Commission Regulations Chapter 52, Section 6.
4. Hall (1981), the taxonomy followed by Armstrong et al. (2011).
5. Maxwell and Brown (1968).

Plains pocket mouse, *Perognathus flavescens*

DESCRIPTION: A small mouse with a tail, not strongly bicolored, around the same length as its head and body. The dorsal pelage is yellowish buff, with many dark hairs intermixed. The pelage just posterior to the pinnae (postauricular patches) is not noticeably light-colored as it is in other members of the genus. The anterior surface of each upper incisor is grooved.[1] Measurements (**Sex** [n] mean (mm, g), *range*): Nebraska, L = **F** [20] 119, *98–134*; **M** [33] 118, *97–136*; TL = **F** [20] 57, *46–67*; **M** [33] 57, *35–89*; HF = **F** [20] 17, *16–19*; **M** [33] 16, *12–18*; EFN = **F** [19] 6, *4–8*; **M** [31] 6, *4–8*; WT = **F** [11] 8.8, *6.0–11.3*; **M** [9] 10.8, *10.1–11.8*.[2] Dental formula: 1/1, 0/0, 1/1, 3/3 = 20.

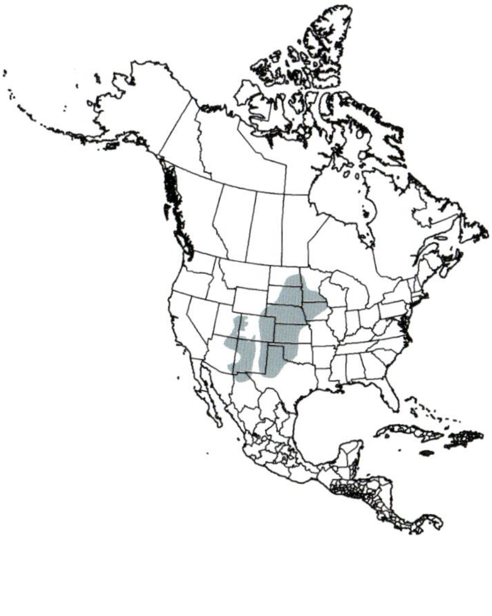

Locations from which the plains pocket mouse has been identified in Wyoming.

STATUS: A protected nongame animal and state Species of Greatest Conservation Need as of 2012.[3]

NOMENCLATURE AND SYSTEMATICS: The species has been recognized since around the turn of the twentieth century when it was raised from a subspecies of *P. fasciatus* to full species status.[4] Of the eight subspecies recognized as of 1996, only *P. f. flavescens* occurs in Wyoming, although a distributional peninsula extends from eastward into Moffat County, Colorado, and reaches less than 90 km (50 mi) south of the Wyoming state line. This part of the geographic range is variously described as *P. f. caryi, P. apache,* or *P. f. flavescens*. The species name is the Latinized "yellowish."[1]

DISTRIBUTION AND HABITAT ASSOCIATIONS: The Great Plains distribution extends into southeastern Wyoming, as far northwest as Johnson County. Habitats are sites with loose, sandy soils and dunes. In Wyoming, the species was shown to occupy sites with various admixtures of grass, sagebrush, and yucca in variable soil types.[5]

ECOLOGY: The diet is generally thought to be seeds and fruits, as inferred from contents of cheek pouches and contents of underground food caches. The plant species represented are diverse and include grasses, forbs, and some fruit-bearing shrubs. The species is absent above ground for extended periods in winter and has been hypothesized to become torporous. Predators are assumed to be a wide range of small carnivores, small raptors, and snakes.[1]

CONSERVATION AND MANAGEMENT: The species is a conservation concern in Wyoming because of its incompletely documented distribution and unknown population status in the state.

NOTES AND REFERENCES

1. Monk and Jones (1996).
2. UNSM specimens.
3. Commission Regulations Chapter 52, Section 11.
4. Osgood (1900) raised this taxon from a subspecies of *P. fasciatus* to the level of species. He also recognized three major groupings of pocket mice, which were subsequently confirmed by chromosomal attributes, and again more recently by molecular methods. The taxonomic status of animals in northwestern Colorado can be contrasted between Hall (1981), Hoffmeister (1986), and Neiswenter and Riddle (2011). The last of these sources shows the complex pattern of divergence in this subgenus and region.
5. Monk and Jones (1996), Maxwell and Brown (1968).

Silky pocket mouse, *Perognathus flavus*

DESCRIPTION: A pocket mouse—small in size for the genus—with darker dorsal pelage and a more prominent postauricular patch than other Wyoming pocket mice. The underside is white or very light, and the anterior surface of each upper incisor is grooved. The sexes are similar in size.[1] Measurements (**Sex** [n] mean (mm, g), *range*): Nebraska, L = **F** [6] 119, *113–125*; **M** [17] 120, *86–160*; TL = **F** [6] 56, *52–60*; **M** [17] 54, *40–64*; HF = **F** [6] 16,

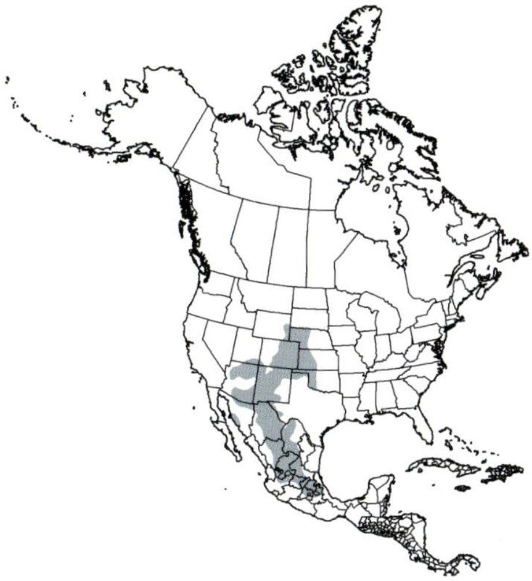

Locations from which the silky pocket mouse has been identified in Wyoming.

15–19; **M** [17] 16, *15–19*; EFN = **F** [6] 6, *5–6*; **M** [17] 6, *5–8*; WT = **F** [6] 9.6, *8.3–11.6*; **M** [15] 9.8, *5.3–13*.[2] Dental formula: 1/1, 0/0, 1/1, 3/3 = 20.

STATUS: A protected nongame animal and a state Species of Greatest Conservation Need as of 2012.[3]

NOMENCLATURE AND SYSTEMATICS: The species was recognized in the mid-nineteenth century, and species-level taxonomy has been stable since then. Of the 14 subspecies recognized as of 1994, only *P. f. piperi* appears to occur in Wyoming, although the putative boundary between *P. f. piperi* and *P. f. bunkeri* is near the southeast corner of the state.[1]

DISTRIBUTION AND HABITAT ASSOCIATIONS: This species of the southern Great Plains and Sonoran and Chihuahuan Deserts occurs in southeastern Wyoming. Habitats in

Wyoming and Colorado include various vegetation associations and soil types, tending towards semiarid grasslands and grass–shrub–yucca admixtures with sandy or sandy-loamy soils. In northeastern Colorado, the species was most common and abundant on sites lacking black-tailed prairie dog activity.[4]

ECOLOGY: The diet of this little-studied species is dominated by seeds, with some insects and scant green plant material. Burrows are constructed at the bases of cacti or shrubs where pocket mice spend days in burrows that they plug, and where they spend extended periods in winter, eating stored seeds. They are suspected of going into at least short-term torpor while in their burrows during winter.[5]

CONSERVATION AND MANAGEMENT: As for other pocket mice, conservation concern is generated by the poor knowledge base regarding distribution, population status, and life needs.

NOTES AND REFERENCES

1. Best and Skupski (1994).
2. UNSM specimens.
3. Commission Regulations Chapter 52, Section 11.
4. Maxwell and Brown (1968), Stapp (2007), Armstrong et al. (2011).
5. Forbes (1964), Maxwell and Brown (1968), Armstrong et al. (2011).

Great Basin pocket mouse, *Perognathus mollipilosus*

DESCRIPTION: The largest-bodied *Perognathus*, this newly recognized species has pelage that is geographically variable. The tail is proportionally longer than in other pocket mice, but this trait likewise is not distinctive.[1] The species is discerned from *P. parvus*, from which it was split in 2014, by nuclear and mitochondrial DNA sequences.[2] Measurements (**Sex** [n] mean (mm, g), *range*): Sweetwater and Lincoln Counties, Wyoming, L = **F** [7] 152, *105–170*; **M** [7] 167, *145–182*; TL = **F** [7] 79, *64–88*; **M** [7] 86, *62–96*; HF = **F** [7] 23, *22–25*; **M** [7] 23, *22–24.5*; EFN = **F** [7] 7, *5–9*; **M** [7] 8, *7–9*; WT = **F** [7] 14.4, *9.5–18.6*; **M** [7] 17.9, *13.6–21.3*.[3] Dental formula: 1/1, 0/0, 1/1, 3/3 = 20.

STATUS: A protected nongame animal and a state Species of Greatest Conservation Need as of 2012.[4]

NOMENCLATURE AND SYSTEMATICS: The species was first described as *P. parvus* in the mid-nineteenth century, and the scientific name was stable from then until 2014.[5] The new species includes populations from the southwestern Wyoming part of the geographic range, although the distribution in southern Idaho is not yet fully mapped. It is morphologically cryptic relative to *P. parvus*.[2]

DISTRIBUTION AND HABITAT ASSOCIATIONS: The Wyoming distribution of this Great Basin species is limited to the southwest corner of the state, including along the Green River as far north as Fontenelle Dam.[6] Habitats are broadly characterized as arid and semiarid grasslands and shrub-steppe, which may be dominated by sagebrush, greasewood, or annual grasses. Preferred soil types are variously described as sandy or as having low amounts of clay.[7]

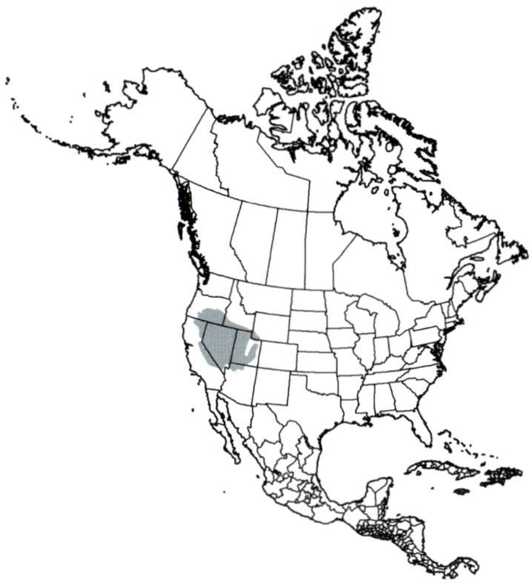

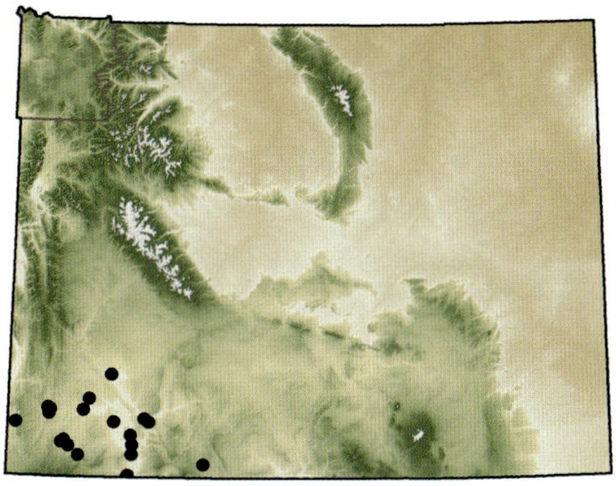

Locations from which the Great Basin pocket mouse has been identified in Wyoming.

ECOLOGY: As for other pocket mice, the diet is dominated by seeds, especially of annual plants, but insects and green material are eaten as well. Burrows are dug at the bases of shrubs, and animals limit aboveground activities to nighttime. Activity above ground is restricted during the cold season; the occurrence of torpor is uncertain.[1] The species is noted for co-occurring with the deer mouse, and for being an important agent of seed mortality and of seed dispersal (via scatter hoarding). This and presumably other species of pocket mice store seeds of singleleaf pinyon in burrows too deep to allow the seeds to germinate and grow to seedlings. But they also scatter-hoard many seeds near the surface.[8]

CONSERVATION AND MANAGEMENT: As for other pocket mice, the species' distribution, status, and life needs are insufficiently known for Wyoming to understand its conservation status.

NOTES AND REFERENCES

1. Verts and Kirkland (1988).
2. Riddle et al. (2014).
3. DMNS, KU specimens.
4. Commission Regulations Chapter 52, Section 11.
5. Cassin (1858), cited in Hall (1981).
6. The continental distribution map is inferred from Riddle et al. (2014).
7. Feldhamer (1979).
8. Vander Wall (1997), Hollander and Vander Wall (2004), Beck and Vander Wall (2010).

Hispid pocket mouse, *Chaetodipus hispidus*

DESCRIPTION: This largest-bodied Wyoming pocket mouse overlaps in length with *P. mollipilosus*. The dorsal pelage is a mixture of orange with black hairs and feels coarse to the touch, the flanks are orange-tinted, and the underside nearly white. The tail is strongly bicolored, with a black middorsal stripe and white underside.[1] Measurements (**Sex** [n] mean (mm, g), *range*): Laramie County, L = **M** [9] 188, *153–220*; TL = **M** [9] 90, *68–105*; HF = **M** [9] 25, *24–27*; EFN = **M** [9] 12, *11–15*; WT = **M** [9] 32.9, *19.8–59.4*.[2] Dental formula: 1/1, 0/0, 1/1, 3/3 = 20.

STATUS: A protected nongame animal and a state Species of Greatest Conservation Need as of 2012.[3]

NOMENCLATURE AND SYSTEMATICS: The species was recognized in the mid-eighteenth century, placed into genus *Perognathus* at the end of that century, and then reclassified to the new genus *Chaetodipus* (previously a subgenus) in 1983. Most subsequent treatments have recognized this classification.[4] Of the four subspecies recognized as of 1988, only *C. h. paradoxus* occurs in Wyoming; a more recent mitochondrial DNA-based analysis has generated

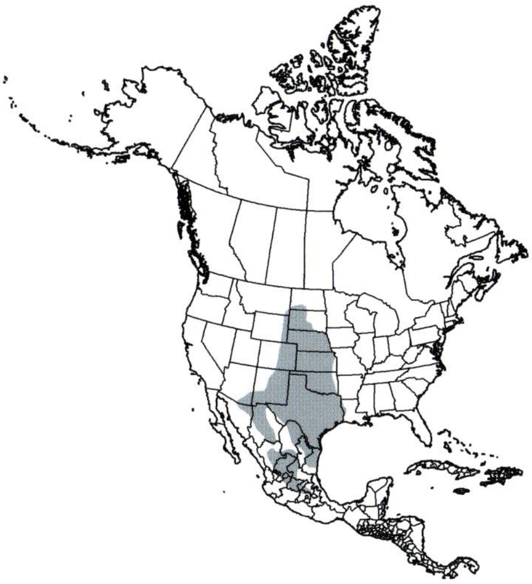

Locations from which the hispid pocket mouse has been identified in Wyoming.

markedly different subspecies boundaries, but the Wyoming portion of the geographic range is unaffected by those findings.[5]

DISTRIBUTION AND HABITAT ASSOCIATIONS: The Wyoming distribution of this Great Plains species includes the easternmost part of the state—the western border of the distribution follows the foothills of the Laramie Mountains north to around Devils Tower. Habitats are arid grasslands, particularly shortgrass admixtures with yucca and shrubs. Soil types occupied in Goshen County and in eastern Nebraska were not highly specialized.[6] In Pawnee National Grassland in northeastern Colorado, hispid pocket mice were most common and abundant on sites outside of black-tailed prairie dog colonies.[7]

ECOLOGY: This poorly understood omnivore feeds mostly on seeds of shrubs and trees. Seeds are carried in the cheek pouches and stored in underground tunnels and chambers to be eaten during winter. Predators include the full suite of avian, reptilian, and mammalian predators of appropriate sizes to kill pocket mice.[4]

CONSERVATION AND MANAGEMENT: Concern about the species is focused on the lack of information regarding distribution, ecology, and population status.

NOTES AND REFERENCES

1. Paulson (1988), Armstrong et al. (2011).
2. KU specimens.
3. Commission Regulations Chapter 52, Section 11.
4. Paulson (1988).
5. Andersen and Light (2012).
6. Maxwell and Brown (1968), Geluso and Wright (2010).
7. Stapp (2007).

POCKET GOPHERS—FAMILY GEOMYIDAE

The pocket gophers are a fossorial lineage restricted in distribution to North and Central America. They are characterized by distal forelimbs that are flattened for burrowing through dirt, strong forelimb claws, cylindrical shape, tiny eyes, and small pinnae. The availability of molecular methods has begun to clarify systematic relationships among species and subspecies, but much remains to be learned.[1]

NOTES AND REFERENCES

1. Wilson and Reeder (2005).

KEY TO SPECIES OF GEOMYIDAE

1a. Upper incisors with one or more longitudinal grooves on the anterior surfaces: . . . . **_Geomys lutescens_**
1b. Upper incisors lacking longitudinal grooves on the anterior surfaces: . . . . **_Thomomys_, 2**

2a. Dark periauricular patch is faint or absent; margin of pinna is buff-colored and lighter than dorsal pelage: . . . . **_Thomomys clusius_**
2b. Dark periauricular patch is distinct; margin of pinna is the same shade or darker than dorsal pelage: 3

3a. Dark periauricular patch encircles pinna, and extends dorsally: . . . . **_Thomomys talpoides_**
3b. Dark periauricular patch is present, but does not encircle pinna and does not extend dorsally: . . . . **_Thomomys idahoensis_**

Wyoming pocket gopher, *Thomomys clusius*

DESCRIPTION: A small rodent of typical fossorial form: rounded body and head, short pinnae, medium-length tail, and long claws on forefeet. Upper incisors lack longitudinal grooves on their anterior surfaces, in contrast to *Geomys*. Pelage on the dorsum is short and grayish or yellowish, compared with the more reddish of sympatric *T. talpoides*. *T. clusius* has an indistinct or no dark periauricular patch, whereas *T. talpoides* has a distinct patch. The hairs on the fringe of the pinna are lighter than the dorsal fur, whereas in *T. talpoides* from the same area they are darker. The most reliable diagnostic feature is chromosome number (2n = 46), in contrast to *T. talpoides* (2n = 48 or 56, depending on subspecies) and *T. idahoensis* (2n = 56 in the Wyoming part of the range). Burrow diameters of Wyoming pocket gophers are smaller than for the northern pocket gopher, but this field character is not reliable without others.[1] Measurements (**Sex** [n] mean (mm, g), *range*): Carbon and Sweet-

Wyoming pocket gopher, showing the absence of a periauricular dark patch.

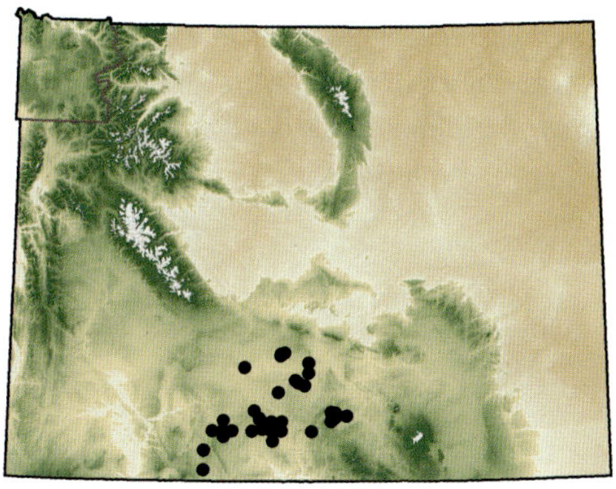

Points from which the Wyoming pocket gopher has been identified in Wyoming.

water Counties, L = **B** [22] 173, *161–184*; BL = **B** [22] 128, *112–134*; HF = **B** [22] 21, *20–22*; WT = **B** [22] 57.9, *44.1–71.5*.[2] Dental formula: 1/1, 0/0, 1/1, 3/3 = 20.

STATUS: A protected nongame wildlife species that may not be taken intentionally.[3] It is also a state Species of Greatest Conservation Need as of 2012.

NOMENCLATURE AND SYSTEMATICS: The species was first collected (one specimen) by Hammond in 1857, and described on the basis of that specimen as *T. clusius* by Elliot Coues in 1875, but later regarded by most taxonomists as *T. talpoides clusius*.[4] Considerable confusion surrounded the taxon's identity, in part because only the single 1857 type specimen existed in collections as of 1948. In the early and mid-twentieth century, many pale-coated *Thomomys* from as far south as central Colorado and as far north as southern Montana were characterized as *T. clusius*.[5] Then, remarkably, Charles Thaeler, Jr. and Landis Hinesley described their "rediscovery" of *T. clusius* in 1979. Their chromosome studies showed that *T. clusius* was indeed a distinct species and that its geographic range scarcely extended beyond the site where originally collected.[2] No subspecies are recognized.

DISTRIBUTION AND HABITAT ASSOCIATIONS: This is the only mammalian species endemic, on the basis of known localities, to Wyoming. The species has been collected at sites across western Carbon and eastern Sweetwater Counties. Habitats are not well known, but tend toward dry, well-drained ridges and slopes with at least some gravel in the soil. This, in contrast with deeper soils of draws and valleys used by the northern pocket gopher, is an apparent segregating mechanism where the two are sympatric. Vegetation associations of the Wyoming pocket gopher tend toward more Gardner's saltbush and winter fat cover, and less yellow rabbitbrush and big sagebrush cover than is the case for sympatric *T. talpoides*.[6]

ECOLOGY: The species is a fossorial herbivore, feeding on plant roots. Being a pocket gopher with a highly restricted distribution, it is assumed to have poor dispersal ability. The species is too poorly known to characterize other aspects of its ecology.

CONSERVATION AND MANAGEMENT: The species is of considerable conservation interest because of its rarity, high endemicity, presumed poor dispersal ability, and occurrence in a region with high levels of oil and gas development.[7]

NOTES AND REFERENCES

1. Thaeler and Hinesley (1979), Keinath et al. (2014). The latter provides a comparison of characters that can be used to distinguish among *Thomomys* species in southern Wyoming.
2. Thaeler and Hinesley (1979).
3. Commission Regulation Chapter 52, Section 11.
4. Hammond was a member of a US Army expedition led by Francis Bryan, and caught the animal near Bridger Pass, around 29 km (18 mi) southwest of Rawlins. The exact type locality of the 1857 specimen has not been relocated since then, and *T. talpoides* collected in the general vicinity were incorrectly labeled *T. clusius* (Thaeler and Hinesley 1979).
5. Long (1965).
6. Keinath et al. (2014).
7. Keinath and Beauvais (2006).

Idaho pocket gopher, *Thomomys idahoensis*

DESCRIPTION: A small-bodied pocket gopher, but with total length overlapping that of *T. clusius*. The form is typical of the family: rounded head and body, cheek pouches that are lined with fine fur, and long foreclaws (although not to the same degree as *Geomys*). Pelage color is variable from yellowish brown with dark-tipped hairs to grayish or dark brown. The species lacks a distinct periauricular patch, and the subspecies in Wyoming has 2n = 56 chromo-

Distribution of the Idaho pocket gopher in Wyoming.

somes.[1] Measurements (**Sex** [n] mean (mm, g), *range*: Uinta County, BL = **F** [7] 125, *117–133*; **M** [4] 126, *123–132*; TL = **F** [8] 58, *50–70*; **M** [4] 53, *50–59*; HF = **F** [8] 21, *21–22*; **M** [4] 21, *21–22*; WT = **F** [6] 51.8, *49.6–55.7*; **M** [4] 51.4, *46.2–63.0*.[1] Dental formula: 1/1, 0/0, 1/1, 3/3 = 20.

STATUS: A nongame animal that may not be intentionally taken and a state Species of Greatest Conservation Need as of 2012.[2] It is common but discontinuous within its geographic range.

NOMENCLATURE AND SYSTEMATICS: The species was recognized in 1972, when two taxa previously recognized at various times as distinct species (*T. pygmaeus*, *T. idahoensis*) or subspecies (*T. talpoides pygmaeus*, *T. t. idahoensis*) were pooled to form *T. idahoensis* on the basis of chromosome number. A further line of evidence was the local sympatry with *T. talpoides* in Uinta County, where no hybrids could be detected.[3] Of the two subspecies recognized, only *T. i. pygmaeus* occurs in Wyoming. This new classification scheme was acknowledged, but not adopted, by Hall (1981).[4]

DISTRIBUTION AND HABITAT ASSOCIATIONS: This species with a highly restricted distribution occurs in Uinta, Lincoln, and Sublette Counties.[5] Habitats are not well known, but appear to be shallower, rockier soils at higher elevations than is the case for the sympatric northern pocket gopher.[6]

ECOLOGY: This is a fossorial herbivore that is poorly known and assumed to have ecological relationships similar to those for *T. talpoides*. Pocket gophers in general have been proposed to serve important functions in soil turnover, water infiltration, suppressing woody plant growth, and forming Mima mounds, the latter particularly in shortgrass prairie.[7] Gophers are important prey for a wide range of mid-sized mammals and raptors, but the limited distribution of this species may make it too scarce to be important in that way.

CONSERVATION AND MANAGEMENT: The taxon is a conservation concern primarily because of its restricted distribution, specialized habitats and our poor knowledge of its distribution and ecology. This species has too restricted a distribution in the state to represent a major conflict with agriculture or landscaping.

NOTES AND REFERENCES

1. Thaeler (1972).
2. Commission Regulations Chapter 52, Section 11.
3. Long (1965), Thaeler (1972).
4. Hall (1981, 1179).
5. The distribution mapped by Thaeler (1972) is erroneous, showing four locations in Teton, Park, and northernmost Fremont Counties that should be in Sublette County.
6. Beauvais and Dark-Smiley (2005).
7. Grant et al. (1980), Reichman and Seabloom (2002).

Northern pocket gopher, *Thomomys talpoides*

DESCRIPTION: A rodent of typical gopher form. Pelage is pale to dark reddish brown, with dark periauricular patches around the small pinnae. Chromosome number varies among (and even within) recognized subspecies in Wyoming, suggesting that this taxon contains more than one biological species.[1] Measurements (**Sex** [n] mean (mm, g), *range*): Albany, Laramie, Carbon, and Teton Counties, L = **F** [9] 196, *184–218*; **M** [5] 179, *128–223*; TL = **F** [9] 57, *48–64*; **M** [5] 54, *43–75*; HF = **F** [9] 26, *23–28*; **M** [5] 26, *24–28*; EFN = **F** [9] 7, *5–11*; **M** [5] 8, *4–16*; WT = **F** [9] 110.8, *83–140*; **M** [5] 94.4, *81.5–113*.[2] Dental formula: 1/1, 0/0, 1/1, 3/3 = 20.

STATUS: A non-protected nongame animal. The species is common or locally abundant in most parts of its geographic range.[3]

NOMENCLATURE AND SYSTEMATICS: The genus has been recognized since the mid-nineteenth century, when *T. talpoides* was the sole named member of the genus. Today, this is one of the most zoogeographically complex North America mammals (the "*Thomomys talpoides* complex"), with as many as 58 subspecies recognized as of 1981, and 54 subspecies still recognized in 2012.[4] The 10 subspecies now believed to occur in Wyoming comprise *T. t. attenuatus, T. t. bridgeri, T. t. bullatus, T. t. caryi, T. t. cheyennensis, T. t. ocius, T. t. meritus, T. t. nebulosus, T. t. pierreicolus,* and *T. t. rostralis*. The subtle morphological differences that are the bases of these putative subspecies can be attributed to the poor dispersal ability of fossorial mammals, for which areas of bedrock near the surface can represent a complete biogeographical barrier. The species epithet (*talpoides*) means mole-like.

DISTRIBUTION AND HABITAT ASSOCIATIONS: This broadly distributed species occurs statewide, in almost all habitats where surface soil with some vegetation is available. It is found in shrub-steppe, grassland, meadow, and forested habitats, sympatric with other pocket gopher species.

ECOLOGY: Pocket gophers in general have been proposed to serve important functions in soil turnover, water infiltration, suppressing woody plant growth, and forming Mima

Northern pocket gopher from south-central Wyoming, showing distinct dark periauricular patch.

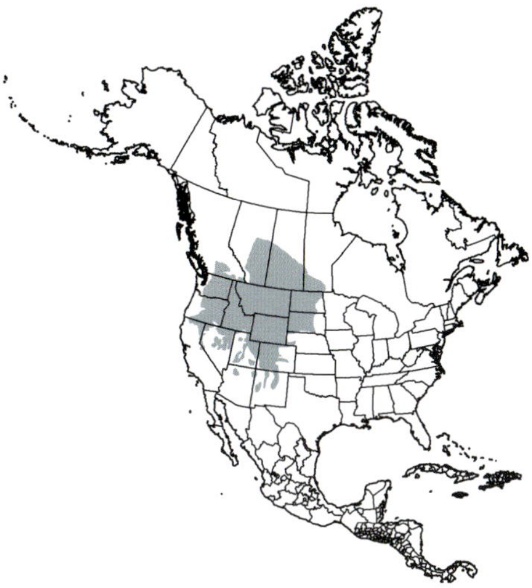

Distribution of the northern pocket gopher in Wyoming.

mounds, which are particularly visible in shortgrass prairie. The role of pocket gophers in the latter formation has been debated for decades. As the most widespread pocket gopher in the state, the species can be credited with most of the ecological effects that are attributed to pocket gophers in Wyoming.[5] They are important prey for many species of mid-sized carnivores and raptors, as a result of coming to the surface more often than moles. Almost nothing is known about their population biology.

CONSERVATION AND MANAGEMENT: The species is not generally considered a conservation concern because of its broad distribution and abundance. It is considered an agricultural pest and is controlled locally.

NOTES AND REFERENCES

1. Thaeler (1980), Verts and Carraway (1999).
2. UWMV, DMNS specimens.
3. Commission Regulations Chapter 52, Section 6.
4. Hall (1981), Wilson and Reeder (2005).
5. Grant et al. (1980), Reichman and Seabloom (2002).

Nebraska pocket gopher, *Geomys lutescens*

DESCRIPTION: Unlike *Thomomys, Geomys* has two longitudinal grooves along the anterior surface of each upper incisor. This species is larger than *Thomomys* gophers, with longer foreclaws. The dorsal fur is paler brown than is typical in *T. talpoides,* where the two species are sympatric.[1] Measurements (**Sex** [n] mean (mm, g), *range*): Goshen and Laramie Counties, L = **F** [12] 230, *210–251*; **M** [9] 248, *198–265*; TL = **F** [12] 72, *67–81*; **M** [9] 77, *61–89*; HF = **F** [12] 31, *29–32*; **M** [9] 32, *28–35*; EFN = **F** [12] 5, *4–6*; **M** [9] 5, *3–6*; WT = **F** [12] 182.4, *117.4–201*; **M** [9] 206.3, *88–264*.[2] Dental formula: 1/1, 0/0/, 1/1, 3/3 = 20.

STATUS: An unprotected nongame animal in Wyoming, which can be taken without a permit, and a state Species of Greatest Conservation Need as of 2012.[3]

NOMENCLATURE AND SYSTEMATICS: This eastern North American genus name dates from over 200 years ago; *G. bursarius*—the earlier species designation for *G. lutescens*—was one of the first New World species given a scientific name. In recent decades, with the

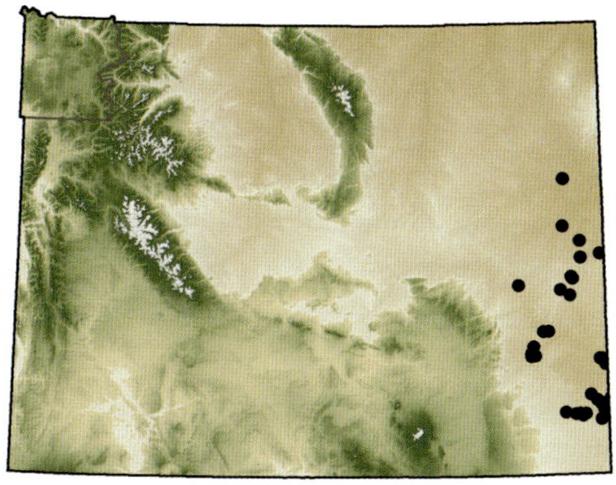

Locations from which the Nebraska pocket gopher has been identified in Wyoming.

combination of molecular and morphological methods, some previous subspecies have been raised to the level of species.[4] The form occurring in Wyoming was earlier referred to as *G. b. lutescens*; now it has been raised to the species level by two recent corroborating studies, and I use that scheme here.[5]

DISTRIBUTION AND HABITAT ASSOCIATIONS: This central Great Plains species occurs along the eastern edge of the state in a north–south band around 100 km (60 mi) in width; the northernmost occurrence in Wyoming is not certain. Habitat is not well known for Wyoming, but elsewhere has been reported to be deep, light soils.[6]

ECOLOGY: Nebraska pocket gophers are solitary, burrowing herbivores that associate with the deep, moist soils characteristic of the central Great Plains. They particularly associate with sandy soils and disturbed sites, and have been found to eat and be spatially associated with forbs in the Sand Hills of Nebraska.[7]

CONSERVATION AND MANAGEMENT: The species is poorly known in the Wyoming portion of its range and is a concern because of its poor dispersal and land use changes within its Wyoming distribution. It is controlled locally because of conflicts with agriculture or gardening.

NOTES AND REFERENCES

1. Keinath et al. (2014).
2. KU specimens.
3. Commission Regulations Chapter 52, Section 6.
4. Hall (1981), Jolley et al. (2000), Sudman et al. (2006).
5. Sudman et al. (2006), Genoways et al. (2008).
6. Miller (1964).
7. Steuter et al. (1995).

NEW WORLD PORCUPINES—FAMILY ERETHIZONTIDAE

The New World porcupines are a group of close relatives to such South American rodents as the tuco-tucos, chinchillas, and guinea pigs. These rodents are, in turn, phylogenetically close to African rodents such as cane rats and Old World porcupines.[1] How an African lineage gave rise to so many extant South American rodents is a continuing biogeographic puzzle, but fossils tell us that the ancestors of this group arrived in South America around 40–45 Ma, when the two continents were separated by a much narrower ocean.[2] Although around 15 species of porcupine in several genera occur in northern South America and Central America, a geographic gap occurs over central Mexico, and the monotypic genus holding the North American porcupine occurs only from northern Mexico northward to the Arctic Ocean. This is by far the most widely distributed of any of the South American forms that dispersed northward during the Great American Biotic Interchange. New World porcupines are more arboreal than their Old World counterparts, and have barbed quills covering most of their bodies. The tropical and subtropical species have prehensile tails and are highly arboreal, seldom coming to the ground. The diet is entirely leaves, stems, flowers, and cambium.

NOTES AND REFERENCES

1. Wilson and Reeder (2005).
2. Antoine et al. (2012).

North American porcupine, *Erethizon dorsatum*

DESCRIPTION: A large, slow-moving rodent with dark brown underfur and highly contrasting quills on the dorsum of head, back, and tail; adults have blond dorsal guard hairs.[1] A musty odor sometimes betrays its presence in enclosed spaces. No vocalizations are made in response to alarm, but several are given during intraspecific interactions. Porcupines are readily distinguished from other large rodents at close range, but at a distance can be confused with various mammals, including light-colored bears. Measurements (**Sex** [n] mean

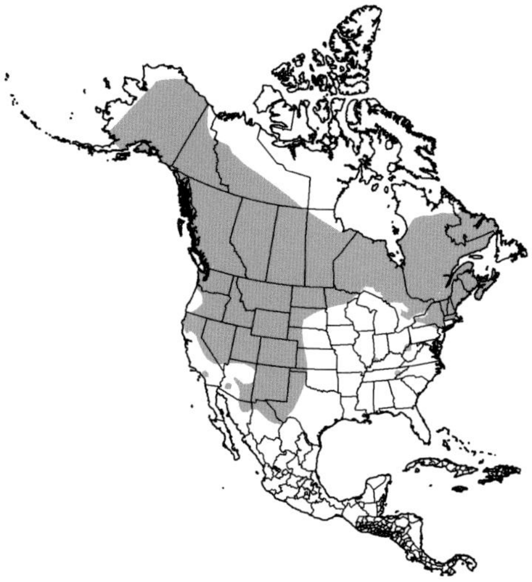

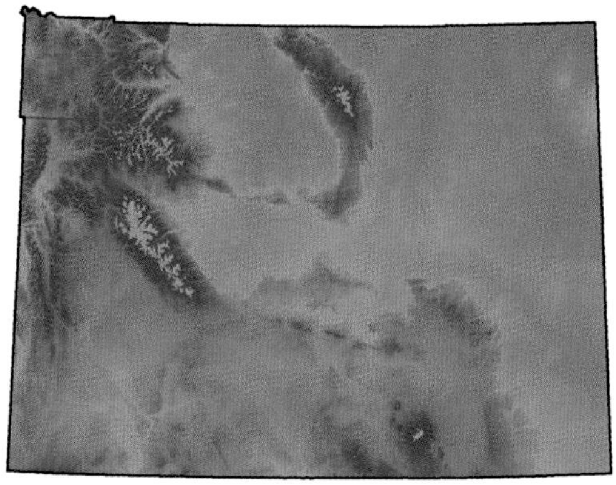

Distribution of the North American porcupine in Wyoming.

(kg), *range*: adults, Albany County, WT = **M** [5] 9.5, *6.6–15.1*.[2] Dental formula: 1/1, 0/0, 1/1, 3/3 = 20.

STATUS: The species is classified as a "predatory animal" under state law.[3] It is common, although highly variable in abundance, across Wyoming.

NOMENCLATURE AND SYSTEMATICS: The genus name has been stable for over a century; the specific epithet is also seen as *E. dorsata*.[4] Six subspecies were recognized as of 1981, two of which occur in Wyoming. The eastern half of the state is represented by *E. d. epixanthus* and the western half by *E. d. bruneri*. No geographical discontinuities correspond with this subspecies boundary. The genus and family are monotypic; this is the only North American representative of a South American lineage (New World porcupines) that diverged from other South American caviomorph rodents around 28 Ma.[5] *Erethizon* crossed into North America during the Great American Biotic Interchange (5–2 Ma), but subsequently went extinct in South America. The earliest fossils from the latitude of Wyoming are from around 1 Ma.

DISTRIBUTION AND HABITAT ASSOCIATIONS: This species of forests and woodlands occurs statewide in Wyoming where woody vegetation is available. Extensive areas of grassland and areas far above elevational limit of trees are not occupied.

ECOLOGY: The species is a strict herbivore, feeding on leaves, stems, and especially cambium of various woody species. The diet in the central Rockies is poorly documented, but ponderosa pine, Douglas-fir, lodgepole pine, limber pine, Engelmann spruce, and deciduous shrubs are commonly eaten winter foods. Winter is typically spent in dens, but some animals may spend the entire winter above the snow surface. During summer, ground foraging for forbs appears important, although studies specific to the Rocky Mountain region are scarce.[6] Population attributes have been studied little, but it is clear that the species has a slow life history, with delayed first reproduction, long gestation, small litters, and high longevity. The oldest known porcupine in the wild was >20 years old.[1] The winter diet is nutrient-poor, and the species is believed to experience high rates of starvation. A number of mammalian predators eat porcupines, notably fishers. The mountain lion is one of the few predators in the temperate West that can effectively kill porcupines, which it does by using its claws to hook the porcupine under the chin and flip it over, exposing the unprotected abdomen. Wolverines are important predators farther north.

CONSERVATION AND MANAGEMENT: The species receives little management attention in Wyoming, in part because of its status as a "predatory animal." This legal classification removes it from the jurisdiction of the Wyoming Game and Fish Department. Porcupines are commonly shot in all seasons, in part because livestock can be injured by quills. The species has become a conservation issue in several parts of the West because of perceived declines in abundance and in some areas, disappearance; the hypothesized (although not demonstrated) mechanism is predation by increasingly abundant mountain lions.[7] In Wyoming, mountain lions are suspected of limiting their abundance, and may drive local populations to low levels.

NOTES AND REFERENCES

1. Roze (2009).
2. Dematteo (1995, 133–136). Four of five adult males gained body weight from autumn 1993 to spring 1994.
3. Wyoming Statutes 23-1-101-a-viii.
4. Wilson and Reeder (2005).
5. Huchon and Douzery (2001).
6. Woods (1973), Roze and Ilse (2003)
7. Sweitzer et al. (1997). Some ranchers in Wyoming report that the appearance of higher abundances of mountain lions in the 1990s coincided with reductions in abundances of porcupines in the same areas.

Pikas, Rabbits, and Hares—Order Lagomorpha

This widely distributed and ecologically important order is characterized by its gnawing dentition, hind limbs adapted for hopping, and digestive morphology. Lagomorphs exhibit little sexual dimorphism, with females the same size as males, or slightly larger, among North American taxa.[1] All species have two incisors on each upper jaw—the second upper one a small peg immediately posterior to the first—and one on each lower jaw. All species are herbivores and hindgut digesters that use bacteria concentrated in the cecum to inoculate the upper digestive tract by reingesting specialized fecal pellets. These reingested bacteria facilitate fermentation of cellulose in the diet. The lineage is believed to have originated in East Asia, with Rodentia its sister lineage, but lagomorphs have spread to become one of the most cosmopolitan mammalian orders. On continental lands, only Australia, the Empty Quarter of the Arabian Peninsula, and the Southern Cone of South America lacked native lagomorphs at the time of European exploration. No lagomorph species hibernates, even those in seasonally cold climates. The phylogenetic relationships among species that occur in Wyoming are illustrated in the figure below.[2]

NOTES AND REFERENCES

1. Orr (1940) for California taxa.
2. Phylogenetic tree modified from Robinson and Matthee (2005).

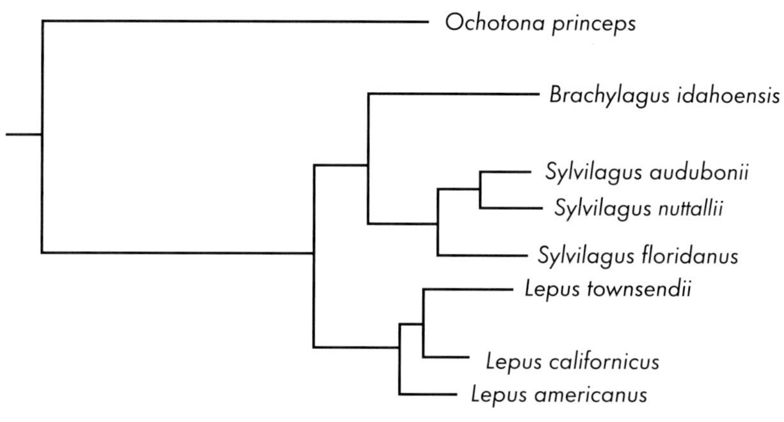

Phylogenetic relationships among members of the Lagomorpha occurring in Wyoming, based on nuclear and mitochondrial DNA analyses. Branch lengths are proportional to degree of molecular divergence.

KEY TO SPECIES OF LAGOMORPHA

1a. Length less than 200 mm: . . . . ***Ochotona princeps***
1b. Length greater than 200 mm: 2

2a. Length greater than 500 mm: . . . . ***Lepus*, 6**
2b. Length less than 500 mm: 3

3a. Ventral tail is buff, rather than white: . . . . ***Brachylagus idahoensis***
3b. Ventral tail is white: . . . . ***Sylvilagus*, 4**

4a. Interior of pinnae noticeably furred; auditory bullae small; some vibrissae partly white (or light): . . . . ***Sylvilagus nuttallii***
4b. Interior of pinnae darker, lightly furred, vibrissae either all black (or dark) or variable: 5

5a. Dorsal pelage relatively brown and dark; adult ear length less than 54 mm (2.1 in): . . . . ***Sylvilagus floridanus***
5b. Dorsal pelage relatively pale gray-brown; adult ear length greater than 54 mm (2.1 in): . . . . ***Sylvilagus audubonii***

6a. Dorsal tail with black longitudinal stripe: . . . . ***Lepus californicus***
6b. Dorsal tail lacking black longitudinal stripe: 7

7a. Ear length less than 80 mm: . . . . ***Lepus americanus***
7b. Ear length greater than 80 mm: . . . . ***Lepus townsendii***

PIKAS—FAMILY OCHOTONIDAE

Pikas occur across the Holarctic from the southern Rockies westward to the Ural Mountains, and the group is characterized by its small body size relative to other lagomorphs, small ears relative to body size, and a skull with only a single small fenestration in the maxilla. Pikas comprise only two New World representatives, both dwellers of mountainous regions, especially—but not exclusively—areas above elevational tree line with interspersed talus fields and meadows.

American pika, *Ochotona princeps*

DESCRIPTION: A small-bodied lagomorph with round pinnae, proportionally shorter than those of rabbits or hares. The body form is similar to that of rabbits, but with shorter hind legs and a more rounded head. The pelage is tan or buff, with a lighter underside and darker underfur showing through the guard hairs. A small cottony tail is not obvious in the field. In alpine zones, pikas are commonly heard before being seen, their nasal beep-like vocalizations audible across the talus fields. Their field sign is distinctive: piles of

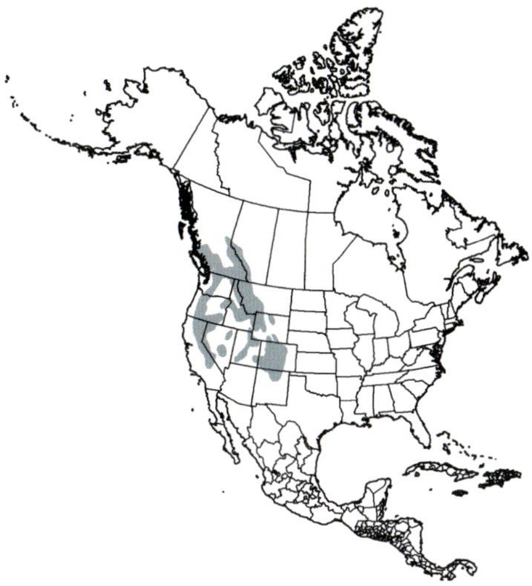

Distribution of the American pika in Wyoming. The smoothed 2880 m (9500 ft) contour approximates the geographic range, but presence within each of these areas is not confirmed.

small-diameter spheroid fecal droppings, and piles of cached green plant leaves being dried to hay for winter use.[1] Measurements (**Sex** [n] mean (mm, g), *RANGE*): Teton, Carbon, Sheridan, and Big Horn Counties, L = **F** [14] 160, *141–179*; **M** [23] 159, *129–174*; HF = **F** [14] 30, *27–33*; **M** [23] 30, *29–33*; EFN = **F** [14] 23, *21–25*; **M** [23] 23, *20–26*; WT = **F** [14] 153.07, *122–186*; **M** [23] 150.0, *102–177*.[2] Dental formula: 2/1, 0/0, 3/2, 2/3 = 26.

STATUS: This is a "protected animal" in Wyoming and a state Species of Greatest Conservation Need as of 2012.[3] It is common in suitable habitats within the geographic range.

NOMENCLATURE AND SYSTEMATICS: The genus and species are well established, and the Late Pleistocene and Holocene biogeography well documented. Recently, allozymes, mitochondrial DNA, nuclear DNA, and vocalization dialects have been used to revise the classification. Only five major subspecific lineages are now recognized; these are believed to have been reinforced through repeated glacial advances that reconnected populations in adjacent—but now isolated—mountain ranges.[4] In this classification, animals from Wyoming are considered *O. p. princeps*, the northern Rocky Mountain lineage that extends from central British Columbia south through Alberta, Montana, Idaho, and Wyoming to Colorado, ending at the Colorado River.

DISTRIBUTION AND HABITAT ASSOCIATIONS: This alpine–subalpine specialist of western mountains was more widely distributed when Europeans arrived (shown) than today. It occurs in Wyoming on the Yellowstone Plateau, Absaroka Mountains, Wind River Mountains, Wyoming Range, Big Horn Mountains, and Medicine Bow/Sierra Madre Mountains. The Uinta Mountains north of the state line are plausibly occupied, but this has not been confirmed. Habitat is cool, rocky areas typically, but not always, above the elevational limit of trees. Pikas occasionally occupy talus fields surrounded by subalpine coniferous forest.

ECOLOGY: Pikas are diurnal, nonhibernating, generalist herbivores. One of the few North American mammals to store food for winter, pikas cut leaves of various plants, dry them to hay in exposed spots, remove them to storage sites beneath talus rocks, and eat them while active in winter.[1] The species is known for its poor dispersal abilities across non-alpine habitats, which is thought to have led to strong population structuring.[5] Pikas are relatively intolerant of high ambient temperatures, perishing if held at temperatures above about 25°C (77°F).[6] Population densities in the Beartooth Mountains of Wyoming in late summer were found to be 2–11/ha (500–2800/mi^2), depending on year and habitat type.[7] The ecological influences of American pikas, aside from herbivory and serving as important prey for weasels and martens, include the creation of nutrient hotspots, where stored but unused hay has decomposed.[8]

CONSERVATION AND MANAGEMENT: This species is one of the most often mentioned in the context of the impacts of climate change on North American mammals, because of shrinking mountaintop islands of cool climate, and tundra habitat. Fossil pika remains on mountains of the Great Basin show that the average elevation at which pikas occur is 783 m (2600 ft) higher today than it was in Late Wisconsinan (40,000–10,000 radiocarbon ybp) and early Holocene times. A number of mountain ranges in the Great Basin have seen pika populations become extinct during the historical period.[9]

NOTES AND REFERENCES

1. Dearing (1997).
2. MSB specimens.
3. Wyoming Statutes 23-1-101-a-ix.
4. Hafner and Smith (2010).
5. Smith (1974a, 1974b).
6. MacArthur and Wang (1974).
7. Kruezer and Huntley (2003). These densities did not include nonresident juveniles.
8. Smith and Weston (1990) for predators, Aho et al. (1998) for hotspots.
9. Beever et al. (2003), Grayson (2005).

RABBITS AND HARES—FAMILY LEPORIDAE

Rabbits and hares are larger than pikas, have proportionately longer ears, and have multiple fenestrations (a bony lattice) in each maxilla. This a species-rich, ubiquitous, and ecologically important group in North America north of Mexico (20 spp.) and in Wyoming (7 spp.).[1] Scarcely a region or habitat in Wyoming does not hold one or several leporid species, and the group represents an important source of herbivory on grasses and forbs, and prey biomass to vertebrate predators. This group is puzzling in terms of our understandings of systematics and ecology. The phylogenetic relationships among leporids have long been obscure, in part because the group is morphologically conservative: physical traits have changed little in the face of speciation and local adaptation. Molecular tools recently have allowed us to understand phylogeny and distributions better. As an example, between 1990 and 2005, the number of named species of Leporidae worldwide increased from 53 to 60, largely based on molecular evidence.[2] Relevant to Wyoming, recent molecular evidence has shown that the three Wyoming *Lepus* species are quite distinct and exhibit no nuclear genetic introgression, snowshoe hares having diverged from the jackrabbit lineage around 2.7 Ma, and the two jackrabbit species from Wyoming having diverged from each other around 1.2 Ma. Interestingly, mitochondrial genes from the black-tailed jackrabbit are believed to have entered Pacific Northwestern snowshoe hare populations around 0.5 Ma.[3]

Wyoming holds four named species of rabbit and three of hares. The rabbits are in two genera: *Sylvilagus* (cottontails) and *Brachylagus* (pygmy rabbit). The three cottontails—desert, mountain, and eastern—have puzzling systematic and community relationships. They are semi-cryptic morphological species, difficult to distinguish reliably at a distance or from field measurements, and all three species have the same number of chromosomes ($2n = 42$).[4] The eastern cottontail's range extends to Florida and occurs in the southeastern corner of Wyoming, where its distribution is parapatric with that of the mountain cottontail. The broadly distributed eastern cottontail is currently suspected of representing at least four biological species. Mountain and desert cottontails are syntopic across much of Wyoming, occurring in the same habitats at the same locations at the same seasons and apparently eating similar foods. Ordinarily, such overlap of co-occurring similar species leads to interspecific competition, character displacement, habitat partitioning, genetic introgression, or some combination of these consequences. But these two cottontail species are not known to exhibit such phenomena in Wyoming. Molecular data show that they are more closely related to each other than to any other cottontails, but they exhibit no evidence of interbreeding. The three hare species—characterized by precocial young—comprise the black-tailed and white-tailed jackrabbits and the snowshoe hare. These three species likewise share the same number of chromosomes ($2n = 44$), but are more easily distinguished morphologically and ecologically than are the cottontails.[2]

NOTES AND REFERENCES

1. Bradley et al. (2014).
2. Robinson and Matthee (2005).
3. Melo-Ferreira et al. (2014).
4. Robinson et al. (1983).

Pygmy rabbit, *Brachylagus idahoensis*

DESCRIPTION: A small rabbit lacking a cottony underside to its tail. Adults can be confused with cottontails if the tail cannot be seen; juvenile pygmy rabbits are tiny (smallest individuals aboveground weighed 71–79 g [~2.6 oz, *n* = 6]).[1] The interior pinnae are more heavily furred than those of cottontails, and the skin of the inner pinna is gray or brown, rather than pink as for cottontails. Females are slightly larger than males (by about 3%).[2] Field sign is distinctive with experience—very small spherical fecal droppings clustered in tall big sagebrush, especially around the entrances of burrows that pygmy rabbits dig. Measurements (**Sex** [n] mean (mm, g), *SD*): Lincoln, Sweetwater, and Fremont Counties; HF = **F** [25] 71.2, *4.6*; **M** [12] 72.4, *4.7*; EFN = **F** [25] 51.1, *2.4*; **M** [13] 51.9, *3.7*; WT = **F** [25] 410, *52*; **M** [14] 412, *86*.[3] Dental formula: 2/1, 0/0, 3/2, 3/3 = 28.

STATUS: A nongame wildlife species that cannot be intentionally taken and a state Species of Greatest Conservation Need.[4] Hunting has been prohibited since 1999.[5] Populations in Wyoming were petitioned for E.S.A. listing in 2003, but found not warranted.[6] The Columbia Basin population was listed as E.S.A. Endangered in 2003.[7]

NOMENCLATURE AND SYSTEMATICS: The genus was earlier considered to be *Sylvilagus*, which is still seen occasionally.[8] Subspecies have not been recognized.

DISTRIBUTION AND HABITAT ASSOCIATIONS: The current distribution of this Great Basin species (shown) in Wyoming reflects profound change over several decades. It was not known in the state in 1965, then found in a small area south of Kemmerer by Tom Campbell and coworkers in 1981.[9] Since then, surveys by various agencies and organizations, compiled by the Wyoming Natural Diversity Database, have shown the species to be widely distributed across southwestern Wyoming, extending south into Colorado in places.[10] This rapid expansion of the known range has been erroneously interpreted by some as evidence of rapid range expansion since 1981. But such a scenario is not plausible; the species is simply cryptic to untrained searchers, and substantial efforts were expended in the last 25 years to map the geographic range. Habitat is typically regarded as shrub-steppe dominated by late-successional big sagebrush and underlain by soils that permit burrowing.[11] Some variation in habitats used is reported, and pygmy rabbits can disperse through various habitat types, as long as they have overhead shrub cover at short intervals.

ECOLOGY: Diets are known from outside Wyoming to be herbaceous leaves, especially grasses, in summer, but almost entirely big sagebrush leaves in winter. Sagebrush plants serve another important function by providing a structural matrix that captures drifting snow in winter. Within this matrix, pygmy rabbits can reach sagebrush stems to browse, and

Pygmy rabbit, showing the well-furred interior pinnae.

Distribution of the pygmy rabbit in Wyoming. The two points mark where the species was first detected in the state in 1981.

emerge from their burrows, climb snowdrifts to shrub branches, and reach leaves at the tops of tall big sagebrush plants.[12] Suitable habitats have been regarded as patchily distributed across the range, and the species believed to exist as isolated subpopulations. But Amanda Thimmayya at the University of Wyoming found the species in 2012 to have almost no spatial genetic structure, showing that pygmy rabbits in Wyoming are able to disperse readily among suitable habitat patches.[13] Illustrating this ability, one pygmy rabbit in Fossil Butte National Monument dispersed several miles through various habitat types, before dying of predation. Young animals have successfully dispersed much farther than that in Idaho.[14] Survival rates are low; at sites in Oregon and Idaho only 0.3–17% of animals survived for one year. Predation accounts for a high proportion of deaths among both juveniles and adults; predators include the coyote, various raptors and owls, and both co-occurring weasel species.[15]

A juvenile pygmy rabbit emerging from a burrow.

CONSERVATION AND MANAGEMENT: The species has been a concern throughout its range for some years, because of the wholesale elimination or conversion of sagebrush-dominated habitats across the Great Basin. In Wyoming, the species was considered peripheral and rare 30 years ago, and intensive searches motivated by conservation concern have clarified that issue. Sagebrush alteration is a major concern for the species because fire and other manipulations—usually for the purpose of improving ungulate forage—can remove pygmy rabbit habitat for decades. Prescribed fire selectively removes the tallest and oldest big sagebrush, the habitats most important to pygmy rabbits.[16] Cattle grazing at moderate intensities in Montana did not produce demonstrable effects on vegetation structure important as cover or on the integrity of burrows.[17]

NOTES AND REFERENCES

1. Estes-Zumpf and Rachlow (2009).
2. Davis and Roth (2008).
3. Thimmayya (2010, Appendix A).
4. Commission Regulation Chapter 52, Section 11.
5. Terry Creekmore, Wyoming Game and Fish Department, Laramie, WY, pers. comm.
6. US Fish and Wildlife Service (2010a).
7. US Fish and Wildlife Service (2003).
8. Wilson and Reeder (2005).
9. Long (1965), Campbell et al. (1982).
10. Purcell (2006) conducted a survey across 10 Wyoming counties; the US Fish and Wildlife Service (2010a) reviewed survey efforts. The first documentation of pygmy rabbits in northernmost Colorado was by Estes-Zumpf et al. (2014).
11. This is one of the few leporids that excavates its own burrows. See Katzner and Parker (1997) for a description of field sign.

12. Green and Flinders (1980).

13. Thimmayya and Buskirk (2012).

14. Katzner and Parker (1998), Estes-Zumpf and Rachlow (2009). The high connectivity of pygmy rabbit populations in Wyoming has existed over the period of time that microsatellite DNA is informative.

15. Crawford et al. (2010).

16. Thimmayya (2010).

17. Camp et al. (2014).

Desert cottontail, *Sylvilagus audubonii*

DESCRIPTION: A medium-sized leporid with buff-colored pelage on the back, rusty throat, and lateral forelimbs. The pinnae are longer than those of mountain cottontails, and the interior pinnae are lightly furred over pink skin. The white cottony ventral tail is prominent when running. The auditory bullae are proportionally larger than is the case for eastern or mountain cottontails. Vibrissae generally are uniformly black or dark.[1] Females are slightly larger than males, by 2–5% in field measurements and some skull dimensions.[2] Measurements (**Sex** [n] mean (mm, g), *RANGE*): various counties; L = **F** [7] 331, *238–410*; **M** [9] 368, *339–390*; TL = **F** [7] 32.6, *17–52*; **M** [9] 42.6, *30–53*; HF = **F** [7] 77.4, *19–52*; **M** [9] 87.2, *80–97*; EFN = **F** [7] 64.6, *46–76*; **M** [9] 63.4, *52–75*.[3] Albany County, WT = **F** [98 measurements over all months] 1152, *907–1540*; **M** [98 measurements over all months] 1034, *900–1280*.[4] Dental formula: 2/1, 0/0, 3/3, 3/3 = 28.

STATUS: Legal classification is as a small game animal. [5] It is common or abundant in suitable habitats within the geographic range.

NOMENCLATURE AND SYSTEMATICS: The scientific and common names (it is also called Aubudon's cottontail) have been stable for many decades, but the species-level classification must be considered tentative until detailed molecular studies are completed for *Sylvilagus*. Of the 12 subspecies recognized as of 1978, only *S. a. baileyi* occurs in Wyoming. The species is most closely related to the mountain cottontail, with which it is sympatric over much of Wyoming.[6] In spite of the close phylogenetic affinity and morphological similarity to the mountain cottontail, the two taxa showed no evidence of introgression in molecular studies

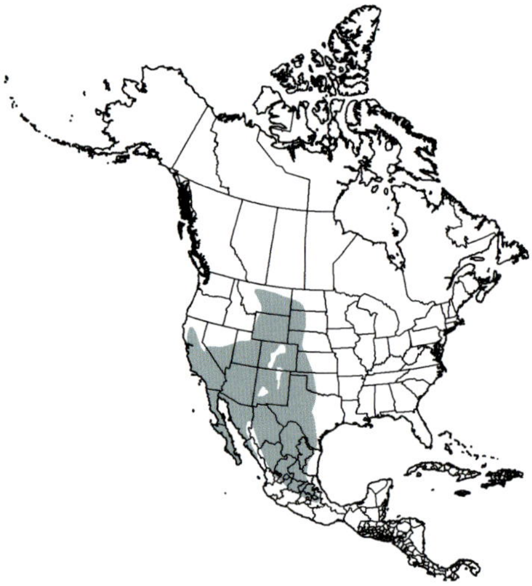

Distribution of the desert cottontail in Wyoming.

in Wyoming.[7] Such a lack of introgression between such closely related sympatric species is unusual, considering the absence of evidence of ecological separation.

DISTRIBUTION AND HABITAT ASSOCIATIONS: In Wyoming, the species occurs statewide in shrub-steppe, grassland, riparian, and some forest habitats. It is broadly sympatric with the mountain cottontail over mid-elevations (2000 m [6500 ft] to 3400 m [7500 ft]), but uncommon in areas of deep winter snow or above elevational treeline.[8] The common name "desert" suggests the species occurs at lower elevations or drier sites than is the case for mountain cottontails, but data associated with museum collections do not support such a distinction. The mean elevation of collection of 198 specimens taken in Wyoming found in various museums was 1811 m (5978 ft), compared with 1935 m (6384 ft) for 110 mountain cottontails. However, both species were reported from sites at elevations greater than 3030 m (10,000 ft), so great overlap in elevations of occurrence is puzzling.[9]

ECOLOGY: The species is poorly studied particularly in and near Wyoming, for such a widespread and common taxon. Most medium-sized carnivores and large raptors in Wyoming are potential predators; bobcats, red foxes, coyotes, red-tailed hawks, great horned owls, and golden eagles are most commonly identified.[10] Population size is variable across seasons and years. Population cycles have been mentioned or assumed to occur for some decades, partly because such cycles are reported in leporids farther north, but only recently have such cycles been supported by data from Wyoming. Cottontails of unknown species in Wyoming undergo population cycles, the cycles have a period of around eight years, and are synchronous with those of greater sage grouse.[11] This synchrony of population cycling between unrelated species was explained by similar habitats, and common predators and parasites.

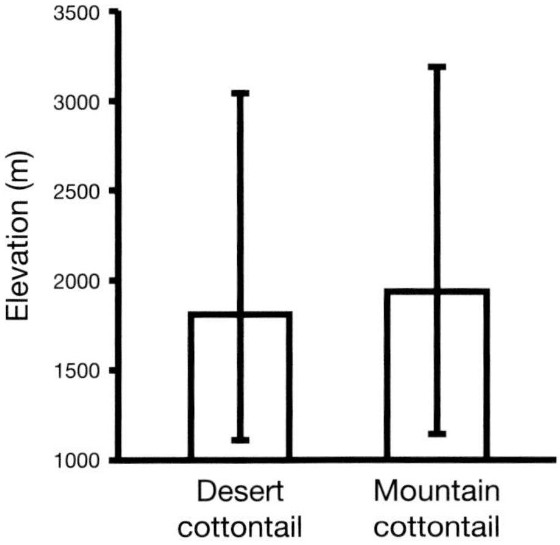

The mean elevations of specimens of desert and mountain cottontails in museum collections show that the two species have broadly overlapping elevational distributions, although mountain cottontails occur at slightly higher (by 124 m [409 ft]) average elevations. Note: Error bars indicate the range of elevations at which 198 desert cottontails and 110 mountain cottontails were collected.

CONSERVATION AND MANAGEMENT: The species is managed as small game, but appears not to be hunted intensively over a large portion of its range. It is common and not a conservation concern.

NOTES AND REFERENCES

1. Chapman and Willner (1978), Hoffmeister (1986).
2. Davis and Roth (2008).
3. UWMV specimens.
4. Krumm (1969, Table 8).
5. Wyoming Statutes 23-1-101-a-xi.
6. Robinson and Matthee (2005).
7. Thimmayya (2010).
8. This differs from the pattern shown for Colorado by Armstrong et al. (2011), who considered desert and mountain cottontails to be parapatric along the Front Range, but sympatric on the Western Slope and in the San Luis Valley.
9. See online appendix for data ($t = 2.79$, df = 306, $p = 0.006$).
10. Chapman and Willner (1978), Zimmerman et al. (1996), Olendorff (1976).
11. Fedy and Doherty (2011).

Eastern cottontail, *Sylvilagus floridanus*

DESCRIPTION: A medium-sized leporid with more brown in the dorsal pelage than for other Wyoming members of the genus, the species has a rusty throat, nape of neck, and lateral forelimbs.[1] The pinnae are as short as for the pygmy rabbit, their interior surfaces lightly furred, as for *S. audubonii*. Most animals have all black (or dark) vibrissae; a few animals show white (or light) banding on some vibrissae. The auditory bullae are smaller than for *S. audubonii*. Sexes do not differ in overall size or in skull dimensions.[2] Measurements (**Sex** [n] mean (mm, g), *RANGE*): Goshen and Laramie Counties, WY, L = **F**

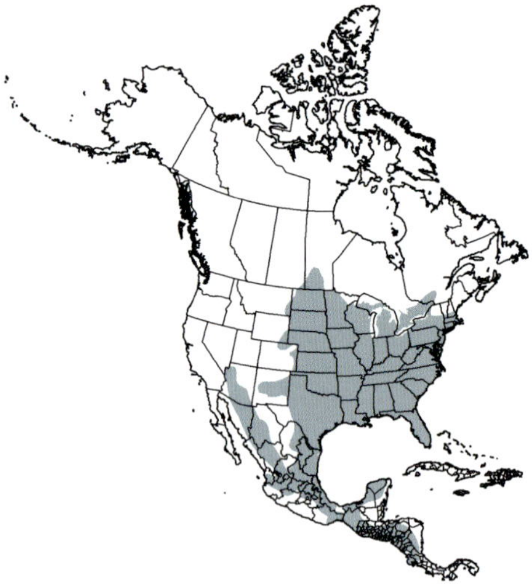

Distribution of the eastern cottontail in Wyoming.

[5] 416, *407–445*, TL = **F** [5] 42, *33–53*; HF = **F** [5] 97, *93–103*; EFN = **F** [5] 51, *48–54*; WT = **F** [5] 1434.60, *1259–1763*.[3] Dental formula: 2/1, 0/0, 3/3, 3/3 = 28.

STATUS: The legal status in the state is that of a small game animal.[4] It is common or abundant in suitable habitats within the geographic range.

NOMENCLATURE AND SYSTEMATICS: The scientific and common names have been stable for many decades, but must be considered tentative until detailed molecular studies throughout the species' range are completed. It is currently believed that this single nominal species likely contains at least four biological species, the other three representing disjunct populations in Mexico and Central America.[5] The nominal subspecies in Wyoming, *S. f. similis*, is one of around 24 named subspecies.[6]

DISTRIBUTION AND HABITAT ASSOCIATIONS: This eastern North American species occupies the southeasternmost corner of Wyoming, mostly in Laramie County, extending slightly northward. The species is sympatric with the desert cottontail here, but roughly parapatric with mountain cottontails. Across the species' range, habitats are highly diverse, including arid grassland, temperate forest, urban parks and yards, and tropical rain forest types. In Wyoming, habitat associations have not been studied specifically, but include mixed-grass prairie and shrub-steppe.[7]

ECOLOGY: The species has been studied extensively over its large geographic range, but little in the central Rocky Mountain states, including the western Great Plains.

CONSERVATION AND MANAGEMENT: The species is managed as a small game species, but hunting mortality is not specifically known, although harvest and hunting effort are estimated annually by the Wyoming Game and Fish Department. Habitats in eastern North America have been managed to attempt to enhance eastern cottontail populations.[8] There is no habitat management specific to the species in Wyoming.

NOTES AND REFERENCES

1. Compare Hoffmeister (1986, 130) with Armstrong et al. (2011, 266).
2. Chapman et al. (1980), Davis and Roth (2008).
3. KU specimens.
4. Wyoming Statutes 23-1-101-a-xi.
5. L. A. Ruedas in Robinson and Matthee (2005).
6. Chapman et al. (1980).
7. Hall (1981).
8. Chapman and Flux (1990).

Mountain cottontail, *Sylvilagus nuttallii*

DESCRIPTION: A medium-sized to large cottontail with relatively short, rounded ears, the inner surfaces of which are more densely haired than is the case for other Wyoming *Sylvilagus*. The pelage is buff-colored on the back, rusty on the throat and lateral forelimbs—not different enough from *S. audubonii* to be a reliable character. The ventral tail is cottony and white. Vibrissae are variously banded black-white (or dark-light), but no animals have all their vibrissae all black (or dark). The auditory bullae are proportionally smaller than is the case for the desert cottontail. Females are slightly larger than males, by 2–5% in field measurements, but not in size of the skull.[1] Fecal droppings are slightly irregular spheres. Measurements (**Sex** [n] mean (mm, g), *RANGE*): California; L = **F** [7] 331, *238–410*; **M** [11] 352, *338–371*; HF = **F** [7] 94, *90–101*; **M** [11] 95, *80–97*; WT = **F** [4] 790, *690–871*; **M** [4] 720, *629–830*.[2] Dental formula: 2/1, 0/0, 3/3, 3/3 = 28.

STATUS: Classified as a small game animal by State Statute.[3] Common or abundant within the geographic range.

NOMENCLATURE AND SYSTEMATICS: The scientific and common names have been stable for many decades. Of the three subspecies recognized as of 1981, *S. n. granger* occurs nearly

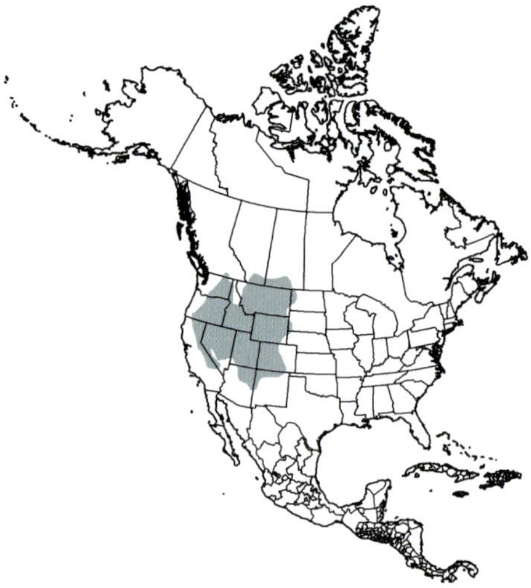

Distribution of the mountain cottontail in Wyoming.

statewide. Wyoming *Sylvilagus* have not been well characterized at the subspecies levels by molecular data, and subspecies now recognized are subject to revision. This species shows extensive local sympatry but no introgression with the sibling species desert cottontail; these two lineages are clearly closely related good biological species.[4] The common name "Nuttall's cottontail" also is used.

DISTRIBUTION AND HABITAT ASSOCIATIONS: The distribution of this temperate western species within Wyoming is poorly understood, but I modify that of Clark and Stromberg (1987) to reflect collections in Nebraska <15 mi (24 km) from the Wyoming border near where

Nebraska, South Dakota, and Wyoming meet.[5] Traditional depictions show this species as parapatric with the eastern cottontail in Platte, Goshen, Laramie, and Crook Counties, and that scheme is followed here.[6] Possible introgression at that zone of contact has not been studied. Habitat is shrub-steppe, grassland, forest, and riparian areas, with some tendency to associate with rocky outcrops. The common name implies a higher elevational distribution than is the case for desert cottontails, but such a distinction is weak, based on current data.[7]

ECOLOGY: The mountain cottontail is a strict herbivore, feeding on a range of grasses and forbs during the warm season, and shrub leaves and small stems (including sagebrush) in winter. On the Snake River Plain of southern Idaho, the species occurred on basalt outcrops, dominated by basin big sagebrush. Fecal pellets contained primarily grasses.[8] It uses hindgut digestion to break down cellulose, which requires inoculation of the gastrointestinal tract by reingesting specialized fecal droppings. The species forages above ground at dusk and at night, but remains stationary most of the day, often in shallow burrows.[9] Populations have been studied elsewhere in the geographic range, but not in Wyoming. As for desert cottontails, population cycles with periods of around eight years have been described, but amplitudes are low.[10] Predators include a wide range of mid-sized mammals and larger raptors.

CONSERVATION AND MANAGEMENT: The species is managed as small game, but no species-specific evidence is available for harvest levels. Hunting cannot be considered to be a major source of mortality statewide for any of the three cottontails, because of their high reproductive rates and the small proportion of the state over which they likely are hunted.

NOTES AND REFERENCES

1. Chapman (1975), Davis and Roth (2008).
2. Orr (1940).
3. Wyoming Statutes 23-1-101-a-xi.
4. Hall (1981), Robinson and Matthee (2005), Thimmayya (2010).
5. Jones (1964), Clark and Stromberg (1987).
6. Long (1965), Hall (1981).
7. See account for desert cottontail.
8. Johnson and Hansen (1979).
9. Chapman (1975).
10. Fedy and Doherty (2011).

Snowshoe hare, *Lepus americanus*

DESCRIPTION: A medium-sized leporid of typical form, with wide, well-furred hind feet and pelage of seasonally varying color. In summer, fur is dull brown with light undersides; in winter, it is white with black margins of pinnae. It can be confused in winter with the white-tailed jackrabbit, which is nearly as white, but lankier, with longer ears. The two species are roughly separated along habitat lines (forest vs. shrub-steppe), but can be nearly syntopic at forest edges. Measurements (**Sex** [n] mean (mm, g), *SD*: Albany County, HF = **F** [7] 141.6, *4.4*; **M** [10] 137.4, *6.9*; EFN = **F** [7] 82.9, *6.9*; **M** [10] 79.8, *7.8*; WT = **F** [7] 1274, *172*; **M** [10] 1169, *139*.[1] Dental formula: 2/1, 0/0, 3/2, 3/3 = 28.

STATUS: The legal status is as a "small game animal" under state statute.[2] It is patchily distributed and common, but not abundant, over most of its range in Wyoming.

NOMENCLATURE AND SYSTEMATICS: The genus and species names have been stable for many decades. Of the 15 subspecies recognized in 1981, two are found in Wyoming. The nominal subspecies in the northwestern mountains, Uinta Mountains and Medicine Bow Mountains is *L. a. bairdii*, which extends from New Mexico to southern Alberta. The disjunct population in the Bighorn Mountains is nominally *L. a. seclusus*, reflecting slight differences in pelage color, body size, and skull morphology for small sample sizes.[3]

Snowshoe hare in winter pelage.

DISTRIBUTION AND HABITAT ASSOCIATIONS: This hare of boreal forests is undergoing range contractions at the southern margin of its distribution (current distribution shown). It occupies suitable habitats in the northwestern, Uinta, Medicine Bow, Bighorn, and Laramie Mountains (near Laramie Peak). It is uncommon and patchily distributed in Yellowstone National Park. The species has not been confirmed from the forested mountains along the Sweetwater River. Habitat is typically coniferous forest or nearby shrub meadows with woody browse accessible from the snow surface in winter. In western Wyoming, preferred habitat has been found to consist of late-seral multistoried forest with a spruce–fir component and dense horizontal cover. Lodgepole pine stands 30–70 years of age with high stem densities were likewise important.[4]

ECOLOGY: The species is well studied in the northern, boreal part of its geographic range, but less so in southern boreo-montane areas, including Wyoming. Diets are highly diverse in this region, concentrating on woody stems and conifer leaves during winter. Lodgepole pine, quaking aspen, willows, Oregon grape, *Vaccinium* sp, *Amelanchier* sp., *Rosa* sp., and *Rubus* sp. are important winter browse; typically leaves and stems <1 cm diameter are eaten. During times of food scarcity, hares are more likely to eat the cambium of larger stems. Food limitation and strong effects of plant secondary compounds on palatability are less common in the central Rockies than in the North. Populations in the central Rockies have not been studied intensively, but are believed to be cyclic, with smaller amplitudes

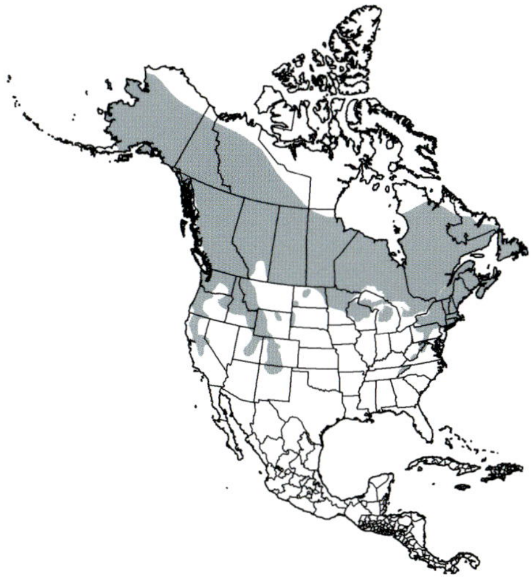

Distribution of the snowshoe hare in Wyoming.

and less predictable (mostly 8–11 years) periods than farther north.[5] Generally, densities in Wyoming are lower than in the North, but higher than in Colorado. In the southern Yellowstone area, density differed by habitat type: 0.03/ha (7.7/mi^2) in multistory spruce–fir, 1.2/ha (310/mi^2) in multistory lodgepole pine–spruce–fir, 2.8/ha (722/mi^2) in 30- to 60-year-old lodgepole pine, and 3.3/ha (851/mi^2) in multistory aspen–spruce–fir.[6] These values are higher than those reported for southern Colorado (0.13/ha [34/mi^2] in lodgepole pine, 0.39/ha [100/mi^2] in spruce–fir), and some of the highest values reported for the contiguous United States or southern Canada.[7] Species-wide, survival rates vary strongly with phase of the population cycle and study methods used. Few studies have been conducted as far south as Wyoming, but in Colorado survival rates (30 days) were around 95%. Predation tends to be the primary cause of mortality, with a wide range of mammalian and avian predators contributing.[8]

CONSERVATION AND MANAGEMENT: The species is considered a conservation concern as the primary prey for the Canada lynx, which was listed as E.S.A. Threatened in the contiguous United States in 2000.[9] Approaches to conservation tend to focus on maintaining forest stand conditions that can support hares. This, in turn, involves maintaining stands or landscapes that provide sustainable sources of small-diameter woody stems accessible from the snow surface in winter. Whether even-aged regenerating stands—particularly of lodgepole pine—or uneven-aged late successional stands are better able to provide these conditions is not clear. One important, unresolved issue is whether regenerating lodgepole pine forest following mountain pine beetle infestation will be able to support hare populations for long periods. Estimated annual statewide harvests from 2008 to 2011 were in the low hundreds. Climate-mediated effects, including warming temperatures, altered snowfall regimes, thawing episodes, and outbreaks of conifer-feeding insects, have the potential to profoundly change the distribution of snowshoe hares in the state, particularly in isolated patches of mesic coniferous forest.

NOTES AND REFERENCES

1. Lawrence (1955).
2. Wyoming Statutes 23-1-101-a-xi.
3. Hall (1981).
4. Dolbeer and Clark (1975), Hodges et al. (2009), Berg et al. (2012).
5. Keith (1983).
6. Hodges (2000a).
7. Berg and Gese (2010). All values are means of 2–7 plots.
8. Zahratka and Shenk (2008), Berg and Gese (2010). Ivan et al. (2014) reported somewhat lower densities for mature spruce–fir in central Colorado. Annual survival was highest (0.14–0.20) in mature spruce–fir.
9. Hodges (2000a, Table 7.3; 2000b).

Black-tailed jackrabbit, *Lepus californicus*

DESCRIPTION: A large, lanky hare with slightly longer ears than the white-tailed jackrabbit. The relatively long tail has a black stripe on its mid-dorsum, even among juveniles. The dorsal fur is gray with brownish tones, and the tips of the pinnae are nearly black. The pelage does not vary strongly in color with season. The skull is nearly indistinguishable from that of a white-tailed jackrabbit, and the dentition is the same except for the enamel folding on the occlusal surface of the upper first incisor.[1] Fecal pellets are the same size and shape as for snowshoe hare. Measurements (**Sex** [n] mean (mm, g), *SE*), adults, northern California; L = **F** [11] 607, *573–630*; **M** [16] 588, *556–610*; TL = **F** [53] 150, *0.9*; **M** [16] 588, *556–610*, HF = **F** [10] 133, *120–143*; **M** [17] 134, *118–144*; EFN = **F** [4] 134, *129–138*, **M** [4] 134, *129–143*; WT = **F** [9] 2870, *2564–3146*, **M** [10] 2454, *1685–2823*.[2] Dental formula: 2/1, 0/0, 3/3, 3/3 = 28.

STATUS: The species is classified by law as a "predatory animal," and is not protected.[3] It is common in shrub-steppe and grassland habitats within the limited geographic range.

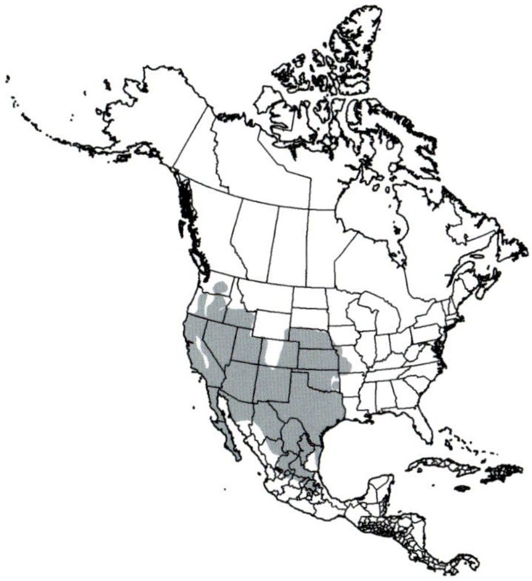

Distribution of the black-tailed jackrabbit in Wyoming.

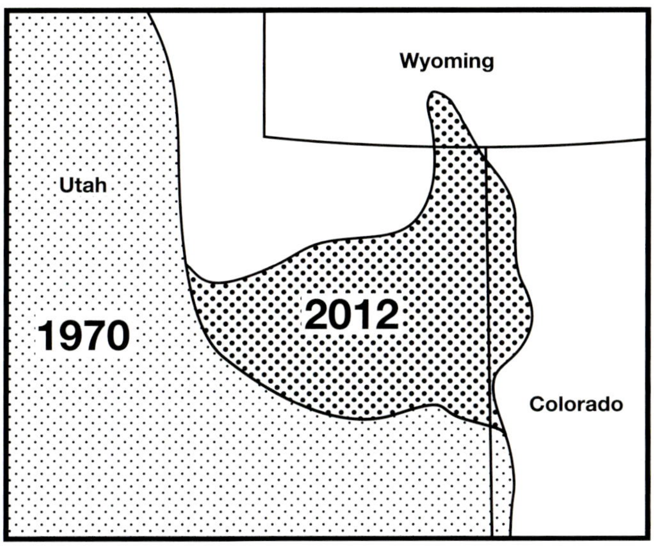

Estimated expansion of the geographic range of the black-tailed jackrabbit into northwestern Colorado and southern Wyoming, 1970–2012.

NOMENCLATURE AND SYSTEMATICS: The genus and species names have been stable for many decades. Of the 17 named subspecies, the one occurring in southeastern counties is *L. c. melanotis*, and that in southern Sweetwater County is not clear.[4]

DISTRIBUTION AND HABITAT ASSOCIATIONS: This species occurs in the southeast corner of Wyoming, particularly Laramie and Goshen Counties. A second area of occurrence in the state was recently recognized along the eastern shore of Flaming Gorge Reservoir, extending from the border with Colorado north to near the impounded confluence of the Blacks Fork and the Green Rivers.[5] The distribution also extends eastward to the vicinity of Adobe Town Rim in southeastern Sweetwater County.[6] This occurrence in Wyoming reflects a rapid range expansion into northwestern Colorado and adjacent Wyoming from the south and west. The species was not known from northwestern Colorado before 1972, but now is found along the entire western border with Utah. The occupied area in Sweetwater County is the vanguard of that northward range expansion, the species having first been detected in Moffat County, Colorado, just south of Sweetwater County, in 1991.[7] Habitats are highly variable across the geographic range, but can include arid and grazed-to-overgrazed sites with little vegetative structure. In the Flaming Gorge area, the species associates with sagebrush and greasewood types, commonly with scattered Utah junipers.

ECOLOGY: This solitary, nocturnally active herbivore associates with a wide range of warm-climate environments across its broad geographic range. Its presence in Wyoming shows that it can persist in shortgrass prairie with little grazing and long, cold winters. The diet in Colorado was found to shift seasonally from grasses in spring to forbs in fall and shrubs in winter. The shrubs eaten in Utah included most species present: big sagebrush, saltbush, and halogeton.[4] Population biology has been studied in northern Utah and southern Idaho, where fluctuations approaching cycles were reported. The cycles appeared to have periods of around eight years, and densities varied between low and high years by a factor of 9, with a range of 0.12–1/ha (31–261/mi^2) in the autumn.[8] Adult sex ratios are near 1:1. Survival rates are low—in northern Utah 20% of animals survive from birth to the following March. Jackrabbits are important prey animals for a large group of grassland and shrub-steppe predators: coyotes, bobcats, wolves, golden eagles, and other raptors. Long-distance dispersal is possible, with movements of up to 35 km (22 mi) recorded.[9] Coyote populations are particularly linked to those of black-tailed jackrabbits, both species fluctuating in unison in one study in northern Utah. There, 64% of all deaths resulted from coyote predation.[10]

CONSERVATION AND MANAGEMENT: The species receives little conservation attention in the state, partly because of its status as a "predator" and partly because of its limited distribution. Paleontological evidence shows that black-tailed and white-tailed jackrabbits in the northern Great Basin have fluctuated inversely in response to climate change during the Holocene.[11] This inverse relationship results from the differing temperature adaptations of the two species, with white-tailed jackrabbits prevailing at high latitudes, high elevations, and during periods of cold climate. It is likely that the recent expansion of the species into northwestern Colorado and southern Wyoming is the result of climatic warming, and it is plausible that ongoing climate warming will result in a much wider distribution for the black-tailed jackrabbit in Wyoming, with consequent displacement of white-tailed jackrabbits. The species is an economic pest in cropland areas, and preferentially consumes barley and alfalfa, but not potato, where available.[12]

NOTES AND REFERENCES

1. Hall (1951), Best (1996). This character is useful for southeastern Wyoming, but not in the Great Basin part of the zone of sympatry.
2. Orr (1940).
3. Wyoming Statutes 23-1-101-a-viii.

4. Sparks (1968), Best (1996).
5. Mark Zornes, Wyoming Game and Fish Department, Green River, pers. comm. He described the occupied area, where he considered the species to be common.
6. Beth Buskirk, 41.291393° N, 108.403708° W, September 1, 2013.
7. Meaney (1991), Armstrong et al. (2011).
8. Gross et al. (1974).
9. Smith et al. (2002).
10. Stoddart (1970), Wagner and Stoddart (1972).
11. Grayson (1977).
12. Fagerstone et al. (1980).

White-tailed jackrabbit, *Lepus townsendii*

DESCRIPTION: A large hare with long pinnae and seasonally variable pelage color—this is a true varying hare in the Wyoming portion of its geographic range. Summer color is light brown with lighter undersides; in winter the pelage is off-white, with rusty brown legs and pinnae. In all seasons, the ventral tail is white, and the amount of gray on the dorsal tail is variable, in some animals slight and limited to the midline. The tips of the ears are black. The sexes are similar in size, and sex determination requires close examination of the genitalia. Fecal pellets are not readily distinguished from those of black-tailed jackrabbits, nor from those of snowshoe hares. Measurements (**Sex** [n] mean (mm, g), *SE*): Lincoln, Uinta, and Sweetwater Counties, HF = **F** [53] 150, *0.9*; **M** [44] 148, *0.9*; EFN = **F** [53] 106, *0.8*, **M** [45] 105, *1.0*; WT = **F** [53] 2900, *30*, **M** [44] 2800, *40*.[1] Dental formula: 2/1, 0/0, 3/2, 3/3 = 28.

STATUS: The species is classified by law as a predatory animal, rather than as wildlife, and receives no management by the Wyoming Game and Fish Department.[2]

NOMENCLATURE AND TAXONOMY: The genus, species, and common names have been stable for decades. The species name honors John Kirk Townsend, a naturalist on Nathaniel Wyeth's 1834 expedition from St. Louis to Oregon, who collected the type specimen. Two subspecies were recognized as of 1981, with the crest of the Rocky Mountains the approximate subspecies boundary between *L. t. campanius* to the east and *L. t. townsendii* to the west. That boundary runs roughly north–south through western Wyoming, but no barriers to movement correspond with this putative boundary.[3]

DISTRIBUTION AND HABITAT ASSOCIATIONS: This north-temperate western species is distributed nearly statewide, with extensive closed-canopy forest the only likely exception. Habitat is typically shrub-steppe or grassland, where it is common, but includes open areas above the elevational limit of trees, at least in summer.[4] The species may undergo seasonal elevational migrations, so that alpine meadows and tundra are not occupied in winter.

ECOLOGY: This solitary herbivore, a hindgut digester with a well-developed cecum, feeds on herbaceous plant leaves during the warm season, shifting toward shrub stems in winter. It remains active all winter, sheltering in shallow depressions or in the lee of rocks or shrubs;

White-tailed jackrabbit in winter pelage.

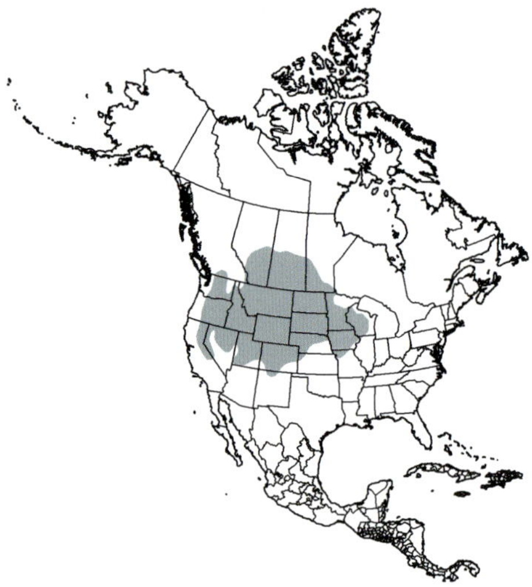

Distribution of the white-tailed jackrabbit in Wyoming.

it does not occupy burrows. Dandelion and western wheatgrass are important warm-season foods. Commonly eaten shrub species in winter and spring include fringed sagewort, big sagebrush, and rabbitbrush. Cultivated plants, especially alfalfa and winter wheat, are readily consumed.[5] Populations have been studied in Uinta County, and densities were remarkably stable from March to October, with mean densities of 7/km^2 (18/mi^2) over three years. On the Pawnee National Grassland of northern Colorado, where they were sympatric with black-tailed jackrabbits, white-tailed jackrabbits had estimated densities of 0.4–2.3/km^2 (1–5.9/mi^2). Sex ratios of adults in southwestern Wyoming were nearly 1:1, and life expectancy at birth was <1 year. Only 12% of animals were >1.5 years old.[6] Jackrabbits are important prey animals for many predators: coyotes, bobcats, Canada lynx, wolves, golden eagles, and other raptors.[7]

CONSERVATION AND MANAGEMENT: Little management attention is directed toward this species in Wyoming, in part because of its legal status as a "predatory animal", and in part because of its presumed commonness. During the 2000s, questions arose about whether the species had become less common in some areas, specifically northern Yellowstone National Park.[8] The putative decline was attributed to increasingly abundant coyotes, which followed the extinction of wolves in the 1920s.[9] This controversy was difficult to resolve with data because the distribution and abundance of jackrabbits had been poorly documented. The species has been noted to have become extinct in at least one other western national park during the historical period.[10] The northward expanding black-tailed jackrabbit is expected to displace this species to some degree.[11]

NOTES AND REFERENCES

1. Rogowitz and Wolfe (1991).
2. Wyoming Statutes 23-1-101-a-viii.
3. Lim (1987).
4. Braun and Streeter (1968).
5. Lim (1987), Rogowitz (1997).
6. Flinders and Hansen (1973, 1975), Rogowitz and Wolfe (1991).
7. Gunther et al. (2009).
8. Berger (2008).
9. Newmark (1995).
10. See discussion at black-tailed jackrabbit.
11. Rogowitz and Wolfe (1991).

Carnivores—Order Carnivora

The carnivores are a diverse and cosmopolitan lineage that originated in the landmass now known as North America around 42 Ma. Shortly after its origin, the group diverged into two lineages recognizable today: the Caniformia, represented in Wyoming by dogs, bears, raccoons, ringtails, weasels, and skunks, and the Feliformia, represented by cats. Traditionally, this order comprised 11 families, but recent molecular and genomic studies have split five new families of carnivores from the existing phylogenetic tree. Order Carnivora is characterized by several morphological features: 4–5 digits per foot, large, pointed canine teeth, rooted cheek teeth that have more (cats) or less (bears) sharp, shearing edges between the upper fourth premolar and the lower first molar, and a diet that tends toward animal material. Carnivores have a baculum, a well-developed zygomatic arch, and a large cranium. They exhibit a number of physiological and ecological consequences of their diets, including simple digestive tracts, large home ranges, and low population densities, compared to herbivores. Of course, the predaceous tendencies of carnivores are variable across families, with some bear species mostly herbivorous, raccoons fairly omnivorous, and cats entirely predaceous.

The roles of carnivores in the ecology, economy, and culture of Wyoming are large and visible; these mammals figure prominently in popular culture and public policy. The roughly two dozen extant species consume many species of birds, mammals, and insects, and limit populations of some of them. They also disperse seeds of berry-producing shrubs and raid seed caches of some rodents.[1] In recent years, carnivores have been shown or proposed to have newly recognized influences on the communities of which they are part, beyond killing prey. These include affecting the habitat use, distribution and behaviors of prey, limiting or reducing abundances of smaller carnivores, and cascading effects on other community members.[2] Most studies of these possible effects have been conducted outside Wyoming, but some have related to wolves, prey, and vegetation in the Greater Yellowstone Area.[3] How reintroduced wolves have affected the Yellowstone community has been controversial, with some claiming strong cascading effects (via wapiti) on vegetation, and others challenging those claims.[4] A 2013 study tested whether wapiti were encountering wolves frequently enough, and reacting to their presence strongly enough, that their opportunity to feed and maintain body condition was impaired. The study showed that wapiti encountered wolves within a distance of 1 km (0.6 mi) only once every nine days for migratory animals, and once every 50 days for nonmigratory ones, so that such indirect effects of wolves on body condition of their prey were not plausible.[5]

NOTES AND REFERENCES

1. See accounts for grizzly bear and Pacific marten.
2. Hambäck et al. (2004), Estes et al. (2011).
3. Berger et al. (2003), Hebblewhite and Smith (2010).
4. Ripple and Beschta (2007), Kauffman et al. (2010).
5. Middleton, Kauffman et al. (2013).

DOGS AND ALLIES—FAMILY CANIDAE

Wild members of the Canidae are characterized by their long rostrums, long well-furred tails, four digits per foot that touch the ground, and a hunting pattern that favors walking or trotting over long distances, rather than lying in ambush. Their body forms are similar, from the smallest fox to the largest wolf; canids differ morphologically mostly in their body size, ranging in Wyoming from the swift fox (~2.3 kg [5 lb]) to the wolf (~57 kg [126 lb]). Species within this family tend to show strong interspecific competition across species, more so than is the case for the Mustelidae or Procyonidae. Thus, wolves influence the distribution and abundance of coyotes, as coyotes do red and swift foxes.

Coyote, *Canis latrans*

DESCRIPTION: A medium-sized carnivore, with typical dog-like form, locomotion, and behavior. Males are slightly larger than females, and the body shape is similar to that of other wild canids. The pelage is seasonally long, variably gray with buff and rufous highlights, and all coyotes have black tips on their tails. Coyote–dog hybrids, rarer in Wyoming than in southeastern states, can be difficult to identify at a glance. Measurements (**Sex** [n] mean (kg), *range*): adults, Kansas, WT = **F** [155] 12, *8–15*; **M** [145] 14, *9–19*.[1] Dental formula: 3/3, 1/1, 4/4, 2/3 = 42.

STATUS: The species is classified as a "predatory animal" under state law.[2] It is common or abundant statewide.

NOMENCLATURE AND SYSTEMATICS: As many as 19 subspecies have been named, including for geographic areas the species did not occupy 150 years ago.[3] Coyotes readily disperse long distances through diverse habitats, producing high gene flow and low levels of genetic structure. So, subspecies endemic to areas colonized since 1850 are of doubtful validity.[4] Geographic variation occurs—especially in body size—across new areas the species has occupied; however, this is more ecotypic than phylogenetic in nature. Hybrids with dogs and introgressed types of uncertain ancestry called "feral *Canis*" are known across the continent, and such animals complicate all taxonomic issues for wild and feralized North American *Canis*.[5]

DISTRIBUTION AND HABITAT ASSOCIATIONS: Currently extending from Costa Rica to the Arctic Ocean, and from the Pacific Ocean to Newfoundland Island, the coyote was originally distributed over a much more restricted area (shown), centered on the Rocky Mountains and extending south to southern Mexico.[6] It occurred over most of Wyoming at the time of European contact, being referred to in some historical accounts as "brush wolves," although likely at lower densities than today, because of much more widespread and abundant wolves then. There is scarcely a habitat or township of land not occupied by coyotes in Wyoming today.

ECOLOGY: The ecology of the coyote is perhaps the best known of any wild carnivoran across North America.[7] The diet is a wide range of animal and plant species. In the arid

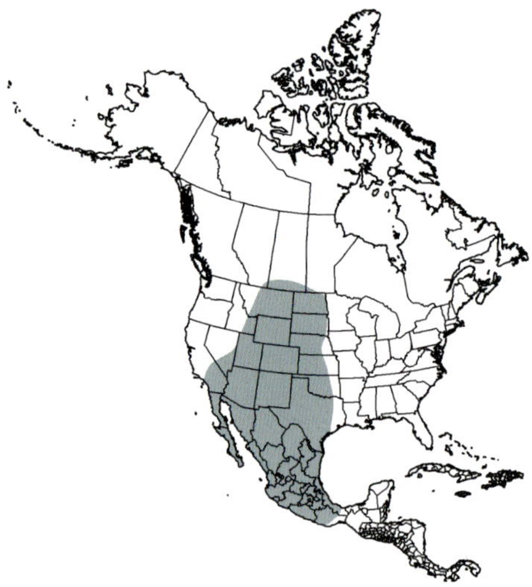

Distribution of the coyote in Wyoming.

West, diets tend to be dominated by leporids (rabbits and hares), with ungulate carrion and small rodents seasonally important. In central Wyoming, coyotes consumed those three major food types in roughly equal proportions.[8] Olaus Murie, working in Grand Teton National Park and the nearby Thorofare country, found the summer diet dominated by pocket gophers, voles, and grasshoppers. His brother Adolph, studying coyotes in Yellowstone National Park, found diets from April to November likewise to be dominated by pocket gophers and voles, although elk made up 15% of food occurrences. On the Yellowstone Northern Range, where elk and deer overwintered, diets were dominated by those species.[9] After wolves returned to Yellowstone, coyote diets became subsidized by carrion from wolf kills; at the same time, ungulate carcasses became the sites of conflict between the two canid species, commonly resulting in harassment or death of coyotes.[10] Coyotes also prey on bird eggs and nestlings, and pronghorn fawns in summer.[11] In the Black Hills, the largest diet item was white-tailed deer (44%), followed by cottontails (16%) and voles (12%).[12] Livestock, both killed by coyotes and scavenged, are doubtless important locally.

A study of population biology on the National Elk Refuge in the 1970s found mean litter size to be 4.5, and mean post-whelping population density to be 1.4/km² (3.6/mi²). Spring population density averaged 0.53/km² (1.4/mi²) across five years.[13] A critical factor in coyote ecology is the abundance of sympatric larger carnivores, particularly the gray wolf. Coyotes are especially vulnerable to interference competition from wolves, which are 2–4 times the weight of coyotes; this size difference leads generally to a high likelihood of interspecific killing of smaller carnivores by larger ones.[14] When the reintroduction of wolves was planned, the demise of coyotes in the recovery area was predicted; by 10 years after the first wolf releases, coyotes had declined by 39–50% in areas near roads in Yellowstone National Park. It is estimated that wolves kill 25–33% of the coyote population yearly. Coyotes adjust territory boundaries, diets, and behaviors in response to the presence of wolves.[15] Where small prey are important, group sizes are 1–2; in Yellowstone and elsewhere where ungulates are preyed upon or scavenged, packs can be as large as 10.[16] Human causes likely account for most coyote deaths in the state. Of 19 deaths of coyotes radio-collared in Grand Teton National Park, human causes accounted for 17.[17]

CONSERVATION AND MANAGEMENT: The coyote is not a conservation concern across Wyoming; the animal is widespread, and likely more abundant than 150 years ago when wolves occurred statewide. Even persistent and widespread efforts at control today are not effective in reducing densities or distribution more than temporarily or locally.

NOTES AND REFERENCES

1. Gier (1968).
2. Wyoming Statutes 23-1-101-a-viii.
3. Hall (1981).
4. Roy et al. (1994).
5. Thurber and Peterson (1991), Lehman et al. (1991).
6. Continental distribution is modified from that by Stanley D. Gehrt, Urban Coyote Program.
7. Bekoff and Gese (2003).
8. Springer and Smith (1981).
9. Murie (1935), Murie (1940).
10. Merkle et al. (2009).
11. Berger and Conner (2008).
12. MacCracken and Uresk (1984). Use of white-tailed deer peaked in winter; relative contributions of predation and scavenging were not known.
13. Camenzind (1978).
14. Donadio and Buskirk (2006).
15. Ballard et al. (2003), Berger and Gese (2007).
16. Gese et al. (1996).
17. Tzilkowski and Knowlton (1979).

Gray wolf, *Canis lupus*

DESCRIPTION: A large-bodied wild dog, with seasonally long fur variable in color from nearly black to pale buff, or even white. The tail is typically carried at a higher angle than that of a coyote while moving. Vocalizations—howls and barks—are recognizable even to untrained observers as not those of coyotes. Measurements (**Sex** [n] mean (mm, g), *range*): >1 year old, Yellowstone National Park, L = **F** [72] 1774, *705–1990*; **M** [91] 1866, *1350–2210*; SH = **F** [69] 770, *580–890*; **M** [89] 831, *650–980*; WT = **F** [115] 41.4, *28–55*; **M** [137] 49.2, *32–65*.[1] Dental formula: 3/3, 1/1, 4/4, 2/3 = 42.

STATUS: The species was listed as endangered under E.S.A. in the contiguous United States in 1974, with both eastern and western subspecies listed in their respective geographic areas. The gray wolf was relisted at the species level in 1978.[2] A recovery plan for the northern Rocky Mountain region was approved in 1987, and the Yellowstone area was named as one recovery area. The species was reintroduced to Yellowstone National Park beginning in 1995 (with the status of Threatened) and simultaneously given a status of a nonessential, experimental population outside national parks.[3] Its legal status in Wyoming outside the national parks from 1995–2012 remained nonessential and experimental. In 2012, the Wyoming portion of the Greater Yellowstone gray wolf population was declared recovered and not in need of further E.S.A. protection. By 2014, the management and legal status of the species in Wyoming remained unstable.

NOMENCLATURE AND SYSTEMATICS: The species was described from Sweden by Linnaeus in 1758, using the same genus-species epithet applied here. Still, the taxonomy at species and subspecies levels is confused, with issues of introgression among coyotes, red wolves, and domestic dogs confounding the biological species concept in this genus. Some regard genus *Canis* as a "superspecies" or "species complex" because of the deep introgression among the various types, domestic and wild. Around 25 subspecies of wolf were recognized in 1981, three of which were mapped as occurring in Wyoming, but these 25 subspecies—some described on the basis of body size, diet (e.g., bison wolf), or major habitat (e.g., plains wolf) have not been challenged with molecular data, and most of the forms earlier found in the contiguous United States no longer occur.[4] Those translocated to Yellowstone from Alberta and British Columbia would have been from the geographic ranges of *C. l. columbianus* or *C. l. occidentalis*. Various common names, including "wolf" and "timber wolf" are also applied to the species.

DISTRIBUTION AND HABITAT ASSOCIATIONS: The species occurred over most of North America before Europeans settled (shown). The distribution was statewide at the time of European contact, shrank to the northwestern mountains by the early twentieth century, and disappeared completely by 1930. A few reports of individual wolves in northwestern Wyoming appeared during 1930–1995.[5] Reintroduced animals were released in three areas in northern Yellowstone in 1995–1996, then the population expanded to the area mapped by around 2010.[6] Most, but not all, of the area marked as the current distribution

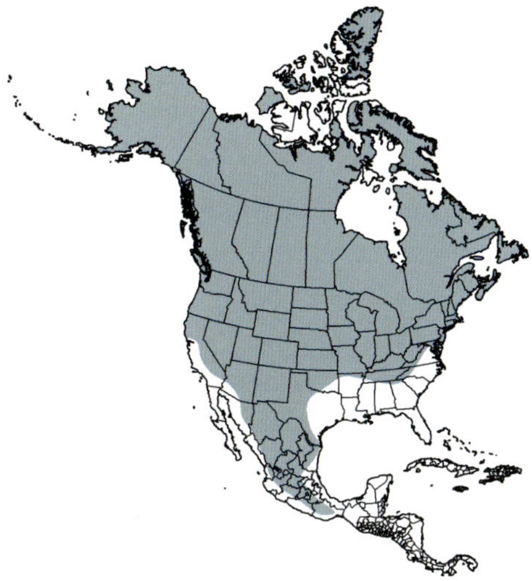

Distribution of known packs of gray wolves in Wyoming in 2012. Individual wolves range variable distances beyond this area.

contains wolf packs. Individual animals are occasionally observed in almost all parts of Wyoming, as well as in Colorado, Utah, and Idaho, with decreasing frequency at increasing distances from the northwestern mountains. The wolf is a habitat generalist, being found in almost all habitats that support ungulates at sufficient densities: plains, shrub-steppe, tundra, and forest. Three centuries ago, it was the most widely distributed mammalian species in the world, spanning almost all of the temperate and subpolar Northern Hemisphere.[7]

ECOLOGY: The wolf is well understood in terms of ecology, behavior, and population biology. Since its reintroduction in 1995, the Yellowstone population has received more research and monitoring attention than perhaps any other Wyoming mammal.[8] The diet in Yellowstone is mostly wapiti (88% of kills, 1995–2003) followed by bison (6%) and other ungulates.[9] Packs are the basis of population organization, with dominant animals typically excluding others from mating. Large males are primarily involved in subduing and killing elk; smaller-bodied animals were more successful at pursuing and isolating targeted elk.[10] Reproductive performance of females increased with body size, and to age 4–5, after which it declined.[11] The estimated number of wolves in Yellowstone National Park and Wyoming outside the park at the end of 2013 was 306, in at least 43 packs.[12] Considerable research and discussion attends the issue of whether wolves have affected the communities into which they were reintroduced. Indeed, Yellowstone has been considered a useful laboratory for investigating the ecological effects of top predators in terrestrial communities—an important issue in contemporary community ecology. Unfortunately, the reintroduction of wolves represents an imperfect experiment, because of confounding environmental factors that changed around the same time that wolves were released: vegetation recovery following the 1988 fires, the onset of a severe multiyear drought, the recovery of the grizzly bear population in distribution and abundance, a severe die-off of whitebark pine, and the introduction of nonnative lake trout to Yellowstone Lake, which led to major declines of cutthroat trout. Each of these factors had the potential to cause some effects that are ecologically linked to wolves or their foods.[13]

CONSERVATION AND MANAGEMENT: Management of the species is controversial and highly political, with debate and discussion featuring prominently in the public media since the early 1990s. Key considerations involve identifying damage to domestic livestock outside Yellowstone National Park and handling individual offending animals or packs. During 2013, 62 wolves were harvested by hunters and 33 removed in agency control efforts. Wolves caused 41 confirmed losses of cattle and 33 of sheep during that year.[12] Economic benefits from the presence of wolves in Yellowstone National Park include visitor spending related to the desire to observe wolves in the three states that make up the Greater Yellowstone Area, estimated to be tens of millions of dollars in 2005.[14]

NOTES AND REFERENCES

1. Unpublished data from National Park Service, Yellowstone Wolf Project, 1995–2010, used by permission.
2. US Fish and Wildlife Service (1978).
3. US Fish and Wildlife Service (1994).
4. Mech (1974), Hall (1981).
5. Fritts et al. (1995).
6. Schullery (1996).
7. Corbet (1978).
8. The grizzly bear is the only possible exception.
9. Stahler et al. (2006).
10. MacNulty et al. (2009).
11. Stahler et al. (2012).
12. Jimenez and Becker (2014).

13. The Yellowstone-based literature on putative effects of wolves on community members has grown rapidly. See discussions in Berger et al. (2003), Ripple and Beschta (2003, 2007), Berger and Gese (2007), and Kauffman et al. (2010).

14. Duffield et al. (2008).

Red fox, *Vulpes fulva*

DESCRIPTION: A medium-sized fox with prominent pinnae, black on their posterior surfaces, and a long, well-furred, white-tipped tail. Sexes are similar in size. The color of adults varies geographically, with age and season, and across individuals within a population. In the Far North, discrete coat color polymorphism is apparent, with distinct red, black, and "cross" color phases common. In Wyoming, color variation is less pronounced—although black forms are observed—with most animals in various shades of red or reddish blond when coats are prime. Animals from northeastern Yellowstone National Park and the Beartooth Plateau have a grayish cast, in common with the high-elevation red foxes of the Cascades and Sierra Nevada. Measurements (**Sex** [n] mean (mm, g), *range*): adults, Iowa, L = **F** [22] 946, *842–1020*; **M** [19] 998, *954–1045*; TL = **F** [22] 337, *294–368*; **M** [19] 359, *320–390*; HF = **F** [22] 145, *127–156*; **M** [19] 157, *140–170*; EFN = **F** [22] 79, *66–85*; **M** [19] 85, *75–93*; WT = **F** [22] 3938, *2951–4585*; **M** [19] 4822, *4131–5675*.[1] Dental formula: 3/3, 1/1, 4/4, 2/3 = 42.

STATUS: It is considered a "predatory animal" under Wyoming statute and is not protected.[2] Animals of the high-elevation type from near the northeastern corner of Yellowstone National Park are not accorded special regulatory status.

NOMENCLATURE AND SYSTEMATICS: Classification at the species and subspecies levels has changed markedly through the decades, along with our understandings of phylogeography across North America. Early taxonomies used *V. fulva* and various subspecific epithets.[3] The use of *V. vulpes* beginning around 1990 reflected the presumed close relatedness of New and Old World forms. Most recently, molecular evidence has shown that North America was colonized by a single event ca. 0.4 Ma, with no detectable subsequent eastward exchange across Beringia, and that North American red foxes are distinct from the Old World ancestral lineage at the species level.[4] Interestingly, for most of the late twentieth century it was believed that red foxes in temperate North America were of two distinct lineages. The first lineage comprised native red foxes, limited to Alaska, boreal Canada, mountains of the Pacific States (Cascades and Sierra Nevada), and the northern Rocky Mountains.[5] Animals on the Beartooth Plateau and adjacent areas of Yellowstone National Park were thought to be such high-elevation types. All of these high-elevation western populations shared a phenotypic trait—a gray cast to the red or blond pelage.[6] The second presumed lineage, which was understood to account for most of the foxes at low elevations across the temperate Midwest and West, were thought to be descendants of animals introduced from Western Europe to the northeastern United States, which had escaped or were released from captivity, and spread eastward across temperate North America in a colonization front.[7]

Red fox in standard pelage.

Red fox with gray-cast pelage, typical of mountain red foxes of the Beartooth Plateau and Cascades.

Recent molecular studies have provided a different interpretation: red foxes in the eastern and Midwestern United States—including the Dakotas and Kansas—are almost entirely of North American ancestry.[8] According to this new understanding, introduced European foxes have made little genetic contribution to the red foxes in the region that includes Wyoming. Such a pattern may reflect previously unrecognized reproductive barriers between native and any possible introduced red foxes. A recent study in California, where introduced European red foxes are known to occur adjacent to native high-elevation foxes, showed that the two lineages had not introgressed, in spite of decades of contact.[9] The same pattern has been found for high-elevation red foxes on the Beartooth Mountains, which showed partial reproductive isolation from nearby low-elevation animals.[6] Although over 40 subspecies have been named worldwide, and 10 in North America, it is difficult to assign any subspecific name to the Wyoming portion of the geographic range. Foxes in the northwestern mountains might be considered *V. f. macroura*, because of their gray pelage and partial genetic isolation from low-elevation forms. But two other subspecies mapped by Hall (1981) approach the southeast corner of Wyoming, and the validity of either of these subspecies—or whether either may have colonized the state—is not clear.[10]

DISTRIBUTION AND HABITAT ASSOCIATIONS: This is arguably the most broadly distributed mammal species in the world today, yet the pre-European distribution in Wyoming and surrounding region is uncertain. Long (1965) depicted the species as occurring statewide, but showed only one specimen collected outside the northwestern mountains. Hall (1981) showed the species as occurring in the northwestern mountains, but absent from most of Wyoming. The plausible scarcity of red foxes over lower elevations of the Intermountain West over most of the twentieth century (shown) has been variously attributed to nonspecific trapping and poisoning of carnivores, or to the presettlement rarity of the red fox in these areas.[11] Skinner noted them to be "a rare animal in" Yellowstone National Park, observing them only twice (in 1919 and 1922) over a 10 year period.[3] They were not mentioned by Col. Dodge among Black Hills mammals in 1875, although he described mountain lions, Canada lynx, gray wolves, and gray foxes. By 1974, it was noted that "*Vulpes vulpes* [sic] was relatively uncommon in the Black Hills five to 10 years ago, but . . . since 1965 it has increased noticeably."[12] On coyote survey traplines conducted at decadal intervals over the period 1940–1970, the red fox was absent during 1940 and 1950, appeared in 1960 (one animal) and grew to 57 animals in 1970.[13] These accounts suggest that red foxes were absent or rare over most of the interior of Wyoming away from cities and towns before the 1960s. Red foxes today appear to be most common near towns and associate with human development in the low-elevation parts of the state; the opposite is true of coyotes, which are a lethal hazard to red foxes.[14] So, red foxes likely avoid areas away from towns because of high coyote densities and the reduced availability of human foods away from towns. Across its North American range, the red fox can be found in a wide range of habitats, from Arctic tundra to southwestern deserts, but definite habitat preferences (e.g., montane forests) and barriers (deserts) are recognized.[15] Red fox habitats in Wyoming have been studied only in Yellowstone National Park, where foxes were found to associate more closely than expected with stands of Douglas-fir and less than expected with

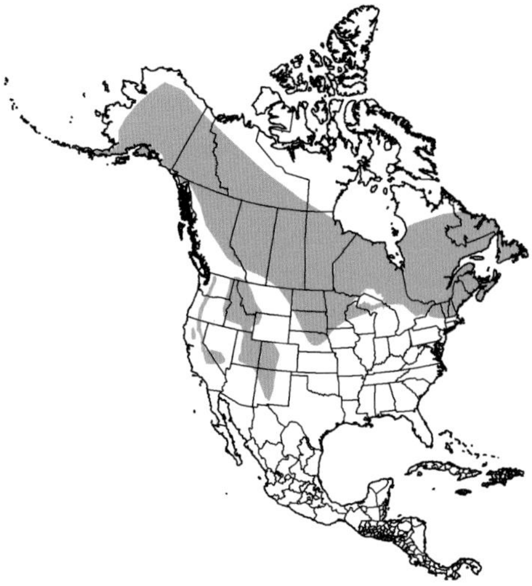

Presettlement geographic range of the red fox in North America.

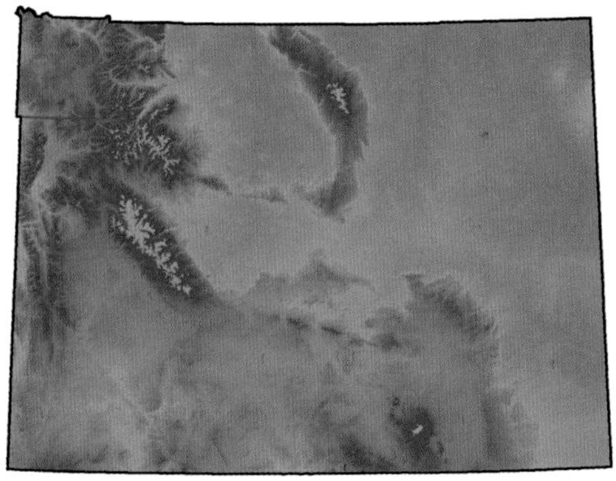

Distribution of the red fox in Wyoming. The species is mostly limited to areas near towns and other developments.

open mesic grasslands. Aggressive behaviors by coyotes toward red foxes were thought to be an important mechanism underlying habitat selection.[16]

ECOLOGY: The species is poorly studied in Wyoming outside of Yellowstone. It is a territorial, seasonally monogamous omnivore. Small, ground-dwelling mammals likely make up most of the diet, but near developed or residential areas red foxes may live mostly on human foods and carrion of road-killed vertebrates. Episodic mortality events can drive populations to low levels over fairly wide areas. Red foxes were common within the cities of Laramie and Fort Collins, Colorado (90 km [50 mi] from Laramie) in the late 2000s, but a die-off in both cities in around 2011—generally attributed to sarcoptic mange—reduced populations to low levels for at least three years.[17] Over most of Wyoming, few native plant foods would be of value to foxes. A study of trapped red foxes in southeastern Wyoming showed that 74% were juveniles (<1 year old).[18]

CONSERVATION AND MANAGEMENT: The species is not generally considered a conservation concern in Wyoming, in part because of its legal classification as a predatory animal, rather than as a furbearer. High-elevation populations from the Pacific states have become conservation concerns because they occupy restricted areas and have gone extinct in some mountain ranges, including the Sierra Nevada.

NOTES AND REFERENCES

1. Storm et al. (1976).
2. Wyoming Statutes 23-1-101-a-viii.
3. Skinner (1927).
4. Statham et al. (2014).
5. Aubry (1983), Aubry et al. (2009).

6. Swanson et al. (2005).

7. Kamler and Ballard (2002, fig. 2) depicted the westward expansion of the red fox as extending across most of Wyoming during the 1950s and 1960s.

8. Statham et al. (2012). Frey (2013) showed that documented translocations from Europe to eastern North America were very few in number, and concluded that the importance of translocations from Europe had been exaggerated.

9. Sacks et al. (2011).

10. Hall (1981).

11. Long (1965), Fichter and Williams (1967), Hall (1981).

12. Dodge (1876), Turner (1974).

13. The total number of trap nights of effort by decade are 9491 (1940–1941), 9681 (1951), 9913 (1960), and 9787 (1970) (Linhart and Robinson 1972). Federal animal damage trapping reports, which are based on damage-causing animals, mirrored these trends, with no red foxes removed from 1948 to 1959. Red foxes first appeared in those data in 1960, and increased until 819 animals were removed in 1975 (Wyoming Game and Fish Department 1982). Trends in fur trapper harvests resemble these patterns, suggesting that red foxes were uncommon or absent in Wyoming during 1940–1959, increased until around 1975, and remained at around that level afterward.

14. Goad et al. (2014).

15. Lariviere and Pasitschniak-Arts (1996).

16. Van Etten et al. (2007).

17. Donal O'Toole, Wyoming State Veterinary Laboratory, and Terry Spraker, Colorado State Veterinary Laboratory, pers. comm.

18. Crowe and Strickland (1975).

Swift fox, *Vulpes velox*

DESCRIPTION: A small wild canid about the size of a small domestic cat with buff or dull orange back, light underside and prominent black tip on a long, well-furred tail. The posterior surfaces of the pinnae are the same color as the dorsum. A dark gray or black patch covers the anterior lateral rostrum. Males are slightly larger than females in overall measurements. Measurements: (**Sex** [n] mean (kg), *range*): Natrona County, WT = **F** [20] 2.26, *2.10–2.55*; **M** [20] 2.39, *1.90–2.90*.[1] Dental formula: 3/3, 1/1, 4/4, 2/3 = 42.

STATUS: A protected nongame animal that may not be taken intentionally, and a state Species of Greatest Conservation Need as of 2012.[2] Uncommon to common within its Wyoming distribution.

NOMENCLATURE AND SYSTEMATICS: The genus and species names have been stable for many decades. The species name is the Latinized "rapid." Some authors earlier considered the swift fox to be conspecific with the more western and southern kit fox (*V. macrotis*); most now consider the two lineages distinct at the species level, in part because of reportedly reduced fitness of hybrids at the main zone of contact in eastern New Mexico.[3] Another zone of contact plausibly could occur in northwestern Colorado; kit foxes have been documented in Delta County, and a small *Vulpes* of unknown species has been reported from Moffat County, just south of Wyoming.[4] Subspecies were designated during the last century and recognized by Hall (1981), but their boundaries do not correspond with population genetic

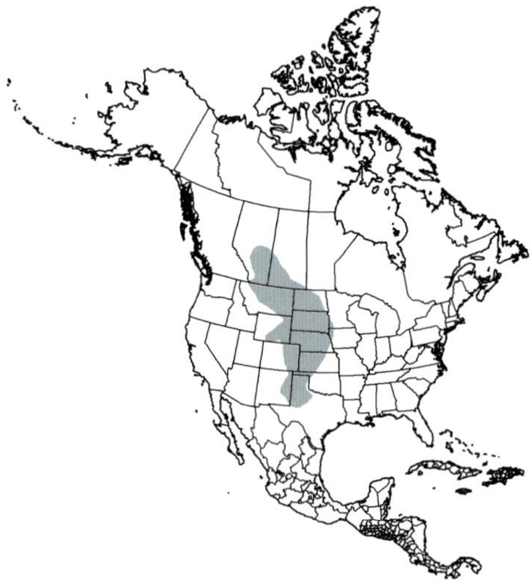

Distribution of the swift fox in Wyoming.

structure based on recent molecular data.[5] Subspecies seem implausible because of the species' strong dispersal ability through shortgrass prairie and the absence of apparent geographic discontinuities at putative subspecies boundaries.

DISTRIBUTION AND HABITAT ASSOCIATIONS: Species-wide, the distribution originally extended from southeastern New Mexico to central Alberta (shown). Today, that area has been reduced by >50%.[6] The distribution in Wyoming covers the eastern half of the state, roughly bounded on the west by the Bighorn Mountains (in the north) and the Green River in the southwest. A natal den near Thermopolis was reported in 2011, and the distribution may extend west of the Green River as well. The known distribution has changed somewhat from that reported in 1965, especially westward, likely in response to extensive surveys conducted during the intervening 50 years; some range expansion may have occurred as well. Records since the mid-1980s occur no farther north than around 43°45′ in Campbell and Weston Counties, and do not occur in Crook County at all. Habitat is short-grass and mixed grass prairie. A shrub component is tolerated as well; the species in Wyoming occupies the grassland–big sagebrush transition and seems to experience high survival rates there.[7]

ECOLOGY: This smallest of canids in the region is a dietary generalist, feeding on small mammals, birds, reptiles, and insects (mostly beetles, grasshoppers, and crickets). Its dry prairie and steppe habitats hold little soft mast that might serve as plant foods. Near Medicine Bow, the summer diet was dominated by mammal material (>95% of fecal droppings with remains), with *Microtus* sp. and *Peromyscus* sp. contributing most material. Insect remains were found in around three-fourths of fecal droppings. The winter diet was similar, but with fewer insects.[8] Parturition occurs in dens, and multiple females with young can share dens at the same time. Density and age structure are poorly known for the species in habitats such as those in Wyoming. Mean annual survival rate in the Bates Hole area just north of Shirley Basin was 58%.[7] The primary cause of death was killing by coyotes (73% of known-cause deaths). Lower survival rates were found in western Kansas where coyotes were likewise the primary cause of death.[9] Additional deaths are caused by vehicle collisions and possibly canine distemper.[7]

CONSERVATION AND MANAGEMENT: The species was proposed for listing under E.S.A. in the 1990s, but found not warranted for such protection in 2001.[10] While range reductions are reported in other states, attributed to habitat loss and fragmentation, habitat in Wyoming is sufficiently intact that range reductions seem to have been minimal.[11]

NOTES AND REFERENCES

1. Olson (2000, Appendix A).
2. Commission Regulations Chapter 52, Section 11.
3. Hall (1981), Cypher (2003), Wilson and Reeder (2005).
4. Link (1995).
5. Hall (1981), Schwalm (2012, Chapter 2).
6. Sovada et al. (2009).
7. Olson and Lindzey (2002).
8. Olson (2000, Table 3).
9. Sovada et al. (1998).
10. US Fish and Wildlife Service (2001).
11. Stephens and Anderson (2005).

Common gray fox, *Urocyon cinereoargenteus*

DESCRIPTION: A medium-sized fox with shorter legs than a red fox. The dorsal pelage is mottled gray; the margins of the dorsum, ventral tail, and posterior ears are pale orange. The lateral rostrum has a black spot, similar to that of the swift fox. The well-furred tail pelage is distinctive: a prominent black stripe follows the dorsal midline. Measurements (**Sex** [n] mean (mm, g), *range*): Alabama, L = **F** [43] 894, *821–940*; **M** [44] 921,

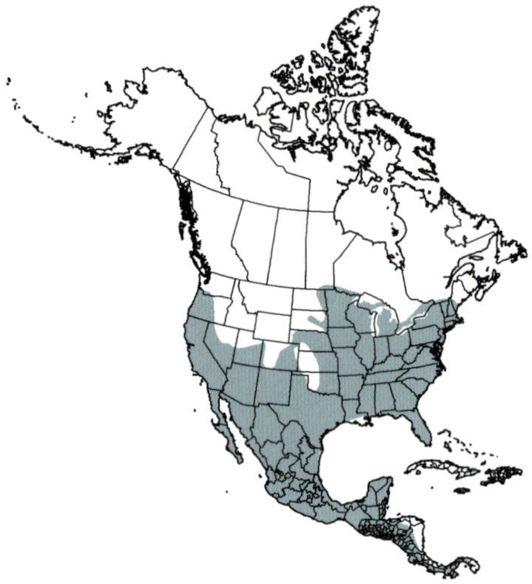

The location of the 1910 specimen of the gray fox, inferred to have originated from Owens Draw, Weston County. No state distribution can be mapped.

820–983; HF = **F** [43] 132, *126–142*; **M** [44] 138, *118–155*; EFN = **F** [43] 68, *62–74*; **M** [44] 69, *63–75*; WT (fall and winter) = **F** [20] 7700, *6200–9200*; **M** [21] 8280, *6800–11,200*.[1] Dental formula: 3/3, 1/1, 4/4, 2/3 = 42.

STATUS: A protected nongame animal, likely extinct in the state.[2] The distribution in the state during the early European period is uncertain. The species underwent review for listing under E.S.A. in late 2012.[3]

NOMENCLATURE AND SYSTEMATICS: The genus and species names have been stable for over 200 years. The specific epithet "*cinereoargenteus*" refers to the silver-gray color. The descriptor "common" is to distinguish it from the island gray fox, *C. g. littoralis*, endemic on the Channel Islands of California.[4]

DISTRIBUTION AND HABITAT ASSOCIATIONS: This south temperate and tropical species extends as far south as Colombia (shown). The single specimen record for Wyoming is a female in the US National Museum collected in Weston County in 1910.[5] No confirmed records are available for the species in Wyoming for the last 100 years. Gray foxes are absent from westernmost South Dakota today, and the South Dakota counties closest to Weston County have gray foxes only in their eastern parts, if at all. The distribution in Nebraska in 1964 included the westernmost counties, but the species is now rare in western Nebraska.[6] The geographic range in the Great Plains was expanding in the decades before 1980; whether this represented recovery of previously occupied range is unclear. The species may have its closest current distribution to Wyoming in northern Colorado, where the geographic range approaches the Wyoming border in Moffat and Larimer Counties.[7] Habitat is wooded, brushy, and rocky habitats, including or especially where low-intensity farming or old fields are interspersed. The species has moderate climbing abilities and can use deciduous trees and rocky, cliffy sites to escape other carnivore species. Coniferous forests with vertical boles lacking easily accessible lower branches seem not to be occupied.

ECOLOGY: The species is a dietary generalist, eating mostly mice and rabbits in winter, but mostly insects in summer. Soft mast, including cultivated fruits, is important in the Midwest during autumn, but few plants in eastern Wyoming would be likely to serve as foods. Like the red fox, this species is affected by interspecific competition from coyotes,[8] which may account for their disappearance from Wyoming over the last century. The close association of the species with physical structure—trees, brush, and cliffy or rocky sites—is consistent with vulnerability to coyote aggression; coyote removal in western Texas was shown to increase abundance of gray foxes.[9]

CONSERVATION AND MANAGEMENT: Although a protected nongame animal, the species is not generally considered a conservation concern, because evidence is lacking of its past presence as a breeding population in the state.

NOTES AND REFERENCES

1. Sullivan (1956).
2. Commission Regulations Chapter 52, Section 11.
3. US Fish and Wildlife Service (2012).
4. Fritzell and Haroldson (1982).
5. USNM Cat. No. 167892, collected by W. O. Randenbush. Long (1965) plots this location as near Newcastle, but the USNM record shows "Owen" as the site of collection. This presumably refers to the Owens Draw Mine, which operated about 5 km (3 mi) east of Upton, Weston County, Wyoming, when the animal was collected. Higgins et al. (2000) showed the nearest South Dakota County to Weston County that has literature and specimen records for gray fox to be Butte County, north of the Black Hills, and showed the species to be uncommon or rare in the western part of the state. Turner (1974) considered the gray fox to be a "species of

uncertain status" in the Black Hills region at the time of his writing, but noted that Dodge (1876, 123) reported that "silver gray" foxes were "numerous" in the Black Hills in 1875. Dodge does not mention the red fox or coyote in his account, but describes "large timber" wolves as "numerous."

6. Jones (1964) showed the species as occurring throughout western Nebraskan counties. Since the requirement that trapped or shot gray foxes be tagged was implemented in 2002, only two gray fox specimens had been tagged in the western one-half of Nebraska by 2012. One specimen was taken in Dawes County in 2008, about 85 km (53 mi) from Wyoming. The other was taken in Frontier County in 2009, much farther east (Sam Wilson, Furbearer Specialist, Nebraska Game and Parks, pers. comm., 2013). The distribution in eastern Wyoming, however extensive it was in decades past, appears to have contracted to the eastward by scores or hundreds of miles.

7. Armstrong et al. (2011).

8. Willingham (2008).

9. Henke and Bryant (1999).

SKUNKS—FAMILY MEPHITIDAE

The skunks, a strictly New World group, were long placed within the Mustelidae, but for several decades were suspected of being closely related to the oriental stink badgers (*Mydaus* sp.).[1] By the 1990s, this suspected relationship was confirmed, and they were placed in the new family Mephitidae.[2] The group is characterized by medium body size, generalized dentition with non-shearing cheek teeth, well-developed anal sacs that are used to spray their contents defensively, and striking black-white patterned pelage. They are solitary and mostly nocturnal.

NOTES AND REFERENCES

1. Radinsky (1973).
2. Dragoo and Honeycutt (1997).

Western spotted skunk, *Spilogale gracilis*, and eastern spotted skunk, *S. putorius*

DESCRIPTION: These two small carnivore species are very similar in appearance, with black pelage marked with symmetrical curved, white bands. The tail is coarsely furred with mostly long white guard hairs (shows eastern species). The stride is a walk or shuffling run. The eastern species is somewhat larger than the western one, but the morphology of the baculum is the only morphological trait that can be used to distinguish between the two forms reliably—not a useful field character. The most useful criteria for identification include molecular markers and reproductive strategies.[1] Measurements (**Sex** [n] mean (mm, g), *range*): *S. gracilis*, northern Arizona, L = **F** [5] 375, *345–402*; **M** [7] 404, *380–426*; TL = **F** [5] 141, *126–156*; **M** [7] 146, *142–160*; HF = **F** [5] 41, *38–47*; **M** [7] 43, *38–47*.[2] Dental formula: 3/3, 1/1, 3/3, 1/2 = 34.

STATUS: Both species are classified as "predatory animals" by state statute.[3] They are uncommon or rare almost anywhere in the state.

NOMENCLATURE AND SYSTEMATICS: The taxonomy of the two species has been complex and unstable over the past century. The genus was recognized in the mid-nineteenth century, but for most of the twentieth century, spotted skunks of the western United States and northern Mexico were considered a subspecies of *S. putorius*: *S. p. gracilis*. By the 1960s evidence began to accumulate that western and eastern forms were distinct at the species level: the morphology of the baculum, timing of the breeding season, and molecular evidence all suggested that the two taxa were distinct biological species in temperate North America.[4] Long (1965) described what were then regarded as two subspecies as occurring parapatrically with little apparent introgression. He stated, "The two kinds seemingly meet without intergrading at Iron Mountain where Mr. Elmer Smith, of Horse Creek, in the course of trapping furbearers in the winter of 1943, took an individual of *S. p. gracilis* and on another day an individual of *S. p. interrupta* in the same [trap] set."[5] More recent molecular studies have suggested that the two lineages of spotted skunk are about as dissimilar as the species of *Mephitis*. Subspecies of *Spilogale* may undergo genetic exchange in Central America, but be fully isolated in Wyoming, in the manner of a ring species.[6] The nominal subspecies of the eastern species found in Wyoming is *S. p. interrupta* and that of the western species is *S. g. gracilis*. A number of regional common names are used to refer to both species, of which "civet cat" is perhaps the most common. Neither "civet" nor "cat" is correct or descriptive.

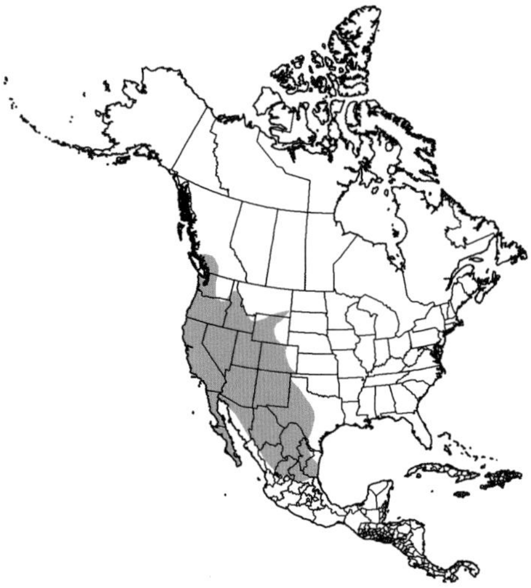

Presettlement geographic range of the western spotted skunk.

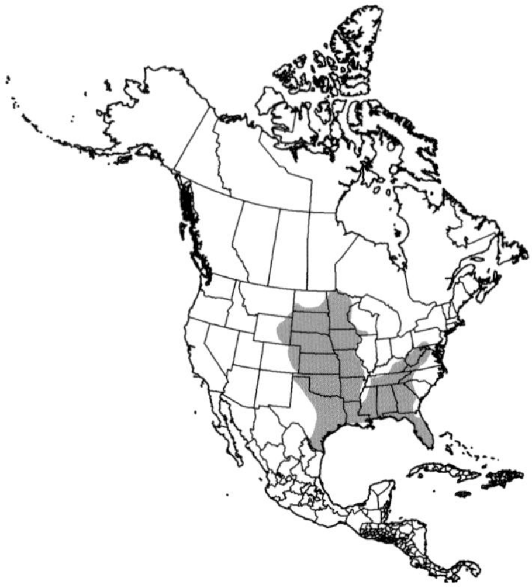

Presettlement geographic range of the eastern spotted skunk.

DISTRIBUTION AND HABITAT ASSOCIATIONS: The two species collectively occupy most of the temperate United States. They occur parapatrically across Wyoming, with few records from the northwestern mountains. Richard Van Gelder (1959) believed that the eastern form had arrived in northern Wyoming during the previous few decades, and attributed the distributional shift to climatic warming over that period.[7] Distributional records are sufficiently patchy across the state that understanding the boundary between species—or where spotted skunks occur at all—is difficult. The putative boundary between the two forms differs across accounts: some have showed the boundary running northwest from Laramie County to the eastern Bighorns, while other authors show the eastern spotted skunk occupying the eastern one-third of the state, or barely entering it, in eastern Laramie and Goshen counties only.[8] Habitats for both species tend to be riparian zones, rocky outcrops, and other habitats with abundant physical structure. Some evidence suggests that the western species is more prone to occupy rocky areas, while the eastern one is more common in plains and level areas.

ECOLOGY: Food habits of these skunks in Wyoming are unknown, but they are omnivorous elsewhere. They depend largely on insects in summer, but also eat birds and their eggs, small mammals, fruits, small reptiles, and carrion. Both species are nocturnal, and neither hibernates in winter. The eastern species mates in spring and immediately enters active gestation; the western one mates in fall, and delays implantation of the blastocyst until spring. The population biology of both species is known fairly well, but for neither species in environments like Wyoming's.[4]

CONSERVATION AND MANAGEMENT: These two species pose interesting examples of the difficulty of understanding the nature and causes of long-term changes in distribution and abundance of uncommon, shy species over large geographic areas. The eastern spotted skunk is generally regarded as having declined in abundance across the Midwest, including Kansas and Nebraska, with most accounts describing a decrease in trapping harvest by a factor of 100–1000, first noted in the 1940s, but continuing through the 1980s.[9] As a result, the species is considered a species of conservation concern in several eastern states—as well as Montana—where it was perceived to be abundant 75 years ago. The subspecies that occurs in Wyoming was reviewed for possible E.S.A. listing in late 2012.[10] A dramatic decline in abundance of eastern spotted skunks began around the end of World War II, and continued for 2–3 decades. No satisfactory explanation has been postulated for such a decline, but evidence for it from Wyoming is lacking, in part because spotted skunks likely never were abundant here. Although abundance was not specifically addressed, a comparison of the distribution of western spotted skunks across the Sierra Nevada from the early 1900s to 2002 showed no obvious change in distribution, although some other mid-sized carnivore species had changed dramatically.[11] Considering its general rarity and status as a "predatory animal," it is not surprising that we have very little information on which to base an assessment of its conservation status.

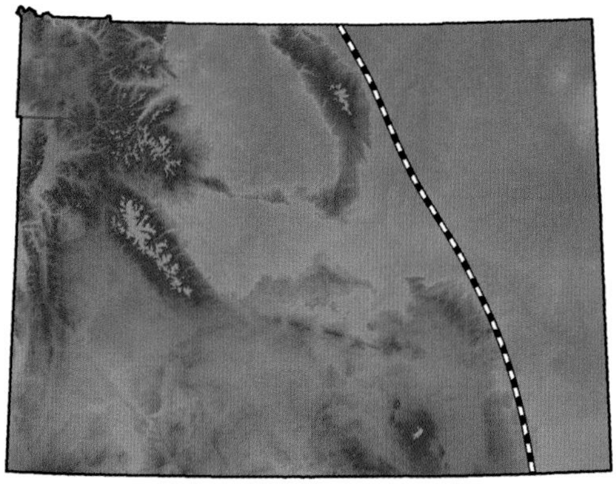

Distribution of the western and eastern spotted skunk in Wyoming, showing the approximate species boundary (dashed line).

NOTES AND REFERENCES

1. Verts et al. (2001). The western species breeds in September and pregnant females undergo embryonic diapause; the eastern species breeds in March or April. Thus, the two species are thought to be reproductively isolated.

2. Hoffmeister (1986).

3. Wyoming Statutes 23-1-101-a-viii.

4. Kinlaw (1995), Verts et al. (2001).

5. Long (1965, 700–701). Horse Creek and Iron Mountain are in the Laramie Mountains northwest of Cheyenne.

6. Dragoo et al. (1993).

7. Van Gelder (1959).

8. Van Gelder (1959), Long (1965), Hall (1981), Verts et al. (2001), Rosatte and Lariviere (2003).

9. Choate et al. (1973), Gompper and Hackett (2005).

10. US Fish and Wildlife Service (2012).

11. Zielinski et al. (2005).

Striped skunk, *Mephitis mephitis*

DESCRIPTION: A medium-sized carnivore with a long, well-furred tail. The fur is black with a prominent white stripe down each side of the dorsal midline, converging on the top of the head, although considerable variation is observed. Measurements (**Sex** [n] mean (mm, g), *range*): Albany County, L = **M** [5] 659, *600–715*; TL = **M** [5] 299, *200–390*; HF = **M** [5] 68.6, *50–78*; EFN = **M** [5] 22.6, *12–30*.[1] Dental formula: 3/3, 1/1, 3/3, 1/2 = 34.

STATUS: The species is classified as a "predatory animal" under Wyoming statute.[2]

NOMENCLATURE AND SYSTEMATICS: The scientific name has been stable for over two centuries; of the 13 subspecies recognized as of 1981, only *M. m. hudsonica* is depicted as occurring in Wyoming.[3] The recent evolution and biogeography of the species is complicated by its generalized habitats, and by patterns of Pleistocene range contraction and expansion. Populations in the northern Rocky Mountains are believed to have originated from a southern refugium (Texas and northern Mexico) during the pre-Illinoian glacial stage, which then expanded into Nevada and Oregon during the Illinoian glaciation. That population, which lay east of the Sierra Nevada, subsequently colonized the Great Basin and northern Rocky Mountains, where it came into secondary contact and introgressed with the original southern lineage. Animals from Wyoming represent a roughly equal mix of these two lineages.[4]

DISTRIBUTION AND HABITAT ASSOCIATIONS: The species is broadly distributed across temperate and southern boreal North America (current distribution shown). In Wyoming, striped skunks occur in suitable habitats nearly statewide. Some anecdotal evidence suggests that the species was rare in or absent from intermountain basins before European settlement. Habitats tend to be grasslands and forest edges, but the strongest associations observed are with water bodies and riparian zones.

ECOLOGY: The species is a trophic generalist, feeding on a wide range of animal and plant foods. Insects (especially beetles, grasshoppers, and crickets) are dominant during summer in grasslands, but this species also eats bird eggs and nestlings, small mammals, soil invertebrates, and berries when available. It dens below ground or beneath human structures in winter for weeks at a time, often with other skunks.[5] Population biology has not been studied in or near Wyoming. Mortality causes include predation by raptors and owls, shooting by hunters and landowners, and vehicle collisions; rabies is too uncommon to affect populations over wide areas or in the long term.[6]

CONSERVATION AND MANAGEMENT: Species-wide, the geographic range appears to be changed little from presettlement times. The species is not generally considered a conservation concern, but is monitored as a potential reservoir and vector for rabies. The virus found in skunks in Wyoming takes several forms, and cases from northeastern and north-central Wyoming are related to the form in Montana and the Dakotas, which has its origin in skunks. Northwestern Wyoming experienced an epizootic during 1988–1993, when 215 of

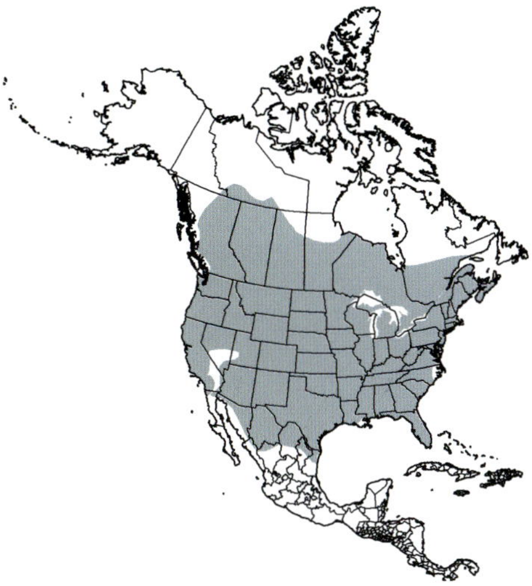

Distribution of the striped skunk in Wyoming.

1015 tested foxes were positive for the disease.[7] Recently, cases of skunk rabies in Laramie County have been identified as a south-central form, derived from canids from Arizona to Louisiana.[8] Cases from far western Wyoming (e.g., the Star Valley in 1991) have been identified as bat rabies that had been recently transmitted by bats.[9]

NOTES AND REFERENCES

1. UWMV specimens.
2. Wyoming Statutes 23-1-101-a-viii.
3. Hall (1981).
4. Barton and Wisely (2012).
5. Renatte and Lariviere (2003).

6. Hansen et al. (2004), Blanton et al. (2011).

7. Ramey et al. (2013).

8. Blanton et al. (2011). In Wyoming in 2010, 20 positive cases of skunk rabies were recorded.

9. Mills et al. (2006). Although a number of humans have been potentially exposed to and treated for skunk rabies in Wyoming, none have died of rabies during the modern era of disease monitoring.

WEASELS AND ALLIES—FAMILY MUSTELIDAE

This family, named for the musky odor emitted by the anal glands of some members, is a cosmopolitan group that features a wide range of body sizes and forms, locomotor styles, types of dentition, habitats occupied, and diets. Some species are semiaquatic, some tree climbers, and others semi-fossorial. All species have five toes per foot that touch the ground, medium to long tails, short rostrums, and most are highly predaceous. Because of their great morphological radiation and ecological specialization, mustelids have lower levels of interspecific competition than do members of the dog and bear families; as many as seven species of mustelid live in sympatry in some areas of North America.

Pacific marten, *Martes caurina*

DESCRIPTION: A mustelid the size of a small domestic cat with a fox-like face, glossy brown fur of medium length, a well-furred tail, and darker limbs. A throat patch varies in size, shape, and seasonally in color (from amber to pale yellow). The throat-patch color fades over winter. Measurements (**Sex** [n] mean (mm, g), *range*): Carbon County, L = **F** [7] 571, *53–64*; **M** [5] 653, *60–74*; TL = **F** [7] 191, *152–267*; **M** [5] 201, *178–216*; HF = **F** [7] 76, *76–76*; **M** [5] 93, *83–102*; EFN = **F** [7] 38, *32–41*; **M** [5] 44, *41–48*; WT = **F** [7] 715, *648–785*; **M** [5] 1044, *860–1245*.[1] Dental formula: 3/3, 1/1, 4/4, 1/2 = 38.

STATUS: A furbearing animal under state law.[2] Uncommon to common in occupied mountain ranges holding mesic coniferous forest. Francis Parkman described trappers

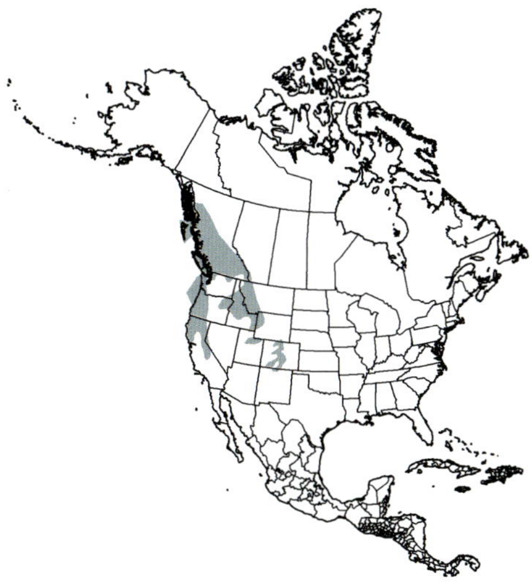

Distribution of the Pacific marten in Wyoming. The species may occur within the Wyoming portions of the Uinta Mountains and Black Hills.

setting pitfalls for "sables and martens" in the Laramie Mountains in the early 1840s, but this population, if it existed then, has gone extinct.[3]

NOMENCLATURE AND SYSTEMATICS: This is a newly recognized North American species of *Martes* (the other is the American marten), the earlier nominal species having been split into as many as six species and 14 subspecies in the early twentieth century. By the 1960s martens in North America were considered a single species (*M. americana*) within which two "subspecies groups" were recognizable, *caurina* and *americana*,[4] with a zone of contact extending from near Prince of Wales Island, Alaska southeast to Montana. These subspecific groupings were based on skull morphology, but have been corroborated by molecular evidence. In 1997, these subspecies groups were proposed to deserve full species status, and subsequent reports have corroborated that view, which I take here.[5] The corresponding common names have converged on Pacific marten (*M. caurina*) and American marten (*M. americana*). The former species descended from a population confined to the Cascades and Sierra Nevada during Pleistocene glaciations, when eastern and western boreal forests were disjunct. Evidence was earlier lacking for a reproductive barrier between the two putative species at the zone of contact. However, it has recently been shown that the two lineages exhibit assortative mating—preferring to mate within, rather across lineages—where parapatric.[6]

DISTRIBUTION AND HABITAT ASSOCIATIONS: This species of western mountains and boreal forest north to Southeast Alaska has contracted its distribution in the Sierra Nevada and a few other areas (current distribution shown). It is found in the large mountain ranges that ring the state: Uinta Mountains, northwestern mountains, Bighorns, and Medicine Bows. Introduced to the Black Hills of South Dakota, martens have established themselves in white-spruce-dominated mesic sites, particularly Spearfish Canyon, but may range as far west as Wyoming.[7] The species has been absent from the Bear Lodge Mountains during the historical period. Suitable habitats in midwinter are late-successional coniferous forest with complex physical structure near the ground. The structure is typically provided by lower branches of living trees and angled or horizontal logs. Such stands tend to be dominated by Engelmann spruce and subalpine fir in Wyoming, but old lodgepole pine and Douglas-fir can produce these conditions as well. This need by Pacific martens for cover is based on predator avoidance, access to spaces beneath the snow for foraging, and thermally protected resting sites during cold weather. During warmer, snow-free seasons, martens occur in a wider range of forest types (e.g., whitebark pine, coniferous-broadleaved mixtures) and woodland or alpine habitats. The latter include krummholz conifers near tree line and talus fields to above 4000 m (13,000 ft). Martens forage for voles and mice in herbaceous meadows during summer. Ponderosa pine-dominated forest is seldom used, presumably due to the typical lack of physical structure near the ground and the scarcity of herbaceous growth that would support abundant small rodents. Optimum habitat is an area of complex physical structure near the ground adjacent to herbaceous openings that support abundant prey, and perhaps talus fields nearby. Martens are patchily distributed in the Bighorn Mountains, although a few have been taken by trappers in the past decade.[8]

ECOLOGY: The generalized diet varies across sites and seasons. Small mammals, particularly voles (southern red-backed vole, *Microtus* spp.), are the primary year-round food in most places, especially the taiga. Red squirrels are less preferred, but martens resort to them in late winter and rest in their underground nests during the cold season. Summer diets are diverse: neonatal mammals, bird eggs and nestlings, insects, and carrion. Fruits are important in late summer and autumn, and martens may serve as important seed dispersal agents for soft-masting species. Stomachs of martens trapped in the Absaroka Mountains in midwinter can be filled with whitebark pine seeds, presumably cached by red squirrels and usurped by martens.[9] How martens extract the seeds from cones is not known, but no cone bracts appear in these stomachs. The species is highly sensitive to the presence of overhead structure, especially in winter, staying near cover for predator avoidance, and has limited dispersal ability across areas below the lower limit of trees. The reproductive strategy is consistent with a "slow" life history, characterized by delayed sexual maturity and high longevity.[10]

CONSERVATION AND MANAGEMENT: Martens remain common over most of their presettlement North American range, and are still trapped as furbearers in Wyoming and elsewhere. Because of their inability to cross large areas below the lower elevational limit of forest, the species in Wyoming occurs as semi-isolated or completely isolated populations in some mountain ranges and opportunities for colonization and exchange among populations are limited. This has implications for the isolated populations in the Bighorn Mountains and Black Hills. Widespread changes in the extent of mesic coniferous forest or the amount of overhead cover can be expected to affect marten populations. How the outbreaks of mountain pine beetle and spruce bark beetle will affect martens is difficult to predict. On the one hand, massive tree death and associated losses of red squirrels and other forest obligate species can be expected to reduce marten abundance. On the other hand, the recruitment of tree boles to horizontal positions near the ground and beneath the snow surface, associated with an increase in herbaceous vegetation, could create compensating effects for martens in the near term. The Bighorn Mountains population is of some concern because it is isolated, and its distribution within the Bighorn Mountains seems to be limited. Gary Beauvais, conducting dissertation research there during 1992–1997, found that martens occupied a small portion of the coniferous forest area.[11] Modeling of climate change effects in the Cascades has yielded predictions that the distribution of Pacific martens will contract vertically in elevation to a marked degree under most climate scenarios by the late twenty-first century.[12]

NOTES AND REFERENCES

1. Author's unpublished data collected, 1989–1990.
2. Wyoming Statutes 23-1-101-a-iii.
3. Parkman (2008, original publication 1847). Ashenfelder Basin, featuring spruce–fir vegetated slopes north of Laramie Peak, is regarded as the likely core of any past marten population in the northern Laramie Mountains.
4. Hagmeier (1958, 1961).
5. Carr and Hicks (1997), Dawson and Cook (2012).
6. Williams and Scribner (2010).
7. The species may have been native to the Black Hills, so that this may have represented a reintroduction; Turner (1974, 153) considered this a species "of uncertain status." The sole early specimen that originated from the Black Hills was taken in 1930 near Custer, South Dakota (USNM 249298).
8. Habitat use in Wyoming has been described by Buskirk et al. (1989), Corn and Raphael (1992), Sherburne and Bissonette (1994), and Ruggiero et al. (1998). Beauvais (1997) described distribution in the Bighorn Mountains.
9. Diets in Wyoming were described by Murie (1961) and reviewed range-wide by Martin (1994). Seed dispersal is described by Hickey et al. (1999).
10. Buskirk et al. (2012).
11. Beauvais (1997).
12. Johnston et al. (2012).

Fisher, *Pekania pennanti*

DESCRIPTION: A medium-sized mustelid carnivore, around three times the weight of a marten, and with a stockier build. The pelage is similar in color to that of martens, but coarser. The face is less fox-like than that of a marten. Males are considerably larger than females. Measurements (**Sex** [n] mean (mm, g), *range*): adults, Southern Sierra Nevada, California, L = **F** [11] 873, *815–920*; **M** [5] 970, *933–1016*; TL = **F** [11] 349, *320–372*; **M** [5] 376, *355–390*; HF = **F** [11] 107, *98–115*; **M** [5] 117, *114–120*; EFN = **F** [11] 35, *31–40*; **M** [5] 39, *31–45*; WT = **F** [11] 1880, *1460–2100*; **M** [5] 3160, *2450–3650*.[1] Dental formula: 3/3, 1/1, 4/4, 1/2 = 38.

STATUS: The legal status is as a "protected animal" and a state Species of Greatest Conservation Need as of 2012.[2] It is occasionally reported from Wyoming, but no specimens have

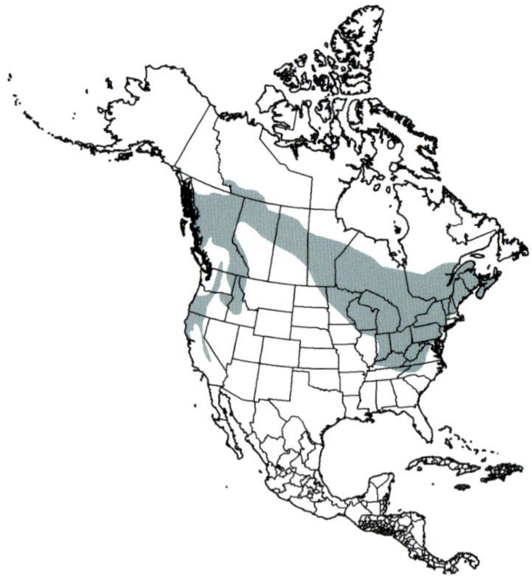

Presettlement geographic range of the fisher.

been collected for over 100 years, and unconfirmed reports have not originated from consistent locations. Such a lack of consistent pattern suggested absence of any breeding populations.[3] One specimen described by Ernest Thompson-Seton in 1929 was attributed to Yellowstone in around 1893.[4] Other reports, all lacking physical evidence, have originated from the Laramie and Medicine Bow Mountains, Jackson Hole, and the Beartooth Plateau. The species cannot be considered a resident species, because no evidence indicates populations in or adjacent to Wyoming, recently or in the historical past.

NOMENCLATURE AND SYSTEMATICS: Although long considered a large marten, the fisher has been reexamined using molecular markers, which place it closer to the wolverine than previously understood. Thus, the genus *Martes* as used over the twentieth century was paraphyletic.[5] To correct this problem, the fisher has now been assigned to its own genus, *Pekania*; it could be reassigned to *Gulo* based on current understandings. The species name has been stable for many decades; Thomas Pennant was a seventeenth-century Welsh naturalist.[6]

DISTRIBUTION AND HABITAT ASSOCIATIONS: The current species distribution is around half of its area at the time of European settlement (shown).[7] No distribution—or even point locations—in Wyoming can be depicted. The anomalous record from near Ucross is far from any fisher habitat, and the 1952 report from the Beartooth Plateau mentioned by Long (1965) is similar to a number of more recent accounts from that region, although no physical evidence has ever emerged to support the assertion that the species occurs on or near the Beartooth Plateau.[8] The nearest location of the species in the northern Rocky Mountains to

Wyoming during the period 1989–2008 was around 240 km (150 mi) west of Wyoming's western margin.[9] Habitat is coniferous and mixed forest; fishers are as averse to entering areas without overhead cover as are Pacific martens. Deep, soft snow constrains fisher locomotion.

ECOLOGY: Like the North American martens, the fisher is a habitat specialist but dietary generalist. It feeds on a wide range of animal foods, specializing in some areas on porcupines or snowshoe hares, both species found, but neither species abundant, in Wyoming mountains.[10] Populations have been studied in New England, boreal Canada, and at several sites in the Pacific states and provinces, but those in the North Rocky Mountains of Idaho and Montana are little studied.

CONSERVATION AND MANAGEMENT: No conservation or management approaches have been applied or proposed in Wyoming, because the species has not been shown to occur in the state, in spite of repeated efforts. The conservation status of the species is highly variable across its geographic range: rare and protected on the Pacific coast, and a nuisance animal in parts of New England. The factors thought to contribute to scarcity and extinction have varied across time and geography, but generally focus on fur trapping and habitat alteration, especially forest conversion.[11]

NOTES AND REFERENCES

1. William J. Zielinski, Arcata, California, unpublished data, US Forest Service, Southern Sierra Fisher Study.
2. Wyoming Statutes 23-1-101-a-ix.
3. Brown (1965) described a specimen taken near Ucross, far from any coniferous forest, with associated evidence of translocation. Smith (1999) reported a visual observation from Grand Teton National Park; no corroboration or physical evidence was associated.
4. Seton (1925, 457) stated, "When I lived in the Yellowstone Park in the summer of 1897, I made all inquiries, and was told by the hunters that the Fisher was unknown; Capt. George S. Anderson told me that about 1893, he had confiscated a skin taken by a poacher in the Park. All hunters assure me that the Fisher is never found in Jackson's Hole."
5. Stone and Cook (2002), Koepfli et al. (2008).
6. Sato et al. (2012).
7. Powell (1993).
8. Long (1965).
9. US Fish and Wildlife Service (2011).
10. Buskirk et al. (1994) and Proulx et al. (1997) provide reviews of species biology.
11. Powell and Zielinski (1994).

Short-tailed weasel, *Mustela erminea*

DESCRIPTION: A small carnivore with seasonally varying pelage, uniformly white in winter and brown on the dorsum with cream undersides in summer. The tail is black-tipped year-round; tail length (ca. one-fourth of total length) is not a good field character. The overall similarity of this and the long-tailed weasel, combined with the lack of clear geographic partitioning between the two species, leaves many field observations in doubt as to species. Age and sex variations are additional confounding factors in species identification in the field. Molecular markers are diagnostic for species, and the number of tail vertebrae and other morphological features are reliable, given a cleaned skeleton. In Wyoming, males are around one and a half times the body length and twice the body weight of females. Same-sex comparisons show the short-tailed weasel to be smaller than the long-tailed weasel by about one-third.[1] Measurements (**Sex** [n] mean (mm, g), *range*): various counties, L = **B** [16] 222, *192–270*; TL = **B** [16] 56, *41–70*; HF = **B** [16] 26, *20–30*; EFN = **B** [16] 12, *7–16*; WT = **B** [6] 50.0, *35.2–62.3*.[2] Dental formula: 3/3, 1/1, 3/3, 1/2 = 34.

STATUS: The species is a furbearing animal by state law.[3] It is common in suitable habitats.

NOMENCLATURE AND SYSTEMATICS: The genus and species names have been stable since the time of Linnaeus.[4] Common names applied to this species also include "ermine," and "stoat" (the latter in Britain, Ireland, and New Zealand). "Ermine" is imprecise because some use that term to denote white-phase weasels of any species. Short-tailed weasels were part of a Eurasian radiation within *Mustela*; *M. erminea* crossed the Bering Land Bridge into North America during the Illinoian glaciation, 300,000–130,000 years ago.[5] Twenty North American subspecies were recognized by Hall (1981), around half of them island endemics; only the nominal *M. e. muricus* was depicted in Wyoming. In contrast, Corbet considered there to be only one subspecies across all of mainland Eurasia and North America.[6]

DISTRIBUTION AND HABITAT ASSOCIATIONS: This is a high-latitude species of tundra, boreal forest, and adjacent areas. It is assumed here to occur in Wyoming's mountains, but to have a limited distribution in arid shrub-steppe. It was considered to be distributed statewide in the 1965 and 1987 faunas, but what few museum species have been collected suggest that most come from forested or mesic lowland areas.[7] The mammal faunas for Colorado, Montana, and Utah consider the species restricted to mountains, and I hypothesize that the Wyoming distribution is likewise limited to mountains or lowlands near water.[8] How competition between weasel species may affect sympatry is not clear. Habitats are highly varied, but the species is most common in grassy and forested areas with abundant voles and mice. Weasels, presumably short-tailed, are common in talus fields in summer, far above the upper elevational limit of trees.

ECOLOGY: This solitary, territorial mammal likely is one of the most common carnivores in Wyoming, but not often observed and poorly understood. The species is highly predaceous and fairly food-specialized, feeding on small rodents, birds, and carrion. In winter, short-tailed weasels in mountains live beneath deep snow, moving via tunnels formed beneath logs, limbs and shrubs, and the snow surface.[9] Populations have not been studied in Wyoming or nearby. Densities are believed to respond primarily to prey abundances. Important mortality causes are not known, but the species is killed by other carnivores, hawks and owls, the avian predators more likely to eat them than mammalian ones.[10] Intraguild competition imposed by the larger long-tailed weasel, including interspecific killing, is believed to be an important factor in local distribution and habitat use.[11]

Short-tailed weasel in winter pelage.

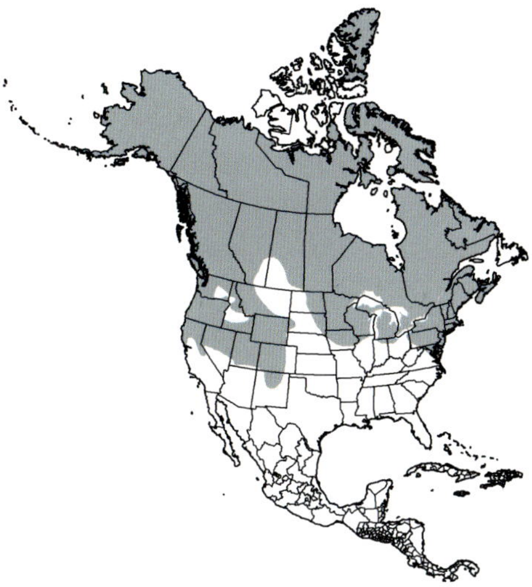

Distribution of the short-tailed weasel in Wyoming.

CONSERVATION AND MANAGEMENT: Short-tailed weasels are seldom trapped for fur because of the low value of their pelts. They are not typically considered a species of conservation concern.

NOTES AND REFERENCES

1. St-Pierre, Ouellet, Dufresne et al. (2006), Elsasser and Parker (2008).
2. UWMV, DMNS specimens, Long (1965).
3. Wyoming Statutes 23-1-101-a-iii.
4. Linnaeus (1758).

5. Harding and Smith (2009).

6. Corbet (1978), Hall (1981).

7. See points on maps by Long (1965) and Clark and Stromberg (1987). Also, see Negus and Findley (1959). So few short-tailed weasel specimens have originated from Wyoming that the elevational distribution of *M. ermine* cannot be confirmed with data. The mean elevation at which 46 short-tailed weasels collected in Wyoming and held in various museum collections was 2307 m (7615 ft), 235 m (776 ft) higher than the corresponding value for long-tailed weasel (online appendix).

8. See Armstrong et al. (2011) for Colorado, Foresman (2012) for Montana, and Durrant (1952) for Utah. The species is completely absent from Nebraska, suggesting a restriction to mountains. The above accounts limit the distribution to mountainous areas with forest or alpine vegetation.

9. King (1983).

10. Squires (2000), Walley (1982).

11. St-Pierre, Ouellet, and Crête (2006).

Long-tailed weasel, *Mustela frenata*

DESCRIPTION: A small carnivore with seasonally varying pelage, white in winter and brown above with buff below in summer. The tail is black-tipped in all seasons; tail length (ca. three-eighth of total length) is not a good field character. Males are roughly twice the weight of females, and the species is larger (length and body weight) than a short-tailed weasel.[1] Identification from morphological measurements has been shown to lead to fairly high rates of error.[2] Measurements (**Sex** [n] mean (mm, g), *range*): various counties, L = **F** [6] 350, *310–432*; **M** [6] 395, *333–442*; TL = **F** [6] 126.2, *110–157*; **M** [6] 143, *113–164*; HF = **F** [6] 45.6, *26–89*; **M** [6] 40.0, *31–45*; EFN = **F** [6] 15.7, *12–19*; **M** [6] 26.2, *14–50*; WT = **M** [6] 219, *127–299*.[3] Dental formula: 3/3, 1/1, 3/3, 1/2 = 34.

STATUS: This is a furbearing animal under Wyoming statute.[4] It is common within the geographic range.

NOMENCLATURE AND SYSTEMATICS: The species name has been stable for nearly two centuries, but the genus name is being reconsidered because of the newly recognized lineage of New World mustelids that includes the American mink, two South American weasel species, and this taxon.[5] The oldest appropriate genus name to apply to a member of this group is *Vison*, which may be applied to this group in the future. The specific name derives from the Latin "bridled," referring to the masked face seen in some animals in the southwestern United States. Four of the 35 nominal subspecies occur in Wyoming, but only *M. f. nevadensis* is widespread in the state.[6]

DISTRIBUTION AND HABITAT ASSOCIATIONS: The species occurs from northern British Columbia to Argentina. In Wyoming, it is found in suitable habitats statewide, particularly grassy sites with high densities of mice and voles. It has undergone range expansions northward in eastern Canada in recent decades.[2] Although commonly mapped as occurring in all temperate habitats, I hypothesize that in Wyoming the species occurs at lower elevations than do short-tailed weasels.[7]

Long-tailed weasel in summer pelage.

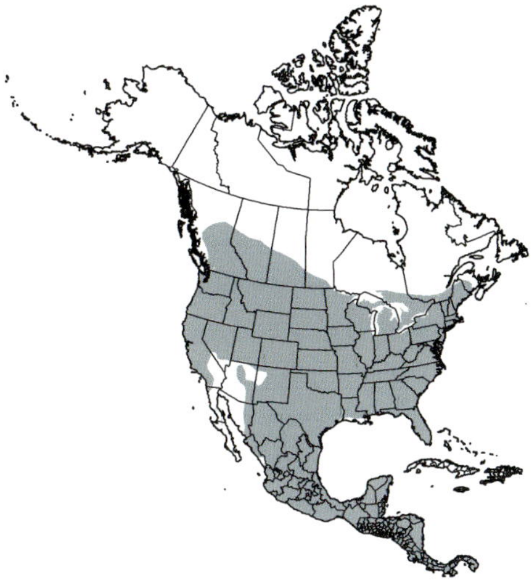

Distribution of the long-tailed weasel in Wyoming. Occurrence of the species above the upper elevational limit of trees is doubtful.

ECOLOGY: The species' ecology is similar to that of *M. erminea*, but its diet is more generalized, and includes small rabbits and birds along with mice, chipmunks, and voles. Carrion is eaten when available. Plant foods are not important. Because of its more generalized diet, the species is thought to not undergo such dramatic population fluctuations as the more specialized least weasel and short-tailed weasel.[8] The species is killed by other carnivores, particularly foxes, and preyed upon by owls and hawks.[9]

CONSERVATION AND MANAGEMENT: The species is trapped for its fur in small numbers; it is not generally considered a conservation concern. Its species-wide distribution has not changed detectably from that in presettlement times.

NOTES AND REFERENCES

1. See discussion of identification at short-tailed weasel.
2. St-Pierre, Ouellet, Dufresne et al. (2006).
3. UWMV specimens.
4. Wyoming Statutes 23-1-101-a-iii.
5. Koepfli et al. (2008), Harding and Smith (2009).
6. Hall (1981).
7. For 112 museum specimens collected in Wyoming, the mean elevation of collection was 2073 m (6840 ft), lower than that for short-tailed weasels.
8. Quick (1951), Sheffield and Thomas (1997).
9. Squires (2000).

Black-footed ferret, *Mustela nigripes*

DESCRIPTION: A weasel-like carnivore the size of a mink or marten. In spite of its distinctive shape and pelage markings, the species is confused in the field with other grassland mammals, especially when viewed briefly or in the beam of a light. Fur on back and flanks is buff or tan; that on forelimbs and hind feet is black. The tail is of medium length and black-tipped. The black facial mask is distinctive in good light, but in poor viewing conditions can be confused with facial markings of prairie dogs, with which it is locally sympatric.[1] Eye-shine is emerald green, similar in color to that of other mammals observed at night on prairie dog towns, but with practice becomes a useful field character. Distinctive field signs include dirt trenches leading from prairie dog burrow entrances, particularly in winter.[2] Measurements (Age class, **Sex** [n] mean (mm, g), *SD*): Carbon and Albany Counties, WT = Juvenile (three months) **F** [61] 662, 59, Adult (≥ 1 year) **F** [70] 691, 39; Juvenile **M** [74] 862, 96; Adult **M** [31] 1018, 77.[3] Dental formula: 3/3, 1/1, 3/3, 1/2 = 34.

STATUS: The species was listed under the first federal endangered species law in 1967, before the enactment of E.S.A. Its legal status in Wyoming is as a "protected animal."[4] It was believed extinct in the wild and in captivity by 1979, but was rediscovered on a ranch west of Meeteetse, west of the known geographic range, in 1981. Intensive fieldwork established an approximate size and distribution of that population by 1984; numbers were estimated at around 130.[5] Following a disease outbreak, all surviving wild animals were taken into captivity during 1985–1987. Successful captive breeding was followed by reintroduction to multiple states. The Shirley Basin population initially showed weak population growth, then signs of increase in the early 2000s. Intensive surveys during 2003–2006 showed exponential growth in the core release area, accompanied by expansion of the geographic range via dispersal.[6]

NOMENCLATURE AND SYSTEMATICS: The taxonomy is stable, and no other scientific names have been used since the early twentieth century. The first description was of a specimen

from Goshen County by Audubon and Bachman (1851).[7] The species arose in Eurasia, diverging from Old World polecat stocks around 1.3 Ma, and made its first appearance in North America around 0.3 Ma. In the Old World, morphologically similar forms are found in the steppe polecat (*M. eversmanii*), of which the Amur River subspecies (*M. e. amurensis*) is most similar.[8] During research on the reproductive biology of Amur River steppe polecats and black-footed ferrets conducted early during the captive breeding program, it was found that crossings of these two taxa produced viable hybrids.[9] Apparently, these two nominal species are part of the same biological species, reflecting a common recent ancestry.

DISTRIBUTION AND HABITAT ASSOCIATIONS: The ferret was formerly widely distributed (shown) across the ranges of the five prairie dog species; reduction of the geographic range followed land conversion to other uses, prairie dog eradication, and the spread of plague. Initial releases back into the wild began in 1991, into Shirley Basin, followed by Badlands National Park and adjacent Buffalo Gap National Grasslands, South Dakota. Ferrets had been reintroduced to 19 sites in seven states, Canada, and Mexico by 2011, although some sites no longer held ferrets by then. Some additional releases onto Shirley Rim and Foote Creek, near Arlington, Wyoming, were conducted as well. The most recent approximate distribution extends from northern Shirley Basin to at least as far south as the Rock River area. The former Meeteetse population and current extended Shirley Basin population both occurred solely in association with white-tailed prairie dogs; reintroductions to sites within the geographic range of the black-tailed prairie dog in Wyoming had not been attempted by 2014.

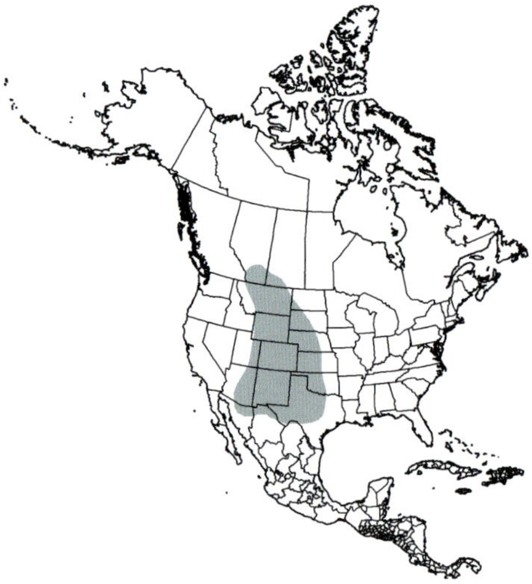

Distribution of the black-footed ferret in Wyoming. The point marks the location of the population discovered in 1981, which served as the basis for all captive breeding and reintroductions that led to the species' recovery to date. The enclosed area was the approximate geographic range of the population reintroduced to the Shirley Basin area as of 2012.

ECOLOGY: The black-footed ferret is one of the most ecologically specialized carnivores in the world, eating almost exclusively prairie dogs (*Cynomys* spp.) and living in their burrows. Although not believed to hibernate, ferrets remain below ground almost continuously in Wyoming in winter. White-tailed prairie dogs are obligate hibernators, so ferrets must find and excavate them from plugged burrows. This involves excavating burrow plugs and killing the torpid occupants.[10] During late summer and early autumn females with young are readily observed above ground at night, traveling together. Average population density on white-tailed prairie dog colonies in Shirley Basin was 0.03/ha (7.6/ mi^2) in late summer.[11] Ferrets are easily able to disperse among prairie dog colonies separated by gaps of several hundred meters. Disease has become an important ecological factor during the last 100 years, because both ferrets and prairie dogs are susceptible to plague, caused by the bacterium *Yersinia pestis*, which spread eastward from California during the first half of the twentieth century. The disease slowed its eastward advance at around 104° W. longitude, near the eastern boundary of Wyoming. Canine distemper, a viral disease of many carnivorans, infected ferrets at Meeteetse in 1986, and is enzootic in various carnivores that occur sympatrically with ferrets.[12] Important predators of ferrets include the coyote, great horned owl, and American badger. Black-footed ferrets have a "fast" life history. They reproduce at one year of age, have brief or no delayed implantation, and can produce very large litters (the largest in captivity to date have been >10). The rate of population change observed in Shirley Basin during the period 2003–2006 was a 35% increase yearly.[13]

CONSERVATION AND MANAGEMENT: The primary factors causing the decline of the ferret in Wyoming, aside from land conversion to crop production, was widespread control of prairie dogs via poisoning during the early and mid-1900s; epizootics of introduced diseases—plague and canine distemper—have been more recent factors. Nationally, conservation and management are now focused on continued captive breeding, additional reintroductions to sites within the original range, and in some cases, attempts to control disease in the wild via either applying pesticides to control flea vectors of plague or vaccinating ferrets for diseases of concern. The Shirley Basin population of ferrets was arguably the most extensive contiguous one as of 2011, although South Dakota holds more ferrets in more disjunct populations than does Wyoming. In Wyoming, recent conservation efforts are directed at monitoring the expansion of existing populations; estimating abundance across the entire >72 km (60 mi) of ferret distribution is not feasible. The Shirley Basin area holds the largest single complex of prairie dogs remaining in North America (>59,000 ha of colony area) and is nearly contiguous with complexes that extend across the southern tier of Wyoming from the southern Laramie Basin to as far west as the Green River. It is plausible that ferrets could recolonize much of this prairie dog habitat with little or no human assistance.

NOTES AND REFERENCES

1. Hillman and Clark (1980).
2. Clark, Campbell et al. (1984), Clark, Richardson et al. (1984).
3. Grenier (2008, fig. 1).
4. Wyoming Statutes 23-1-101-a-vii.
5. Forrest et al. (1988).
6. Miller et al. (1996), Grenier et al. (2007).
7. Audubon and Bachman (1851).
8. Anderson et al. (1986).
9. Elizabeth S. Williams, University of Wyoming, pers. comm.
10. Campbell et al. (1987), Richardson et al. (1987).
11. $n = 24$ colonies, including some with 0 ferrets detected (Grenier 2008, Table 2).
12. Williams et al. (1988, 1994).
13. Grenier et al. (2007). Reports of litter sizes are from Paul Marinari, US Fish and Wildlife Service, Colorado, pers. comm.

Least weasel, *Mustela nivalis*

DESCRIPTION: A tiny carnivore, the weight of a vole, but longer and more slender. Pelage in summer is medium brown on the dorsum and off-white on the undersides, and solid white in winter in Wyoming. The tail is very short (ca. one-fifth of total length) and lacks a black tip, although a few dark hairs on the tail tip may be noted. Males are larger than females, but by how much in North America is not clear. Measurements (**Sex** [n] mean (mm, g), *range*): Nebraska and Saskatchewan, L = **F** [6] 192, *184–225*; **M** [6] 202.0, *188–208*; TL = **F** [6] 35, *28–40*; **M** [6] 32.5, *31.5–34*; HF = **F** [6] 23.0, *20.5–26*; **M** [6] 22.8, *21–24*.[1] Dental formula: 3/3, 1/1, 3/3, 1/2 = 34.

STATUS: As a weasel it is classified as a furbearing animal, but few, if any, animals likely are trapped for fur, and none intentionally.[2] It is also a state Species of Greatest Conservation

Need as of 2012. The species is too poorly known to characterize its biological status. It is rare in field notes and in museum collections, even from the core of its North American geographic range.

NOMENCLATURE AND SYSTEMATICS: This is a Eurasian species that crossed Bering Strait into North America, perhaps as recently as the Wisconsin Glaciation, which peaked 18,000 years ago. The genus name is stable, but uncertainty has existed as to whether this species and the very similar Old World *M. rixosa* are synonyms, subspecies, or biological species. Molecular studies have shown that populations in North America have greater genetic structure than is the case for the short-tailed weasel.[3] Such a finding is usually indicative of habitat specialization, and limited dispersal ability. Of the four North American subspecies recognized as of 1981, the geographic ranges of two (*M. n. rixosa* and *M. n. campestris*) occur near the northeastern corner of Wyoming.[4]

Least weasel in summer pelage.

DISTRIBUTION AND HABITAT ASSOCIATIONS: The species occurs across arctic and boreal North America and in Eurasia, to as far southwest as North Africa. Wyoming is outside most mapped distributions for the species. Occurrences in Wyoming, including poorly documented possible recent records from southern Crook County, extend in an arc from the foothills of the Bighorns, in Sheridan County, eastward an unknown distance.

ECOLOGY: Arguably the most poorly known of North American carnivores, the least weasel is a mouse and vole specialist. The species is most difficult to study in the wild, but from captive studies, we know that it does not exhibit embryonic diapause, first conceiving at around 11 months and first giving birth at one year of age.[5]

CONSERVATION AND MANAGEMENT: The species is considered a conservation concern, in part because it is so rare and so little information about its biology or distribution is available.

NOTES AND REFERENCES

1. Hall (1951).
2. Wyoming Statutes 23-1-101-a-iii.
3. Kurose et al. (2005).
4. Hall (1981), Sheffield and King (1994).
5. Sheffield and King (1994).

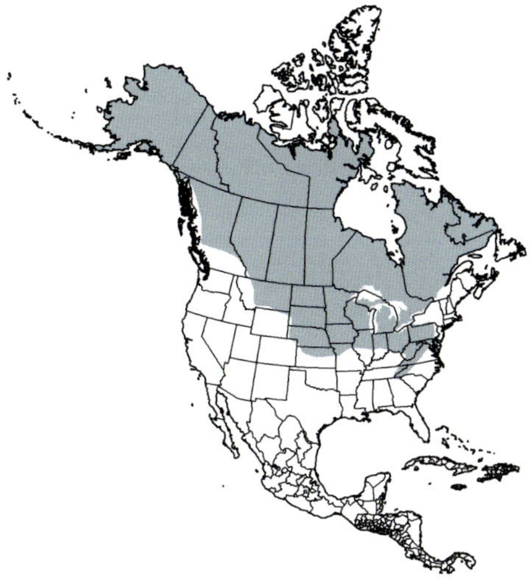

Distribution of the least weasel in Wyoming.

American mink, *Neovison vison*

DESCRIPTION: A medium-sized, semiaquatic weasel with short limbs and a medium-length tail. Pelage is unmarked medium to dark glossy brown, with a slightly darker tail tip. Fur is typically wet when observed in the field. Small, irregular white patches and streaks mark the chin and throat, extending to the abdominal midline in some animals. Measurements (**Sex** [n] mean (mm, g), *range*): various counties, L = **F** [10] 461, *472–551*; TL = **F** [10] 165, *143–183*; HF = **F** [10] 58, *52–64*; EFN = **F** [8] 21, *16–26*.[1] Dental formula: 3/3, 1/1, 3/3, 1/2 = 34.

STATUS: This is a "furbearing animal," by state statute.[2] Minks are uncommon to common in suitable habitats.

NOMENCLATURE AND SYSTEMATICS: Although the species' validity has been unquestioned for over a century, its location in a mustelid genus has been and continues to be unstable.[3] The current genus name may soon revert to the earlier *Mustela* or to the proposed *Vison*, the latter being the oldest appropriate name assigned to the North American lineage of which it is part (see account for long-tailed weasel).[4] The American mink was once regarded as a sibling species to the superficially similar European mink (either *Mustela* or *Lutreola*), but the two lineages are now regarded as more distantly related. The most recent interpretation places the American mink with three other New World taxa (*M. felipei*, *M. africana*, and *M. frenata*) as an American weasel lineage.[3] Two subspecies were recognized as occurring in Wyoming as of 1999; *N. v. energumenos* occurs west of the crest of the Bighorns and Medicine Bows, whereas *N. v. letifera* is east of that line.

DISTRIBUTION AND HABITAT ASSOCIATIONS: This broadly distributed species occurs in suitable habitats across Wyoming: streams, river sloughs, wetlands, marshes, and other

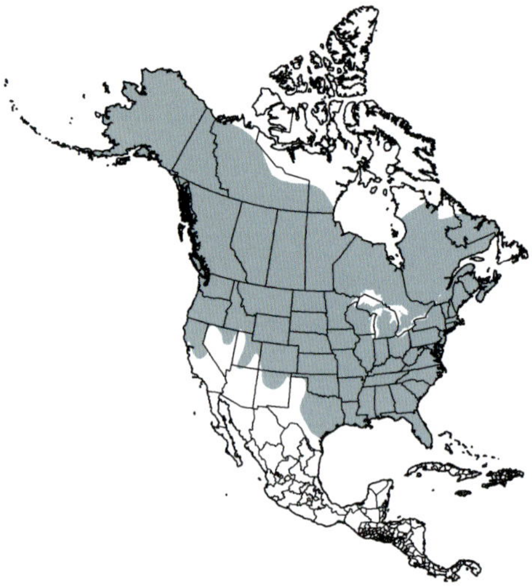

Distribution of the American mink in Wyoming.

slow-moving water bodies with irregular shorelines. Its continental distribution at the time of European exploration (shown) has changed little. Minks are also common near irrigated hay meadows, and travel hundreds of meters from water to seek food or shelter.[5]

ECOLOGY: American minks are semiaquatic predators, feeding on a wide range of aquatic and terrestrial animals, depending on setting. The ecology of the species is not well known from this region; most detailed studies have been done in Midwestern, Canadian, or coastal environments, where minks are more abundant. The primary food is fish, particularly slow-moving small-bodied species; minks also eat frogs, crayfish, aquatic insects, muskrats, and small mammals.[4] In a low-gradient stream in the Wyoming Range, at least 8% of radio-tagged cutthroat trout and at least 28% of tagged brook trout were believed to have been killed by mink.[6]

CONSERVATION AND MANAGEMENT: Managed as furbearers, minks are not trapped in large numbers because of the general scarcity of aquatic and wetland habitats in Wyoming. They are not considered a conservation concern in most parts of their native range. However, they are implicated in ecological webs that concentrate contaminants, especially metals, in high trophic levels in aquatic systems. As such, they have been considered to be the potential foci of contaminant concentration processes.[7]

NOTES AND REFERENCES

1. UWMV, KU specimens.
2. Wyoming Statutes 23-1-101-a-iii.
3. See Harding and Smith (2009) for review.
4. Bradley et al. (2014) prefer *Vison*, but are not consistent in following Harding and Smith (2009) for the long-tailed weasel.
5. Lariviere (1999).
6. Lindstrom and Hubert (2004).
7. See Harding et al. (1998) for review.

Wolverine, *Gulo gulo*

DESCRIPTION: A carnivore slightly larger than a badger, with long, dark brown fur. A band of lighter fur (gold or pale) runs longitudinally along the flanks, meeting on the dorsal midline over the shoulder and over the base of the tail. The face between the eyes and ears is light in some animals. The chest features a light patch, variable in size and irregular in shape, often visible from the front. The stride is distinctively mustelid—an undulating bound. Wolverines are often confused in the field with other, more common mammals: marmots, young bears, and even porcupines. Measurements (**Sex** [n] mean (mm, kg), *range*): >1 year old, central Idaho, L = **F** [7] 875, *840–895*; **M** [10] 972, *920–1000*; TL = **F** [7] 183, *170–200*; **M** [10] 191, *170–203*; HF = **F** [6] 157, *152–160*; **M** [10] 168, *145–181*; WT = **F** [7] 7.9, *6.5–8.8*; **M** [9] 12.7, *11.1–14.7*.[1] Dental formula: 3/3, 1/1, 4/4, 1/2 = 38.

STATUS: The species is a "protected animal" under Wyoming statute and a state Species of Greatest Conservation Need as of 2012.[2] It was found warranted for listing in 2010, and was proposed for listing as Threatened in the contiguous United States in 2013.[3] The proposal to list was withdrawn in 2014.[4] It is uncommon even within the occupied geographic range.

NOMENCLATURE AND SYSTEMATICS: The genus name has been used for nearly two centuries; the specific epithet is sometimes presented as *G. luscus*, which is more often used only for the New World form.[5] Recent molecular studies show that wolverines are phylogenetically positioned more closely to the martens (six species of *Martes*) than is the fisher.[6] Four subspecies

in North America were recognized in 1981, the one in Wyoming (*G. g. luscus*) being the same as that across most of Canada and Alaska. Any subspecific designations are of doubtful validity because of the extraordinary dispersal ability of this animal.[5]

DISTRIBUTION AND HABITAT ASSOCIATIONS: In North America the wolverine historically extended as far south as southern Colorado and the southern Sierra Nevada, and as far north as the Canadian arctic islands (shown). The presettlement Wyoming distribution can only be inferred: it likely occurred in all mountain ranges with extensive areas of tundra and subalpine forest, although it is not clear that small mountain ranges could sustain populations for multiple generations. Wolverines in the northwestern mountains have been well documented in recent years. Radio-collared animals have recently shown the extraordinary mobility of this species, during both routine movements and dispersal. Both sexes in Wyoming have shown the ability to disperse as far as 170 km (100 mi). One male captured in the Teton Mountains in 2001 when 11 months old traveled a minimum distance of 874 km (542 mi) over 42 days.[7] Another young male captured south of Togwotee Pass traveled at least 942 km (585 mi) in a 2009 dispersal that ended south of Rocky Mountain National Park in Colorado, where it still lived in 2013.[8] Typical habitat is tundra and subalpine coniferous forest, although dispersal across shrub-steppe and desert occurs.[9] Wolverines den beneath late spring snow fields and in talus fields and rock crevices above tree line.[10] In central Idaho they have been found to occur at higher elevations—in association with whitebark pine—in summer, but in winter often associate with lower-elevation Douglas-fir and lodgepole pine.[11] In Wyoming wolverines preferred elevations >2150 m (7100 ft) in winter as well as summer, showing that they remain mostly in the forest zone or above, year-round.

ECOLOGY: The species is a solitary predator and scavenger, primarily the former during summer and the latter during winter. Summer foods are highly varied, including birds and

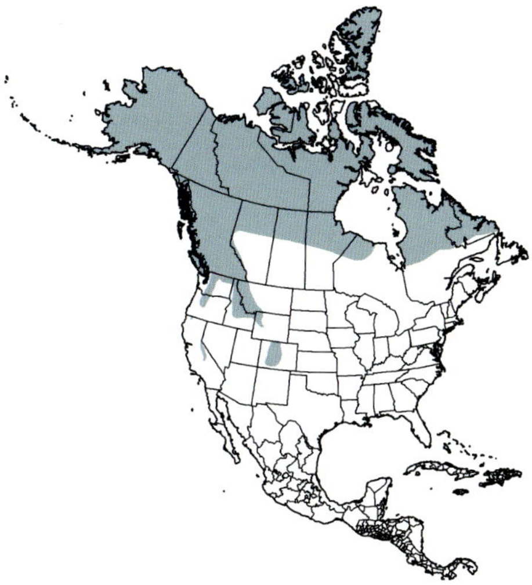

Distribution of the wolverine in Wyoming. The species is known to traverse shrub-steppe below the lower elevational limit of coniferous forest during dispersal.

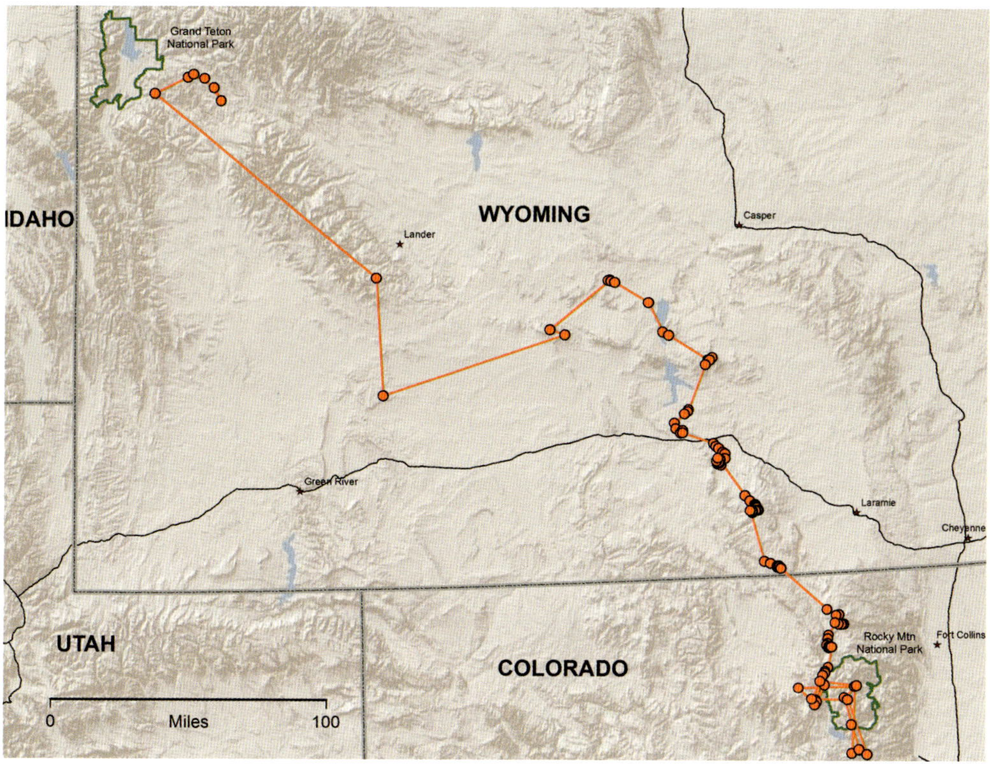

Movement of a young male wolverine from December 21, 2008 to August 6, 2009. Most of the long-distance movement occurred during April 6–June 5, 2009, tracked by the Wildlife Conservation Society's Greater Yellowstone Wolverine Program.

small and medium-sized mammals. This is a species of low reproductive capacity as a result of embryonic diapause and small litter sizes. Population biology has not been studied in detail anywhere; however, patterns of mortality were examined among 36 study animals in western Montana. Accidental harvest from licensed trapping of other species accounted for 9 of 14 mortalities. A similar pattern has been reported across the species' North American range.[12] Population density has been estimated recently for areas above 2150 m (7100 ft) elevation in northwestern Wyoming: 3.5/1000 km^2 (9/1000 mi^2).[7]

CONSERVATION AND MANAGEMENT: Some authors have speculated that the species became extinct in the contiguous United States around 1900; current populations descended from colonizers from the north, according to this view.[13] Other evidence suggests that Yellowstone National Park served as a refugium during the earliest twentieth century, when trapping and poisoning were common outside the park. Milton Skinner quoted Vernon Bailey as stating in 1920 that "trappers around the borders obtain more specimens than in any other part of the United States. The park evidently serves as a breeding and recruiting ground that has kept this interesting and rare animal from local extinction."[14] Within the state, the species was confined to remote parts of the northwestern mountains over most of the twentieth century. A 1982 compilation of observations concluded that wolverines likely had undergone a recent increase in abundance in areas near Yellowstone National Park.[15] In 2010 the US Fish and Wildlife Service ruled that the wolverine in the contiguous United States was a Distinct Population Segment. In 2013 it proposed the species for listing as Threatened in the contiguous United States, a proposal that was withdrawn in 2014. By late 2014, the E.S.A. status of the species was in litigation. Wolverines require large areas of high elevation wilderness in the contiguous United States to persist, and intensive snowmobile use and trapping in these areas are concerns.[16] The primary concern given in the 2013 proposal to list the species was declining snow depths in response to climate warming.

NOTES AND REFERENCES

1. Copeland (1996).
2. Wyoming Statutes 23-1-101-a-ix.
3. US Fish and Wildlife Service (2010b, 2013).
4. US Fish and Wildlife Service (2014).
5. Hall (1981).
6. Koepfli et al. (2008).
7. Inman et al. (2004, 2011). The project which captured, radio-collared, and tracked this wolverine was the result of collaboration between the Wildlife Conservation Society; Alta, Wyoming 4-H Club; Grand Targhee Ski Resort; Wyoming Game and Fish Department; Idaho Department of Fish and Game; USDA Forest Service; Mountain Air Research; Grand Teton National Park; and The Wolverine Foundation. Its success was the impetus for further wolverine studies in the region.
8. Inman et al. (2009). This animal's closely monitored dispersal into Colorado was the first documentation of the wolverine in that state since 1919.
9. A dispersing young male was accidentally trapped on upper Horse Creek on the east side of the Laramie Mountains in 1996 (Wyoming Game and Fish Department, unpublished data).
10. Aubry et al. (2007).

11. Copeland et al. (2007).
12. Banci (1994), Squires et al. (2007).
13. McKelvey et al. (2014).
14. Skinner (1927, 195).
15. Hoak et al. (1982).
16. Ruggiero et al. (2007), US Fish and Wildlife Service (2013).

American badger, *Taxidea taxus*

DESCRIPTION: A medium-sized carnivore with low profile, short legs, shuffling gait, long claws on the forelimbs, and a medium-length tail. The pelage is distinctive: strong dark-light banding on the face and head, mottled dark and tan over the back, buff to pale orange on the chest, and a prominent white band down the middorsal line from nose to shoulders. The dorsal fur on the thorax and abdomen is long enough in winter that it nearly touches the ground, making the animal look dorsoventrally flattened. Measurements (**Sex** [n] mean (mm, kg)): Arapaho County, Colorado, L = **F** [10] 727; **M** [10] 777; TL = **F** [10] 125; **M** [10] 130; HF = **F** [10] 105; **M** [10] 105; WT = **F** [10] 6.4; **M** [9] 8.5.[1] Dental formula: 3/3, 1/1, 3/3, 1/2 = 34.

STATUS: The species is classified as a "furbearing animal" under state law.[2] It is common within the geographic range.

NOMENCLATURE AND SYSTEMATICS: The genus and species names have been stable for over 200 years. Badger species on other continents occupy other genera; therefore, the common name "badger" does not describe a monophyletic group. Subspecies were revised in 1972 to recognize four, two of them represented as occurring in Wyoming: *T. t. jeffersoni* west of a line extending south from the eastern foothills of the Bighorn Mountains to the Laramie Basin, and *T. t. taxus* occurs east of that line.[3]

DISTRIBUTION AND HABITAT ASSOCIATIONS: This broadly distributed plains and desert species has disappeared from parts of the Canadian prairies and the Midwest since European settlement (original distribution shown). It occurs in grasslands and shrub-steppe habitats statewide. It is absent from closed forests, but occurs in alpine tundra, above the limit of trees, during summer. This suggests that it may disperse through forests to reach seasonally occupied habitats.[4]

ECOLOGY: The species is a solitary, territorial predator that specializes on fossorial mammals. Ground squirrels, including prairie dogs, are the preferred prey species in Wyoming, and are mostly captured by excavation. Badgers have been noted in Wyoming to hunt cooperatively with coyotes, with coyotes driving ground squirrels down burrows, and badgers driving them to the surface.[5] Mammals captured above ground include rabbits and mice.[6] The species remains active above ground during autumn and spring, but becomes inactive during midwinter, and can remain below ground for as long as 70 consecutive days. While underground in winter, badgers repeatedly enter brief periods of torpor.[7] Populations have not been well studied, but minimum densities have been estimated at 0.8–1.1/km^2 (2–2.8/mi^2) where white-tailed prairie dogs were the primary prey in Carbon County, and

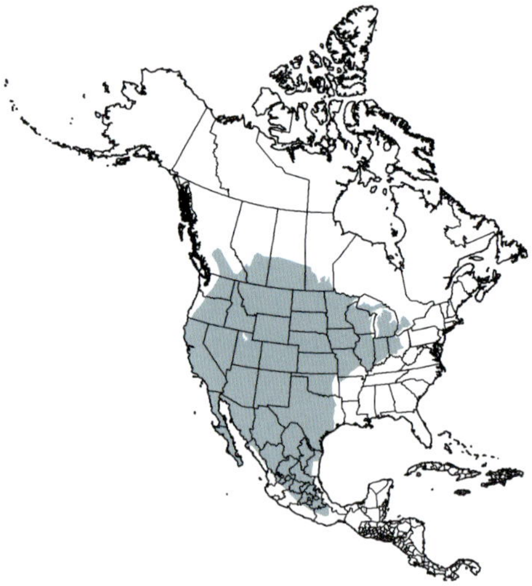

Distribution of the American badger in Wyoming.

only 0.4/km² (1/mi²) in northern Utah and southern Idaho, where smaller prey dominated.[8] Of badgers in trapper harvests from southeastern Wyoming, 48% were animals less than one year old and 28% aged 1–3 years. The oldest badger among the sample of 80 was 13 years old.[9] Badgers have been proposed to influence shrub-steppe ecosystems as a result of their burrowing and mounding activities. Badger-built mounds were found to represent 5–8% of the surface of the landscape, and mounded soil to represent 23 metric tons/ha (6660 tons/mi²) in an area of west-central Idaho.[10] Burrows provide pathways for water percolation and structures that can be used by other species, although most badger burrows represent enlarged rodent burrows.

CONSERVATION AND MANAGEMENT: The species is considered a conservation concern in Canada, where its geographic range has contracted markedly. In the Midwest, its range contracted during the settlement period, but has since expanded, at least in some areas. In Wyoming, its range is apparently unchanged since presettlement times. It is seldom trapped for fur and is not generally considered a conservation concern in Wyoming.

NOTES AND REFERENCES

1. Hoff (1998).
2. Wyoming Statutes 23-1-101-a-iii.
3. Long (1972).
4. Lindzey (2003).
5. Minta et al. (1992). This mutualism was first reported by Samuel Aughey, who conducted geological surveys in Wyoming Territory in 1883, but had heard of it from locals earlier (Aughey 1884).
6. Goodrich and Buskirk (1998), Lindzey (2003).
7. Harlow (1981).
8. Lindzey (1971), Goodrich and Buskirk (1998).
9. Crowe and Strickland (1975)
10. Eldridge (2004).

Northern river otter, *Lontra canadensis*

DESCRIPTION: This medium-sized semiaquatic carnivore is long, slender, and short-legged, and has a conspicuously arched lumbar spine while running. Its relatively short, dense pelage, usually viewed while wet, is medium brown on the dorsum, but lighter on the ventral face, throat, and abdomen. While swimming, otters can be confused with beavers, minks, and muskrats. Typically silent when traveling alone, they vocalize with bird-like whistles and chirps to communicate with group members. Measurements (**Sex** [n] mean (mm, kg), *range*): Yellowstone Lake, L = **B** [5] 860, *800–1030*; TL = **B** [5] 482, *466–514*; HF = **B** [4] 131, *124–137*; WT = **B** [5] 8.6, *8.0–10.3*.[1] Dental formula: 3/3, 1/1, 4/3, 1/2 = 36.

STATUS: The species has been a "protected animal" under Wyoming statute since 1973; the trapping season was closed beginning in 1952–1953.[2] It also a state Species of Greatest Conservation Need as of 2012. The biological status over the state is difficult to assess.

NOMENCLATURE AND SYSTEMATICS: The genus name underwent revision during the last 40 years. Before 1970, the species was considered to be congeneric with the superficially similar Eurasian river otter, *L. lutra*. But the late C. G. Van Zyll de Jong, in a series of reports from 1972 to 1987 based on skeletal morphology, concluded that all New World river land otters, except for the giant otter (*Pteronura*), constituted a monophyletic genus (*Lontra*) distinct from the Old World *Lutra*.[3] Subsequently, molecular analyses confirmed this finding.[4] Seven subspecies were recognized as of 1981; only *Lontra canadensis pacifica* is considered to occur in Wyoming; this subspecies extends to Hudson Bay and northwestern Alaska.[5]

DISTRIBUTION AND HABITAT ASSOCIATIONS: The species' presettlement distribution (shown) was remarkably broad, from the Arctic Ocean to the southern tip of Florida. This is among the

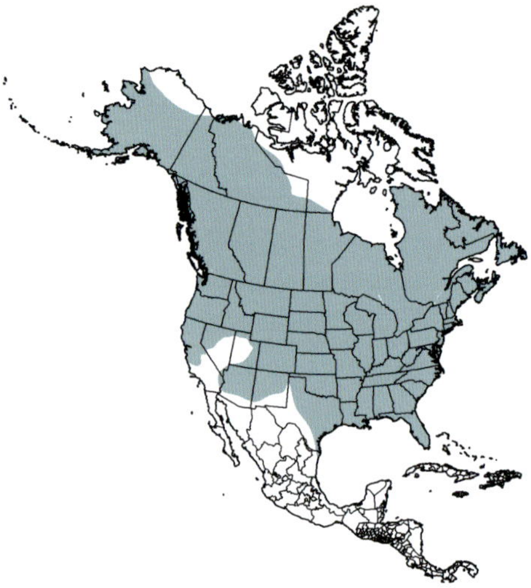

Distribution of the northern river otter in Wyoming.

Wyoming mammals with the most poorly documented distributions. Charles Long's (1965) account is based on examination of a single specimen from Yellowstone.[6] Some otters are accidentally taken by Wyoming trappers and a few of those result in curated specimens. Credible accounts of otters have originated from various water bodies in recent decades. Some of these reports seem reliable, but others lack specificity and could be mink, or they are from anomalous sites, and seem not to be part of any pattern of occurrence. It is common to receive puzzling oral reports of otters from river reaches that seem to be isolated from any known otter populations. The species underwent major distributional declines across the southern portion of its range, including in eastern Wyoming, during the last 180 years. It was driven to ecological or complete extinction on the Great Plains states from North Dakota south to Oklahoma and in the Southwest.[5] The species has been extensively reintroduced in the contiguous United States over the last 40 years.[7] In Wyoming, otters have become more widespread in the state over around the same period, although the details of this presumed recovery are not understood. Yellowstone National Park may have served as a refugium for otters during the early twentieth century, from which they recolonized areas to the south and east.[8] No translocations of otters have made to or within Wyoming, but they have occurred in northernmost Colorado, including into the Laramie River drainage which flows northward into Albany County.[9] Otters are most consistently observed in Yellowstone Lake and near the Oxbow Bend of the Snake River in Grand Teton National Park. They are observed only sporadically in the easternmost counties. Habitat is streams, rivers and lakes, and riparian vegetation nearby. Otters are typically associated with slow-moving sloughs, slow-moving stretches of otherwise fast rivers, and shallow lakes. Coarse woody debris, including logjams in rivers, has been found an important habitat feature for resting and denning.[5] Some areas of Yellowstone Lake are particularly favored because hydrothermal vents create openings in lake ice for much of the winter; continuous surface ice excludes otters from lakes in winter.

ECOLOGY: River otters are strictly predaceous, eating bottom fish, trout (especially while spawning in shallow streams), small mammals, aquatic birds, and invertebrates. Around Yellowstone Lake, the diet has shifted away from cutthroat trout toward amphibians over the last decade.[10] No diet studies have been conducted elsewhere in the state, but in neighboring states the diet was over 90% fish, with crayfish found infrequently. The fish species eaten in this region tend to be dominated by suckers, whitefish, and other slow-moving bottom feeders.[4] Populations are not well studied in the region; in Yellowstone annualized survival rates averaged 0.67 during the period of declining cutthroat trout.[10] Otters have ecological effects aside from removing some fish. In recent years they have been shown to transport nutrients from aquatic to terrestrial systems as a result of foraging in water and defecating, urinating, and leaving orts on land. Where otters create latrines, soil is nitrogen-enriched, and plant growth stimulated.[11]

CONSERVATION AND MANAGEMENT: The drastic decline of otters over the temperate portion of the range has been the basis of much conservation concern. Otters are negatively affected by accidental trapping, some river impoundments, loss of stream-bank vegetation and structure, loss of in-stream flow, water pollution, and reductions in prey abundance.[5,9] A new and potentially serious problem in the conservation of otters in and near Yellowstone Lake is the decline of cutthroat trout in response to rapid increases in the abundance of lake trout, introduced during the 1990s. Cutthroat trout captured in spawning streams made up much of the diet of otters during spawning season.[12]

NOTES AND REFERENCES

1. Crait and Ben-David (2006), Jamie R. Crait, University of Wyoming, unpublished data.
2. Wyoming Statutes 23-1-101-a-ix.

3. See Van Zyll de Jong (1987) for review.
4. Koepfli and Wayne (1998).
5. Melquist et al. (2003).
6. Long (1965).
7. Raesly (2001).
8. Thomas (1954, 51).
9. Boyle (2006).
10. Crait (2013).
11. Crait and Ben-David (2007).
12. Crait and Ben-David (2006), Wengeler et al. (2010).

RACCOONS AND ALLIES—FAMILY PROCYONIDAE

The raccoons and allies are a North American lineage that spread to South America when the continents became connected 9–5 Ma; most extant species occur from Mexico southwards. They are medium-sized, have generalized dentition without shearing cheek teeth, climb trees readily, and occur in subtropical or temperate forested or riparian habitats. The North American species have strikingly banded or marked pelage, especially on the face and tail. The diet is fairly omnivorous, in the various raccoon species emphasizing aquatic invertebrates.

Ringtail, *Bassariscus astutus*

DESCRIPTION: A slender medium-sized carnivore with large pinnae and a very long tail—nearly the length of the head and body. The dorsal pelage is buff-gray with lighter undersides. The facial shape and body size are similar to that of a marten, but the long, dark-light banded tail is distinctive. Measurements (**Sex** [n] mean (mm, g), *range*): eastern Arizona, L = **F** [5] 728, *700–750*; **M** [5] 764, *723–793*; TL = **F** [5] 369, *330–390*; **M** [5] 386, *358–411*; HF = **F** [5] 66, *63–69*; **M** [5] 69, *63–75*; EFN = **F** [5] 59.86, *50–66*; **M** [5] 51.4, *38–65*.[1] Dental formula: 3/3, 1/1, 4/4, 2/2 = 40.

STATUS: This is a protected nongame animal, which may not be intentionally taken.[2] It is uncommon or rare within its geographic range in Wyoming. Occasional visual reports are received from sites in southern Wyoming far from the known range; these reports are difficult to evaluate with respect to whether they represent natural occurrences, misidentifications, or released pets.[3]

NOMENCLATURE AND SYSTEMATICS: The scientific name has been in use since the early nineteenth century, and few alternatives have been proposed, none of them used widely. Of the 14 subspecies recognized as of 1988, two are depicted in or near Wyoming: *B. a. arizonensis* and *B. a. nevadensis*.[4]

DISTRIBUTION AND HABITAT ASSOCIATIONS: This species of southwestern and Pacific coast canyons and riparian zones is best documented in Wyoming from the southern border near Flaming Gorge, but reports have originated from the Red Desert and other anomalous sites.[5]

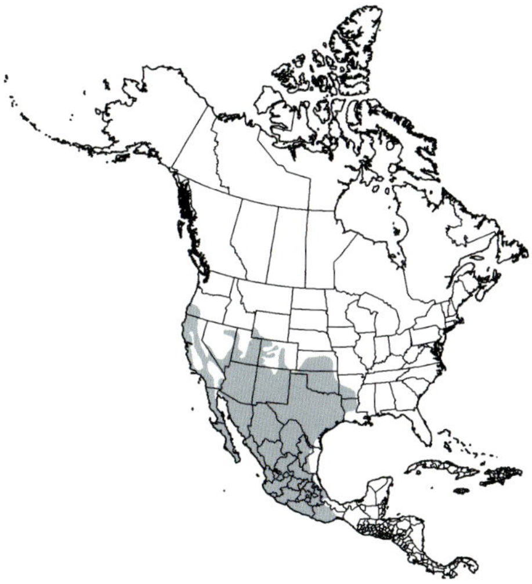

Geographic range of the ringtail.

Distribution of the ringtail in Wyoming.

One animal found dead where I-80 crosses the Blacks Fork of the Green River in 2006 is displayed in the Wyoming Game and Fish Department Regional Office in Green River.[6] Habitats are varied, but tend in Colorado toward arid canyon country and riparian woodlands. Rocky areas with dense forest cover are avoided. In Colorado, the species has been noted to occur up to 2700 m (9000 ft) in elevation.[7]

ECOLOGY: The species is poorly known from Wyoming and Colorado; elsewhere it is a dietary generalist, eating fruits, invertebrates, small vertebrates, and bird eggs. Juniper berries can be important foods and may allow ringtails to range widely across Utah juniper woodlands around Flaming Gorge Reservoir.[8] They readily eat human foods and tolerate human presence, living at times in attics and other human structures.[1]

CONSERVATION AND MANAGEMENT: The species is too poorly known and its status too uncertain to receive much consideration in Wyoming. It is not often mentioned as a conservation concern in the state, being at the limit of its geographic range here.

NOTES AND REFERENCES

1. Hoffmeister (1986).
2. Commission Regulations Chapter 52, Section 11.
3. Considerable information is available online regarding ringtails as pets. Such animals presumably escape or are released into the wild, and could account for some observations outside the known geographic range.
4. Pogalayen-Neuwall and Toweill (1988).
5. Long (1965). Crowe (1986) showed two records in southwestern Sweetwater County, and one in the southwestern corner of Lincoln County, but no information is associated with those plotted points.
6. The animal was found dead near the Blacks Fork of the Green River (41.5342° N, 109.7191° W).
7. Hoffmeister (1986), Galvin and Richards (1993).
8. Pogalayen-Neuwall and Toweill (1988), Armstrong et al. (2011).

Northern raccoon, *Procyon lotor*

DESCRIPTION: A stout mid-sized carnivore with gray, medium-length fur on the back, a tail ringed with wide dark and light bands, a black facial mask, and typically ambling gait. Measurements (**Sex** [n] mean (mm, g), *range*): various counties, L = **F** [5] 735, *650–780*; **M** [5] 727, *650–820*; TL = **F** [5] 254.60, *225–300*; **M** [5] 205, *160–250*; HF = **F** [5] 111.40, *70–170*; **M** [5] 98.0, *84–110*; EFN = **F** [5] 60, *50–66*; **M** [5] 51, *38–65*.[1] Dental formula: 3/3, 1/1, 4/4, 2/2 = 40.

STATUS: It is classified as a "predatory animal" under state statute and is common in suitable habitats.[2]

NOMENCLATURE AND SYSTEMATICS: The genus and species names have been stable for over 200 years; "*lotor*" refers to the lotic or moving-water habitats with which the species associates. The family Procyonidae is restricted to the New World today, but originated in western Eurasia during the Late Eocene or Oligocene. Raccoons are closely related to ringtails.[3] Of the 25 subspecies recognized as of 1981, three occur in or near Wyoming: *P. l. hirtus*, *P. l. excelsus*, and *P. l. pallidus*.[4]

DISTRIBUTION AND HABITAT ASSOCIATIONS: This broadly distributed temperate species has undergone arguably the most dramatic distribution change over the last 200 years of any Wyoming mammal, without direct human intervention. It was apparently not or barely present in the state at the time of European settlement (shown), but occurs in suitable habitats statewide today. The species was mentioned in few or no accounts of explorers in or near Wyoming as recently as 1876, but moved up the Missouri River bottoms so that it occupied the "eastern one-fifth" of the state by 1965.[5] The range expansion westward

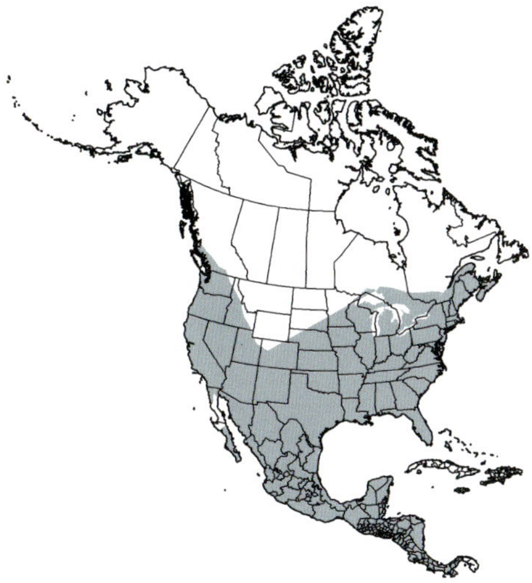

Distribution of the northern raccoon in Wyoming.

continued through the 1980s, extending to westernmost Wyoming, including Lincoln and Sublette Counties.[6] Rabies monitoring by the Wyoming State Veterinary Laboratory during the period 1985–1989 (150 specimens) showed raccoons originating from every county except Lincoln and Sublette, and for the period 2000–2011 (552 specimens) showed raccoons from every county except Teton.[7] The historical-period range expansion in North America has been attributed to human land use and to the ecological extinction of wolves during the early 20th century.[8] Habitats tend to be riparian-associated or developed areas; areas far from surface water or moist vegetation are unlikely to be unoccupied. The species has been noted at elevations over 2700 m (9000 ft) in Colorado.[9]

ECOLOGY: The species has not been studied in Wyoming. In other areas, it is a dietary generalist, feeding on crayfish, small vertebrates, berries, carrion, agricultural crops, and, in developed areas, human foods. In some Wyoming towns, raccoons move around via storm sewers and emerge from storm drains at night to forage in alleyways and backyards. Population biology is not well understood in or near Wyoming. Humans likely account for much mortality via opportunistic shooting and deaths on highways. Although raccoons are important reservoirs and vectors for rabies on the East Coast, positive diagnoses in Wyoming are rare.[10]

CONSERVATION AND MANAGEMENT: Although legally a "predatory animal," raccoons are trapped for their fur, which has considerable value. The species is more often considered a pest than a conservation concern, because of its expanding distribution and commensal relationship with humans. It is plausible that the expanded distribution and abundance of raccoons may influence other community members on which raccoons prey, and which may be conservation concerns themselves.

NOTES AND REFERENCES

1. UWMV specimens, Long (1965).
2. Wyoming Statutes 23-1-101-a-viii.
3. Koepfli et al. (2007).
4. Lotze and Anderson (1979).
5. Long (1965), Turner (1974), Finley (1995).
6. Hart (1982).
7. Finley (1995) for data 1985–1989; Wyoming State Veterinary Laboratory, unpublished data for the period 2000–2011.
8. Finley (1995).
9. Armstrong et al. (2011).
10. The last positive diagnosis of raccoon rabies from Wyoming was from Natrona County in 1998 (Wyoming State Veterinary Laboratory, unpublished data).

BEARS—FAMILY URSIDAE

The eight species in this family make up a large-bodied Northern Hemisphere lineage that has a long rostrum, short tail, five toes per foot that touch the ground, and cheek teeth that have rounded surfaces, rather than shearing edges. All three North American species undergo seasonal torpor, and have diets that are variably omnivorous. As a group, and partly because of their large bodies and low population densities, bears present some the world's most serious mammalian conservation issues, with various taxa recognized under national and international protection systems.

American black bear, *Ursus americanus*

DESCRIPTION: A dark-colored large carnivore with rounded form, rostrum pelage lighter colored than the rest of the body, and no hump over the shoulders. Color is typically black, but chocolate or cinnamon-colored animals are common. There is a light colored mark on the chest of some animals, visible from the front, but this is also seen in some young grizzly bears. The size is smaller than for sympatric grizzly bears, but difficult to

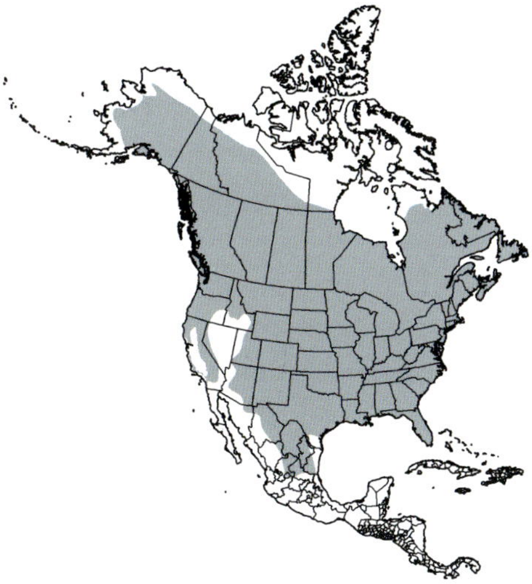

Distribution of the American black bear in Wyoming.

evaluate in the field without experience. The rostrum of a black bear is more sloped than that of a grizzly bear (more dished), and it has larger ears in proportion to its head. Black bears in the Medicine Bow Mountains are similar in size to or smaller than those in adjacent states.[1] Measurements (**Sex** [n] mean (mm, kg), *range*): adults, Park, Teton, Fremont, and Sublette Counties, L = **F** [25] 1247, *555–1470*; **M** [52] 1408, *1120–1680*; TL = **M** [11] 116, *75–140*; HF = **F** [25] 184, *157–207*; **M** [51] 212, *175–245*; WT = **F** [25] 76.0, *44.4–124.7*; **M** [50] 102.0, *37.2–173.3*.[2] Dental formula: 3/3, 1/1, 4/4, 2/3 = 42.

STATUS: The species is a "trophy game animal" under state statute.[3] It is common within some parts of its Wyoming geographic range.

NOMENCLATURE AND SYSTEMATICS: The genus name *Euarctos* was applied by some taxonomists in the early and mid-twentieth century. Four species are now placed within *Ursus*; of these, the Asiatic black bear appears most closely related to the American black bear. Of the 16 subspecies recognized in 1981, two are shown as occurring in Wyoming, but Long believed that only *U. a. cinnamomum* occurred in the state.[4] Population genetic data show that populations in the northwestern mountains, Bighorn Mountains, and Medicine Bow Mountains show evidence of genetic isolation over long periods.[5] This genetic isolation reflects infrequent dispersal across shrub-steppe habitats, likely due to grizzly bears occupying those non-forested habitats over the millennia. Black bears were restricted to forested mountains where they could escape grizzlies by climbing trees. Even the loss of grizzly bears from the Wyoming plains over the last 130 years seems not to have resulted in measurable genetic interchange among black bear populations.

DISTRIBUTION AND HABITAT ASSOCIATIONS: This species occurs from Florida to northern Alaska (current distribution shown). In Wyoming, the current geographic range is depicted variously by different authors: statewide, absent from prairie of the eastern part of the state, or only in mountain ranges.[3] The pre-European distribution likely was limited to forested habitats; today, in the absence of grizzly bears in shrub-steppe habitats, black bears occasionally disperse among mountain ranges, crossing large unforested areas. The species was common in the Black Hills in the late nineteenth century.[6] Habitat in Wyoming is forest, woodland, and areas nearby. In northwestern Wyoming black bears are less likely than grizzlies to use open areas.[7] In the Medicine Bows where grizzlies do not occur, non-vegetated areas and lodgepole pine were used more than expected based on availability. Shrublands and ponderosa pine were used less than expected.[1]

ECOLOGY: The species is a solitary, territorial, dietary generalist. Black bears eat almost every form of available, digestible plant and animal material: grasses, forbs, roots, tree cambium, fruits, insects (e.g., grasshoppers, ants), spawning cutthroat trout, small mammals, carrion, and neonatal ungulates.[8] Whitebark pine nuts are an important food in the northwestern mountains, and black bears use these nuts much as do grizzlies, raiding red squirrel caches repeatedly during the year, in some cases at the same seasons and locations as grizzly bears.[9] Ants appear in many black bear scats in dry forest types of southern Wyoming, and human foods were important during the twentieth century, particularly before modern landfill management. Stable isotopic analyses showed that in the Yellowstone area the terrestrial meat component of black bear diets was similar to that of female and subadult male grizzly bears, 40–60%.[10] Similarly, analyses of fecal droppings from the Greys River area of far western Wyoming showed that around 60% of the May–September diet was herbaceous plants.[11] Interference competition from grizzlies is a key limitation on black bears where the two species are sympatric, affecting habitat associations and access to some resources. Black bears are more likely than grizzlies to be active in daytime, reducing the chance of contact with dangerous adult male grizzlies.[12] Black bears undergo seasonal torpor in dens that are often simple overhangs, rock crevices, or piles of logs. Populations have been studied little, relative to other geographic areas and relative to grizzly bears in Wyoming. Mean population density for the Medicine Bow Mountains was 2.5/100 km^2 (6.4/100 mi^2). Values for other mountainous areas of western North America tend to be 10–20 times higher.[13]

Black bears have ecological effects that are diverse and difficult to assess. Much of their diet overlaps with that of the grizzly bear, and the two species are often pooled in reports where field sign of both species can be confused. Therefore, ecological contributions of the two species are difficult to parse. Black bears accounted for only 3% of predation mortality and 1.5% of total mortality among Yellowstone elk in the 1990s.[14] The species also disperses fruit seeds, removes some mammal carrion, and causes some tree death among Engelmann spruce and lodgepole pine.[15]

CONSERVATION AND MANAGEMENT: The species is managed as a trophy game species outside protected areas and is hunted in spring and fall hunting seasons. Harvest is managed by a quota for the several hunt areas; around 300 animals per year were killed by hunters during 2005–2009.[16] Most bear habitat conservation actions tend to focus on grizzlies, but a diversity of stand types and successional stages has been proposed to address the seasonally diverse diets of black bears.[11]

NOTES AND REFERENCES

1. Grogan (1997).
2. Wyoming Game and Fish Department, Trophy Game Section, unpublished data, 2013.
3. Wyoming Statutes 23-1-101-a-xii.
4. Long (1965), Hall (1981).
5. See chapter on Zoogeography. Genetic distances are from Cachelin and McDonald, unpublished data.
6. Turner (1974).
7. Holm et al. (1999).
8. Murie (1937), Reinhart and Mattson (1990), Pelton (2003).
9. Kendall (1983).
10. Jacoby et al. (1999).
11. Irwin and Hammond (1985).
12. Mattson et al. (1992), Holm et al. (1999), Gunther et al. (2002).
13. 95% confidence interval: 2.1–3.0, Grogan and Lindzey (1999).
14. Singer et al. (1997).
15. Bailey (1930).
16. Wyoming Game and Fish Department (2010b).

Grizzly bear, *Ursus arctos*

DESCRIPTION: Grizzlies are large bears with fur of seasonally variable length and color, but seldom coal black. In northwestern Wyoming, most adults have some degree of banding or grizzling of the dorsal guard hairs at some time of year. The rostrum is not as consistently or uniformly tan as in black bears, but tends to be lighter than the other facial pelage. A hump over the shoulder is visible in most adults, and the feet, long foreclaws, and behavior are distinctive with experience. The rostrum in adults is dished in profile in contrast to that of a black bear, and the ears are small in proportion to the head.

Grizzly bear (adult male).

Measurements (**Sex** [n] mean (mm, g), *range*): adults, northwestern counties, L = **F** [67] 1466 *1230–1730*; **M** [117] 1622, *1320–1860*; HF = **F** [67] 217, *180–245*; **M** [119] 246, *115–285*; WT = **F** [67] 119, *70.3–168*; **M** [119] 180, *58.5–302*.[1] Dental formula: 3/3, 1/1, 4/4, 2/3 = 42.

STATUS: This is a "trophy game animal" under state statute.[2] The species was listed as "threatened" in the contiguous United States under E.S.A. in 1975. In 2007, the Yellowstone Distinct Population Segment was found to no longer meet the E.S.A. definition of Threatened or Endangered, and E.S.A. protections were removed. However, a 2009 court decision reversed that decision and reinstated the Yellowstone population as Threatened.[3]

NOMENCLATURE AND SYSTEMATICS: This genus name has been in use since the mid-eighteenth century; the specific epithet has been revised multiple times. Merriam's (1918) classification, repeated in Hall (1981), identified at least 86 "forms" of brown-grizzly bears, of which no fewer than seven putative species were named in and near Wyoming.[4] Today, only a single Holarctic brown bear species is recognized: *Ursus arctos*. It is likely that subspecies names will not again be assigned by systematists to Wyoming forms, because even the original species names were not assigned with regard to valid biological species, and subspecies names were assigned even less cautiously. A further problem is that most of the geographic range over which these many taxa were distributed was depopulated of grizzly bears by the early twentieth century. It would be difficult to correctly assign a subspecies name to the population that has expanded out of Yellowstone National Park.[5] The population has experienced some loss of genetic variability over the nearly 100 years of isolation from other populations, but the loss is less severe than was long hypothesized, and not believed capable of affecting population persistence over the next few decades.[6] The common name "brown bear" is correctly applied to all members of the species in North America and Eurasia.

DISTRIBUTION AND HABITAT ASSOCIATIONS: This species was once broadly distributed (shown) west of the Mississippi River, and occurred statewide when Europeans arrived. Historical accounts show the presence of the species in virtually every habitat statewide: shrub-steppe, forests, and alpine tundra; densities would have been low in the Red Desert. With human hunting and settlement, grizzlies were soon limited to the northwestern mountains and by the mid-twentieth century to only a portion of Yellowstone National Park. Subsequent recovery has expanded the geographic range to beyond the borders of Yellowstone National Park into Jackson Hole, the Teton and Wind River Mountains, and the Wind River Indian Reservation.[7]

ECOLOGY: The ecology of grizzlies in the Yellowstone area is well documented. This is a solitary omnivore that in northwestern Wyoming seasonally shifts among highly disparate food sources: carrion, herbaceous plant leaves and flowers, berries, spawning trout, neonatal ungulates, whitebark pine seeds, and aggregated insects.[8] This species has one of the slowest life histories of any North American land vertebrate, with mean age of first reproduction in Yellowstone of 5.7 years, reproductive interval among females thereafter of 2.6–3.2 years, and litters averaging around two cubs. So, the typical female grizzly bear in Yellowstone has produced two cubs by the age of 8.5 years. The number of animals in the Yellowstone area has increased from as few as 136 animals in the mid-1970s to more than 700 animals as of 2013.[9] Cubs living within Yellowstone National Park have an 83% probability of living to one year; those outside the Grizzly Bear Recovery Zone have a corresponding 12% survival rate.[10] About 85% of all deaths of independent grizzlies in the Yellowstone population are human-caused.[11] Population density in the Yellowstone population has been estimated at 14–18/1000 km^2 (36–46/1000 mi^2).[12] Grizzlies have a number of ecological effects, aside from participating in energy and biomass cycling. They are important seed predators on whitebark pine, likely are important seed dispersers for berry-producing shrubs, and have been hypothesized to be effective predators on neonatal elk in the Yellowstone area.[13]

CONSERVATION AND MANAGEMENT: A long history, controversial at times, describes efforts to conserve the greater Yellowstone population of grizzlies. Steps have included hunting prohibitions; educating recreational users, grazing lessees, and landowners about reducing conflicts; eliminating open garbage dumps; removing and securing other food attractants where possible; and adopting nonlethal control methods.[14] In response, the distribution and abundance of grizzlies in northwestern Wyoming have expanded markedly since the 1970s. Grizzlies have a long-standing tendency to depredate livestock, and nearly all sheep grazing leases and some cattle grazing leases on public land in northwestern Wyoming have been canceled to reduce conflicts. It is unlikely that the Yellowstone population will grow much further in size or farther in area because of social and economic constraints. Grizzlies within Yellowstone National Park began to face new dietary limitations in the 1990s, with declines of spawning Yellowstone cutthroat trout and whitebark pine nuts. Pine nuts had been a protein source second only to meat before the major die-off of whitebark pine. And the loss of cutthroat trout spawning has resulted in bears shifting their distribution to areas where elk are calving during that season.[15] Whether elk calves will prove to be

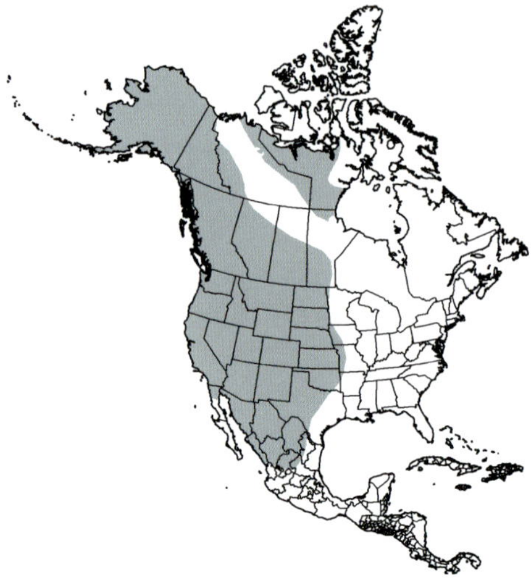

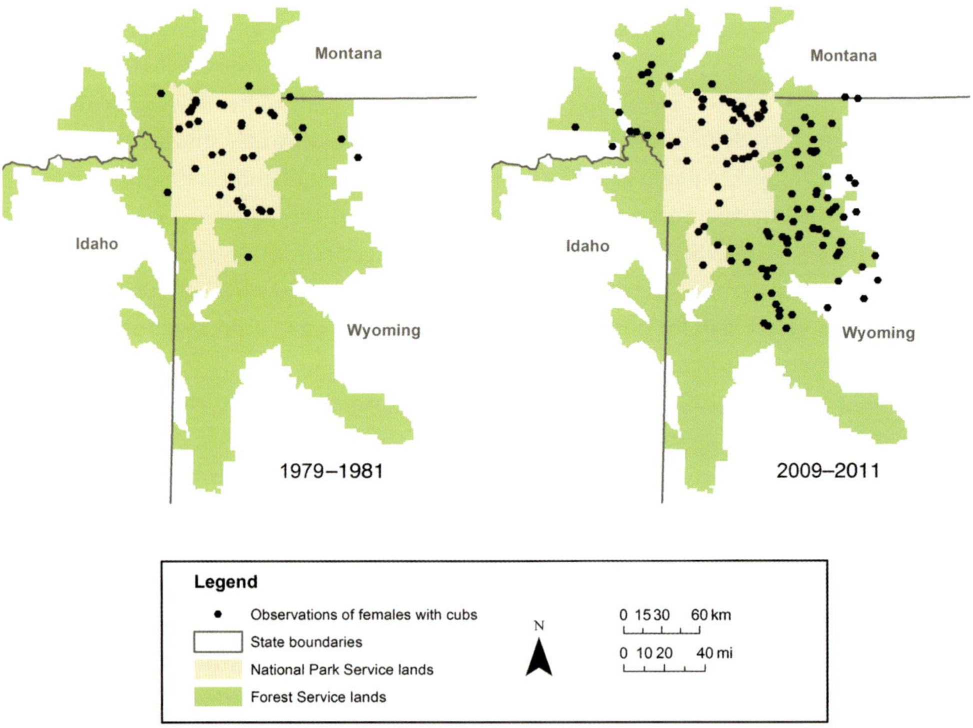

Distribution of observations of female grizzly bears with cubs of the year gathered by state and federal agencies and compiled by the Interagency Grizzly Bear Study Team for two periods, 1979–1981 and 2009–2011. Each point shows the initial observation of a female-cub group for each year during the interval. Much of the increase in numbers of female grizzly bears with cubs has occurred via range expansion to areas outside Yellowstone National Park.

a food resource equivalent to trout and to what degree hunter-taken ungulate carrion will offset the loss of whitebark pine nuts in the autumn seasons is not yet known.

NOTES AND REFERENCES

1. Wyoming Game and Fish Department, Trophy Game Section unpublished data 2013.
2. Wyoming Statutes 23-1-101-a-xii.
3. See US Fish and Wildlife Service (2010c) for a review of regulatory and judicial actions.
4. Merriam (1918) described *Ursus horribilis, U. idahoensis, U. mirus, U. rogersi rogersi,* and *U. washake* from the Greater Yellowstone Area; *U. absarokus* from the Bighorns; and *U. rogersi bisonophagus* from the Black Hills area. Hall (1981).
5. The US Fish and Wildlife Service (2010c) refers to the Yellowstone population as *U. arctos horribilis,* although four other species or subspecies were named from the same population during the early 1900s.
6. Miller and Waits (2003).
7. Schwartz, Haroldson, Gunther et al. (2006). Location data used by permission of the Interagency Grizzly Bear Study Team.
8. Mattson et al. (1991), French et al. (1994), Green et al. (1997), Mattson (1997), Schwartz et al. (2003).
9. Van Manen et al. (2013).
10. Schwartz, Haroldson, and White (2006).
11. Haroldson et al. (2006).
12. Schwartz et al. (2003).
13. Middleton, Morrison et al. (2013).
14. Murie (1948) described conflicts between grizzlies and cattle grazing. Anderson et al. (2002), Schwartz et al. (2003).
15. Felicetti et al. (2003), Haroldson et al. (2005), Koel et al. (2005).

CATS—FAMILY FELIDAE

The cats are a broadly distributed group of medium-sized to large carnivores with short rostrums, four digits per foot that touch the ground, retractile claws in most species, shearing cheek teeth, and highly predacious foraging. Short- and long-tailed species occur. The first upper molar is a vestigial nub, and the second lower molar, posterior to the shearing carnassial tooth, is absent. This reflects the unimportance of plants foods, which must be thoroughly chewed, in the diets of cats. The group tends toward stalking or ambush as a means of prey capture. Taxonomic structure below the family level has been unstable for decades, with various subfamily and genus-level organizations proposed.

Mountain lion, *Puma concolor*

DESCRIPTION: A distinctive large cat with a short rostrum, short fur, and long, agile tail. Pelage is uniformly buff or reddish tan. Measurements (**Sex** [n] mean (cm, kg), *range*): Bighorn Mountains, **L** = F [9] 200, *187–208*; **M** [8] 208, *170–223*; TL = **F** [9] 79, *71–85*; **M** [7] 80, *75–84*; WT = **F** [9] 41.9, *36.3–45.4*; **M** [8] 60.7, *45.4–74.8*.[1] Dental formula: 3/3, 1/1, 3/2, 1/1 = 30.

Mountain lion mother and young.

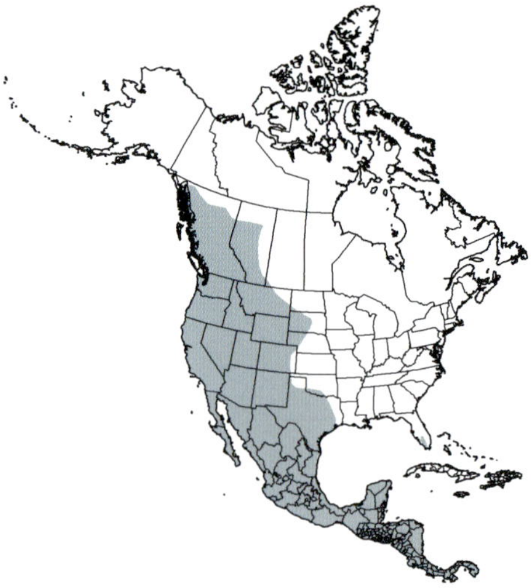

Distribution of the mountain lion in Wyoming.

STATUS: The legal status in the state is as a "trophy game animal." It is uncommon to common, but shy and hidden in daytime, so sightings are not a reliable indicator of its abundance.[2]

NOMENCLATURE AND SYSTEMATICS: The species was first described from a Brazilian specimen in 1648, but named *Felis concolor* by Linnaeus in 1771. The genus name underwent change to *Puma* in 2000; the species epithet means "uniform in color." The common names "cougar," "puma," and "panther" are used variously in parts of the Americas. A total of

30 subspecies were recognized based on morphology as of 1983, but a 2000 molecular analysis showed that the 15 previously recognized North American subspecies were a single lineage (*P. c. cougar*), which was homogeneous relative to the highly divergent various South American lineages. These data suggested the species underwent a massive die-off in North America coincident with the late-Pleistocene extinction spasm, followed by recolonization by relatively few animals via Central America.[3] Within the Rocky Mountain region, recent molecular analyses showed that populations in the mountain ranges surrounding Wyoming were highly interconnected via dispersal. Mountain ranges that are close together have higher rates of genetic interchange than more distant ones; the species exhibits isolation by distance. The migration rate between the northwestern mountains and the Bighorn Mountains was estimated at 30 animals/generation. Between the Bighorns and the more distant Snowy Range the rate was only eight animals/generation.[4]

DISTRIBUTION AND HABITAT ASSOCIATIONS: This is the most widely distributed mammalian species in the Western Hemisphere, found from Patagonia to Yukon Territory. Its presettlement distribution (shown) was greatly reduced during the first half of the twentieth century, but has since recovered somewhat in the Midwest. Long (1965) reported the species to be very rare, but it occurs nearly statewide today, although sightings in the northeastern corner and on the eastern plains remain uncommon. Expansion of the species to the northeastern corner has occurred over the last 20 years or so.[5] Habitats are quite varied, but tend toward areas with sufficient cover for hiding and for ambushing prey.[6] In the Bighorn Mountains, cougars were found to use mixed coniferous and mountain shrub communities more than expected, and sagebrush-grass less than expected. Mountain lions have difficulty ambushing prey in areas of short grass.[7]

ECOLOGY: Mountain lions are solitary, territorial, stalking, and ambush predators. Their large body size makes ungulates their obligate prey, but they eat a wide range of other animal foods. Studies using telemetry in southeastern Wyoming have shown large prey to include deer (almost entirely mule deer, 65% of carcasses), wapiti (22%), pronghorn (9%), livestock (3%), and moose (1%). There was a tendency for female cougars to kill mule deer and males to kill the larger wapiti. Among smaller prey, porcupines were commonly taken.[8] Of course, other ungulates, including bighorn sheep, are important in some areas, and leporids are taken in summer. Populations have been studied in various regions, and density estimated for Wyoming and western states. In the Bighorn Mountains, density was estimated to be 1.5 residents/100 km^2 (3.8/100 mi^2). For lions of all residency statuses, the value was 3.9/100 km^2 (10/100 mi^2).[9] For the Medicine Bow Mountains, estimated density was somewhat higher: 2.4–3.4 independent lions/100 km^2 (6–8.7/100 mi^2).[10] Population size in Wyoming has been estimated via molecular methods at 1800–4500 animals.[4] Mortality causes are varied, including human hunting, killing by other lions, and disease.[11] Mountain lions from California carry antibodies to various diseases of felids, but which ones cause population-level effects is not known.[12] Mountain lions are believed to influence prey communities by limiting abundances. Specifically the species has been found to limit or lower densities of prey alternate to deer and wapiti: wild horses, bighorn sheep, and porcupines. In contrast, most population scientists doubt whether cougars can be a primary limiting factor in the abundance of mule deer, which are the primary prey in Wyoming and adjacent states.[13]

CONSERVATION AND MANAGEMENT: The species is managed via quotas for several hunt areas within the state, with seasons and bag limits modified when populations exceed or fall below the stated objective. The conservation status of and human attitudes toward cougars have changed drastically in the West over the last 50 years. This has corresponded with a decline in the abundance of domestic sheep and the economic importance of the sheep industry. While the species is now widespread in the state and cannot be considered a conservation concern, it has important ecological effects on other vertebrates, populations of which can be reduced under high cougar densities.[14]

NOTES AND REFERENCES

1. Logan (1983, Appendices B and C).
2. Wyoming Statutes 23-1-101-a-xii. Thomas (1954) described mountain lions as "so scarce in Wyoming that they have little significance," whereas Long (1965) referred to them as "nearly extirpated in Wyoming," including all of Yellowstone National Park.
3. Hall (1981), Culver et al. (2000).
4. Anderson et al. (2004).
5. Berg et al. (1983).
6. Lindzey (1987).
7. Logan and Irwin (1985).
8. Anderson and Lindzey (2003).
9. Logan (1983).
10. Anderson and Lindzey (2005). The value for independent young animals is roughly comparable.
11. Logan et al. (1986).
12. Paul-Murphy et al. (1994).
13. Pierce and Bleich (2003).
14. See account for North American porcupine for suspected effects of high cougar predation.

Canada lynx, *Lynx canadensis*

DESCRIPTION: A medium-sized cat with a short rostrum on the skull, a tail slightly longer than that of bobcat, long limbs, and large feet. The rear half of the body can seem taller than the front in some views. Adult lynx are generally lighter in weight than sympatric bobcats. Pelage is variable in color, tending toward gray with dark highlights. The ear tufts are more pronounced than those of the bobcat, and the tail has black bands that extend to the ventral side. Measurements (**Sex** [n] mean (kg), *range*): >1 year of age, animals born in Colorado, WT = **F** [9] 8.1, *6.0–9.5*; **M** [9] 9.7, *7.8–11.8*.[1] Dental formula: 3/3, 1/1, 2/2, 1/1 = 28.

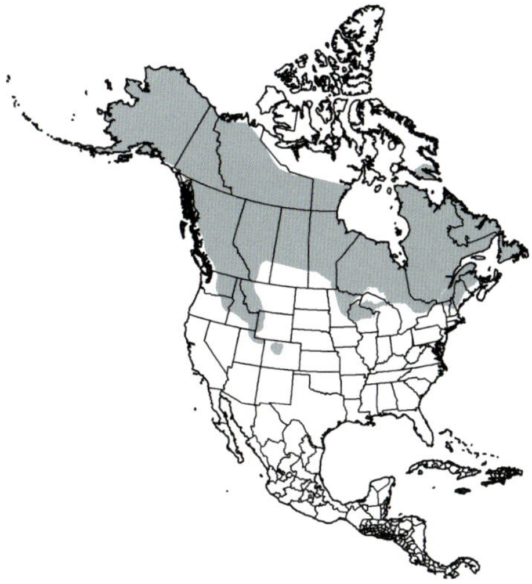

STATUS: The legal status in Wyoming is as a "protected animal," and in the contiguous United States as E.S.A. Threatened since 2000.[2] It is also a state Species of Greatest Conservation Need as of 2012.

NOMENCLATURE AND SYSTEMATICS: The genus has been considered by a few authors to be *Felis*.[3] The species is distinct from the Old World boreal lynx (*L. lynx*); and the two taxa are morphologically and otherwise distinct. Only two subspecies have been recognized, one of them (*L. c. atrata*) isolated on Newfoundland Island; the mainland subspecies is *L. c. canadensis*.[4] Animals are morphologically and genetically similar across the vast mainland distribution, reflecting frequent long-distance movements, typically triggered by the periodic population crashes of the snowshoe hare, the primary prey.[5]

DISTRIBUTION AND HABITAT ASSOCIATIONS: Range-wide, the boreal forest distribution corresponds closely with that of the snowshoe hare, with considerable range contraction at the southern margin over the last 150 years (presettlement distribution shown). In Wyoming, records have been concentrated in the Wind River Mountains (early 1900s), Hoback Rim (1940), Salt River and Wyoming Ranges (1949–1955, 1990s), the upper Hoback River (1954), and southwestern Teton County (1957).[6] Other records have come from eastern

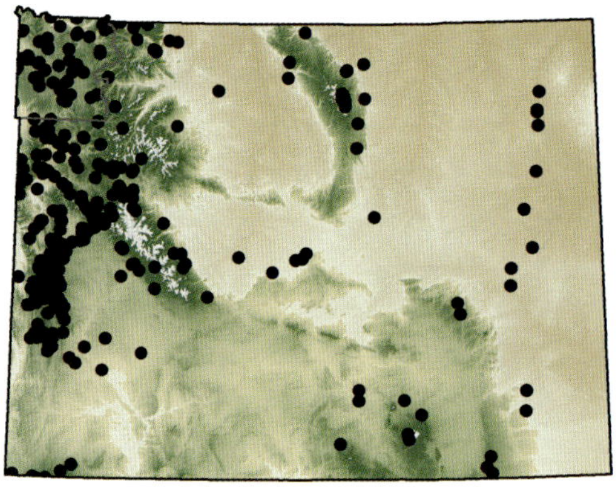

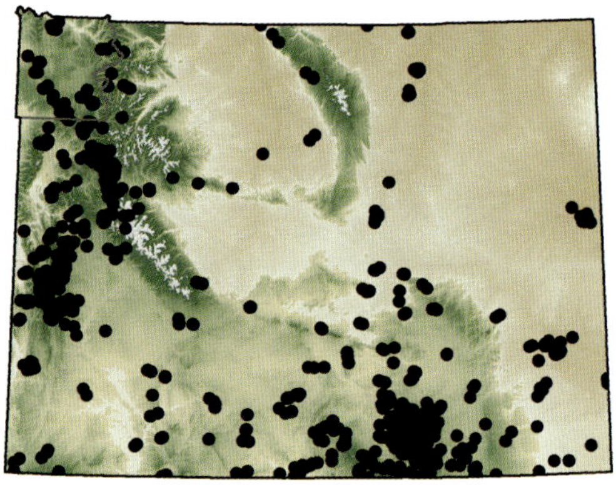

[Left] Distribution of locations of Canada lynx in Wyoming before 1999, collected by the US Forest Service Lynx Science Team. The greatest concentration is in the Wyoming Range, extending to the northern Wind River Mountains.

[Right] Locations of radio-collared Canada lynx observed by the Colorado Lynx Reintroduction Project during 1999–2010. The locations were of animals translocated to or born in Colorado. Many locations are in low-quality lynx habitat where snowshoe hares do not occur, reflecting lynx attempting to return to their place of capture. The concentration in the Medicine Bow Mountains likely represents northward moving animals that were blocked by arid shrub-steppe to the north. The concentration of locations in the Wyoming Range and northern Wind River Mountains is consistent with the pattern in the map on left, showing the high habitat quality in that area. Source: Wyoming Statutes 23-1-101-a-ix; US Fish and Wildlife Service (2000).

Albany County (1963) and southern Converse County (1983).[7] Intensive searches in the mid-1990s led to telemetric studies of two lynx in the Wyoming Range; a litter was observed in 1998.[8] The species was rare on the Yellowstone Plateau during the period 2001–2004, being detected mostly in the eastern part of the park.[9] The reintroduction of Canada lynx into southern Colorado beginning in 1999 led to some animals moving northward to (and in some cases through) Wyoming; the locations of these radio-collared animals show many occurrences in unsuitable habitats, but also concentrations where lynx have historically been abundant.[10] The greatest concentrations of locations occurred in the Medicine Bow Mountains and the Wyoming Range and northern Wind River Mountains. The Medicine Bow Mountains likely are well represented because dispersing lynx followed the Front Range northward to a peninsula of suitable habitat, where their movements were blocked by arid lands to the north. Relatively few lynx were observed in the Bighorn Mountains, in spite of snowshoe hares occurring there. Habitat is mesic coniferous forest and habitats nearby. Habitat quality is strongly influenced by snowshoe hare abundance. Snowshoe hares are associated with small-diameter woody stems accessible from the snow surface in winter.[11]

ECOLOGY: The species is well studied, especially with regard to populations, in the boreal core of its geographic range, but less so in Wyoming. Lynx populations are tied to their obligate prey, snowshoe hares; populations at the southern limit of the geographic range, including Wyoming, are believed to not undergo cycles.[12] Populations closer to the core boreal range—for example, in Minnesota—show periodic population increases, which lag three years after northern cyclic peaks and are associated with lynx dispersing southward.[13] Wyoming likely is too distant from the sources of these dispersals to be so affected. The population density of snowshoe hares required to sustain lynx populations is variable, but tends to converge on 0.5/ha (129/mi^2).[14] Besides snowshoe hares, Canada lynx prey on red squirrels, white-tailed jackrabbits, cottontails, and grouse; these alternate prey likely are more important in Wyoming than in Alaska and Canada. Causes of death are not known for Wyoming, but the fates of translocated and native-born lynx in Colorado suggest the range of mortality causes. Of 65 known-cause deaths, 29% were known or suspected of being shot, 20% were hit by vehicles, and 14% died of illness, mostly plague.[9] The remarkable lack of aggressiveness in the species may render it more vulnerable to interspecific killing than other carnivores.

CONSERVATION AND MANAGEMENT: Protected from legal trapping in Wyoming since 1973, the distribution and abundance of lynx are influenced primarily by habitat quality and community interactions. Although snowshoe hares at sufficient densities are a necessary habitat feature, they alone are not sufficient to sustain viable lynx populations. Lynx also require places to hunt (where hares are vulnerable), sites for denning, and freedom from carnivore competition and human disturbance. Managing habitat for lynx in the contiguous United States has come to focus on forest structural stages, regeneration of small-diameter woody stems, road densities, and snowmobile use.[12]

NOTES AND REFERENCES

1. Jake Ivan, Colorado Parks and Wildlife, unpublished data.
2. Wyoming Statutes 23-1-101-a-ix, US Fish and Wildlife Service (2000).
3. Wilson and Reeder (2005).
4. Hall (1981).
5. Schwartz et al. (2002).
6. Map data used by permission of USDA Forest Service.
7. McKelvey et al. (2000), Halloran and Blanchard (1959), Reeve et al. (1986),
8. Squires and Laurion (2000).
9. Murphy et al. (2006).
10. Devineau et al. (2010). Map data used by permission of Colorado Parks and Wildlife.
11. Hodges (2000a, 2000b).
12. Buskirk et al. (2000).
13. Ruggiero et al. (2000).
14. This is approximately the density of hares at the low phase of population cycles in the North (Hodges 2000b).

Bobcat, *Lynx rufus*

DESCRIPTION: A medium-sized cat, characterized by a short rostrum, medium-length limbs, and a short tail. Pelage is variable in color, with a tan background, variable dark spots, rufous flanks, and black tufts on pinnae. The tail has dark bands on its dorsum. The winter pelage is grayer and more easily confused than is the summer with that of a Canada lynx. Adult bobcats are more variable in size, and commonly larger, than lynx from the same

Distribution of the bobcat in Wyoming.

region. Measurements (**Sex** [n] mean (mm), *range*): >1 year of age, various counties, L = **F** [5] 878, *832–927*; **M** [9] 960, *921–997*; TL = **F** [5] 155, *143–178*; **M** [9] 165, *137–184*; HF = **F** [5] 184, *165–197*; **M** [4] 198, *191–213*; EFN = **F** [5] 78, *76–83*; **M** [9] 82, *76–89*; WT = **F** [41] 8.5, *6.4–12.7*; **M** [43] 11.5, *8.2–14.5*.[1] Dental formula: 3/3, 1/1, 2/2, 1/1 = 28.

STATUS: The species is a "furbearing animal" by state statute.[2] It is common in suitable habitats statewide, although not commonly observed.

NOMENCLATURE AND SYSTEMATICS: The species has long been recognized, although the genus name is occasionally given as *Felis*. Although the species is morphologically distinct from the Canada lynx, hybrids have been noted in the upper Midwest and Maine; the fertility of these animals is not known.[3] Of the 12 subspecies recognized as of 1997, only *L. r. pallescens* occurs in Wyoming.[4]

DISTRIBUTION AND HABITAT ASSOCIATIONS: This desert, temperate, and subboreal species occurs in suitable habitats statewide, although areas above the upper limit of trees tend to be avoided. The geographic range for the species (current distribution shown) has changed little over the historical period. Habitats are varied, but particularly shrubby, rocky areas, woodlands with dense understory, and other areas with abundant physical structure near the ground. Shortgrass prairie and croplands tend to be avoided.

ECOLOGY: Bobcats are territorial, solitary, ambush, or stalking predators. They feed mainly on leporids, but also consume prairie dogs, ground squirrels, rats, mice, and the occasional ungulate. Fresh carrion is eaten when available. Plant foods are consumed only incidentally to ingesting prey. Trapper harvests in Wyoming tend to be dominated by animals 0–2 years old (60–70% of males, 58–64% of females); the oldest animals observed were 10–11 years old. Mean litter size in Wyoming was 2.4.[5]

CONSERVATION AND MANAGEMENT: The species is managed as a furbearer, with trapping seasons set statewide. Managers regulate trapping intensity at a local level (e.g., quotas by management area), although trapping pressure can vary sharply with pelt prices, which averaged several hundred dollars in 2013.[6] During the last 12 years or so, bobcat populations are believed to have increased in most of United States where they occur, including Wyoming.[7] The species' geographic range today is similar to that at the time of European contact.

NOTES AND REFERENCES

1. Crowe (1974).
2. Wyoming Statutes 23-1-101-a-iii.
3. F^1 hybrids were identified using DNA markers by Schwartz et al. (2004), and described morphologically by Homyack et al. (2008).
4. Lariviere and Walton (1997).
5. Crowe (1974, 1975), Crowe and Strickland (1975).
6. Average price to trappers via fur auction houses was $400–500 (Edward Crain, Walla Walla, Wash., pers. comm.).
7. Roberts and Crimmins (2010b).

Even-Toed Ungulates—Order Artiodactyla

The even-toed ungulates are a widespread lineage that first appears in the fossil record 55 Ma.[1] Extant artiodactyls are characterized by mostly medium to large body size and two or four digits, with two touching the ground in Wyoming forms. The skull is elongated, with no upper incisors, a long dental diastema, and open-rooted cheek teeth that have enamel crescents on their occlusal surfaces. Head ornamentation, in the form of horns or antlers growing from the frontal bones, is present in males and some females of all North American forms. The digestive tract is highly modified for foregut digestion of plant foods, with a compartmented stomach. Artiodactyls make up the largest-bodied, most economically important, and most conspicuous wild Wyoming mammals, with several species easily observed on a drive across the state. They are the obligate prey of wolves and mountain lions, and important prey for several other species and important agents of herbivory, and provide many birds and mammals with carrion food sources.

NOTES AND REFERENCES

1. Gingerich (2006).

DEER AND ALLIES—FAMILY CERVIDAE

Members of the deer family are distributed across the Old and New Worlds except for sub-Saharan Africa, and are characterized by antlers in most species. Antlers are seasonally grown branched extensions of the frontal bone and shed yearly. Each antler is covered with finely furred skin while growing; "velvet" is shed in late summer and the antler is used in courtship battles among males during fall; antlers are shed in winter or early spring. This group is the most economically important mammalian family in the state, with hunting for deer, wapiti, and moose the source of most revenue for game management.

Wapiti, *Cervus canadensis*

DESCRIPTION: A large ungulate with brown fur, tan rump, and seasonally deciduous antlers in adult males. The head and neck are darker than the rest of the body. Pelage on the dorsal trunk is dark brown when newly grown, but fades to pale brown by early winter. Gender-specific vocalizations are distinctive. Measurements (**Sex** [n] mean (mm, kg), *range*): Yellowstone, L = **F** [55] 2423, *2310–2510*; **M** [12] 2272, *2083–2483*; TL = **F** [55] 136, *133–159*; **M** [12] 143, *102–178*; HF = **F** [55] 663, *603–670*; **M** [12] 669, *641–692*; EFN = **F** [55] 203, *184–222*; **M** [12] 212, *191–229*; WT = **F** [11] 255, *236–292*; **M** [10] 331, *298–373*.[1] Dental formula: 0/3, 1/1, 3/3, 3/3 = 34.

Wapiti mother and young.

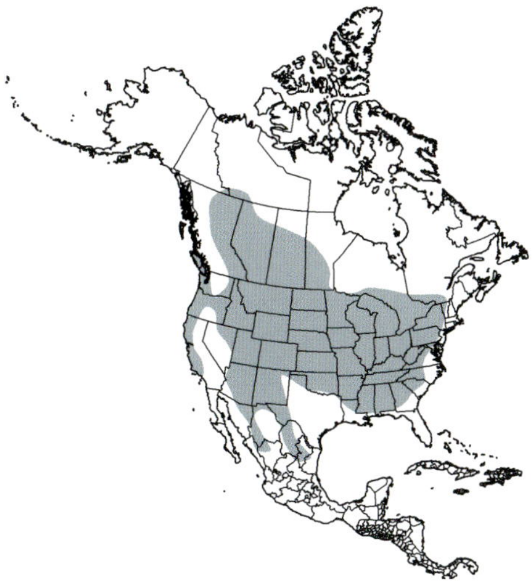

Distribution of the elk in Wyoming.

STATUS: Classified as a "big game animal" under state statute.[2] It is common or abundant statewide.

NOMENCLATURE AND SYSTEMATICS: The genus name has been stable for many decades, but the accepted species epithet has alternated between *C. elaphus* and *C. canadensis*. Use of the former denotes that New World and Old World forms are conspecific. Use of the latter for North America reflects the view that the two forms have diverged sufficiently to attain separate species status. The New World form is locally universally called "elk," although this term is used in Europe to denote *Alces alces* (the sibling species to North American moose). A further complication arose in response to a 2002 phylogenetic comparison of Old World and New World forms—as well as the closely related sika deer (*C. nippon*)—using mitochondrial DNA. That analysis showed that red deer and wapiti are genetically distinct, and the species status of *C. canadensis* should be reaffirmed. Further, the authors found that the appropriate species boundary between *C. elaphus* and *C. canadensis* is not Bering Strait, but a north–south line across central Asia. In other words, animals from Mongolia are more closely related to those in Wyoming than they are to those across the Ural Mountains, in western Eurasia.[3] This analysis and others support the view that all North American *C. canadensis* should be considered the same subspecies (*C. c. canadensis*) and not split into the six subspecies recognized traditionally.[4]

DISTRIBUTION AND HABITAT ASSOCIATIONS: The species occurs nearly statewide. Early European explorers found them in most temperate parts of the continent and state (shown), including shortgrass prairie and sagebrush-steppe habitats. Olaus Murie (1951) reviewed the distribution of elk in the state during the period of European exploration and identified concentrations around all forested mountain ranges.[5] The species also occurs in the Red Desert where surface water is nearby. Overhunting and land-use changes reduced the distribution, especially in plains habitats, and later in forests of the northwestern mountains.[6] Through hunting restrictions and translocations from Yellowstone National Park, by the mid-twentieth century elk were approaching their pre-European distribution in the state, although local distributions and migration routes had been much changed by agricultural development in valley bottoms and other land uses. Habitat is highly varied, particularly during summer and in the absence of human disturbance. In a non-forested setting in southwestern Wyoming, elk were found to associate in summer with higher elevations, with greater vegetative diversity, with sites near shrubs, with northerly aspects, on moderate slopes, and away from roads. Habitats selected in winter were similar, except for a shift to windblown sites with southerly aspects.[6] Habitat differences between the sexes have been described from the Bighorn Mountains in spring and early summer, with bulls associating with more timbered habitats and higher elevations than cows and calves.[7]

ECOLOGY: Diets comprise woody plant stems and grasses in winter, shifting to grasses during spring green-up, forbs in early summer, and grasses again in autumn. Elk eat aspen bark in fall and early winter, and scarred aspen boles are a conspicuous sign of elk presence, although moose leave similar marks slightly higher on boles. Competition between elk and other wild ungulates has been difficult to demonstrate in the Central Rocky Mountains, but that with livestock has been documented.[8] The northern Yellowstone elk herd—the most intensively studied of Wyoming populations—numbered around 17,000 in 1934, 5500 in 1968, 25,000 in 1987, and 4200 in early 2012.[9] The decline from 1995 to 2012 has been attributed to changed predator communities, to increased hunter harvests, and to drought.[10] Elk are seasonally social, with those in the Gros Ventre area commonly observed in groups of 30–50 in summer. Groups are smaller in spring and late fall.[11] Survival rates are highly variable across sexes, ages, years, and hunting regimes. Animals in and around Jackson Hole were ear-tagged on feed grounds for many years beginning around 1950; the mean annual survival rate for adult females during 1962–1966 was 0.68, and the rate for female calves was 0.61. Bulls had lower survival rates: 0.54 for the period 1958–1960; male calves had a mean rate of 0.66.[12] Statewide the hunter take of elk for 2010 was around 25% of the estimated population.[13]

Elk in Wyoming are infected with various parasites and infectious diseases, some of them ecologically important. Scabies, caused by mites of the genus *Psoroptes*, can cause fur loss, cold stress, and, in severe cases, death. Infections with the bacterium *Fusobacterium necrophorum*, which causes necrotic stomatitis and foot rot, have been epizootic on the National Elk Refuge in Jackson Hole during some years (for example, 1943, when nearly 10% of animals succumbed).[14] Brucellosis is a bacterial disease transmitted between elk and cattle, which causes abortion of fetuses and is common in the Yellowstone area. Migratory behaviors of elk are well known, although incompletely documented in Wyoming. Those behaviors are closely linked to the need of animals to track spatial patterns of vegetation phenology, and are susceptible to interruptions by human activities and land use, with subsequent effects on elk numbers.

CONSERVATION AND MANAGEMENT: The primary management approach is hunting on public and private land under permits issued by the Wyoming Game and Fish Department. Regional issues surrounding elk have been the impetus for many resource management programs, including reintroductions, winter feeding, disease monitoring, depopulation, habitat improvements, and the reintroduction of the wolf to northwestern Wyoming. The interruption of elk migration routes by developments in southern Jackson Hole, leading to starvation of elk within sight of the early town of Jackson, was the reason for the establishment of the National Elk Refuge in 1912.[15] Statewide abundance at its nadir in the early 1900s is not well documented, but by around 1923, 22,600 were estimated to live in the state. This increased to an estimated 65,000 in 1976 and 104,000 in 2010, compared to a state population management objective of 83,640.[16] In 2009, 22,971 animals were legally harvested by hunters who spent an estimated $43,000,000, making elk the single most important game species in terms of total economic activity generated. The integrity of elk migration routes in various parts of the state is an emerging conservation concern. Some elk associate with low-elevation hay meadows seasonally, often in areas that were historical winter range, and this use results in landowner claims of crop or hay losses. Twenty-three feed grounds are maintained in northwestern Wyoming, and the hunting seasons in that area are intended to limit conflicts between elk and agricultural operations. Important conservation questions have been discussed for decades about the relationships among winter feeding of elk, brucellosis transmission between elk on feed grounds, and brucellosis transmission between cattle and elk.[17] The prevalence of brucellosis is much higher among animals that use feed grounds (20–30%) than among those that do not (2–3%).[18] Chronic wasting disease is another emerging issue. This lethal and contagious disease of Cervidae, including elk, has moved northwestward across Wyoming over the past 30 years and was detected in the Cody area as of 2011. It seems likely that it will cross the mountains into Jackson Hole and the Yellowstone Plateau in coming decades.

NOTES AND REFERENCES

1. Quimby and Johnson (1951).
2. Wyoming Statutes 23-1-101-a-i.
3. Polziehn and Strobeck (2002).
4. Schonewald (1994), Polziehn et al. (2000).
5. Murie (1951).
6. Oedekoven and Lindzey (1987), Sawyer et al. (2007).
7. Sawyer and Lindzey (2001).
8. Hobbs et al. (1996).
9. Eberhardt et al. (2007), Northern Yellowstone Cooperative Wildlife Working Group (unpublished data).
10. Middleton (2012).
11. Long et al. (1980).
12. Sauer and Boyce (1983).
13. Wyoming Game and Fish Department (2011).
14. Boyce (1989).
15. Smith et al. (2004).
16. Bryant and Maser (1982), Blair (1987, 59), Wyoming Game and Fish Department (2011).
17. Scurlock and Edwards (2010).
18. Cross et al. (2010).

Mule deer, *Odocoileus hemionus*

DESCRIPTION: A medium-sized ungulate with dorsal pelage that varies seasonally from reddish tan in midsummer to gray in the fall. The underside is whitish. The rump is white, the tip of the narrow tail black, the forehead variably dark, and the pinnae proportionately larger than for other members of the genus. A white band crosses the throat. Seasonally shed antlers grow out of pedicels on the frontal bones of males. The antlers increase in size with age and nutritional condition and tend to branch in a series of even forks, without an apparent "main beam" distal to the first bifurcation. Small brow tines project vertically from the main beam near its base. Males are considerably larger than females, especially in late adulthood. When alarmed, mule deer flee by means of a four-legged hop, a "stott," which is less swift than the running of a white-tailed deer, but may allow clearing shrubs better and communicate alarm to conspecifics. Measurements (**Sex** [n] mean (mm, g), *range*): Colorado for dimensions, Sublette County for female weight, L = **F** [>89] 1424, *1250–1560*; **M** [51] 1523, *1260–1675*; TL = **F** [>89] 172, *120–220*; **M** [50] 180, *135–220*; HF = **F** [>89] 465, *419–500*; **M** [50] 487, *444–530*; EFN = **F** [>89] 203, *175–220*; **M** [50] 210, *165–230*; WT = **F** [94] 6349, *4472–7511*.[1] Dental formula: 0/3, 0/1, 3/3, 3/3 = 32.

STATUS: The legal status is that of a "big game animal."[2] This species is common and conspicuous over most of the state.

NOMENCLATURE AND SYSTEMATICS: The species was first described in 1817 and has carried its current scientific name since the late nineteenth century. It has undergone relatively little revision, aside from the naming of many subspecies in the early twentieth century. The genus name *Dama* is no longer applied.[3] The main challenges to the validity of the species

Mule deer male.

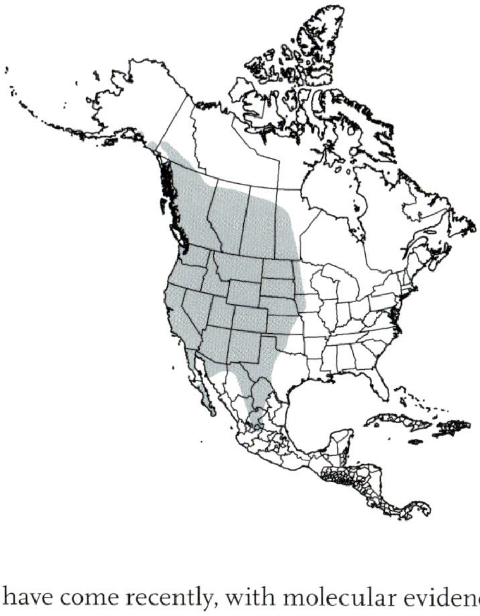

Distribution of the mule deer in Wyoming.

have come recently, with molecular evidence suggesting extensive introgression with white-tailed deer in some regions.[4] Of the nine subspecies recognized as of 1984, two belong to a coastal subspecies group that comprises *O. h. columbianus* and *O. h. sitkensis*: the black-tailed deer group. All Wyoming animals belong to subspecies *O. h. hemionus*.

DISTRIBUTION AND HABITAT ASSOCIATIONS: This western deer occurs almost statewide, including in alpine tundra in summer, in all but the densest conifer forest types, and in shortgrass prairie associated with breaks, shrubby areas, or riparian zones. The species is most commonly associated with mountain shrub communities interspersed with meadows and woodlands. Uneven topography is a common feature of mule deer habitats, which may facilitate the growth of mountain shrubs—important winter forage—rather than being important for their elevational relief per se. Wintering habitats over much of the state tend to be sagebrush with little snow.[5]

ECOLOGY: The species is a ruminating foregut fermenter, which eats forbs and shrubs from spring until the first snow, then shifts almost entirely to shrubs for most of the winter. A study conducted in southeast Wyoming showed that browse composed 93% of the winter diet and 76% of the summer diet. The major browse species were big sagebrush in all seasons, antelope bitterbrush and alderleaf mountain mahogany in winter, and russet buffaloberry in summer.[6] Seasonal migrations are common, and some in Wyoming are the longest recorded any mammal in the contiguous United States. For 171 mule deer radio-collared in western Wyoming, the mean seasonal migration distance was 84 km (50 mi), with some deer migrating as far as 241 km (150 mi).[7] Interactions—including hybridization—with white-tailed deer have become an increasing concern, as the latter species has become more widely distributed and abundant over the last 75 years. Hybrids have been described with either species as the mother, and some hybrids have distinctive behavioral or morphological traits, while others are only detectable via molecular markers. Hybrid animals are considered by some to suffer reduced fitness because of behavioral or biomechanical maladaptations. Specifically, hybrids are thought to be unable to stott like a mule deer, or to run like a white-tailed deer.[8] In 2010, an estimated 427,000 mule deer occupied the state; the hunter harvest was 34,500.[9] Predation is often supposed to be important to mule deer populations, and the main predators of mule deer are mountain lions and coyotes, the latter primarily killing neonates. Little evidence supports the notion that predators can suppress mule deer densities, however, and they are most likely to do so when deer populations are well below carrying capacity of the habitat.[10] An emerging concern over the last three decades has been chronic wasting disease, a transmissible spongiform encephalopathy, caused by an infectious prion protein that infects Cervidae. The disease is invariably fatal after a prolonged course and can have population-level effects through a number of mechanisms.[11] For example, infected deer may be selectively preyed upon by mountain lions, and infection could affect the ability of females to care for fawns.[12]

CONSERVATION AND MANAGEMENT: Mule deer have undergone multiple past declines and population irruptions, but as of 2012 are in a decline that is variously attributed to habitat deterioration, drought, exurban development, loss of winter range, disturbance on the winter range, and other factors. Some consider the expansion of white-tailed deer to be a factor.[13] Of particular concern in Wyoming are the barriers to seasonal migration and disturbance to winter ranges caused by various forms of development.[14] Control of coyote numbers is at times asserted to increase mule deer recruitment or population densities, but experimental studies have shown little effect of coyote or mountain lion reduction alone on recruitment or abundance of mule deer.[15] Similar to elk and white-tailed deer, mule deer associate with agricultural croplands in some areas and at some times. They consume several crops, including alfalfa, hay, and sugar beets, often in sufficient quantities to result in claims of damage by landowners.

NOTES AND REFERENCES

1. Anderson et al. (1974) for dimensions, K. L. Monteith, University of Wyoming, unpublished data for female weights.

2. Wyoming Statutes 23-1-101-a-i.

3. Hall (1981), Anderson and Wallmo (1984).

4. Bradley et al. (2003).

5. Oedekoven and Lindzey (1987).

6. Goodwin (1975).

7. Sawyer et al. (2005), Wyoming Migration Initiative (http://migrationinitiative.org).

8. Lingle (1993).

9. Wyoming Game and Fish Department (2011).

10. Ballard et al. (2001).

11. Statewide, infection rates are difficult to estimate because sampling has been conducted selectively at the presumed front of the expansion wave of the epizootic. The hunt area with the highest infection rate over the period 1997–2011 was Area 65 (northeastern Albany and northwestern Platte Counties), where 32% of 834 mule deer sampled were CWD-positive. For 2011 only, the value was 57% ($n = 30$) (Wyoming Game and Fish Department, unpublished data).

12. Krumm et al (2010), Dulberger et al. (2010).

13. Kramer (1973).

14. Sawyer et al. (2002, 2006, 2012).

15. Hurley et al. (2010), Brown and Conover (2011).

White-tailed deer, *Odocoileus virginianus*

DESCRIPTION: A medium-sized ungulate with medium brown dorsal pelage, reddish tan in some animals in early summer, but commonly faded to nearly gray by midwinter. The underside is white. A white band crosses the throat, as for mule deer, but the tail is wider, white on the underside and flanks, lacking any black marks, and carried vertically when alarmed. The pinnae are smaller than for mule deer. The seasonally shed antlers in males have a single main beam identifiable from the pedicel to the tip. Measurements (**Sex** [n] mean (kg), *SE*): late autumn, eastern Montana; WT = **F** fawns [14] 33, *1*; **M** fawns [13] 34, *1*; **F** 1 year [18] 53, *1*; **M** 1 year [25] 65, *1*; **F** 2 years [7] 51, *2*; **M** 2 years [13] 76, *2*; **F** ≥ 3 years [22] 58, *1*; **M** ≥ 3 years [17] 87, *4*.[1] Dental formula: 0/3, 0/1, 3/3, 3/3 = 32.

STATUS: This species is a "big game animal" under Wyoming law.[2]

NOMENCLATURE AND SYSTEMATICS: The species was first described in 1780, and the current scientific name has been in use since around the turn of the twentieth century. This species has undergone more taxonomic splitting and revision than perhaps any other North American ungulate, with dozens of genus and species epithets, and even more subspecies names. As of 1991, around 38 putative subspecies were still recognized, of which three (*O. v. ochrourus, O. v. dacotensis*, and *O. v. texanus*) are represented in Wyoming.[3]

DISTRIBUTION AND HABITAT ASSOCIATIONS: The distribution (current, shown) is transcontinental, extending north to Yukon Territory and south to South America. The distribution in Wyoming at the time of European exploration is difficult to reconstruct, but likely included Jackson Hole, northern Yellowstone National Park, the Bighorn Basin, eastern foothills of the

White-tailed deer female, with tail in the alarm position.

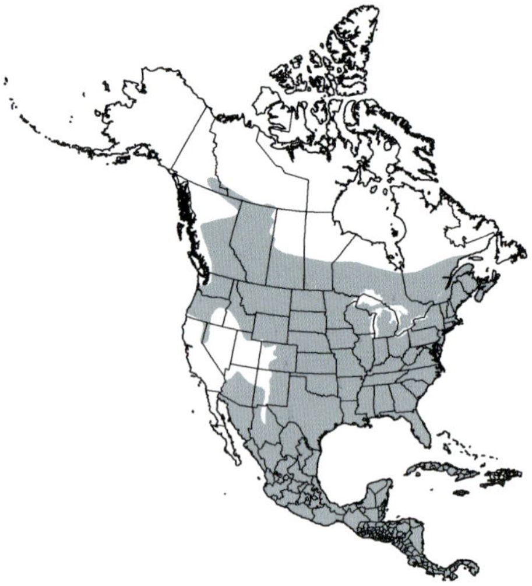

Distribution of the white-tailed deer in Wyoming.

Bighorn Mountains, Powder River Basin, Black Hills and Bear Lodge Mountains, and the North Platte River and its tributaries. The species was rare or absent over much of southwestern Wyoming.[4] White-tailed deer underwent a dramatic decline around the turn of the twentieth century, and did not reappear in numbers until the 1940s or 1950s. Some of this recovery was due to effective enforcement of game laws, but some resulted from translocations during 1949–1953. Flood control promoted the growth of dense riparian vegetation, important to white-tailed deer. Translocations originated from the Black Hills area and targeted Teton, Platte, Sweetwater, Johnson, Albany, Fremont, and Park Counties.[5] As of 2012, the species is widely distributed in suitable habitats across the state, with the areas of lowest occupancy in southwestern

Wyoming and other arid areas. The highest densities occur in the Black Hills area and the east slope and foothills of the Bighorn Mountains. Some possibly isolated populations are found outside the mapped area, including near Hoback Junction and along the Green River. Habitats over most of the Wyoming portion of the range are riparian-associated meadows, shrublands and woodlands, and irrigated row crops or alfalfa. In the Black Hills region, the species is much more widely distributed in upland habitats, including various admixtures of ponderosa pine, quaking aspen, white spruce, and chokecherry. During winter, the species tends to occupy southern and western exposures and grassy understories.[6]

ECOLOGY: The species is a ruminating herbivore that consumes woody vegetation as well as grasses and forbs. Although arguably the most intensively studied North American ungulate, the species has received relatively little research attention in Wyoming. In southeastern Wyoming, it ate mostly forbs and crops (86%, primarily alfalfa) in summer, but almost entirely (94%) browse and shrubs in winter.[7] Important browse species in the Black Hills area in winter include kinnikinnick, Oregon grape, and western snowberry.[8] Survival is highly variable across the geographic range, but in the central Black Hills of South Dakota deaths of females were primarily caused by predation, killing by domestic dogs, starvation, and disease (71%), followed by hunting (23%). Annual survival rates ranged from 50% to 62%, depending on year.[9] In that area seasonal migrations were typical then, with animals moving down in elevation when snow depths precluded browsing on low shrubs.[10] During summer in northwestern Colorado, the species was sedentary, and seasonal migrations, if they occurred, were inconsistent across years.[11] White-tailed deer are capable of higher rates of production than mule deer, and are believed capable of ecologically displacing them. The mechanisms underlying the displacement are not well understood, but sufficient correlational data exist to suggest such an effect.[12] Like mule deer, white-tailed deer are vulnerable to chronic wasting disease, which, during 2001–2010, reached its highest infection rates in Hunt Area 65 between Casper and Douglas.[13]

CONSERVATION AND MANAGEMENT: White-tailed deer have been increasing in numbers and distribution for several decades. It is not clear whether these increases are merely returns to pre-1880 levels or community shifts at the expense of mule deer. White-tailed deer are often found in agricultural croplands and use those foods enough that depredation claims by landowners are frequent. Population estimates of white-tailed deer tend to be less reliable than for mule deer, because white-tailed deer tend to concentrate less in winter. The replacement of mule deer by white-tailed deer, via either competition or habitat changes and other anthropogenic causes, is a consideration in some hunting-season-setting decisions. In some areas, such as along the eastern slope of the Bighorn Mountains, the species is sufficiently abundant to be regarded as a nuisance. Similar to mule deer and other cervids, chronic wasting disease is an emerging concern, and management options are being explored.

NOTES AND REFERENCES

1. Dusek et al. (1989)
2. Wyoming Statutes 23-1-101-a-i.

3. Smith (1991).

4. For occurrence in Jackson Hole see Murie (1940), for Yellowstone National Park see Bailey (1930), for the Bighorn Basin see King (1992), for southwestern Wyoming see Dorn (1986), and for the Black Hills area see Turner (1974).

5. Pauley and Lindzey (1993). A total of 347 animals were moved in 19 translocations.

6. Stefanich (1995).

7. Sawyer and Lindzey (2000).

8. Hill (1946), Schneeweis et al. (1972).

9. DePerno et al. (2000).

10. Stefanich (1995), DePerno et al. (2002).

11. Garrott et al. (1987).

12. Kramer (1973).

13. The area where infection rates are highest is the same as for mule deer: Hunt Area 65 on the east slope of the Laramie Mountains, where incidence in all sex–age groups peaks at 65% in females during their fifth year of life (Edmunds 2013, 76). In that area the annual survival rate for juvenile and yearling white-tailed deer of both sexes was 0.73 for CWD-negative animals and 0.25 for CWD-positive animals. The strongest effects were observed in yearling males, for which the survival rate for CWD-negative animals was 0.79, whereas that for CWD-positive males was 0.20. Estimated λ for the population was 0.90, representing a 10% decline yearly (Edmunds 2013).

Moose, *Alces americana*

DESCRIPTION: A large deer with a bulbous nose, long legs, and palmate antlers in adult males. The pelage is seasonally variable: dark brown to nearly black on undersides in fall, lighter brown on the dorsum, but faded and blotchy by late spring. Females typically have a triangular patch of white fur below the tail, widening toward the dorsal (vulvar) edge. Both species vocalize; those of bulls are typically limited to around the time of rut. Measurements (**Sex** [n] mean (mm), *range*): Sublette County, L = **F** [65] 2676, *2290–2960*; HF = **F** [65] 555, *500–590*.[1] Dental formula: 0/3, 0/1, 3/3, 3/3 = 32.

STATUS: The species is classified as a "big game animal" under state statute and a state Species of Greatest Conservation Need.[2]

NOMENCLATURE AND SYSTEMATICS: The genus name has been in use since the early nineteenth century. Early twentieth-century taxonomists separated the Old World species (*A. alces*) from that of the New World, some authors naming as many as three New World species (*A. gigas*, *A. shirasi*, and *A. americana*) that now correspond to three of the four currently recognized subspecies (the fourth is *A. a. andersoni*). After several decades of the unification of world *Alces* under a single species, analysis of chromosome number showed that there were indeed Old World (2N = 68) and New World (2N = 70) forms. Curiously, the boundary between these two types is not Bering Strait, but a line dividing east from west Eurasia at around the Yenisei River (86° E).[3] Wyoming animals are considered to be *A. americana shirasi*, although the entire geographic range of that putative subspecies did not hold any moose 250 years ago.[4] George Shiras III was a lawmaker and naturalist who surveyed Yellowstone in 1908–1910 and observed many moose along the south side of Yellowstone Lake. E.W. Nelson subsequently named the subspecies of moose in the northern Rockies in his honor.[5]

Moose, showing white vulvar patch characteristic of adult females.

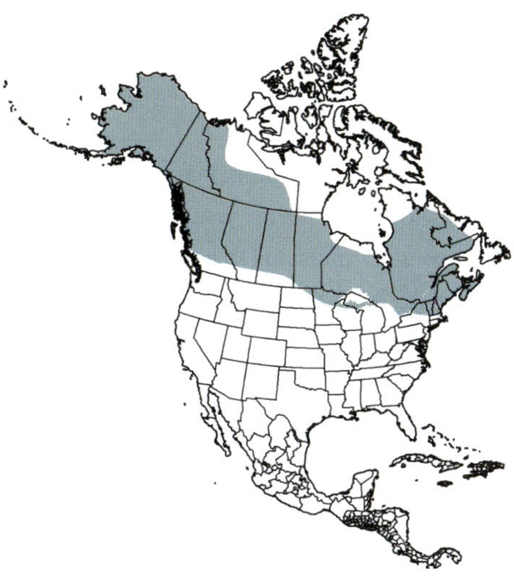

Distribution of the moose in Wyoming.

DISTRIBUTION AND HABITAT ASSOCIATIONS: The distribution within Wyoming (current distribution) is much more extensive than that 150 years ago. The earliest chroniclers of wildlife in the state do not mention moose, although they describe other large mammal species in detail.[6] Moose were apparently absent in the Rockies south of 49° until 140 years ago. Famously, the Lewis and Clark expedition, traveling north of Wyoming during 1805–1806, did not mention observing moose during their journey, and Osborne Russell's detailed journal of travels in northern Wyoming during 1834–1843 did not mention moose, even though he described the large mammals he observed in detail.[7] Moose seem to have appeared in Wyoming in the Yellowstone area during the mid-nineteenth century, spread to Jackson Hole around the turn of the twentieth century, arrived in the Wind River Mountains by the 1930s, and increased in commonness and slightly in extent over the following few decades.[8] The species disperses well, and lone moose have been found beyond the documented population advance, generating anecdotes of moose farther south and east in Wyoming. The 1950s seem to mark the end of the process of natural expansion of the moose in Wyoming, but the species was translocated to northern Colorado in the late 1970s. It thrived there and spread north to the Medicine Bow and Laramie Mountains, where it is now common in the Pole Mountain area and occurs in small numbers as far north as Casper Mountain. The species was introduced to the Bighorn Mountains from Jackson Hole, beginning in 1948, and with additional augmentations in 1950, 1974, and 1987. From these founders, that population has expanded to nearly the entire forested area of that mountain range, as far south as southern Johnson County. The Wind River population has spread to the Sweetwater River to its confluence with the North Platte.[9] Across the continental range (presettlement distribution, shown) the species occupies a broad range of habitats, including shrub tundra, coniferous forest, and other types. In Wyoming and adjacent states, the range of habitats occupied is more limited, with riparian floodplains commonly used. Adjacent coniferous forests are important winter habitats in some cases.[10] Aspens are sufficiently extensive to be important in a few areas, and moose occasionally eat aspen bark, leaving distinctive marks higher on the bole than wapiti can reach. Moose populations on the Northern Yellowstone Range had been stable for several decades before the 1988 fires, but quickly declined by 75% following them, and had not recovered by 2002. This has been attributed to the loss of late-successional coniferous forest, which is important for wintering.[11]

ECOLOGY: Moose represent several ecological paradoxes. They are important, even iconic members of boreal mammalian communities, yet are ephemeral on a decadal and centurial scale, even in regions thought to represent typical habitat. They are large-bodied, yet are capable of high reproductive output, including high twinning rates and first reproduction at two years of age, under optimal conditions. They are generally associated with shrubs and would seem likely to be benefited by post-disturbance early successional stages, but their response to vegetation changes following fire is inconsistent across the geographic range. Diets are primarily, but not exclusively, leaves and stems of woody plants, particularly deciduous shrubs. Aquatic plants are important and sought out, sometimes requiring moose to submerge themselves completely. McMillan (1953) described food plants from Yellowstone;

they comprised 23 taxa, only two of which were woody. But these two willows (*Salix geyeriana* and *S. wolfii*) represented 89% of observations of use; aquatic herbs were most of the remainder.[12] Today a much broader diet is recognized, even though not specifically documented in Wyoming. In Montana, summer diets were dominated by forbs.[13] Leaves of subalpine fir can be important in winter. Populations are routinely monitored in Wyoming before and following the hunting season. Typical postseason calf:cow ratios are 31–45:100, and the statewide bull:cow ratio in 2001 was 63:100. Twinning rates in Wyoming are unusually low, <5%, and first reproduction here occurs at 2–4 years, depending on nutritional condition. Mortality results from various causes, especially starvation in winter, predation on neonates, and hunting. The main predators are grizzly bears (for the first few weeks of life) and wolves, both of which are restricted to the northwestern mountains. Age structure has been documented for hunter-taken animals; 89% of antlered males were ≤7 years old, with the oldest male 12 years of age. Antlerless moose (excluding calves) were slightly older: 84% were ≤7 years of age, with the oldest 18 years.[8]

CONSERVATION AND MANAGEMENT: From dispersers that colonized the Yellowstone area during the mid-nineteenth century, populations grew in size while expanding in area to reach an estimated statewide total of 13,700 animals in 2001. This number had declined to 7439 by 2010.[8] The species has been managed as a big game animal for decades, with an average of 581 animals per year taken by hunters during 2006–2010. Since the 1970s, the older populations in the northwestern part of the state have shown signs of decline, while the younger populations in the Bighorns and the southeastern part of the state continued to increase in number and expand in range. The decline in northwestern Wyoming is gradual and resembles similar ones in other areas of the northern Rockies. The growth and expansion that have occurred in southern Wyoming and Colorado are generally attributed to recent arrival and use of previously unexploited browse. The causes of the gradual declines farther north are multiple and confounded by site-specific factors. They include time since arrival, degree of browse utilization, climate (particularly recent drought and thermal stress on the animals), the 1988 Yellowstone fires, and possibly predation, although how strongly Wyoming predators affect moose populations has yet to be quantified. The current outbreaks of bark beetles likely will further affect moose by removing thermal cover, and possibly browse. Another potential population factor is infection with carotid artery worm (*Elaeophora schneideri*), a nematode that has recently appeared in Wyoming. It shows highly variable infection rates in hunter-killed animals, and some infected animals may die of it.[14]

NOTES AND REFERENCES

1. K. L. Monteith, University of Wyoming, unpublished data, 2013.
2. Wyoming Statutes 23-1-101-a-i.
3. Boyeskorov (1999), Groves and Grubb (2011). This boundary coincides approximately with the boundary between *Cervus elaphus* and *C. canadensis* inferred from molecular evidence.
4. Franzman (1981).
5. Nelson (1914).
6. Karns (2007).

7. Russell (1921).
8. Houston (1967), Brimeyer and Thomas (2004).
9. Brimeyer and Thomas (2004), Wyoming Game and Fish Department (2011).
10. Peek (2007, Table 26).
11. Tyers (2006).
12. McMillan (1953).
13. Knowlton (1960).
14. Henningsen et al. (2012).

PRONGHORNS—FAMILY ANTILOCAPRIDAE

This group radiated in North America over the late Cenozoic and was viewed traditionally as a stand-alone family related to, or as a subfamily within the Bovidae. More recently, the antilocaprids have been shown to be closely related to giraffes and musk deer.[1] In contrast to the Bovidae, pronghorns shed the keratinaceous sheaths of their horns yearly. Miocene and Pliocene pronghorn species were diverse and mostly smaller-bodied than the extant pronghorn. They had elaborately shaped horns with as many as six main shafts. Modern pronghorns from Sweetwater County occasionally are reported with two horn shafts on each side. In addition to these differences from other artiodactyls, pronghorn horns are not simple geometric curves or spirals, as for the bovids. Pronghorns possess a diverse set of scent glands used for communication.

NOTES AND REFERENCES

1. Decker et al. (2009).

Pronghorn, *Antilocapra americana*

DESCRIPTION: A small ungulate by North American standards, with sexually dimorphic pelage and horns. The eyes are proportionally larger than for other western ungulates. Horn sheaths of agglutinated hair grow over a bony core, have forward-pointing spurs, or prongs, and are black, with light tips in older males. The sheaths are shed yearly in October and November, with small growing black sheaths underlying and replacing the older ones.[1] Pelage is reddish tan on the back and white on the undersides and rump, with complex banding patterns on the neck and head. The dorsal snout, rostrum between eyes, and cheeks of males are black. Male horns are generally longer than their ears, and female horns shorter. The alarm vocalization is a sharp cough-like bark. Scent glands, more numerous on males than on females, are found on various parts of the body, including between the digits, near the angle of the jaw, on the rump, and on the lumbar middorsal line. Measurements (**Sex** [n] mean (kg)): Adult (≥2 years), Carbon County, WT = **F** [76] 48; **M** [52] 49.[2] Dental formula: 0/3, 0/1, 3/3, 3/3 = 32.

STATUS: Classified as a "big game animal" under state law.[3] It is common or abundant within the geographic range.

Pronghorn male (foreground) and female.

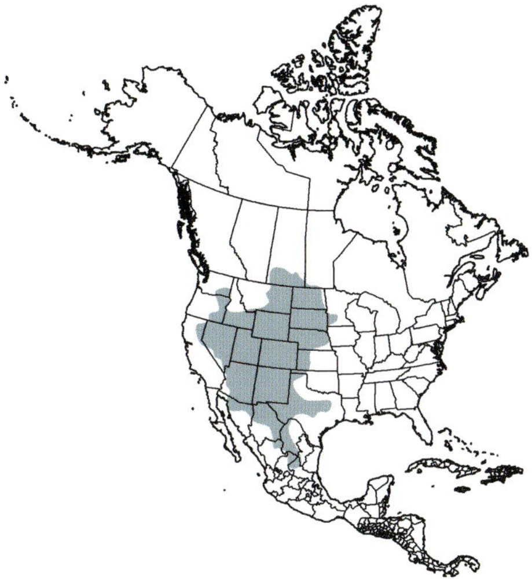

Distribution of the pronghorn in Wyoming.

NOMENCLATURE AND SYSTEMATICS: The genus and species names have been stable for nearly two centuries. The monotypic genus was moved from Bovidae back to the Antilocapridae—a sister family, according to understandings of the time—in the 1980s. Recent molecular studies place Antilocapridae between giraffes and musk deer, quite distant from the Bovidae.[4] Five subspecies were recognized as of 1981; animals in Wyoming and all adjacent states

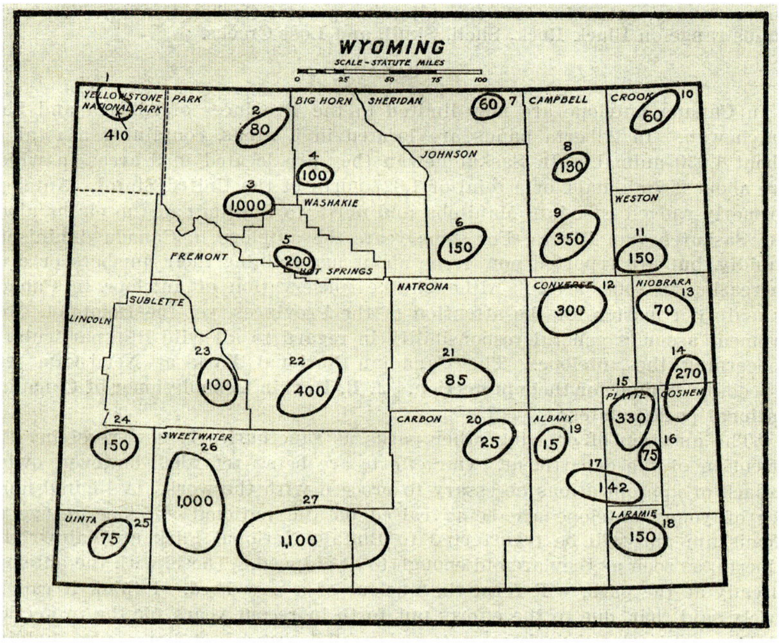

The distribution of pronghorns in Wyoming in 1922–1923.

are of the nominal *A. a. americana*.[5] The common name "pronghorn antelope" is also used, but "antelope" is misleading, because most so-named animals are bovids.

DISTRIBUTION AND HABITAT ASSOCIATIONS: At European exploration, the species occurred from central Saskatchewan south to Hidalgo province, Mexico, and from the eastern Great Plains to coastal California (shown). The distribution of the species range-wide and in Wyoming reached its lowest point around 1920.[6] Today, the species' geographic range is expanding in various directions, including eastward into eastern South Dakota, but has not reached its pre-European extent.[7] In Wyoming, the species now occurs nearly statewide below the elevation of coniferous forest. A small population lives in northern Yellowstone National Park and adjacent lands. Habitat is shrub-steppe, but pronghorns enter woodland and approach forest edges often. Many Wyoming towns have pronghorns in their parks and undeveloped lots, especially in winter. Animals that migrate between the upper Green River Basin and Grand Teton National Park maneuver around timbered sites, staying in sage draws and other open areas as much as possible.

ECOLOGY: Pronghorns are browsers and grazers that migrate varying distances. Energy budgets are a primary consideration in the species' ecology, and reflect quantity and quality of the diet, food availability in winter, ability to migrate to winter range, rut, gestation, and lactation. The highest densities are in grasslands in the east-central part of the state.[8] Episodes of severe winter weather can interact with nutritional condition and barriers to movement (e.g., highways, towns, and fences) to produce dramatic mortality events.

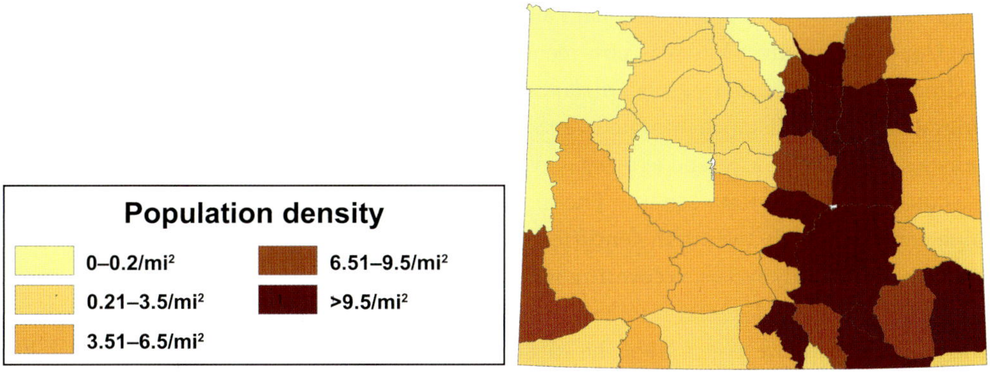

Pronghorn populations attain their highest densities in herd units that form a north–south band in the east-central part of the state. The easternmost part of the state has undergone more land conversion to crop agriculture.

The pronghorn population in the Chain Lakes area northwest of Rawlins suffered an estimated 62% die-off over the winter of 1971–1972, in part because of recently constructed fences that blocked migration routes.[9] Another such mortality event in response to a severe winter occurred near Red Rim, southeast of Rawlins, during the winter of 1983–1984.[10]

Foregut fermenting herbivores, pronghorns require suitable plant forage. The diet is diverse and opportunistic, reflecting availability, season, and plant phenology. In grasslands with mild winters, grass species dominate year-round. In shrub-steppe with severe winters, shrubs are year-round staples. In central and western Wyoming, big sagebrush tends to dominate.[11] Pronghorns are attracted to fields of winter wheat and alfalfa within native rangeland, especially when native foods are drought-stricken. Pronghorns are somewhat migratory, with animals moving seasonally in response to nutritional characteristics of plants. The species has high twinning rates and females produce young in consecutive years, which allows populations to grow rapidly in response to favorable conditions. Mortality is highest during the first year—especially the first 25 days—of life.[12] Sex-biased survival rates are typical in most years and for most ages, so that about 50% of females, but only 15% of males, lived to the age of four in the Red Desert.[13] The importance of predation depends on location and season. Predation was the single largest mortality cause in Yellowstone National Park, and coyotes were the single most important predator (56% of adult deaths; 79% of deaths of fawns), even in the presence of mountain lions, wolves, and golden eagles.[14] In southeastern Wyoming, 91% of known-cause deaths were from predation, with coyotes contributing 67% of predation-caused deaths. Golden eagles caused 23% and red foxes 6% of deaths.[15] Pronghorns are more vulnerable to severe winter weather than larger-bodied Wyoming ungulates. Their small body size shortens fasting-endurance times, and the strong winds that accompany low temperatures increase convective heat loss. Wind also redistributes snow, in some cases packing snow around branches of shrubs so that their stems are not available at the season when pronghorns need them most.[11] The seasonal migration from Grand Teton National Park through the Gros Ventre Mountains and down the Upper Green River Valley is now recognized as the longest described for the species.[16] Animals in recent years migrated distances of 120–260 km (72–156 mi) in spring and fall migrations (mean distance migrated = 178 km [107 mi]). These migrations are threatened by vehicles on roads, fences, and exurban developments.[16]

CONSERVATION AND MANAGEMENT: The species was depleted by overhunting during the late nineteenth and early twentieth centuries; numbers in Wyoming were estimated at 7000–20,000 in the early 1920s, recovering to around 200,000 by 1970, and to 527,000, more than half of the species total numbers, in 2010.[17] This strong long-term recovery speaks to the high reproductive potential of the species (e.g., the high twinning rate), the availability of large areas of unaltered habitat, and scientific management practices. The basic management approach is that of a harvested big game species, with annual surveys of abundance, sex composition, and fawn production. Seasons and quotas are set by hunt areas (117 statewide), and hunting removes 5–12% of the total state population yearly. Severe winters have accounted for massive die-offs on multiple occasions during the historical period. The one around Chain Lakes in the early 1970s led to the establishment of Chain Lakes Wildlife Habitat Management Area, an interagency designation intended to maintain pronghorn habitat and minimize fence obstacles to winter movements. Another such area, the Red Rim-Daley Wildlife Habitat Management Area, was established southwest of Rawlins in 1991 in response to major die-offs of pronghorns during the early 1980s. The integrity of migration routes, particularly in the Upper Green River Basin, has become a concern, and two wildlife overpasses have been constructed to ease the movement of animals across State Highways 189 and 191. Wyoming, including Yellowstone National Park, has been the source for many translocations of pronghorns that have been used to repopulate the former geographic range in Idaho, Utah, Texas, Arizona, New Mexico, Nevada, South Dakota, Kansas, and Oklahoma. Animals were translocated from Yellowstone to the National Bison Range in Montana in the 1950s, from which further translocations were made to various states in the Great Plains and the West.[18]

NOTES AND REFERENCES

1. O'Gara and Matson (1975).
2. Buck (1947), Edwards (1958), O'Gara (2004a). The latter cites William Hepworth, WGFD, unpublished data, which originated mostly from the Seminoe Reservoir area.
3. Wyoming Statutes 23-1-101-a-i.
4. Wilson and Reeder (2005), Agnarsson and May-Collado (2008).
5. Hall (1981).
6. Reproduced from Nelson (1925), Sundstrom et al. (1973). The low estimate presented is from Nelson. The estimate reported by Commissioner Frank Smith to Governor William Ross for the same period was 13,900 (Blair 1987, 59).
7. Yoakum (2004a).
8. Abundances for herd units from Wyoming Game and Fish Department, Wildlife Division, for 2010. Abundances for Jackson Hole from National Park Service (2010) and for Yellowstone from National Park Service (2011).

9. Oakley and Riddle (1974).
10. Guenzel (1986).
11. Yoakum (2004b).
12. Zimmer (2004).
13. O'Gara (2004b).
14. Barnowe-Meyer et al. (2009).
15. Zimmer (2004, 47).
16. Sawyer et al. (2005).
17. Nelson (1925), Sundstrom et al. (1973), Wyoming Game and Fish Department (2011). The low estimate presented is from Nelson. The estimate reported by Commissioner Frank Smith to Governor William Ross for the same period was 13,900 (Blair 1987, 59).
18. Scott (2004), Wyoming Game and Fish Department, Wildlife Division, unpublished data.

ANTELOPE AND ALLIES—FAMILY BOVIDAE

This family comprises even-toed ungulates with horny extensions of the frontal bones, other than pronghorns. The horny sheaths are not shed yearly, and in North American forms are not shed at any time during life. The group includes members ranging in size from the 5 kg (12 lb) dik-dik (*Madoqua* spp.) to buffalo (e.g. *Syncerus* spp.) that weigh over 900 kg (2000 lb); members occur naturally on all continents except Australia and South America. Bovids are primarily grazers, and attain their greatest numbers of species in sub-Saharan Africa. The North American types, at the time of European exploration, were most species rich in the arid, mountainous western half of the continent.

American bison, *Bos bison*

DESCRIPTION: The largest North American ungulate, this species has brown, seasonally variable fur, longer over the dorsal thorax than the abdomen. The slender tail is intermediate in length and lightly haired, with a tuft near the tip. Bony horn cores on the frontal bones are covered by skin-derived, horny sheaths, black and perennial. The body is narrow but tall, the height accentuated by the shoulder hump, which is supported by long dorsal processes of the thoracic vertebrae. Moderate sexual dimorphism of body size, horn size, and facial shape is typical. Measurements (**Sex** [n] *range* (mm, kg)): Yellowstone, L = **F** [?] *2130–3180*; **M** [?] *3040–3800*; TL = **F** [?] *300–510*; **M** [?] *330–910*; HF = **F** [?] *500–530*; **M** [?] *580–680*; WT = **F** [?] *318–545*; **M** [?] *544–907*.[1] Dental formula: 0/3, 0/1, 3/3, 3/3 = 32.

STATUS: A native wild ungulate, bison (including genetic admixtures with cattle) are now held in semi-captivity at various sites in the state. As a result, bison in various locations and circumstances are variously classified as "privately owned," "running at large," or "wildlife," by joint regulation of the Wyoming Game and Fish Department and the Wyoming Livestock Board.[2] Free-ranging bison (wildlife) are currently found in Yellowstone National Park, Jackson Hole (in various land jurisdictions), and on the adjoining Shoshone and Bridger-Teton National Forests.

American bison male.

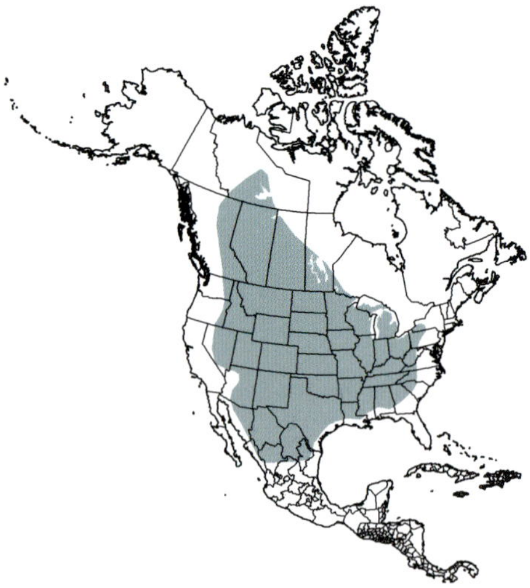

Distribution of the American bison Wyoming. The presettlement range was statewide.

NOMENCLATURE AND SYSTEMATICS: The genus name *Bos* was that used by Linnaeus (1758); *Bison* was employed from 1827 until the twentieth century. The two generic epithets are considered synonyms by some taxonomists.[3] Another modern wild bison—*B. bonasus*—occurs in Europe. The earliest specimen assigned to *B. bison* is around 9700 years old; other Holocene bison (e.g., *B. priscus, B. latifrons,* and *B. antiquus*) have unclear phylogenetic relationships to modern American bison; the last of these extinct forms is thought to have undergone rapid change in western Canada in the last 15,000 years to give rise to modern bison.[4] The animals in Yellowstone National Park—along with several other conservation herds in the Rockies and Great Plains—are pure-strain American bison.[5] Those in Grand Teton

National Park are suspected of having slight cattle introgression, based on the contribution of one founder herd.[6] Only two subspecies of bison were recognized as of 1981—although several others had been named earlier. Both presumptive subspecies occurred in Wyoming in pre-European times: *B. b. athabascae* of the Rocky Mountains (including the Medicine Bow Mountains) and northwestern North America, and *B. b. bison* of the Great Plains and eastern forest.[1] The common name "buffalo" is common in the popular literature, but also refers to various African and Eurasian species of Bovidae.

DISTRIBUTION AND HABITAT ASSOCIATIONS: Bison occurred broadly across North America when Europeans arrived (shown). In Wyoming during the pre-European Holocene Epoch, bison occurred almost statewide, although at geographically variable densities. The species occurred at high elevations in summer: remains have been found near the summit of Medicine Bow Peak (3530 m, 11,650 ft), on Hunt Mountain in the Bighorn Mountains (3090 m, 10,197 ft) and at 3840 m (12,600 ft) in the Uinta Mountains, just south the Wyoming state line.[7] Such records are consistent with seasonal elevational migrations; bison are unlikely to have remained so high through winter because of deep snow. Bison doubtless occurred at higher densities in the mixed-grass regions of the state, particularly the northeastern basins, than in shrub-steppe of southwestern Wyoming. The disappearance of bison from parts of Wyoming occurred sufficiently early that documentation is scarce. Osborn Russell, writing in the early 1840s, referred to the rapid disappearance of bison from the Green River Basin, Bear River drainage, and headwaters of the Snake River.[8] They were absent from the Laramie Basin after 1845 and reduced to isolated pockets statewide by 1875. Various accounts refer to a remnant population that persisted in the Red Desert into the latest nineteenth or early twentieth century.[9] The current distribution includes the northwestern mountains, mostly on federal lands. Habitats are primarily non-forested areas, although bison enter forests for shade, to escape blizzard conditions and to avoid disturbance. Forests were traversed during seasonal migrations.[5]

ECOLOGY: Bison are ruminant grazers that feed primarily on grasses; sedges are preferred where available. They typically drink water or eat snow on a daily basis.[5] The current state population size depends on how animals that migrate seasonally to Montana are included. Seasonal migrations are typical of many bison herds, and current migration routes tend to be riparian zones connecting grasslands. Roads in Yellowstone are believed to facilitate bison movements in winters with deep snow.[10] Estimates of causes and rates of mortality specific to Wyoming are likewise difficult to calculate because of animals that seasonally migrate to Montana where they are legally hunted. The annual survival rate for unhunted females in Yellowstone National Park was 92% and constant over time.[11] The Yellowstone and Jackson Hole populations have tended to grow and are managed via hunting, translocations to quarantine facilities, and capture slaughter in the case of the Northern Yellowstone Range. Severe winters, such as occurred in the mid-1990s and late 2000s, can stimulate migration and cause some winterkill. Wolves are known to kill bison—particularly calves—elsewhere, but instances of predation in Yellowstone following reintroduction have been rare.[12]

CONSERVATION AND MANAGEMENT: The overhunting of bison to the brink of extinction, followed by the protection and fostering of the last remaining wild bison in and around Yellowstone National Park is one of the iconic stories of wildlife conservation in the United States.[13] In 1902, the last 20 or so free-ranging bison in Yellowstone National Park were augmented by 21 bison brought from private herds elsewhere. The captive herd grew to >1000 animals, and the free-ranging herd increased slowly as well. As of 2012, Yellowstone National Park managed to achieve a population size of 2500–4500. The Jackson Hole population was estimated at 927 in 2012, and the population objective was 500.[14] Brucellosis, an Old World bacterial disease of bison, cattle and elk, was discovered in Yellowstone in 1917, and all bison in the state have subsequently been managed so as to reduce possible contact with domestic cattle. Of 199 animals from the Jackson Hole herd tested for brucellosis during 2000–2008, 40–83% were positive, depending on year.[15]

NOTES AND REFERENCES

1. Meagher (1986).
2. Commission Regulation 41.
3. Wilson and Reeder (2005).
4. Wilson (1996).
5. Halbert and Derr (2007).
6. Hedrick (2009).
7. Fryxell (1926, 1928), Cannon (2007), author for Hunt Mountain.
8. Russell (1921).
9. Hornaday (1887).
10. Bruggeman et al. (2007).
11. Fuller et al. (2007).
12. Laundré et al. (2001).
13. Plumb and Sucec (2006).
14. Wyoming Game and Fish Department (2011).
15. Baskin (1998), Scurlock and Edwards (2010).

Mountain goat, *Oreamnos americanus*

DESCRIPTION: A medium-sized even-toed ungulate with all-white pelage; pelage is short and pure white in late summer, and shaggy and soiled by late spring. Males and females have slender, slightly curved, pointed horns, black in color, that grow throughout life. Sexual differences in horn size and body shape are subtle except to an experienced observer. Measurements (**Sex** [n] mean (kg), *range*): ≥4 years, Olympic National Park, Wash.; WT = **F** [57] 61, *50–74*; **M** [15] 96, *50–117*.[1] Dental formula: 0/3, 0/1, 3/3, 3/3 = 32.

STATUS: The paleontological record is puzzling, with a few specimens occurring in sites in Wyoming far from the geographic range at the time of European settlement. A few early historical records suggest the presence of individuals, but not breeding populations within the state. The populations currently in the state are naturalized nonnatives, having expanded

Mountain goat, Fremont Peak, 2009.

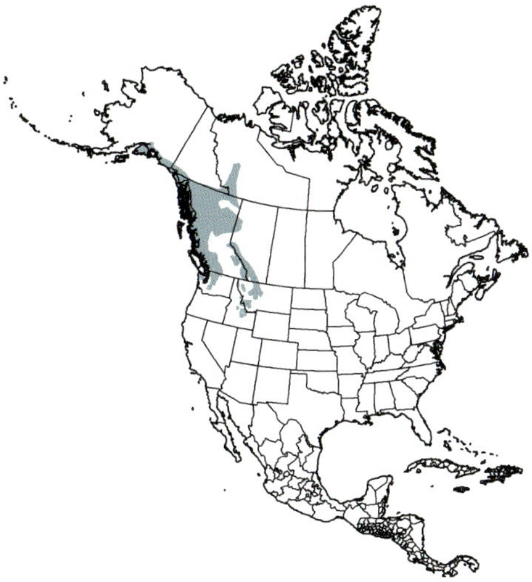

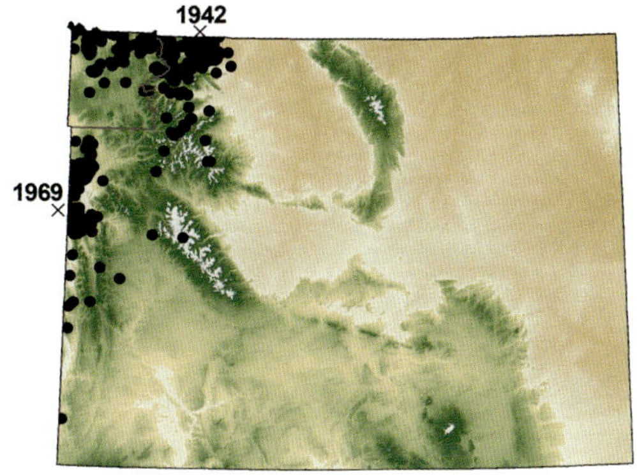

Locations from which the mountain goat ($n = 5988$ observations) has been identified from western Wyoming during 1946–2011, collected by the Greater Yellowstone Area Mountain Ungulate Project. Areas of concentration and sites where solitary dispersing animals were observed are apparent. The southeasternmost point in the Wind River Range is the 2009 sighting on Fremont Peak (see photograph). The two dated ×'s were the sites of introduction that were closest to the state line, in both cases less than 10 km (6 mi).

into Wyoming from sites of introduction in adjacent Montana and Idaho.[2] The species is a "big game animal" under state law.[3]

NOMENCLATURE AND SYSTEMATICS: The genus name has been in use since the early twentieth century; the specific epithet is 100 years older. Of the four subspecies recognized as of 1975, the one that occurred closest to Wyoming before translocations was *O. a. missoulae*. The species is closely related to the European chamois, *Rupicapra*, and first appears in North American deposits 125,000–75,000 years ago.[4] Fragmentation of populations by glaciation during the Pleistocene is thought to account for low levels of genetic variability observed in interior populations.[5]

DISTRIBUTION AND HABITAT ASSOCIATIONS: The species was native to mountains of the northwest coast and northern Rocky Mountains (shown) before being translocated widely. Fossil remains of the species from the Late Pleistocene have been described from Wyoming, and historical anecdotes from westernmost Wyoming date from the 1840s to 1914.[6] The species has been introduced to South Dakota (first releases in 1924) and Colorado (first releases in 1948), but only animals from sites in Montana (first released in 1942) and Idaho (first released in 1969) have colonized Wyoming. The first recorded occurrence of the species in Wyoming was in the Beartooth Mountains in 1946. The first appearance in Yellowstone National Park was in the 1960s, with marked range expansion in the 1990s.[7] Reports from 2011 placed the species near Jenny Lake and Death Canyon in Grand Teton National Park.[8] Habitats are alpine meadows or areas near the upper limit of trees with cliffs or rocky slopes nearby for escape from predators.

ECOLOGY: The species has been studied little in Wyoming or Yellowstone National Park, but elsewhere has been found to eat a mixture of grasses, forbs, and shrubs that is similar to the diet of bighorn sheep. Across studies conducted in summer, mean grass component was 52%, forbs were 30%, and browse was 16%. In winter, the respective values were 60%, 8%, and 32%.[9] Mineral licks are important to the species and can result in travel across dangerous terrain to sites where goats concentrate.[10] Seasonal shifts in elevation are common but variable across studies; goats in southeastern British Columbia used sites in winter that were restricted in area, more rugged, with more southerly exposures and farther from trees than in summer.[11] Twinning is rare, and females produce their first young at 3–7 years, so population productivity is low.[12] Predation is likely the primary cause of death in native populations, with wolves, bears, and mountain lions contributing.[8] Deaths attributed to accidental falls and avalanches also have been reported.[13]

CONSERVATION AND MANAGEMENT: Considerable discussion attends the issue of whether mountain goats were present in Wyoming (or Colorado) at the time of settlement, and before the time of translocations to near the border. Accordingly, there is some uncertainty as to whether the species should be managed as a native or an introduced species. This is especially relevant in Yellowstone and Grand Teton National Parks, where nonnative mammals are regarded as nonnatural members of the animal community, which also includes bighorn sheep.[14] The overpopulation of Olympic National Park by introduced mountain goats in the late twentieth century, followed by population reduction and attendant public controversy, is regarded by many as an undesirable outcome.[15] In the case of nonnational park lands, the question is further complicated by the fact that mountain goats are appealing to the public, and desirable to some as a big game species (21 killed by hunters in Wyoming in 2009). The estimated number of animals in Yellowstone National Park was 175–225 as of 2011; in Wyoming outside of Yellowstone and Grand Teton, it was 280 as of 2009.[16]

NOTES AND REFERENCES

1. Estimated from Houston et al. (1989, fig. 2).
2. Schullery and Whittlesey (2001), McWhirter (2004).
3. Wyoming Statutes 23-1-101-a-i.
4. Rideout and Hoffman (1975), Kurtén and Anderson (1980).
5. Mainguy et al. (2007).
6. Schullery and Whittlesey (2001).
7. Lemke (2004), McWhirter (2004).
8. Dewey (2011).
9. Laundré (1994).
10. Côté and Festa Bianchet (2003).
11. Poole et al. (2009).
12. Hamel et al. (2006).
13. Chadwick (1983).
14. Cain (2012).
15. Scheffer (1993), Laundré (1994).
16. Wyoming Game and Fish Department (2010b), Yellowstone Center for Resources unpublished data (March 2011).

Bighorn sheep, *Ovis canadensis*

DESCRIPTION: A medium-sized ungulate with well-developed horns. Body size and proportions differ markedly between sexes, with adult males larger, having better-developed neck muscles, visible scrotal testes as adults, and horns that grow, with age, longer and more massive than those of females. Fur is reddish to dark brown on the back, with white muzzle, belly, rump, and hind legs. Measurements (**Sex** [n] mean (kg), *range*): western Alta.; WT = **F** ≥ 4 years [65] 72, *54–91*; **M** ≥ 7 years [15] 108, *101–120*.[1] Dental formula: 0/3, 0/1, 3/3, 3/3 = 32.

STATUS: A "big game animal" by state statute, this is a state Species of Greatest Conservation Need as of 2012.[2] It is rare to common in occupied habitats statewide.

NOMENCLATURE AND SYSTEMATICS: One of two North American species of wild sheep, the genus and species names have been stable through time. Of the seven subspecies recognized in 1981, two occurred in Wyoming in presettlement times: *O. canadensis canadensis* over most of the state and *O. c. auduboni* along the eastern margin, including Inyan Kara Mountain.[3] This eastern subspecies (Audubon's bighorn), which occupied the Badlands and Missouri Breaks of eastern Montana and the Dakotas, and the Black Hills, is now extinct. Bighorns have been translocated in recent decades into Wyoming from Nevada, Idaho,

Montana, and Oregon. Those originating from southern Nevada are thought to have died out, but the remainder have persisted and interbred with native animals to various degrees.[4] So, the taxonomic identity of Wyoming bighorns can best be described as admixtures of various named subspecies, particularly the original *O. c. canadensis* and some *O. c. californiana*, the nominal subspecies found in Oregon and Idaho. The common name "mountain sheep" is sometimes used, but makes no distinction among the many species of sheep-like bovids that dwell in mountains.

DISTRIBUTION AND HABITAT ASSOCIATIONS: Broadly, although discontinuously distributed across the mountainous West 200 years ago (shown), the distribution has changed drastically over the last 150 years. European explorers and migrants described bighorns ("ibex" or "argali") in many areas where they are less abundant, rare, or absent today: in the mountain ranges of central Wyoming (Granite Hills, Ferris, Seminoe, and Shirley Mountains) and the Bighorn, Wind River, Medicine Bow, and Laramie Mountains. Low-elevation sites were occupied as well: along the North Platte River into Nebraska, along Poison Spider Creek, and in various canyons at the foot of the Bighorn and Absaroka Mountains. They occurred on steep outcrops in the Red Desert and Bighorn Basin.[5] Bighorns also occurred near the Green River in areas now inundated by Flaming Gorge Reservoir. Non-Wyoming bighorns (a total of 144 animals in six introductions) have been released into the Sweetwater Rocks, Shell Canyon, Devils Canyon, Laramie Peak, and Seminoe Mountains.[4] These translocations of animals from low elevation, nonmigratory herds in other states were intended to increase success in habitats and conditions similar to the animals' points of origin. Previous translocations of bighorns to these sites from higher-elevation, migratory herds had not been successful.

Some bighorn groups seasonally migrated over distances of dozens of miles, from high-elevation summer ranges to wintering areas along rivers and in basins. Recorded migrations include from the Absaroka Mountains to cliffy wintering sites in the Bighorn Basin. Sites where bighorns once spent the summer include Browns Peak in the Medicine Bow Mountains, Hunt Mountain in the Bighorn Mountains, and the northern Laramie Mountains. Some of these migrations were irregular, not followed in every year; others were more yearly. Many migration routes have been abandoned, and the collective herd memory of them has been lost. This likely has had the effect of reducing carrying capacity for sheep and of making the reestablishment of some migration routes across unsuitable habitat or past human-constructed barriers unlikely. At the same time, single rams and ewe bands episodically explore widely and disperse to new areas. These movements have the potential for recolonizing former ranges, but also of facilitating disease transmission between domestic and bighorn sheep.

Typical habitat is unforested slopes, particularly herbaceous meadows or grasslands near escape cover. Sheep can escape in steep terrain with or without rock outcrops. Where sheep winter at high elevations, they require wind-blown sites where forage is available. Bighorns will tolerate a certain amount of low shrub cover, but are sensitive to overhead cover, which blocks their visual field. In areas where wolves have been absent for many generations and visual cover is scarce, bighorns will venture into even, open terrain.

Bighorn sheep adult male.

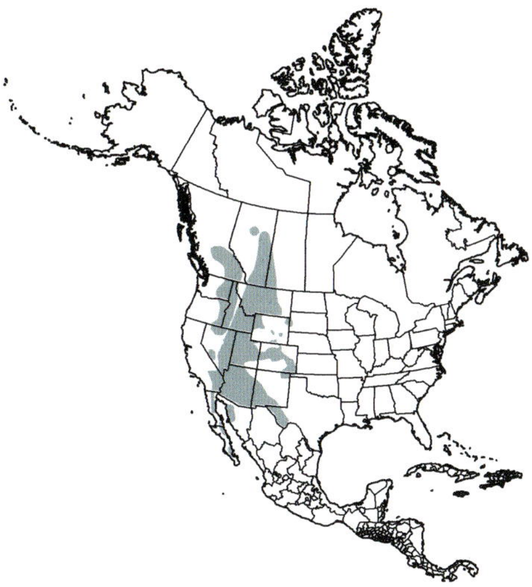

Distribution of the bighorn sheep in Wyoming.

ECOLOGY: The diet is grasses and forbs in summer, tending toward grass–forb–shrub in winter. Bighorns are sexually segregated in summer, with adult rams remaining distant from ewes, lambs, and young rams.[6] Several ecological forces are central to understanding the distribution and abundance of the species. First, the species was more migratory in presettlement times than today, moving from summer ranges in mountains to low-elevation sites in winter. This allowed animals to gain access to nutritious forage year-round. Second, bighorns are grazers rather than browsers; shrub and tree encroachment—for example, in response to fire suppression—lowers habitat quality. Third, the species is highly vulnerable to a range of livestock diseases, particularly those of domestic sheep.[7] A major region-wide die-off in the early 1880s was attributed to "scab," and more recently outbreaks of lungworm-pneumonia complex are a recurring phenomenon. An unusual epizootic of keratoconjunctivitis in Yellowstone in the early 1980s killed an estimated 60% of animals on the northern range.[8] Fourth, the species is generally thought to be vulnerable to genetic effects of small population size and population isolation.[9] This concern is heightened by undesirable genetic effects of selective removal of large-horned rams by hunting. Some have reported, while others have disputed, the effects of selective removal of large-horned rams on genetic variability, horn size, and vigor.[10] Effective predators of bighorns include mountain lions and to a lesser extent wolves.[11] Coyotes and golden eagles take some lambs, but are not generally considered important in limiting population size.[12]

CONSERVATION AND MANAGEMENT: Of the native ungulates that were reduced to low levels or nearly eradicated early in the twentieth century, the bighorn is one of two species that have not staged dramatic recoveries.[13] Numbers in presettlement times are not known accurately, but tens of thousands were described as occupying locations where animals now number only in the hundreds. Statewide abundance was estimated at 2000 in 1960 and at 5000–6000 in 2009–2010, an increase likely attributable to improved survey methods.[14] The basic conservation strategy over the past several decades has been to harvest at low levels via sport hunting, with only males of three-fourths curl or greater horn length allowed to be taken. Recently, hunting of any ram in most hunt areas and any sheep of either sex in one hunt area has become the harvest paradigm.

Translocations have been conducted from within the state and elsewhere; perhaps the largest single source has been the Whiskey Basin Herd near Dubois. Since the 1930s, in 66 separate actions, 1551 bighorns have been translocated to such sites as the Bighorn Mountains (Paintrock Canyon, Porcupine Creek, Middle Fork of Powder River), Laramie Mountains (Laramie Peak, La Bonte Canyon), Medicine Bow Mountains (Douglas Creek and Encampment River), Seminoe Mountains, Ferris Mountains, and Sinks Canyon.[4] This herd also has been an important source for translocations to sites outside Wyoming. A total of 399 animals have been translocated to five western states: South Dakota, Utah, New Mexico, Nevada, and Idaho. Unfortunately, the Whiskey Mountain herd underwent dramatic declines in the early 1990s as a result of lungworm and bacterial pneumonia. The population size dropped by about half, and had not fully recovered by 2012.

Two bighorn sheep are blocked by a woven wire fence bordering the National Elk Refuge, just north of Jackson.

Habitat management typically includes conversion of encroaching shrubs and trees to herbaceous types through prescribed fire. Approaches to reducing contact between bighorn and domestic sheep have included the retirement, by agencies, of domestic sheep grazing leases and lease buyouts by conservation organizations where they have potential to affect bighorn populations. Eliminating contact between domestic and bighorn sheep is a challenge, because both species are capable of wandering long distances in mountains, carrying pathogens between populations. On the positive side, the number of domestic sheep in Wyoming has declined drastically over the twentieth century; 4,000,000 sheep and lambs were estimated to be in Wyoming in 1932, compared with 375,000 in 2013.[15] New emphasis is being given to maintaining and restoring migration routes that had been interrupted by roads and fences, so that the occupied area can be increased, and deaths by disease and human causes reduced. Reductions in forest area and dense shrubs resulting from frequent fires and forest beetle epidemics might improve habitat conditions in coming decades, although other possible climate-mediated effects are not yet understood.

NOTES AND REFERENCES

1. Blood et al. (1970).
2. Wyoming Statutes 23-1-101-a-i.
3. Hall (1981).
4. Hurley (2011).
5. Honess and Frost (1942), Buechner (1960).
6. Woolf et al. (1970).

7. Foreyt and Jessup (1982).

8. Honess and Frost (1942), Shackleton (1985), Meagher et al. (1992).

9. FitzSimmons et al. (1997), Epps et al. (2005)

10. Harris et al. (2002), DiBattista (2008).

11. Sawyer and Lindzey (2002), Rominger et al. (2004), White et al. (2008).

12. Buechner (1960), Shackleton (1985).

13. Beecham et al. (2007). The other species is the bison, recovery of which is limited by economic and disease-related factors.

14. Wyoming Game and Fish Department (2009, 2010b).

15. US Department of Agriculture (2013).

APPENDIX

Plant Species and Non-mammalian Animal Species

Common name[1]	Scientific name
PLANTS	
Subalpine fir	*Abies lasiocarpa*
Rice grass	*Achnatherum* sp.
Indian rice grass	*Achnatherum hymenoides*
Pale agoseris	*Agoseris glauca*
Crested wheatgrass	*Agropyron cristatum*
Alder	*Alnus* sp.
Thinleaf alder	*Alnus incana*
Serviceberry	*Amelanchier* sp.
Saskatoon serviceberry	*Amelanchier alnifolia*
Utah serviceberry	*Amelanchier utahensis*
Bluestem	*Andropogon* sp.
Bearberry, kinnikinnick	*Arctostaphylos uva-ursi*
Three-awn	*Aristida* sp.
Arnica	*Arnica* sp.
Sagebrush	*Artemisia* sp.
Little or low sagebrush	*Artemisia arbuscula*
Silver sagebrush	*Artemisia cana*
Fringed or prairie sagewort	*Artemisia frigida*
Black sagebrush	*Artemisia nova*
Birdsfoot sagewort	*Artemisia pedatifida*
Big sagebrush	*Artemisia tridentata* or *A. t.* ssp. *spiciformis*
Basin big sagebrush	*A. t.* ssp. *tridentata*
Mountain big sagebrush	*A. t.* ssp. *vaseyana*
Wyoming big sagebrush	*A. t.* ssp. *wyomingensis*
American milkvetch	*Astragalus americanus*
Saltbush	*Atriplex* sp.
Four-wing saltbush	*Atriplex canescens*
Shadscale	*Atriplex confertifolia*
Gardner's saltbush	*Atriplex gardneri*

(continued)

Common name[1]	Scientific name
Birch	*Betula* sp.
Dwarf birch	*Betula glandulosa, B. nana*
Paper birch	*Betula papyrifera*
Grama	*Bouteloua* sp.
Buffalograss	*Bouteloua dactyloides*
Blue grama	*Bouteloua gracilis*
Brome, bromus	*Bromus* sp.
Sandreed	*Calamovilfa* sp.
Prairie sandreed	*Calamovilfa longifolia*
Small evening primrose	*Camissonia minor*
Sedge	*Carex* sp.
Indian paintbrush	*Castilleja* sp.
Mountain mahogany	*Cercocarpus* sp.
Curl-leaf mountain mahogany	*Cercocarpus ledifolius*
Alderleaf mountain mahogany	*Cercocarpus montanus*
Rabbitbrush	*Chrysothamnus* sp., *Ericameria* sp.
Thistle	*Cirsium* sp.
Tiny trumpet	*Collomia linearis*
Horseweed	*Conyza* sp.
Redosier dogwood	*Cornus sericea*
Hawthorn	*Crataegus* sp.
Cryptantha	*Cryptantha* sp.
Shrubby cinquefoil	*Dasiphora fruticosa*
Dyssodia	*Dyssodia* sp.
Russian olive	*Elaeagnus angustifolia*
Common spikerush	*Eleocharis palustris*
Wheat grass	*Elymus* sp., *Pascopyrum smithii*
Heather	several low-growing members of Ericaceae
Alpine-forget-me-not	*Eritrichium* sp.
Fescue	*Festuca* sp.
Halogeton saltlover	*Halogeton glomeratus*
Needle-and-thread grass, needle grass, stipa	*Hesperostipa comata*
Rush	*Juncus* sp.
Juniper	*Juniperus* sp.
Utah juniper	*Juniperus osteosperma*
Rocky Mountain juniper	*Juniperus scopulorum*
Winter fat	*Krascheninnikovia lanata*
Lupine	*Lupinus* sp.
Creeping barberry, Oregon-grape	*Mahonia repens*
Alfalfa	*Medicago* sp.
Prickly-pear cactus	*Opuntia* sp.
Mountain sorrel	*Oxyria digyna*
Switchgrass	*Panicum virgatum*
Western wheatgrass	*Pascopyrum smithii*
Spruce	*Picea* sp.
Engelmann spruce	*Picea engelmannii*

White spruce	*Picea glauca*
Blue spruce	*Picea pungens*
Whitebark pine	*Pinus albicaulis*
Bristlecone-pine	*Pinus aristata*
Lodgepole pine	*Pinus contorta*
Pinyon pine	*Pinus edulis*
Limber pine	*Pinus flexilis*
Singleleaf pinyon	*Pinus monophylla*
ponderosa-pine	*Pinus ponderosa*
Sandberg's-bluegrass	*Poa secunda*
Alpine-bistort	*Polygonum viviparum*
Narrowleaf cottonwood	*Populus angustifolia*
Aspen, quaking aspen	*Populus tremuloides*
Chokecherry	*Prunus virginiana*
Douglas-fir	*Pseudotsuga menziesii*
Antelope bitterbrush	*Purshia tridentata*
Oak	*Quercus* sp.
Gambel's oak	*Quercus gambelii*
Bur oak	*Quercus macrocarpa*
High northern buttercup	*Ranunculus hyperboreus*
Skunk bush sumac	*Rhus trilobata*
Currant	*Ribes* sp.
Rose, wild rose	*Rosa* sp.
Western thimbleberry	*Rubus parviflorus*
Red swampfire	*Salicornia rubra*
Willow	*Salix* sp.
Booth's willow	*Salix boothii*
Drummond's willow	*Salix drummondiana*
Geyer's willow	*Salix geyeriana*
Sandbar willow	*Salix interior*
Wolf's willow	*Salix wolfii*
Russian thistle	*Salsola* sp.
Greasewood	*Sarcobatus vermiculatus*
Purple mountain saxifrage	*Saxifraga oppositifolia*
Bulrush	*Schoenoplectus* sp., *Scirpus* sp.
Silver buffaloberry	*Shepherdia argentea*
Russet buffaloberry, soapberry	*Shepherdia canadensis*
Moss campion	*Silene acaulis*
Scarlet globemallow	*Sphaeralcea coccinea*
Dropseed	*Sporobolus* sp.
Alkali sacaton	*Sporobolus airoides*
Snowberry	*Symphoricarpos* sp.
Western snowberry	*Symphoricarpos occidentalis*
Tamarisk	*Tamarix chinensis*
Dandelion	*Taraxacum* sp.
Common dandelion	*Taraxacum officinale*
Wheat, winter wheat	*Triticum* sp.
Cattail	*Typha* sp.

(continued)

Common name[1]	Scientific name
Blueberry, cranberry	*Vaccinium* sp.
Yucca	*Yucca* sp.
INVERTEBRATES	
Western balsam bark beetle	*Dryocoetes confusus*
Mountain pine beetle	*Dendroctonus ponderosae*
Douglas-fir beetle	*Dendroctonus pseudotusgae*
Spruce beetle	*Dendroctonus rufipennis*
FISH	
Cutthroat trout	*Oncorhynchus clarki*
Lake trout	*Salvelinus namaycush*
Brown trout	*Salmo trutta*
REPTILES	
Garter snake	*Thamnophis* sp.
BIRDS	
Boreal owl	*Aegolius funereus*
Burrowing owl	*Athene cunicularia*
Golden eagle	*Aquila chrysaetos*
Great horned owl	*Bubo virginianus*
Greater sage grouse	*Centrocercus urophasianus*
Mountain plover	*Charadrius montanus*
Northern goshawk	*Accipiter gentilis*
Red-tailed hawk	*Buteo jamaicensis*

1. Scientific and common nomenclature follows Kartesz and Thieret (1991), US Department of Agriculture, NRCS (2013), and Chesser et al. (2013).

GLOSSARY

agglutinated	Clumped or glued together.
allopatric	Referring to the pattern of two or more species' geographic ranges that do not overlap and do not come into contact.
altricial	(Of neonates) poorly developed at birth and requiring parental care.
aspect	The compass direction faced by a slope.
baculum	The os penis bone found in male mammals of various orders.
bolide	A generic large crater-forming extraterrestrial body of unspecified composition.
boreomontane	Referring to the boreal forest and southern extensions of it along mountain ranges.
calcar	A spur of cartilage attached to the medial side of the ankle and running along the rear margin of the interfemoral membrane of a bat. A keeled calcar has a small wing-like flap of skin along the posterior margin.
cecum	A blind pouch at the junction of the small and large intestine.
clast	A fragment of a preexisting rock, such as is ejected from volcanoes.
cline	A continuous ecological gradient, lacking discrete steps in variation.
commensal	A relationship between two kinds of organisms in which one kind benefits at no cost or benefit to the other.
concordant	Occurring together; occupying the same place.
condylobasal length	The length of the skull from the anterior edge of the premaxilla to the posteriormost part of the occipital condyles.
conspecific	Belonging to the same species.

cryptic morphological species (also "morphospecies")	Distinct biological species that are difficult or impossible to distinguish on the basis of morphology, at some level of examination. What may be cryptic morphospecies in the field may be reliably distinguished with a cleaned skeleton.
dentary bone	The lower jaw bone of a modern mammal, in which the lower teeth are rooted.
diastema	A space or gap between teeth.
diel	Daily, referring to a 24-hour cycle.
distal	Farther from the main part of the body (cf. proximal).
downlist	Reduce the level of protection under the Endangered Species Act, as from Endangered to Threatened, or from Threatened to Not Listed.
ecological extinction	Reduced in abundance to the point of being ecologically uninfluential.
ecotypic	Referring to local adaptation in response to ecological conditions, but less pronounced than phylogenetic variation.
endemic	Restricted or unique to a place or region.
enzootic	Describing a disease of animals that occurs at some low or moderate frequency in the population, without causing major mortality events or disappearing entirely.
epizootic	Describing a disease of animals that occurs at high frequency in the population, possibly causing large-scale illness or death.
eutherian	Referring to a mammal that develops a true chorioallantoic placenta during pregnancy.
extralimital	Referring to an occurrence or location beyond the known geographic range of a taxon.
fenestration	A small opening in a skull, without nerves or vessels passing through.
fossorial	Referring to living beneath the soil surface or burrowing.
genetic drift	Random changes in the frequency of alleles, especially uncommon ones, in small populations. It is one of the major sources of evolutionary change.
hypogeous	Referring to growing underground, as in the fruiting bodies of some fungi.
insolation	The solar radiation incident on a given surface area, particularly the earth's surface.
interference competition	A direct form of interaction between individuals of the same or different species in which one animal acts aggressively or

	threateningly toward the other. Forms of interference competition include territorial behavior, aggressiveness, and killing.
introgression	Gene flow between two populations or nominal species, particularly beyond first-generation hybridization. Genetic mixing involving repeated backcrossing.
keratinaceous	Characterized by the growth of the tough, resistant protein keratin, which grows in the epidermis, forms nails and claws, and makes up the horn sheath of Antilocapridae and Bovidae.
keratoconjunctivitis	Inflammation of the cornea and conjunctiva, found in many mammalian species.
Ma	Mega annum, one million years, or one million years before present, depending on context.
maxilla	The facial bone that makes up the upper jaw in which the canine teeth, premolars, and molars are rooted.
mesic	Moist, but not featuring standing water.
Mima mound	Low, regularly spaced, roughly circular, naturally occurring mounds found in various areas of the western United States.
multituberculata	A group of rodent-like mammals that existed for around 120 Ma, the longest fossil history of any mammalian lineage.
mycophagy	The practice of eating fungi.
occlusal	Referring to the surface of a tooth that makes contact with other teeth during chewing.
orts	The remains of a dead animal, a carcass, or parts of it.
parapatric	Having geographic ranges that abut, but do not overlap.
pedicel	A stem or stalk, such as that projects from the frontal bone and supports the antlers of the Cervidae.
periauricular	Referring to the area surrounding the pinna.
phenology	The seasonal state of growth of a plant.
phenotypic traits	All of the physical, behavioral, and physiological attributes of an organism. The expression of its genes.
phylogenetic tree	A branching diagram intended to depict the pattern of evolutionary relationships among living, and in some cases extinct, taxa.
phylogeography	The study of the relationship between evolutionary relatedness, generally as inferred from molecular evidence, and geographic distribution.
pinna	The visible flap of the external ear.
proximal	Nearer to the main part of the body (cf. distal).

rostrum	The part of the skull projecting anteriorly from the braincase, and encasing the nasal cavity. It is made up of the prenasal, nasal, premaxilla, and maxilla.
saltatorial	Referring to a pattern of locomotion emphasizing hopping on the hind limbs, although the front limbs may contact the substrate.
scansorial	Adapted for climbing, but not fully arboreal.
squamosal bone	The bone making up the lower lateral portion of the braincase, including the housing for the ear, and composing the proximal part of the jaw joint in modern mammals.
sylvatic plague	The bacterial disease in wild animals caused by *Yersinia pestis*.
sympatric	Referring to the pattern of two species occupying the same or broadly overlapping geographic areas.
syntopic	The relationship between related species that occur together closely. Sympatric and occurring in the same habitats.
taxon	The members of one taxonomic level, such as a subspecies, species, genus, or family.
transmissible spongiform encephalopathy	A group of progressive diseases, caused by prions, or misshapen proteins, that affect the brain and nervous system. Tiny holes appear in the cortex, causing it to appear like a sponge (hence, "spongiform").
tubercle	A round, rough nodule or warty outgrowth on the skin.
turpene	Aromatic or aliphatic hydrocarbon chemicals, typically with strong odor, that are produced by various plants and serve a protective function, deterring herbivores.
unicuspid	Referring to teeth that have a single crown or projection, for example, canine teeth of mammals.
ventrum	The lower surface of a quadrupedal animal; the underside.
vibrissa	A long, stiff mammalian hair on the rostrum specially adapted for tactile sensation.
zoonosis (adj. zoonotic)	A disease of nonhuman animals that can be transmitted to humans.
zygomatic arch	The skull structure that forms the lower margin of the orbit in some mammals, typically composed of the jugal bone anteriorly and the squamosal posteriorly.

REFERENCES CITED

Abel, B., and M. Grenier. 2012. Inventory of Bats in Forest of Southeastern Wyoming: Mist Surveys. Unpublished Completion Report. Wyoming Game and Fish Department, Nongame Section, Lander.

Abel, B., and M. Grenier. n.d. A Strategic Plan for White-Nose Syndrome in Wyoming. Unpublished Report. Wyoming Game and Fish Department, Lander.

Adams, R. A. 2003. Bats of the Rocky Mountain West: Natural History, Ecology, and Conservation. University Press of Colorado, Boulder.

Agnarsson, I., and L. J. May-Collado. 2008. The phylogeny of Cetartiodactyla: the importance of dense taxon sampling, missing data, and the remarkable promise of cytochrome b to provide reliable species-level phylogenies. Molecular Phylogenetics and Evolution 48:964–985.

Agnew, W., D. W. Uresk, and R. M. Hansen. 1988. Arthropod consumption by small mammals on prairie dog colonies and adjacent ungrazed mixed grass prairie in western South Dakota. Proceedings of the 8th Great Plains Wildlife Damage Control Workshop. USDA Forest Service General Technical Report on Range Management 154:81–87.

Agosta, S. J. 2002. Habitat use, diet and roost selection by the big brown bat (*Eptesicus fuscus*) in North America: a case for conserving an abundant species. Mammal Review 32:179–198.

Aho, K., N. Huntly, J. Moen, and T. Oksanen. 1998. Pikas (*Ochotona princeps*: Lagomorpha) as allogenic engineers in an alpine ecosystem. Oecologia 114:405–409.

Alagona, P. 2013. After the Grizzly: Endangered Species and the Politics of Place in California. University of California Press, Berkeley.

Alexander, L. F., and B. R. Riddle. 2005. Phylogenetics of the New World rodent family Heteromyidae. Journal of Mammalogy 86:366–379.

Allred, S. 2010. The Natural History of Tassel-Eared Squirrels. University of New Mexico Press, Albuquerque.

Alroy, J. 1998. Cope's rule and the dynamics of body mass evolution in North American fossil mammals. Science 280:731–734.

Alroy, J. 1999. The fossil record of North American mammals: evidence for a Paleocene evolutionary radiation. Systematic Biology 48:107–118.

Altmann, M. 1952. Social behavior of elk, *Cervus canadensis nelsoni*, in the Jackson Hole area of Wyoming. Behaviour 4:116–143.

Andersen, D. C., J. A. McMahon, and M. L. Wolfe. 1980. Herbivorous mammals along a montane sere: community structure and energetics. Journal of Mammalogy 61:500–519.

Andersen, J. J., and J. E. Light. 2012. Phylogeography and subspecies revision of the hispid pocket mouse, *Chaetodipus hispidus* (Rodentia: Heteromyidae). Journal of Mammalogy 93:1195–1215.

Anderson, A. E., D. E. Medin, and D. C. Bowden. 1974. Growth and morphometry of the carcass, selected bones, organs, and glands of mule deer. Wildlife Monographs 39:1–122.

Anderson, A. E., and O. C. Wallmo. 1984. *Odocoileus hemionus*. Mammalian Species 219:1–9.

Anderson, C. R., Jr., and F. G. Lindzey. 2003. Estimating cougar predation rates from GPS location clusters. Journal of Wildlife Management 67:307–316.

Anderson, C. R., Jr., and F. G. Lindzey. 2005. Experimental evaluation of population trend and harvest composition in a Wyoming cougar population. Wildlife Society Bulletin 33:179–188.

Anderson, C. R., Jr., F. G. Lindzey, and D. B. McDonald. 2004. Genetic structure of cougar populations across the Wyoming Basin: metapopulation or megapopulation? Journal of Mammalogy 85:1207–1214.

Anderson, C. R., Jr., M. A. Ternent, and D. S. Moody. 2002. Grizzly bear-cattle interactions on two grazing allotments in northwest Wyoming. Ursus 13:247–256.

Anderson, E. 1968. Fauna of the Little Box Elder Cave, Converse County, Wyoming. University of Colorado Studies, Series in Earth Sciences 6:1–59.

Anderson, E. 1984. Who's who in the Pleistocene: a mammalian bestiary. Pp. 40–89. In P. S. Martin and R. G. Klein (eds). Quaternary Extinctions: A Prehistoric Revolution. University of Arizona Press, Tucson.

Anderson, E., S. C. Forrest, T. W. Clark, and L. Richardson. 1986. Paleobiology, biogeography, and systematics of the black-footed ferret, *Mustela nigripes* (Audubon and Bachman), 1851. Great Basin Naturalist Memoirs 8:11–62.

Antoine, P.-O., L. Marivaux, D. A. Croft, G. Billet, M. Ganerød, C. Jaramillo, T. Martin et al. 2012. Middle Eocene rodents from Peruvian Amazonia reveal the pattern and timing of caviomorph origins and biogeography. Proceedings of the Royal Society B 279:1319–1326.

Armitage, K. B. 1961. Frequency of melanism in the golden-mantled marmot. Journal of Mammalogy 42:100–101.

Armitage, K. B. 1991. Dynamics of yellow-bellied marmots: results from long-term research. Annual Review of Ecology and Systematics 22:379–407.

Armstrong, D. M., R. A. Adams, and J. Freeman. 1994. Distribution and ecology of bats in Colorado. University of Colorado Museum, Natural History Inventory 15:1–82.

Armstrong, D. M., J. R. Choate, and J. K. Jones, Jr. 1986. Distributional patterns of mammals in the plains states. Occasional Papers, the Museum of Texas Tech University 105:1–27.

Armstrong, D. M., J. P. Fitzgerald, and C. A. Meaney. 2011. Mammals of Colorado. 2nd ed. University Press of Colorado, Boulder.

Armstrong, D. M., and J. K. Jones. 1971. *Sorex merriami*. Mammalian Species 2:1–2.

Arnett, E. B., W. K. Brown, W. P. Erickson, J. K. Fiedler, B. L. Hamilton, T. H. Henry, A. Jain et al. 2008. Patterns of bat fatalities at wind energy facilities in North America. Journal of Wildlife Management 72:61–78.

Arnold, B. D. 2007. Population structuring and sex-biased dispersal in the forest dwelling vespertilionid bat, *Myotis septentrionalis*. American Midland Naturalist 157:374–384.

Aubry, K. B. 1983. The Cascade red fox: distribution, morphology, zoogeography and ecology. PhD dissertation, University of Washington, Seattle.

Aubry, K. B., K. S. McKelvey, and J. P. Copeland. 2007. Distribution and broadscale habitat relations of the wolverine in the contiguous United States. Journal of Wildlife Management 71:2147–2158.

Aubry, K.B., M.J. Statham, B.N. Sacks, J.D. Perrine, and S.M. Wisely. 2009. Phylogeography of the North American red fox: vicariance in Pleistocene forest refugia. Molecular Ecology 18:2668–2686.
Audubon, J.J., and J. Bachman. 1851. The Viviparous Quadrupeds of North America. Vol. 2. V.G. Audubon, New York.
Aughey, S. 1884. Curious companionship of the coyote and badger. American Naturalist 18:645.
Augustine, D.J., and S.K. Skagen. 2014. Mountain plover next survival in relation to prairie dog and fire dynamics in shortgrass steppe. Journal of Wildlife Management 78:595–602.
Bacon, C.D., D. Silvestro, C. Jaramillo, B.T. Smith, P. Chakrabarty, and A. Antonelli. 2015. Biological evidence supports an early and complex emergence of the Isthmus of Panama. Proceedings of the National Academy of Sciences 112:6110–6115.
Bailey, V. 1930. Animal Life of Yellowstone National Park. Charles C. Thomas Publisher, Springfield, IL.
Bailey, V., and C.C. Sperry. 1929. Life history and habits of the grasshopper mice, genus *Onychomys*. US Department of Agriculture Technical Bulletin 145:1–9.
Baird, S.F. 1859. Mammals of North America: The Description of Species Based Chiefly on the Collections in the Museum of the Smithsonian Institution. Part I. General Report upon the Mammals of the Several Pacific Railroad Routes. J.B. Lippincott, Philadelphia, PA.
Baker, B.W., and E.P. Hill. 2003. Beaver. Pp. 288–310. In G.A. Feldhamer, B.C. Thompson, and J.A. Chapman (eds). Wild Mammals of North America. 2nd ed. Johns Hopkins University Press, Baltimore, MD.
Baker, M.D., M.J. Lacki, G.A. Falxa, P.L. Droppelman, R.A. Slack, and S.A. Slankard. 2008. Habitat use of pallid bats in coniferous forests of northern California. Northwest Science 82:269–275.
Baker, R.J., and R.D. Bradley. 2006. Speciation in mammals and the genetic species concept. Journal of Mammalogy 87:643–662.
Ballard, W.B., L.N. Carbyn, and D.W. Smith. 2003. Wolf interactions with non-prey. Pp. 259–271. In L.D. Mech and L. Boitani (eds). Wolves: Behavior, Ecology, and Conservation. University of Chicago Press, Chicago, IL.
Ballard, W.B., D. Lutz, T.W. Keegan, L.H. Carpenter, and J.C. deVos, Jr. 2001. Deer-predator relationships: a review of recent North American studies with emphasis on mule and black-tailed deer. Wildlife Society Bulletin 29:99–115.
Banbury, J.L., and G.S. Spicer. 2007. Molecular systematics of chipmunks (*Neotamias*) inferred by mitochondrial control region sequences. Journal of Mammalian Evolution 14:149–162.
Banci, V. 1994. Wolverine. Pp. 99–127. In L.F. Ruggiero, K.B. Aubry, S.W. Buskirk, L.J. Lyon, and W.J. Zielinski (eds). The Scientific Basis of Conserving Forest Carnivores: American Marten, Fisher, Lynx, and Wolverine in the Western United States. USDA Forest Service General Technical Report RM-254.
Banfield, A.W.F. 1974. The Mammals of Canada. University of Toronto Press, Toronto, ON.
Barbour, R.W., and W.H. Davis. 1969. Bats of America. University Press of Kentucky, Lexington.
Barclay, R.M.R., E.F. Baerwald, and J.C. Gruver. 2007. Variation in bat and bird fatalities at wind energy facilities: assessing the effects of rotor size and tower height. Canadian Journal of Zoology 85:381–387.
Barnowe-Meyer, K.K., P.J. White, T.L. Davis, and J.A. Byers. 2009. Predator-specific mortality of pronghorn on Yellowstone's northern range. Western North American Naturalist 69:186–194.
Bartels, M.A., and D.P. Thompson. 1993. *Spermophilus lateralis*. Mammalian Species 440:1–8.

Barton, H.D., and S.M. Wisely. 2012. Phylogeography of striped skunks (*Mephitis mephitis*) in North America: Pleistocene dispersal and contemporary population structure. Journal of Mammalogy 93:38–51.
Baskin, Y. 1998. Home on the range. BioScience 48:245–251.
Bean, M.J. 1983. The evolution of national wildlife law. Praeger Publishers, New York.
Beauvais, G.P. 1997. Mammals in fragmented forests in the Rocky Mountains: community structure, habitat selection, and individual fitness. PhD dissertation, University of Wyoming, Laramie.
Beauvais, G.P., and D.N. Dark-Smiley. 2005. Species Assessment for Idaho Pocket Gopher (*Thomomys idahoensis*) in Wyoming. Unpublished Report. Wyoming Natural Diversity Database, University of Wyoming, Laramie.
Beauvais, G.P., and J. McCumber. 2006. Pygmy shrew (*Sorex hoyi*): A Technical Conservation Assessment. USDA Forest Service, Rocky Mountain Region, Species Conservation Project.
Beck, M.J., and S.B. Vander Wall. 2010. Seed dispersal by scatter-hoarding rodents in arid environments. Journal of Ecology 98:1300–1309.
Beecham, J.J., C.P. Collins, and T.D. Reynolds. 2007. Rocky Mountain Bighorn Sheep (*Ovis canadensis*): A Technical Conservation Assessment. Prepared for the USDA Forest Service, Rocky Mountain Region, Species Conservation Project. USDA Forest Service, Lakewood, CO.
Beever, E.A., P.F. Brussard, and J. Berger. 2003. Patterns of apparent extirpation among isolated populations of pikas (*Ochotona princeps*) in the Great Basin. Journal of Mammalogy 84:37–54.
Bekoff, M. and E.M. Gese. 2003. Coyote. Pp. 467–481. In G.A. Feldhamer, B.C. Thompson, and J.A. Chapman (eds). Wild Mammals of North America. 2nd ed. Johns Hopkins University Press, Baltimore, MD.
Belk, M.C., C.L. Pritchett, and H.D. Smith. 1990. Patterns of habitat use by *Sorex monticolus* in summer. Great Basin Naturalist 50:387–389.
Benedict, R.A., H.H. Genoways, and P.W. Freeman. 2000. Shifting distributional patterns of mammals in Nebraska. Transactions of the Nebraska Academy of Sciences 26:55–84.
Beneski, J.T., and D.W. Stinson. 1987. *Sorex palustris*. Mammalian Species 296:1–6.
Benkman, C.W. 1995. The impact of tree squirrels (*Tamiasciurus*) on limber pine seed dispersal adaptations. Evolution 49:585–592.
Benkman, C.W., T.L. Parchman, and E. Mezquida. 2010. Patterns of coevolution in the adaptive radiation of crossbills. Annals of the New York Academy of Sciences 1206:1–16.
Bentz, B.J., J. Régnière, C.J. Fettig, E.M. Hansen, J.L. Hayes, J.A. Hicke, R.G. Kelsey, J.F. Negrón, and S.J. Seybold. 2010. Climate change and bark beetles of the western United States and Canada: direct and indirect effects. BioScience 60:602–613.
Berg, N.D., and E.M. Gese. 2010. Relationship between fecal pellet counts and snowshoe hare density in western Wyoming. Journal of Wildlife Management 74:1745–1751.
Berg, N.D., E.M. Gese, J.R. Squires, and L.M. Aubry. 2012. Influence of forest structure on the abundance of snowshoe hares in western Wyoming. Journal of Wildlife Management 76:1480–1488.
Berg, R.L., L.L. McDonald, and M.D. Strickland. 1983. Distribution of mountain lions in Wyoming as determined by mail questionnaire. Wildlife Society Bulletin 11:265–268.
Berger, J. 2004. The last mile: how to sustain long-distance migration in mammals. Conservation Biology 18:320–331.
Berger, J. 2008. Undetected species losses, food webs, and ecological baselines: a cautionary tale from the Greater Yellowstone Ecosystem. Oryx 42(1):1–4.

Berger, J., P. B. Stacey, L. Bellis, and M. P. Johnson. 2003. A mammalian predator–prey imbalance: grizzly bear and wolf extinction affect avian Neotropical migrants. Ecological Applications 11:947–960.

Berger, K. M., and M. M. Conner. 2008. Recolonizing wolves and mesopredator suppression of coyotes: impacts on pronghorn population dynamics. Ecological Applications 18:599–612.

Berger, K. M., and E. M. Gese. 2007. Does interference competition with wolves limit the distribution and abundance of coyotes? Journal of Animal Ecology 76:1075–1085.

Bergstrom, B. J., and R. S. Hoffman. 1991. Distribution and diagnosis of three species of chipmunks (*Tamias*) in the Front Range of Colorado. Southwestern Naturalist 36:14–28.

Berman, J., T. McCay, and P. Scull. 2007. Spatial analysis of species richness of shrews (Soricomorpha: Soricidae) in North America north of Mexico. Acta Theriologica 52:151–158.

Best, T. L. 1996. *Lepus californicus*. Mammalian Species 530:1–10.

Best, T. L., and M. P. Skupski. 1994. *Perognathus flavus*. Mammalian Species 471:1–10.

Betancourt, J. L., T. R. Van Devender, and P. S. Martin, editors. 1990. Packrat Middens: The Last 40,000 Years of Biotic Change. University of Arizona Press, Tucson.

Betke, M., D. E. Hirsh, N. C. Makris, G. F. McCracken, M. Procopio, N.I. Hristov, S. Tang et al. 2008. Thermal imaging reveals significantly smaller Brazilian free-tailed bat colonies than previously estimated. Journal of Mammalogy 89:18–24.

Beuchat, C. A. 1990. Body size, medullary thickness, and urine concentrating ability in mammals. American Journal of Physiology – Regulatory, Integrative and Comparative Physiology 258:R298–R308.

Bilyeu, D. M., D. J. Cooper, and N. T. Hobbs. 2008. Water tables constrain height recovery of willow on Yellowstone's Northern Range. Ecological Applications 18:80–92.

Black, G. 2012. Empire of Shadows: The Epic Story of Yellowstone. St. Martin's Press, New York.

Blair, N. 1987. The History of Wildlife Management in Wyoming. Wyoming Game and Fish Department, Cheyenne.

Blanga-Kanfi, S., H. Miranda, O. Penn, T. Pupko, R.W. Debry, and D. Huchon. 2009. Rodent phylogeny revised: analysis of six nuclear genes from all major rodent clades. BMC Evolutionary Biology 9:71.

Blanton, J. D., D. Palmer, J. Dyer, and C. E. Rupprecht. 2011. Rabies surveillance in the United States during 2010. Journal of the American Veterinary Medical Association 239:773–783.

Blood, D. A., D. R. Flook, and W. D. Wishart. 1970. Weights and growth of Rocky Mountain bighorn sheep in western Alberta. Journal of Wildlife Management 34:451–455.

Blumenthal, E. M., and G. L. Kirkland, Jr. 1976. The biology of the opossum, *Didelphis virginiana*, in southcentral Pennsylvania. Proceedings of the Pennsylvania Academy of Science 50:81–85.

Blumstein, D. T. 1989. Food habits of red-tailed hawks in Boulder County, Colorado. Journal of Raptor Research 23:53–55.

Blumstein, D. T., S. Im, A. Nicodemus, and C. Zugmeyer. 2004. Yellow-bellied marmots (*Marmota flaviventris*) hibernate socially. Journal of Mammalogy 85:25–29.

Bogan, M. A., and P. M. Cryan. 2000. The bats of Wyoming. Pp. 71–94. In Reflections of a Naturalist: Papers Honoring Professor Eugene D. Fleharty. Fort Hays Studies, Special Issue 1, Fort Hays State University, Fort Hays, KS.

Borchgrevink, M. B., J. D. Derner, T. R. Weston, R. A. Olson, G.E. Schuman, and B.W. Hess. 2010. Small mammal and plant community responses to mechanical disturbance and rest in Wyoming big sagebrush grassland. Arid Land Research and Management 24:57–67.

Bowe, A., and G. P. Beauvais. 2012. An Assessment of Species and Subspecies of *Zapus* in Wyoming. Unpublished Report. Wyoming Natural Diversity Database, University of Wyoming, Laramie.

Boyce, M. S. 1978. Variability and body size variation in the muskrats (*Ondatra zibethicus*) of North America. Oecologia 36:1–19.

Boyce, M. S. 1980. First record of the fringe-tailed bat, *Myotis thysanodes*, from southeastern Wyoming. Southwestern Naturalist 25:114–115.

Boyce, M. S. 1989. The Jackson Elk Herd: Intensive Wildlife Management in North America. Cambridge University Press, Cambridge, UK.

Boyd, D. K., and D. H. Pletscher. 1999. Characteristics of dispersal in a colonizing wolf population in the central Rocky Mountains. Journal of Wildlife Management 63:1094–1108.

Boyeskorov, G. 1999. New data on moose (*Alces*, Artiodactyla) systematics. Säugetierkundliche Mitteilungen 44:3–13.

Boyle, S. 2006. North American River Otter (*Lontra canadensis*): A Technical Conservation Assessment. USDA Forest Service, Rocky Mountain Region. Available at http://www.fs.fed.us/r2/prijects/scp/assessments/northamericanriverotter.pdf. Accessed December 6, 2011.

Bradley, R. D., L. K. Ammerman, R. J. Baker, L. C. Bradley, J.A. Cook, R.C. Dowler, C. Jones et al. 2014. Revised checklist of North American mammals north of Mexico, 2014. Occasional Papers, Museum of Texas Tech University 327:1–28.

Bradley, R. D., F. C. Bryant, L. C. Bradley, M. L. Haynie, and R.J. Baker. 2003. Implications of hybridization between white-tailed deer and mule deer. Southwestern Naturalist 48:654–660.

Braun, C. E., and R. G. Streeter. 1968. Observations on the occurrence of white-tailed jackrabbits in the alpine zone. Journal of Mammalogy 49:160–161.

Brimeyer, D. G., and T. P. Thomas. 2004. History of moose management in Wyoming and recent trends in Jackson Hole. Alces 40:133–143.

Brown, D. E., and M. R. Conover. 2011. Effects of large-scale removal of coyotes on pronghorn and mule deer productivity and abundance. Journal of Wildlife Management 75:876–882.

Brown, H. L. 1946. Rodent activity in a mixed prairie near Hays, Kansas. Transactions of the Kansas Academy of Science 48:448–456.

Brown, L. N. 1965. A fisher, *Martes pennanti*, in Sheridan County, Wyoming. Southwestern Naturalist 10:143.

Brown, L. N. 1967a. Ecological distribution of six species of shrews and comparison of sampling methods in the central Rocky Mountains. Journal of Mammalogy 48:617–623.

Brown, L. N. 1967b. Seasonal activity patterns and breeding of the western jumping mouse (*Zapus princeps*) in Wyoming. American Midland Naturalist 78:460–470.

Brown, L. N. 1967c. Ecological distribution of mice in the Medicine Bow Mountains of Wyoming. Ecology 48:677–679.

Brown, L. N. 1970. Population dynamics of the western jumping mouse (*Zapus princeps*) during a four-year study. Journal of Mammalogy 51:651–658.

Bruggeman, J. E., R. A. Garrott, P. J. White, F. G. R. Watson, and R. Wallen. 2007. Covariates affecting spatial variability in bison travel behavior in Yellowstone National Park. Ecological Applications 17:1411–1423.

Brugger, K. A. 2006. Late Pleistocene climate inferred from the reconstruction of the Taylor River glacier complex, southern Sawatch Range, Colorado. Geomorphology 75:318–329.

Bryant, L. D., and C. Maser. 1982. Classification and distribution. Pp. 1–59. In J. W. Thomas and D. E. Toweill (eds). Elk of North America: Ecology and Management. Stackpole Books, Harrisburg, PA.

Buck, P. D. 1947. The biology of the antelope (*Antilocapra americana*) in Montana. MS thesis, Montana State College, Bozeman.

Buechner, H. K. 1960. The bighorn sheep in the United States, its past, present and future. Wildlife Monographs 4:1–174.

Buskirk, S.W., J. Bowman, and J.H. Gilbert. 2012. Population biology and matrix demographic modeling of American martens and fishers. Pp. 77–92. In K.B. Aubry, W.J. Zielinski, M.G. Raphael, G. Proulx, and S.W. Buskirk (eds). Biology and Conservation of Martens, Sables, and Fishers: A New Synthesis. Cornell University Press, Ithaca, NY.

Buskirk, S.W., S.C. Forrest, M.G. Raphael, and H.J. Harlow. 1989. Winter resting site ecology of marten in the central Rocky Mountains. Journal of Wildlife Management 53:191–196.

Buskirk, S.W., A.S. Harestad, M.G. Raphael, and R.A. Powell, eds. 1994. Martens, Sables and Fishers: Biology and Conservation. Cornell University Press, Ithaca, NY.

Buskirk, S.W., L.F. Ruggiero, K.B. Aubry, D.E. Pearson, J.R. Squires, and K.S. McKelvey. 2000. Comparative ecology of lynx in North America. Pp. 397–417. In L.F. Ruggiero, K.B. Aubry, S.W. Buskirk, G.M. Koehler, C.J. Krebs, K.S. McKelvey, and J.R. Squires (eds). Ecology and Conservation of Lynx in the United States. University Press of Colorado, Boulder.

Caceres, M.C., and R.M.R. Barclay. 2000. *Myotis septentrionalis*. Mammalian Species 634:1–4.

Cain, S. 2012. Teton Range bighorn sheep face new threats. Teton Stewards (Grand Teton National Park Foundation) Fall–winter 2012–2013:1–2.

Calder, W.A. 1969. Temperature relations and underwater endurance of the smallest homeothermic diver, the water shrew. Comparative Biochemistry and Physiology 30:1075–1082.

Camenzind, F.J. 1978. Behavioral ecology of coyotes (*Canis latrans*) on the National Elk Refuge, Jackson, Wyoming. PhD dissertation, University of Wyoming, Laramie.

Camp, M.J., J.L. Rachlow, L.A. Shipley, T.R. Johnson, and K.D. Bockting. 2014. Grazing in sagebrush rangelands in western North America: implications for habitat quality for a sagebrush specialist, the pygmy rabbit. Rangeland Journal 36:151–159.

Campbell, L.A., J.G. Hallett, and M.A. O'Connell. 1996. Conservation of bats in managed forests: use of roosts by *Lasionycteris noctivagans*. Journal of Mammalogy 77:976–984.

Campbell, T.M., III, and T.W. Clark. 1980. Short-term effects of logging on red-backed voles and deer mice. Great Basin Naturalist 40:183–189.

Campbell, T.M. III, T.W. Clark, and C.R. Groves. 1982. First record of pygmy rabbits (*Brachylagus idahoensis*) in Wyoming. Great Basin Naturalist 42:100.

Campbell, T.M. III, T.W. Clark, L. Richardson, S.C. Forrest, and B.R. Houston. 1987. Food habits of Wyoming black-footed ferrets. American Midland Naturalist 117:208–210.

Cannon, K.P. 2007. "They went as high as they choose:" what an isolated skull can tell us about the biogeography of high-altitude bison. Arctic, Antarctic, and Alpine Research 39:44–56.

Carey, A.B. 2000. Ecology of northern flying squirrels: implications for ecosystem management in the Pacific Northwest, USA. Pp. 45–61. In R.L. Goldingay and J.S. Scheibe (eds). Biology of Gliding Squirrels. Filander Verlag, Fürth, Germany.

Carleton, M.D., A.L. Gardner, I.Y. Pavlinov, and G.G. Musser. 2014. The valid generic name for red-backed voles (Muroidea: Cricetidae: Arvincolinae): restatement of the case for *Myodes* Pallas, 1811. Journal of Mammalogy 95:943–959.

Carr, S.M., and S.A. Hicks. 1997. Are there two species of marten in North America? Genetic and evolutionary relationships within *Martes*. Pp. 15–28. In G. Proulx, H.N. Bryant, and P.M. Woodard (eds). *Martes*: Taxonomy, Ecology, Techniques, and Management. Provincial Museum of Alberta, Edmonton, AB.

Carroll, L.E., and H.H. Genoways. 1980. *Lagurus curtatus*. Mammalian Species 124:1–6.

Cary, M. 1911. A biological survey of Colorado. North American Fauna 33:1–256.

Cary, M. 1917. Life zone investigations in Wyoming. North American Fauna 42:1–95.

Cassin, J. 1858. Mammalogy and Ornithology. J.B. Lippencott, Philadelphia, PA.

Catania, K.C. 2013. The neurobiology and behavior of the American water shrew (*Sorex palustris*). Journal of Comparative Physiology A 119:545–554.

Cerovski, A.O., M. Grenier, B. Oakleaf, L. Van Fleet, L., and S. Patla. 2004. Atlas of Birds, Mammals, Amphibians, and Reptiles in Wyoming. Wyoming Game and Fish Department, Lander.
Chadwick, D.H. 1983. A Beast the Color of Winter: The Mountain Goat Observed. Sierra Club Books, San Francisco, CA.
Chambers, C.L., M.J. Herder, K. Yasuda, D.G. Mikesic, S.M. Dewhurst, W.M. Masters, and D. Vleck. 2011. Roosts and home ranges of spotted bats (*Euderma maculatum*) in northern Arizona. Canadian Journal of Zoology 89:1256–1267.
Chapman, J.A. 1975. *Sylvilagus nuttallii*. Mammalian Species 56:1–3.
Chapman, J.A., and J.E.C. Flux. 1990. Rabbits, Hares and Pikas: Status Survey and Conservation Action Plan. World Conservation Union, Gland, Switzerland.
Chapman, J.A., J.G. Hockman, and M.M. Ojeda C. 1980. *Sylvilagus floridanus*. Mammalian Species 136:1–8.
Chapman, J.A., and G.R. Willner. 1978. *Sylvilagus audubonii*. Mammalian Species 106:1–4.
Chesser, R.T., R.C. Banks, F.K. Barker, C. Cicero, J.L. Dunn, A.W. Kratter, I.J. Lovette et al. 2013. Fifty-fourth supplement to the American Ornithologists' Union Checklist of North American Birds. Auk 130:558–571.
Choate, J.R., E.D. Fleharty, and R.J. Little. 1973. Status of the spotted skunk, *Spilogale putorius*, in Kansas. Transactions of the Kansas Academy of Science 76:226–233.
Clark, D.R., Jr. 2001. DDT and the decline of free-tailed bats (*Tadarida brasiliensis*) at Carlsbad Cavern, New Mexico. Archives of Environmental Contaminants and Toxicology 40:537–543.
Clark, T.W. 1970. Richardson's ground squirrel (*Spermophilus richardsonii*) in the Laramie Basin, Wyoming. Great Basin Naturalist 30:55–70.
Clark, T.W. 1971a. Notes on the biology of the thirteen-lined ground squirrel in the Laramie Plains, Wyoming. Southwestern Naturalist 15:499–502.
Clark, T.W. 1971b. Ecology of the western jumping mouse in Grand Teton National Park, Wyoming. Northwest Science 45:229–238.
Clark, T.W. 1973. Local distribution and interspecies interactions in microtines, Grand Teton National Park, Wyoming. Great Basin Naturalist 33:205–217.
Clark, T.W. 1975. Ecological notes on deer mice in Grand Teton National Park, Wyoming. Northwest Science 49:14–16.
Clark, T.W. 1977. Preliminary mammal survey of Fossil Butte National Monument, Wyoming. Great Basin Naturalist 37:91–94.
Clark, T.W., T.M. Campbell, III, M.H. Schroeder, and L. Richardson. 1984. Handbook of Methods for Locating Black-Footed Ferrets. Wildlife Technical Bulletin No. 1. Wyoming Bureau of Land Management, Cheyenne.
Clark, T.W., R.S. Hoffman, and C.F. Nadler. 1971. *Cynomys leucurus*. Mammalian Species 7:1–4.
Clark, T.W., L. Richardson, D. Casey, T.M. Campbell III, and S.C. Forrest. 1984. Seasonality of black-footed ferret diggings and prairie dog burrow plugging. Journal of Wildlife Management 48:1441–1444.
Clark, T.W., V.A. Saab, and D. Casey. 1980. A partial bibliography of Wyoming mammals. Northwest Science 54:55–67.
Clark, T.W., and M.R. Stromberg. 1987. Mammals in Wyoming. Public Education Series No. 10. University of Kansas, Museum of Natural History, Lawrence, KS.
Clarke, L.F. 1955. Jackson Hole Biological Research Station. A.I.B.S. Bulletin 5(2):27–28.
Collen, P., and R.J. Gibson. 2001. The general ecology of beavers (*Castor* spp.), as related to their influence on stream ecosystems and riparian habitats, and the subsequent effects on fish: a review. Reviews in Fish Biology and Fisheries 10:439–461.

Conner, M. M., and T. M. Shenk. 2003. Distinguishing *Zapus hudsonius preblei* from *Zapus princeps princeps* by using repeated cranial measurements. Journal of Mammalogy 84:1456–1463.

Conroy, C. J., and J. A. Cook. 2000. Phylogeography of a post-glacial colonizer: *Microtus longicaudis* (Rodentia: Muridae). Molecular Ecology 9:165–175.

Cope, E. D. 1892. One a new genus of Mammalia from the Laramie Formation. American Naturalist 26:758–762.

Copeland, J. P. 1996. Biology of the wolverine in central Idaho. MS thesis, University of Idaho, Moscow.

Copeland, J. P., J. M. Peek, C. R. Groves, W. E. Melquist, K. S. McKelvey, G. W. McDaniel, C. D. Long, and C. E. Harris. 2007. Seasonal habitat associations of the wolverine in central Idaho. Journal of Wildlife Management 71:2201–2212.

Corbet, G. B. 1978. The Mammals of the Palearctic Region: A Taxonomic Review. Cornell University Press, Ithaca, NY.

Cork, S. J., and G. J. Kenagy. 1989. Nutritional value of hypogeous fungus for a forest-dwelling ground squirrel. Ecology 70:577–586.

Corn, J. G., and M. G. Raphael. 1992. Habitat characteristics at marten subnivean access sites. Journal of Wildlife Management 56:442–448.

Cornely, J. E., L. N. Carraway, and B. J. Verts. 1992. *Sorex preblei*. Mammalian Species 416:1–3.

Côté, S., and M. Festa Bianchet. 2003. Mountain goat. Pp. 1061–1075. In G. A. Feldhamer, B. C. Thompson, and J. A. Chapman (eds). Wild Mammals of North America. 2nd ed. Johns Hopkins University Press, Baltimore, MD.

Cotner, L. A., and R. L. Schooley. 2011. Habitat occupancy by riparian muskrats reveals tolerance to urbanization and invasive vegetation. Journal of Wildlife Management 75:1637–1645.

Craighead, J. J., J. S. Sumner, and J. A. Mitchell. 1995. The Grizzly Bears of Yellowstone: Their Ecology in the Yellowstone Ecosystem, 1959–1992. Island Press, Washington, DC.

Crait, J. R. 2013. The impact of nonnative species on river otters and aquatic–terrestrial linkages in Yellowstone National Park. PhD dissertation, University of Wyoming, Laramie.

Crait, J. R., and M. Ben-David. 2006. River otters in Yellowstone National Park depend on a declining cutthroat trout population. Journal of Mammalogy 87:485–494.

Crait, J. R., and M. Ben-David. 2007. Effects of river otter activity on terrestrial plants in trophically altered Yellowstone Lake. Ecology 88:1040–1052.

Crawford, J. A., R. G. Anthony, J. T. Forbes, and G. A. Lorton. 2010. Survival and causes of mortality for pygmy rabbits (*Brachylagus idahoensis*) in Oregon and Nevada. Journal of Mammalogy 91:838–847.

Cross, P. C., E. K. Cole, A. P. Dobson, W. H. Edwards, K.L. Hamlin, G. Luikart, A.D. Middleton, B.M. Scurlock, and P.J. White. 2010. Probable causes of increasing brucellosis in free-ranging elk of the Greater Yellowstone Ecosystem. Ecological Applications 20:278–288.

Crowe, D. M. 1974. Some aspects of reproduction and population dynamics of bobcats in Wyoming. PhD dissertation, University of Wyoming, Laramie.

Crowe, D. M. 1975. A model for exploited bobcat populations in Wyoming. Journal of Wildlife Management 39:408–415.

Crowe, D. M. 1986. Furbearers of Wyoming. Wyoming Game and Fish Department, Cheyenne.

Crowe, D. M. 1990. The First Century: A Hundred Years of Wildlife Conservation in Wyoming. Wyoming Game and Fish Department in cooperation with Falcon Press, Cheyenne.

Crowe, D. M., and D. Strickland. 1975. Population structures of some mammalian predators in southeastern Wyoming. Journal of Wildlife Management 39:449–450.

Cryan, P.M. 2003. Seasonal distribution of migratory tree bats (*Lasiurus* and *Lasionycteris*) in North America. Journal of Mammalogy 84:579–593.

Cryan, P.M., M.A. Bogan, and J.S. Altenbach. 2000. Effect of elevation on distribution of female bats in the Black Hills, South Dakota. Journal of Mammalogy 81:719–725.

Cryan, P.M., M.A. Bogan, and G.M. Yanega. 2001. Roosting habits of four bat species in the Black Hills of South Dakota. Acta Chiropterologica 3:43–52.

Cudworth, N., S. Johnson, and M. Grenier. 2011. Inventories of forest bats in northeastern Wyoming: mist netting. Pp. 119–145. In M.B. Grenier (ed.). Threatened, Endangered, and Nongame Bird and Mammal Investigations. Unpublished Annual Completion Report, Wyoming Game and Fish Department, Nongame Program, Lander.

Culver, M., W.E. Johnson, J. Pecon-Slattery, and S.J. O'Brien. 2000. Genomic ancestry of the American puma (*Puma concolor*). Journal of Heredity 91:186–197.

Curry, A. 2012. Ancient migration: coming to America. Nature 485:30–32.

Cypher, B.L. 2003. Foxes. Pp. 511–546. In G.A. Feldhamer, B.C. Thompson, and J.A. Chapman (eds). Wild Mammals of North America. 2nd ed. Johns Hopkins University Press, Baltimore, MD.

Davis, C.M., and V.L. Roth. 2008. The evolution of sexual size dimorphism in cottontail rabbits (*Sylvilagus*, Leporidae). Biological Journal of the Linnean Society 95:141–156.

Davis, R.B., C.F. Herreid, II, and H.L. Short. 1962. Mexican free-tailed bats in Texas. Ecological Monographs 32:311–326.

Dawson, N.G., and J.A. Cook. 2012. Behind the genes: diversification of North American martens (*Martes americana* and *M. caurina*). Pp. 23–38. In K.B. Aubry, W.J. Zielinski, M.G. Raphael, G. Proulx, and S.W. Buskirk (eds). Biology and Conservation of Martens, Sables, and Fishers: A New Synthesis. Cornell University Press, Ithaca, NY.

Dearing, M.D. 1997. The function of haypiles of pikas (*Ochotona princeps*). Journal of Mammalogy 78:1156–1163.

Decker, J.E., J.C. Pires, G.C. Conant, S.D. McKay, M.P. Heaton, K. Chen, A. Cooper et al. 2009. Resolving the evolution of extant and extinct ruminants with high-throughput phylogenomics. Proceedings of the National Academy of Sciences 106:18644–18649.

DeMatteo, K.E. 1995. The physiological response of the North American porcupine (*Erethizon dorsatum bruneri*) to decreasing ambient temperature and increasing wind speed. MS thesis, University of Wyoming, Laramie.

Demboski, J.R., and J.A. Cook. 2001. Phylogeography of the dusky shrew, *Sorex monticolus* (Insectivora, Soricidae): insight into deep and shallow history in northwestern North America. Molecular Ecology 10:1227–1240.

Demboski, J.R., and J. Sullivan. 2003. Extensive mtDNA variation within the yellow-pine chipmunk, *Tamias amoenus* (Rodentia: Sciuridae), and phylogeographic inferences for northwest North America. Molecular Phylogenetics and Evolution 26:389–408.

DePerno, C.S., J.A. Jenks, S.L. Griffin, and L.A. Rice. 2000. Female survival rates in a declining white-tailed deer population. Wildlife Society Bulletin 28:1030–1037.

DePerno, C.S., J.A. Jenks, S.L. Griffin, L.A. Rice, and K.F. Higgins. 2002. White-tailed deer habitats in the central Black Hills. Journal of Range Management 55:242–252.

Derge, K.L., and R.H. Yahner. 2000. Ecology of sympatric fox squirrels (*Sciurus niger*) and gray squirrels (*S. carolinensis*) at forest–farmland interfaces of Pennsylvania. American Midland Naturalist 143:355–369.

Devineau, O., T.M. Shenk, G.C. White, P.F. Doherty, Jr., and R.H. Kahn. 2010. Evaluating the Canada lynx reintroduction programme in Colorado: patterns in mortality. Journal of Applied Ecology 47:524–531.

Dewey, S. 2011. Wildlife monitoring: mountain goats. Pp. 20–21. In S. L. Cain (ed.). Wildlife Conservation, Management, and Research 2011. National Park Service, Grand Teton National Park, Moose, WY.
Dibattista, J. D. 2008. Patterns of genetic variation in anthropogenically impacted populations. Conservation Genetics 9:141–156.
Diersing, V. E. 1980. Systematics and evolution of the pygmy shrews (subgenus Microsorex) of North America. Journal of Mammalogy 61:76–101.
Diersing, V. E., and D. F. Hoffmeister. 1977. Revision of the shrews *Sorex merriami* and a description of a new species of the subgenus *Sorex*. Journal of Mammalogy 58:321–333.
Dixon, M. D. 2011. Population genetic structure and natal philopatry in the widespread North American bat *Myotis lucifugus*. Journal of Mammalogy 92:1343–1351.
Dodge, R. I. 1876. The Black Hills: A Minute Description of the Routes, Scenery, Soil, Climate, Timber, Gold, Geology, Zoology, Etc. James Miller Publishers, New York.
Dolbeer, R. A., and W. R. Clark. 1975. Population ecology of snowshoe hares in the central Rocky Mountains. Journal of Wildlife Management 39:535–549.
Donadio, E., and S. W. Buskirk. 2006. Diet, morphology, and interspecific killing in Carnivora. American Naturalist 136:524–536.
Dorn, R. D. 1986. The Wyoming Landscape 1805–1878. Mountain West, Cheyenne, WY.
Douglas, R. B., V. T. Parker, and K. W. Cullings. 2005. Belowground ectomycorrhizal community structure of mature lodgepole pine and mixed conifer stands in Yellowstone National Park. Forest Ecology and Management 208:303–317.
Dragoo, J. W., R. D. Bradley, R. L. Honeycutt, and J. W. Templeton. 1993. Phylogenetic relationships among the skunks: a molecular perspective. Journal of Mammalian Evolution 1:255–267.
Dragoo, J. W., and R. L. Honeycutt. 1997. Systematics of mustelid-like carnivores. Journal of Mammalogy 78:426–443.
Dragoo, J. W., J. A. Lackey, K. E. Moore, E. P. Lessa, J. A. Cook, and T. L. Yates. 2006. Phylogeography of the deer mouse (*Peromyscus maniculatus*) provides a predictive framework for research on hantaviruses. Journal of General Virology 87:1997–2003.
Dubay, S. A. 2000. Mycophagy as a nutritional strategy for small mammals in the Rocky Mountains. PhD dissertation, University of Wyoming, Laramie.
Dubay, S. A., G. D. Hayward, and C. Martinez del Rio. 2008. Nutritional value and diet preference of arboreal lichens and hypogeous fungi for small mammals in the Rocky Mountains. Canadian Journal of Zoology 86:851–862.
Ducummon, S. L. 1999. The North American bats and mines project: a cooperative approach for integrating bat conservation and mine-land reclamation. Proceedings of the 23rd Annual British Columbia Mine Reclamation Symposium, Kamloops, BC.
Duffield, J. W., C. F. Neher, and D. A. Patterson. 2008. Wolf recovery in Yellowstone: park visitor attitudes, expenditures, and economic impacts. George Wright Forum 25(1):13–19.
Dulberger, J., N. T. Hobbs, H. M. Swanson, C. J. Bishop, and M. W. Miller. 2010. Estimating chronic wasting disease effects on mule deer recruitment and population growth. Journal of Wildlife Diseases 46:1086–1095.
Durrant, S. D. 1952. Mammals of Utah. University of Kansas Publications, Museum of Natural History 6:1–549.
Durrant, S. D., and M. R. Lee. 1955. Rare shrews from Utah and Wyoming. Journal of Mammalogy 36:560–561.
Dusek, G. L., R. J. Mackie, J. D. Herriges, Jr., and B. B. Compton. 1989. Population ecology of white-tailed deer along the lower Yellowstone River. Wildlife Monographs 104:1–68.

Dzal, Y., L. P. McGuire, N. Veselka, and M. B. Fenton. 2011. Going, going, gone: the impact of white-nose syndrome on the summer activity of the little brown bat (*Myotis lucifugus*). Biology Letters 7:392–394.

Easterla, D. A., and J. O. Whitaker, Jr. 1972. Food habits of some bats from Big Bend National Park, Texas. Journal of Mammalogy 53:887–890.

Eberhardt, L. L., P. J. White, R. A. Garrott, and D. B. Houston. 2007. A seventy-year history of trends in Yellowstone's northern elk herd. Journal of Wildlife Management 71:594–602.

Eberle, J. J., and D. R. Greenwood. 2012. Life at the top of the greenhouse Eocene world: a review of the Eocene flora and vertebrate fauna from Canada's High Arctic. Geological Society of America Bulletin 124(1–2):3–23.

Edmunds, D. R. 2013. Chronic wasting disease ecology and epidemiology of white-tailed deer in Wyoming. PhD dissertation, University of Wyoming, Laramie.

Edwards, C. W., and R. D. Bradley. 2001. Molecular phylogenetics of the *Neotoma floridana* species group. Journal of Mammalogy 82:791–798.

Edwards, W. C. 1958. A study of the reproductive potential and ovarian structures in the pronghorn antelope, *Antilocapra americana* (Ord.) in Wyoming. MS thesis, University of Wyoming, Laramie.

Egoscue, H. J. 1960. Laboratory and field studies of the northern grasshopper mouse. Journal of Mammalogy 41:99–110.

Elbroch, M. 2003. Mammal Tracks and Sign: A Guide to North American Species. Stackpole Books, Mechanicsburg, PA.

Eldridge, D. J. 2004. Mounds of the American badger (*Taxidea taxus*): significant features of North American shrub-steppe ecosystems. Journal of Mammalogy 85:1060–1067.

Elias, S. A., S. K. Short, C. H. Nelson, and H. H. Birks. 1996. Life and times of the Bering land bridge. Nature 382:60–63.

Elsasser, S. C., and G. H. Parker. 2008. Morphometric criteria for distinguishing species and age-cohorts of ermine (*Mustela erminea*) and long-tailed weasel (*M. frenata*). Acta Zoologica Academiae Scientiarum Hungaricae 54:75–88.

Engstrom, M. D., and J. R. Choate. 1979. Systematics of the northern grasshopper mouse (*Onychomys leucogaster*). Journal of Mammalogy 60:723–739.

Epps, C. W., P. J. Palsbøll, J. D. Wehausen, G. K. Roderick, R. R. Ramey II, and D. R. McCullough. 2005. Highways block gene flow and cause a rapid decline in genetic diversity of desert bighorn sheep. Ecology Letters 8:1029–1038.

Eshelman, B. D., and C. S. Sonnemann. 2000. *Spermophilus armatus*. Mammalian Species 637:1–6.

Estes, J. A., J. Terborgh, J. S. Brashares, M. E. Power, J. Berger, W.J. Bond, S.R. Carpenter et al. 2011. Trophic downgrading of planet Earth. Science 333:301–306.

Estes-Zumpf, W. A., and J. L. Rachlow. 2009. Natal dispersal by the pygmy rabbit (*Brachylagus idahoensis*). Journal of Mammalogy 90:363–372.

Estes-Zumpf, W. A., S. E. Zumpf, J. L. Rachlow, J. R. Adams, and L.P. Waits. 2014. Genetic evidence confirms the presence of pygmy rabbits in Colorado. Journal of Fish and Wildlife Management 5:118–123.

Fagerstone, K. A., G. K. Lavoie, and R. E. Griffith, Jr. 1980. Black-tailed jackrabbit diet and density on rangeland and near agricultural crops. Journal of Range Management 33:229–233.

Fahnestock, J. T., and J. K. Detling. 2002. Bison–prairie dog–plant interactions in a North American mixed-grass prairie. Oecologia 132:86–95.

Farentinos, R. C. 1972. Observations on the ecology of the tassel-eared squirrel. Journal of Wildlife Management 36:1234–1239.

Fauteaux, D., L. Imbeau, P. Drapeau, and M. J. Mazerolle. 2012. Small mammal responses to coarse woody debris distribution at different spatial scales in managed and unmanaged boreal forests. Forest Ecology and Management 266:194–205.
Fedy, B. C., and K. E. Doherty. 2011. Population cycles are highly correlated over long time series and large spatial scales in two unrelated species: greater sage-grouse and cottontail rabbits. Oecologia 165:915–924.
Feldhamer, G. A. 1979. Vegetative and edaphic factors affecting abundance and distribution of small mammals in southeast Oregon. Great Basin Naturalist 39:207–218.
Felicetti, L. A., C. C. Schwartz, R. O. Rye, M. A. Haroldson, K.A. Gunther, D.L. Phillips, and C.T. Robbins 2003. Use of sulfur and nitrogen stable isotopes to determine the importance of whitebark pine nuts to Yellowstone grizzly bears. Canadian Journal of Zoology 81:763–770.
Fenton, M. B., and R. M. R. Barclay. 1980. *Myotis lucifugus*. Mammalian Species 142:1–8.
Ferner, J. W. 1974. Habitat relationships of *Tamiasciurus hudsonicus* and *Sciurus aberti* in the Rocky Mountains. Southwestern Naturalist 18:470–473.
Fichter, E., and R. Williams. 1967. Distribution and status of the red fox in Idaho. Journal of Mammalogy 48:219–230.
Fiedel, S. 2009. Sudden deaths: the chronology of terminal Pleistocene megafaunal extinction. Pp. 21–37. In G. Haynes (ed.). American Megafaunal Extinctions at the End of the Pleistocene. Springer Science + Business Media, New York.
Findley, J. S., and S. Anderson. 1956. Zoogeography of the montane mammals of Colorado. Journal of Mammalogy 37:80–82.
Findley, J. S., C. Jones, H. H. Genoways, E. C. Birney, and R. J. Baker. 1996. Obituary: J. Knox Jones, Jr., 1929–1992. Mammalogy Papers. University of Nebraska State Museum, Lincoln.
Finley, R. B., Jr. 1990. Woodrat ecology and behavior and the interpretation of paleomiddens. Pp. 28–42. In J. L. Betancourt, T. R. Van Devender, and P. S. Martin (eds). 1990. Packrat Middens: The Last 40,000 Years of Biotic Change. University of Arizona Press, Tucson.
Finley, R. B., Jr. 1995. The spread of raccoons (*Procyon lotor hirtus*) into the Colorado Plateau from the northern Great Plains. Proceedings of the Denver Museum of Natural History, Series 3, 11:1–6.
Fitzpatrick, J. W., M. Lammertink, M. D. Luneau, T. W. Gallagher, B.R. Harrison, G.M. Sparling, K.V. Rosenberg et al. 2005. Ivory-billed woodpecker (*Campephilus principalis*) persists in continental North America. Science 308:1460–1462.
FitzSimmons, N. N., S. W. Buskirk, and M. H. Smith. 1997. Genetic changes in reintroduced Rocky Mountain bighorn sheep populations. Journal of Wildlife Management 61:863–872.
Flake, L. D. 1973. Food habits of four species of rodents on a short-grass prairie in Colorado. Journal of Mammalogy 54:636–647.
Flinders, J. T., and R. M. Hansen. 1973. Abundance and dispersion of leporids within a shortgrass ecosystem. Journal of Mammalogy 54:287–291.
Flinders, J. T., and R. M. Hansen. 1975. Spring population responses of cottontails and jackrabbits to cattle grazing shortgrass prairie. Journal of Range Management 28:290–293.
Floyd, C. H. 2004. Marmot distribution and habitat associations in the Great Basin. Western North American Naturalist 64:471–481.
Flynn, R. W., and T. V. Schumacher. 2009. Temporal changes in population dynamics of American martens. Journal of Wildlife Management 73:1269–1281.
Forbes, R. B. 1964. Some aspects of the life history of the silky pocket mouse, *Perognathus flavus*. American Midland Naturalist 72:438–443.

Foresman, K.R. 2012. Mammals of Montana. 2nd ed. Mountain Press, Missoula, MT.

Foresman, K.R., and R.E. Jensen, Jr. 1992. A morphometric comparison of the masked shrew (*Sorex cinereus*) and Hayden's shrew (*Sorex haydeni*) in Montana and bordering states. Northwestern Naturalist 73:15–21.

Foreyt, W.J., and D.A. Jessup. 1982. Fatal pneumonia of bighorn sheep following association with domestic sheep. Journal of Wildlife Diseases 18:163–168.

Forrest, S.C., D.E. Biggins, L. Richardson, T.W. Clark, T.M. Campbell, III, K.A. Fagerstone, and E.T. Thorne. 1988. Population attributes for the black-footed ferret (*Mustela nigripes*) at Meeteetse, Wyoming, 1981–1985. Journal of Mammalogy 69:261–273.

Frank, J.L., S. Barry, and D. Southworth. 2006. Mammal mycophagy and dispersal of mycorrhizal inoculum in Oregon white oak woodlands. Northwest Science 80:264–273.

Franklin, L.S. 2004. Adolph Murie: Denali's wilderness conscience. MS thesis, University of Alaska, Fairbanks.

Franzman, A.W. 1981. *Alces alces*. Mammalian Species 154:1–7.

Frase, B.A., and R.S. Hoffman. 1980. *Marmota flaviventris*. Mammalian Species 135:1–8.

French, S.P., M.G. French, and R.R. Knight. 1994. Grizzly bear use of army cutworm moths in the Yellowstone ecosystem. Bears: Their Biology and Management 9:14–24.

Frey, J.K. 2009. Distinguishing range expansions from previously undocumented populations using background data from museum records. Diversity and Distributions 15:183–187.

Frey, J.K. 2013. Re-evaluation of the evidence for the importation of red foxes from Europe to colonial America: origins of the southeastern red fox (*Vulpes vulpes fulva*). Biological Conservation 158:74–79.

Frick, W.F., J.F. Pollock, A.C. Hicks, K.E. Langwig, D.S. Reynolds, G.G. Turner, C.M. Butchkoski, and T.H. Kunz. 2010. An emerging disease causes regional population collapse of a common North American bat species. Science 329:679–682.

Fritts, S.H., E.E. Bangs, J.A. Fontaine, W.G. Brewster, and J.F. Gore. 1995. Restoring wolves to the northern Rocky Mountains of the United States. Pp. 107–126. In L.N. Carbyn, S.H. Fritts, and D.R. Seip (eds). Ecology and Conservation of Wolves in a Changing World. Canadian Circumpolar Institute, Occasional Publication 35. Edmonton, AB.

Fritzell, E.K., and K.J. Haroldson. 1982. *Urocyon cinereoargenteus*. Mammalian Species 189:1–8.

Frost, D.R., and R.M. Timm. 1992. Phylogeny of plecotine bats (Chiroptera: "Vespertilionidae"): summary of the evidence and proposal of a logically consistent taxonomy. American Museum Novitates 3034:1–16.

Fryxell, F.M. 1926. A new high altitude limit for the American bison. Journal of Mammalogy 7:102–109.

Fryxell, F.M. 1928. The former range of the bison in the Rocky Mountains. Journal of Mammalogy 9:129–139.

Fujita, M.S., and T.H. Kunz. 1984. *Pipistrellus subflavus*. Mammalian Species 228:1–6.

Fuller, J.A., R.A. Garrott, P.J. White, K.E. Aune, T.J. Roffe, and J.C. Rhyan. 2007. Reproduction and survival of Yellowstone bison. Journal of Wildlife Management 71:2365–2372.

Fuzessery, Z.M., P. Buttenhoff, B. Andrews, and J.M. Kennedy. 1993. Passive sound localization of prey by the pallid bat (*Antrozous p. pallidus*). Journal of Comparative Physiology A 171:767–777.

Galvin, B.R., and R.E. Richards. 1993. Distribution update for Colorado ringtails (Carnivora: *Bassariscus astutus*). Journal of the Colorado-Wyoming Academy of Science 25:30.

Gardner, A.L., and M.E. Sunquist. 2003. Opossum. Pp. 3–29. In G.A. Feldhamer, B.C. Thompson, and J.A. Chapman (eds). Wild Mammals of North America. 2nd ed. Johns Hopkins University Press, Baltimore, MD.

Garrison, T. E., and T. L. Best. 1990. *Dipodomys ordii*. Mammalian Species 353:1–10.
Garrott, R. A., G. C. White, R. M. Bartmann, L. H. Carpenter, and A.W. Alldredge. 1987. Movements of female mule deer in northwest Colorado. Journal of Wildlife Management 51:634–643.
Geluso, K. 2008. Winter activity of Brazilian free-tailed bats (*Tadarida brasiliensis*) at Carlsbad Cavern, New Mexico. Southwestern Naturalist 53:243–272.
Geluso, K., J. J. Heubschman, and K. N. Geluso. 2013. Bats of the Wildcat Hills and surrounding areas in western Nebraska. Monographs of the Western North American Naturalist 6:20–42.
Geluso, K., J. J. Huebschman, J. A. White, and M. A. Bogan. 2004. Reproduction and seasonal activity of silver-haired bats (*Lasionycteris noctivagans*) in western Nebraska. Western North American Naturalist 64:353–358.
Geluso, K., T. R. Mollhagen, J. M. Tigner, and M. A. Bogan. 2005. Westward expansion of the eastern pipistrelle (*Pipistrellus subflavus*) in the United States, including new records from New Mexico, South Dakota, and Texas. Western North American Naturalist 65:405–409.
Geluso, K., J. A. White, and M. A. Bogan. 2004. Least shrew (*Cryptotis parva*) in extreme west-central Nebraska. Transactions of the Nebraska Academy of Sciences 29:29–32.
Geluso, K., and G. D. Wright. 2010. Hispid pocket mouse (*Chaetodipus hispidus*) in east-central and northeastern Nebraska. Western North American Naturalist 70:126–129.
Geluso, K. N., J. S. Altenbach, and D. E. Wilson. 1976. Bat mortality: pesticide poisoning and migratory stress. Science 194:184–186.
Genoways, H. H., P. W. Freeman, and C. Grell. 2000. Extralimital records of the Mexican free-tailed bat (*Tadarida brasiliensis mexicana*) in the central United States and their biological significance. Transactions of the Nebraska Academy of Sciences 26:85–96.
Genoways, H. H., M. J. Hamilton, D. M. Bell, R. R. Chambers, and R. D. Bradley. 2008. Hybrid zones, genetic isolation, and systematics of pocket gophers (Genus *Geomys*) in Nebraska. Journal of Mammalogy 89:826–836.
Gese, E. M., R. L. Ruff, and R. L. Crabtree. 1996. Foraging ecology of coyotes (*Canis latrans*): the influence of extrinsic factors and a dominance hierarchy. Canadian Journal of Zoology 74:769–783.
Getz, L. L., C. S. Carter, and L. Gavish. 1981. The mating system of the prairie vole, *Microtus ochrogaster*: field and laboratory evidence for pair-bonding. Behavioral Ecology and Sociobiology 8:189–194.
Gier, H. T. 1968. Coyotes in Kansas. Kansas State College Agricultural Experiment Station Bulletin 393:1–118.
Gilbert, B. M., and L. D. Martin. 1984. Late Pleistocene fossils of Natural Trap Cave, Wyoming, and the climatic model of extinction. Pp. 138–147. In P. S. Martin and R. G. Klein (eds). Quaternary Extinctions: A Prehistoric Revolution. University of Arizona Press, Tucson.
Gillihan, S. W., and K. R. Foresman. 2004. *Sorex vagrans*. Mammalian Species 744:1–5.
Gingerich, P. D. 2003. Mammalian responses to climate change at the Paleocene–Eocene boundary: Polecat Bench record in the northern Bighorn Basin, Wyoming. Geological Society of America Special Paper 369:463–478.
Gingerich, P. D. 2006. Environment and evolution through the Paleocene–Eocene thermal maximum. Trends in Ecology and Evolution 21:246–253.
Goad, E. H., L. Pejchar, S. E. Reed, and R. L. Knight. 2014. Habitat use by mammals varies along an exurban development gradient in northern Colorado. Biological Conservation 176:172–182.

Goin, O.B. 1944. A new race of the canyon mouse. Journal of Mammalogy 25:189–191.
Gompper, M.E., and H.M. Hackett. 2005. The long-term, range-wide decline of a once common carnivore: the eastern spotted skunk (*Spilogale putorius*). Animal Conservation 8:195–201.
Good, J.M., S. Hird, N. Reid, J.R. Demboski, S.J. Steppan, T.R. Martin-Nims, and J. Sullivan. 2008. Ancient hybridization and mitochondrial capture between two species of chipmunks. Molecular Ecology 17:1313–1327.
Goodrich, J.M., and S.W. Buskirk. 1998. Spacing and ecology of North American badgers (*Taxidea taxus*) in a prairie-dog (*Cynomys leucurus*) complex. Journal of Mammalogy 79:171–179.
Goodwin, G.A. 1975. Seasonal food habits of mule deer in southeastern Wyoming. USDA Forest service, Research Note RM-287.
Gordus, A.G., and J.H. Theis. 1993. Isolation of *Borrelia burgdorferi* from the blood of a bushy-tailed wood rat in California. Journal of Wildlife Diseases 29:478–480.
Grant, T.A., E.M. Madden, T.L. Shaffer, P.J. Pietz, G.B. Berkey, and N.J. Kadrmas. 2006. Nest survival of clay-colored and vesper sparrows in relation to woodland edge in mixed-grass prairies. Journal of Wildlife Management 70:691–701.
Grant, W.E., N.R. French, and L.J. Flose, Jr. 1980. Effects of pocket gopher mounds on plant production in shortgrass prairie ecosystems. Southwestern Naturalist 25:215–224.
Grayson, D.K. 1977. On the Holocene history of some northern Great Basin lagomorphs. Journal of Mammalogy 58:507–513.
Grayson, D.K. 1984. Nineteenth-century explanations of Pleistocene extinctions: a review and analysis. PP. 5–39. In P.S. Martin and R.G. Klein (eds). Quaternary Extinctions: A Prehistoric Revolution. University of Arizona Press, Tucson.
Grayson, D.K. 2005. A brief history of Great Basin pikas. Journal of Biogeography 32:2103–2111.
Grayson, D.K. 2006. The Late Quaternary biogeographic histories of some Great Basin mammals (western USA). Quaternary Science Reviews 25:2964–2991.
Green, G.I., D.J. Mattson, and J.M. Peek. 1997. Spring feeding on ungulate carcasses by grizzly bears in Yellowstone National Park. Journal of Wildlife Management 61:1040–1055.
Green, J.S., and J.T. Flinders. 1980. *Brachylagus idahoensis*. Mammalian Species 125:1–4.
Grenier, M.B. 2008. Population biology of the black-footed ferret reintroduced into Shirley Basin, Wyoming. MS thesis, University of Wyoming, Laramie.
Grenier, M.B., D.B., McDonald, and S.W. Buskirk. 2007. Rapid population growth of a critically endangered carnivore. Science 317:779.
Grinnell, J. 1915. The Tennessee opossum has arrived in California. California Fish and Game 1:114–116
Grogan, R.G. 1997. Black bear ecology in southeast Wyoming: the Snowy Range. MS thesis, University of Wyoming, Laramie.
Grogan, R.G., and F.G. Lindzey. 1999. Estimating population size of a low-density black bear population using capture-resight. Ursus 11:117–122.
Gross, J.E., L.C. Stoddart, and F.H. Wagner. 1974. Demographic analysis of a northern Utah jackrabbit population. Wildlife Monographs 40:3–68.
Groves, C., and P. Grubb. 2011. Ungulate Taxonomy. Johns Hopkins University Press, Baltimore, MD.
Gruver, J.C. 2002. Assessment of bat community structure and roosting habitat preferences for the hoary bat (*Lasiurus cinereus*) near Foote Creek Rim, Wyoming. MS thesis, University of Wyoming, Laramie.
Gruver, J.C., and D.A. Keinath. 2006. Townsend's Big-Eared Bat (*Corynorhinus townsendii*): A Technical Conservation Assessment. Prepared for the USDA Forest Service, Rocky Mountain Region, Species Conservation Project.

Guenzel, R. J. 1986. Pronghorn ecology in southcentral Wyoming. MS thesis, University of Wyoming, Laramie.
Gunnell, G. F., and N. B. Simmons. 2005. Fossil evidence and the origin of bats. Journal of Mammalian Evolution 12:209–246.
Gunther, K. A., M. J. Biel, N. Anderson, and L. Waits. 2002. Probable grizzly bear predation on an American black bear in Yellowstone National Park. Ursus 13:372–374.
Gunther, K. A., R. A. Renkin, J. C. Halfpenny, S. M. Gunther, T. Davis, P. Schullery, and L. Whittlesey 2009. Presence and distribution of white-tailed jackrabbits in Yellowstone National Park. Yellowstone Science 17(1):24–32.
Gusztak, R. W., R. A. MacArthur, and K. L. Campbell. 2005. Bioenergetics and thermal physiology of American water shrews (*Sorex palustris*). Journal of Comparative Physiology B 175:87–95.
Guthrie, R. D. 1984. Mosaics, allelochemics, and nutrients. Pp. 259–298. In P. S. Martin and R. G. Klein (eds). Quaternary Extinctions: A Prehistoric Revolution. University of Arizona Press, Tucson.
Hafner, D. J., and A. T. Smith. 2010. Revision of the subspecies of the American pika, *Ochotona princeps* (Lagomorpha: Ochotonidae). Journal of Mammalogy 91:401–417.
Hafner, J. C., J. E. Light, D. J. Hafner, M. S. Hafner, E. Reddington, D.S. Rogers, and B.R. Riddle. 2007. Basal clades and molecular systematics of heteromyid rodents. Journal of Mammalogy 88:1129–1145.
Hager, M. W. 1972. A Late Wisconsin-Recent vertebrate fauna from the Chimney Rock animal trap, Larimer County, Colorado. Contributions to Geology, University of Wyoming 11:63–71.
Hagmeier, E. M. 1958. Inapplicability of the subspecies concept to North American marten. Systematic Zoology 7:1–7.
Hagmeier, E. M. 1961. Variation and relationships in North American marten. Canadian Field-Naturalist 75:122–138.
Halbert, N. D., and J. N. Derr. 2007. A comprehensive evaluation of cattle introgression into US federal bison herds. Journal of Heredity 98:1–12.
Hall, E. R. 1943. Intergradation versus hybridization in ground squirrels of the western United States. American Midland Naturalist 29:375–378.
Hall, E. R. 1951. A synopsis of North American Lagomorpha. University of Kansas Publications, Museum of Natural History 5:119–202.
Hall, E. R. 1981. The Mammals of North America. John Wiley, New York.
Hall, E. R., and D. F. Hoffmeister. 1942. Geographic variation in the canyon mouse, *Peromyscus crinitus*. Journal of Mammalogy 23:51–65.
Hall, E. R., and K. R. Kelson. 1959. The Mammals of North America. Ronald Press, New York.
Halloran, A. F., and W. E. Blanchard. 1959. Lynx from western Wyoming. Journal of Mammalogy 40:450–451.
Hambäck, P. A., L. Okansen, P. Ekerholm, A. Lindgren, T. Oksanen, and M. Schneider. 2004. Predators indirectly protect tundra plants by reducing herbivore abundance. Oikos 106:85–92.
Hamel, S., S. D. Côté, D. G. Smith, and M. Festa-Bianchet. 2006. Population dynamics and harvest potential of mountain goat herds in Alberta. Journal of Wildlife Management 70:1044–1053.
Hancock, D. C., and D. J. Nash. 1982. Dorsal hair length and coat color in Abert's squirrel. Great Basin Naturalist 42:597–598.
Hansen, L. A., N. E. Mathews, B. A. Vander Lee, and R. S. Lutz. 2004. Population characteristics, survival rates, and causes of mortality of striped skunks (*Mephitis mephitis*) on the southern high plains, Texas. Southwestern Naturalist 49:54–60.

Harding, L. E., M. L. Harris, and J. E. Elliot. 1998. Heavy and trace metals in wild mink (*Mustela vison*) and river otter (*Lontra canadensis*) captured on rivers receiving metals discharges. Bulletin of Environmental Contamination and Toxicology 61:600–607.

Harding, L. E., and F. A. Smith. 2009. *Mustela* or *Vison*? Evidence for the taxonomic status of the American mink and a distinct biogeographic radiation of American weasels. Molecular Phylogenetics and Evolution 52:632–642.

Harlow, H. J. 1981. Torpor and other physiological adaptations of the badger (*Taxidea taxus*) to cold environments. Physiological Zoology 54:267–275.

Harlow, H. J., and G. E. Menkens. 1986. A comparison of hibernation in the black-tailed prairie dog, white-tailed prairie dog, and Wyoming ground squirrel. Canadian Journal of Zoology 64:793–796.

Haroldson, M. A., K. A. Gunther, D. P. Reinhart, S. R. Podruzny, C. Cegelski, L. Waits, T. Wyman, and J. Smith. 2005. Changing numbers of spawning cutthroat trout in tributary streams of Yellowstone Lake and estimates of grizzly bears visiting streams from DNA. Ursus 16:167–180.

Haroldson, M. A., C. G. Schwartz, and G. C. White. 2006. Survival of independent grizzly bears in the Greater Yellowstone Ecosystem, 1983–2001. Wildlife Monographs 161:33–43.

Harris, R. B., W. A. Wall, and F. W. Allendorf. 2002. Genetic consequences of hunting: what do we know and what should we do? Wildlife Society Bulletin 30:634–643.

Harrison, R. G., S. M. Bogdanowicz, R. S. Hoffmann, E. Yensen, and P. W. Sherman. 2003. Phylogeny and evolutionary history of the ground squirrels (Rodentia: Marmotinae). Journal of Mammalian Evolution 10:249–276.

Hart, E. B. 1982. The raccoon, *Procyon lotor*, in Wyoming. Great Basin Naturalist 42:599–600.

Hart, E. B. 1992. *Tamias dorsalis*. Mammalian Species 399:1–6.

Hart, E. B., M. C. Belk, E. Jordan, and M. W. Gonzalez. 2004. *Zapus princeps*. Mammalian Species 749:1–7.

Harvey, M. W. T. 1994. A Symbol of Wilderness: Echo Park and the American Conservation Movement. University of New Mexico Press, Albuquerque.

Hasenyager, R. N. 1980. Bats of Utah. Publication of the Utah Division of Wildlife Resources 80–15:1–109.

Haynes, C. V. 1969. The earliest Americans. Science 166:709–715.

Hayward, G. D., S. H. Henry, and L. F. Ruggiero. 1999. Response of red-backed voles to recent patch cutting in subalpine forest. Conservation Biology 13:168–176.

Hebblewhite, M., and D. W. Smith. 2010. Wolf community ecology: ecosystem effects of recovering wolves in Banff and Yellowstone National Parks. Pp. 69–120. In M. Musiani, L. Boitani, and P. Paquet (eds). The World of Wolves: New Perspectives on Ecology, Behaviour and Management. University of Calgary Press, Calgary, BC.

Hedrick, P. W. 2009. Conservation genetics and North American bison (*Bison bison*). Journal of Heredity 100:411–420.

Helgen, K. M., F. R. Cole, L. E. Helgen, and D. E. Wilson. 2009. Generic revision in the Holarctic ground squirrel genus *Spermophilus*. Journal of Mammalogy 90:270–305.

Henderson, C. B. 1990. The influence of seed apparency, nutrient content and chemical defenses on dietary preference in *Dipodomys ordii*. Oecologia 82:333–341.

Henderson, L. E., L. J. Farrow, and H. G. Broders. 2008. Intra-specific effects of forest loss on the distribution of the forest-dependent northern long-eared bat (*Myotis septentrionalis*). Biological Conservation 141:1819–1828.

Henke, S. E., and F. C. Bryant. 1999. Effects of coyote removal on the faunal community in western Texas. Journal of Wildlife Management 63:1066–1081.

Hennings, D., and R. S. Hoffman. 1977. A review of the taxonomy of the *Sorex vagrans* species complex from western North America. Occasional Papers of the Museum of Natural History, University of Kansas 68:1–35.
Henningsen, J. C., A. L. Williams, C. M. Tate, S. A. Kilpatrick, S.A. Kilpatrick, and W.D. Walter. 2012. Distribution and prevalence of *Eleaophora schneideri* in moose in Wyoming. Alces 48:35–44.
Henterly, A. C., K. E. Mabry, N. G. Solomon, and A. S. Chesh. 2011. Comparison of morphological versus molecular characters for discriminating between sympatric meadow and prairie voles. American Midland Naturalist 165:412–420.
Hermanson, J. W., and T. J. O'Shea. 1983. *Antrozous pallidus.* Mammalian Species 213:1–8.
Hester, S. G., and M. B. Grenier. 2005. A Conservation Plan for Bats in Wyoming. Unpublished Report. Wyoming Game and Fish Department, Nongame Program, Lander.
Hibbard, E. A. 1956. Range and spread of the gray and the fox squirrels in North Dakota. Journal of Mammalogy 37:525–531.
Hickey, J. R., R. W. Flynn, S. W. Buskirk, and M. Willson. 1999. An evaluation of a mammalian predator, the American marten, *Martes americana,* as a disperser of seeds. Oikos 87:499–508.
Higgins, K. F., E. D. Stukel, J. M. Goulet, and D. C. Backlund. 2000. Wild Mammals of South Dakota. South Dakota Department of Game, Fish and Parks, Pierre.
Higgins, L. C., and P. Stapp. 1997. Abundance of thirteen-lined ground squirrels in shortgrass prairie. Prairie Naturalist 29:25–37.
Hill, E. P. 1982. Beaver. Pp. 256–281. In J. A. Chapman and G. A. Feldhamer (eds). Wild Mammals of North America: Biology, Management, and Economics. 1st ed. Johns Hopkins University Press, Baltimore, MD.
Hill, R. R. 1946. Palatability ratings of black hills plants for white-tailed deer. Journal of Wildlife Management 10:47–54.
Hillman, C. N., and T. W. Clark. 1980. *Mustela nigripes.* Mammalian Species 126:1–3.
Himes, C. M., and G. J. Kenagy. 2010. Influence of montane isolation and refugia on population structure of *Sorex palustris* in western North America. Journal of Mammalogy 91:1000–1010.
Hoak, J. H., J. L. Weaver, and T. W. Clark. 1982. Wolverines in western Wyoming. Northwest Science 56:159–161.
Hobbs, N. T., D. L. Baker, G. D. Bear, and D. C. Bowden. 1996. Ungulate grazing in sagebrush grassland: mechanisms of resource competition. Ecological Applications 5:200–217.
Hodges, K. E. 2000a. Ecology of snowshoe hares in southern boreal and montane forests. Pp. 163–206. In L. F. Ruggiero, K. B. Aubry, S. W. Buskirk, G. M. Koehler, C. J. Krebs, K. S. McKelvey, and J. R. Squires (eds). Ecology and Conservation of Lynx in the United States. University Press of Colorado, Boulder.
Hodges, K. E. 2000b. The ecology of snowshoe hares in northern boreal forests. Pp. 117–161. In L. F. Ruggiero, K. B. Aubry, S. W. Buskirk, G. M. Koehler, C. J. Krebs, K. S. McKelvey, and J. R. Squires. 2000. Ecology and Conservation of Lynx in the United States. University Press of Colorado, Boulder.
Hodges, K. E., L. S. Mills, and K. M. Murphy. 2009. Distribution and abundance of snowshoe hares in Yellowstone National Park. Journal of Mammalogy 90:870–878.
Hoff, D. J. 1998. Integrated laboratory and field investigations assessing contaminant risk to American badgers (*Taxidea taxus*) on the Rocky Mountain Arsenal National Wildlife Refuge. PhD dissertation, Clemson University, Clemson, SC.

Hoffman, J. D., and H. H. Genoways. 2008. Characterization of a contact zone between two subspecies of the big brown bat (*Eptesicus fuscus*) in Nebraska. Western North American Naturalist 68:36–45.

Hoffman, R. S. 1974. Terrestrial vertebrates. Pp. 475–568. In J. D. Ives and R. G. Barry (eds). Arctic and Alpine Environments. Methuen, London, UK.

Hoffman, R. S., and R. D. Fisher. 1978. Additional distributional records of Preble's shrew (*Sorex preblei*). Journal of Mammalogy 59:883–884.

Hoffman, R. S., and J. G. Owen. 1980. *Sorex tenellus* and *Sorex nanus*. Mammalian Species 131:1–4.

Hoffman, R. S., and R. D. Taber. 1960. Notes on *Sorex* in the northern Rocky Mountain alpine zone. Journal of Mammalogy 41:230–234.

Hoffmeister, D. F. 1951. A taxonomic and evolutionary study of the piñon mouse, *Peromyscus truei*. Illinois Biological Monographs 21(4):1–104.

Hoffmeister, D. F. 1981. *Peromyscus truei*. Mammalian Species 161:1–5.

Hoffmeister, D. F. 1986. Mammals of Arizona. University of Arizona Press, Tucson.

Hoffmeister, D. F., and V. E. Diersing. 1978. Review of the tassel-eared squirrels of the subgenus *Otosciurus*. *Journal of Mammalogy* 59:402–413.

Hollander, J. L., and S. B. Vander Wall. 2004. Effectiveness of six species of rodents as dispersers of singleleaf piñon pine (*Pinus monophylla*). Oecologia 138:57–65.

Holloway, G. L., and R. M. R. Barclay. 2001. *Myotis ciliolabrum*. Mammalian Species 670:1–5.

Holm, G. W., F. G. Lindzey, and D. S. Moody. 1999. Interactions of sympatric black and grizzly bears in northwest Wyoming. Ursus 11:99–108.

Holmes, B. E., and B. E. Neubaum. 2013. First record of the northern flying squirrel (*Glaucomys sabrinus*) in Colorado. Northwestern Naturalist 94:84–87.

Homyack, J. A., J. H. Vashon, C. Libby, E. L. Lindquist, S. Loch, D. F. McAlpine, and K. L. Pilgrim. 2008. Canada lynx-bobcat (*Lynx canadensis* × *L. rufus*) hybrids at the southern periphery of lynx range in Maine, Minnesota and New Brunswick. American Midland Naturalist 159:504–508.

Honess, R. F., and N. M. Frost. 1942. A Wyoming bighorn sheep study. Wyoming Game and Fish Department Bulletin 1:1–125.

Hoofer, S. R., R. A. Van Den Bussche, and I. Horacek. 2006. Generic status of the American pipistrelles (Vespertilionidae) with description of a new genus. Journal of Mammalogy 87:981–992.

Hoogland, J. L. 1995. The Black-Tailed Prairie Dog: Social Life of a Burrowing Mammal. University of Chicago Press, Chicago, IL.

Hoogland, J. L. 1996. *Cynomys ludovicianus*. Mammalian Species 535:1–10.

Hoogland, J. L. 2003. Sexual dimorphism of prairie dogs. Journal of Mammalogy 84:1254–1266.

Hooven, E. F. 1973. Notes on the water vole in Oregon. Journal of Mammalogy 54:751–752.

Hope, A. G., N. Panter, J. A. Cook, S. L. Talbot, and D.W. Nagorsen. 2014. Multilocus phylogeography and systematic revision of North American water shrews (genus: *Sorex*). Journal of Mammalogy 95:722–738.

Hope, A. G., K. A. Speer, J. R. Demboski, S. L. Talbot, and J. A. Cook. 2012. A climate for speciation: rapid spatial diversification with the *Sorex cinereus* complex of shrews. Molecular Phylogenetics and Evolution 64:671–684.

Hornaday, W. T. 1887. The extermination of the American bison: with a sketch of its discovery and life history. US National Museum Annual Report, 1887. Government Printing Office, Washington, DC.

House, W. A. 1964. Food habits of Richardson's ground squirrel in south-central Wyoming. MS thesis, Colorado State University, Fort Collins.

Houston, D. B. 1967. The Shiras moose in Jackson Hole, Wyoming. PhD dissertation, University of Wyoming, Laramie.
Houston, D. B. 1982. The Northern Yellowstone Elk: Ecology and Management. Macmillan, New York.
Houston, D. B., C. T. Robbins, and V. Stevens. 1989. Growth in wild and captive mountain goats. Journal of Mammalogy 70:412–416.
Howe, H. F., B. Zorn-Arnold, A. Sullivan, and J. S. Brown. 2006. Massive and distinctive effects of meadow voles on grassland vegetation. Ecology 87:3007–3013.
Howell, A. H. 1918. Revision of the American flying squirrels. North American Fauna 44:1–64.
Howell, A. H. 1938. Revision of the North American ground squirrels, with a classification of the North American Sciuridae. North American Fauna 56:1–256.
Huchon, D., and E. J. P. Douzery. 2001. From the Old World to the New World: a molecular chronicle of the phylogeny and biogeography of hystricognath rodents. Molecular Phylogenetics and Evolution 20:238–251.
Hultén, E. 1968. Flora of Alaska and Neighboring Territories. Stanford University Press, Stanford, CA.
Hurley, K. 2011. Wyoming's bighorn sheep transplant program: 77 years and counting. Wyoming Outdoors Fall:44–45.
Hurley, M. A., J. W. Unsworth, P. Zager, M. Hebblewhite, E.O. Garten, D.M. Montgomery, J.R. Skalski, and C.L. Maycock. 2010. Demographic response of mule deer to experimental reduction of coyotes and mountain lions in southeastern Idaho. Wildlife Monograph 178:1–33.
Inman, R. M., K. H. Inman, M. L. Packila, and A. J. McCue. 2007. Wolverine reproductive rates and maternal habitat in Greater Yellowstone. Pp. 65–84. In Greater Yellowstone Wolverine Study, Cumulative Report, May 2007. Wildlife Conservation Society, North America Program, General Technical Report, Bozeman, MT.
Inman, R. M., M. L. Packila, K. H. Inman, B. Aber, R. Spence, and D. McCauley. 2009. Greater Yellowstone Wolverine Program, Progress Report, December 2009. Wildlife Conservation Society, North America Program, General Report, Bozeman, MT.
Inman, R. M., M. L. Packila, K. H. Inman, A. J. McCue, G.C. White, J. Persson, B.C. Aber et al. 2011. Spatial ecology of wolverines at the southern periphery of distribution. Journal of Wildlife Management 76:778–792.
Inman, R. M., R. R. Wigglesworth, K. H. Inman, M. K. Schwartz, B. L. Brock, and J. D. Rieck. 2004. Wolverine makes extensive movements in the Greater Yellowstone Ecosystem. Northwest Science 78:261–266.
Inouye, D. W., B. Barr, K. B. Armitage, and B. D. Inouye. 2000. Climate change is affecting altitudinal migrants and hibernating species. Proceedings of the National Academy of Sciences 97:1630–1633.
Irwin, L. L., and F. M. Hammond. 1985. Managing black bear habitats for food items in Wyoming. Wildlife Society Bulletin 13:477–483.
Ivan, J. S., G. C. White, and T. M. Shenk. 2014. Density and demography of snowshoe hares in central Colorado. Journal of Wildlife Management 78:580–594.
Jacoby, M. E., G. V. Hilderbrand, C. Servheen, C. C. Schwartz, S. M. Arthur, T. A. Hanley, C. T. Robbins, and R. Michener. 1999. Trophic relations of brown and black bears in several western North American ecosystems. Journal of Wildlife Management 63:921–929.
Jennings, F. 1975. The Invasion of America: Indians, Colonialism, and the Cant of Conquest. University of North Carolina Press, Chapel Hill.

Jimenez, M.D. and S.A. Becker, eds. 2014. Northern Rocky Mountain Wolf Recovery Program 2014 Interagency Annual Report. US Fish and Wildlife Service, Ecological Services, Helena, MT.
Johnson, D.W., and D.M. Armstrong. 1987. *Peromyscus crinitus*. Mammalian Species 287:1–8.
Johnson, J.S., M.J. Lacki, and M.D. Baker. 2007. Foraging ecology of long-legged myotis (*Myotis volans*) in north-central Idaho. Journal of Mammalogy 88:1261–1270.
Johnson, M. K, and R.M. Hansen. 1979. Foods of cottontails and woodrats in south-central Idaho. Journal of Mammalogy 60:213–215.
Johnson, T.N., S.W. Buskirk, G.D. Hayward, and M.G. Raphael. 2014. Tree mortality in response to synchronized forest insect outbreaks: effects of tree species, age, and cutting history. Forest Ecology and Management 319:10–17.
Johnston, D.S., and M.B. Fenton. 2001. Individual and population-level variability in diets of pallid bats (*Antrozous pallidus*). Journal of Mammalogy 82:362–373.
Johnston, K.M., K.A. Freund, and O.J. Schmitz. 2012. Projected range shifting by montane mammals under climate change: implications for Cascadia's national parks. Ecosphere 3:97.
Jolley, T.W., R.L. Honeycutt, and R.D. Bradley. 2000. Phylogenetic relationships of pocket gophers (Genus *Geomys*) based on the mitochondrial 12SrRNA gene. Journal of Mammalogy 81:1025–1034.
Jones, J.K., Jr. 1964. Distribution and taxonomy of mammals of Nebraska. University of Kansas Publications, Museum of Natural History 16:1–356.
Jones, J.K., Jr., and R.W. Manning. 1992. Illustrated Keys to Skulls of Genera of North American Land Mammals. Texas Tech University Press, Lubbock.
Kamler, J.F., and W.B. Ballard. 2002. A review of native and nonnative red foxes in North America. Wildlife Society Bulletin 30:370–379.
Karns, P.D. 2007. Population distribution, density and trends. Pp. 125–139. In A.W. Franzman and C.C. Schwartz (eds). Ecology and Management of the North American Moose. 2nd ed. University Press of Colorado, Boulder.
Kartesz, J.T., and J.W. Thieret. 1991. Common names for vascular plants: guidelines for use and application. Sida 14:421–434.
Katzner, T.E. and K.L. Parker. 1997. Vegetative characteristics and size of home ranges used by pygmy rabbits (*Brachylagus idahoensis*) during winter in southwestern Wyoming. Journal of Mammalogy 78:1063–1072.
Katzner, T.E. and K.L. Parker. 1998. Long distance movements from established burrow sites by pygmy rabbits (*Brachylagus idahoensis*) in southwestern Wyoming. Northwestern Naturalist 79:72–74.
Kauffman, M.J., J.F. Brodie, and E.S. Jules. 2010. Are wolves saving Yellowstone's aspen? A landscape-level test of a behaviorally mediated trophic cascade. Ecology 91:2742–2755.
Kaufman, G.A., D.W. Kaufman, and E.J. Finck. 1988. Influence of fire and topography on habitat selection by *Peromyscus maniculatus* and *Reithrodontomys megalotis* in ungrazed tallgrass prairie. Journal of Mammalogy 69:342–352.
Keinath, D.A. 2004. Fringed Myotis (*Myotis thysanodes*): A Technical Conservation Assessment. Unpublished Report. Prepared for the USDA Forest Service, Rocky Mountain Region, Species Conservation Project.
Keinath, D.A., and G.P. Beauvais. 2006. Wyoming Pocket Gopher (*Thomomys clusius*): A Technical Conservation Assessment. Unpublished Report. Wyoming Natural Diversity Database, University of Wyoming, Laramie.
Keinath, D.A., H.R. Griscom, and M.D. Andersen. 2014. Habitat and distribution of the Wyoming pocket gopher (*Thomomys clusius*). Journal of Mammalogy 95:803–813.

Keinath, D. A., and G. D. Hayward. 2003. Red-backed vole (*Clethrionomys gapperi*) response to disturbance in subalpine forests: use of regenerating patches. Journal of Mammalogy 84:956–966.

Keith, J. O. 2003. The Abert's Squirrel (*Sciurus aberti*): A Technical Conservation Assessment. Prepared for the USDA Forest Service, Rocky Mountain Region. Available at www.fs.fed.us/r2/projects/scp/assessments/abertsquirrel.pdf. Accessed February 24, 2012.

Keith, L. B. 1983. Role of food in hare population cycles. Oikos 40:385–395.

Kenagy, G. J. 1981. Effects of day length, temperature, and endogenous control on annual rhythms of reproduction and hibernation in chipmunks (*Eutamias* spp.). Journal of Comparative Physiology 141:369–378.

Kendall, K. C. 1983. Use of pine nuts by grizzly and black bears in the Yellowstone area. Proceedings of the 5th International Conference on Bear Research and Management 5:166–173.

Kielan-Jaworowska, Z., R. L. Cifelli, and Z-X. Luo. 2004. Mammals from the Age of Dinosaurs. Columbia University Press, New York.

Kiesow, A. M. 2008. Genetic structure of northern flying squirrel (*Glaucomys sabrinus*) and red squirrel (*Tamiasciurus hudsonicus*) populations in the Black Hills. PhD dissertation, University of South Dakota, Vermillion.

Kiesow, A. M., E. M. Monroe, and H. B. Britten. 2012. Genetic structure of the arboreal squirrels (*Glaucomys sabrinus* and *Tamiasciurus hudsonicus*) in the North American Black Hills. Canadian Journal of Zoology 90:1191–1200.

King, C. 1992. History of Wildlife in the Bighorn Basin of Wyoming. Pioneer Printing and Stationery, Cheyenne, WY.

King, C. M. 1983. *Mustela erminea*. Mammalian Species 195:1–8.

King, J. A., ed. 1968. Biology of *Peromyscus* (Rodentia). Special Publication No. 2. American Society of Mammalogists, Stillwater, OK.

King, T. L., M. S. Eackles, and C. C. Young. 2006. Microsatellite DNA markers for assessing phylogeographic and population structure in Preble's meadow jumping mice (*Zapus hudsonius preblei*) and cross-amplification among neighbouring taxa. Molecular Ecology Notes 6:670–673.

King, T. L., J. F. Switzer, C. L. Morrison, M. S. Eackles, C.C. Yound, B.A. Lubinski, and P. Cryan. 2006. Comprehensive genetic analyses reveal evolutionary distinction of a mouse (*Zapus hudsonius preblei*) proposed for delisting from the US Endangered Species Act. Molecular Ecology 15:4331–4359.

Kingham, D. 1990. Wyoming rancher battles state over plans for exotic animals. Los Angeles Times, July 1.

Kinlaw, A. 1995. *Spilogale putorius*. Mammalian Species 511:1–7.

Kirkland, G. L., Jr., R. R. Parmenter, and R. E. Skoog. 1997. A five-species assemblage of shrews from the sagebrush steppe of Wyoming. Journal of Mammalogy 78:83–89.

Klaus, M. 2003. The status, habitat, and response to grazing of water vole populations in the Big Horn Mountains of Wyoming, U.S.A. Arctic, Antarctic and Alpine Research 35:100–109.

Klaus, M., and G. P. Beauvais. 2004. Water Vole (*Microtus richardsoni*): A Technical Conservation Assessment. Prepared for USDA Forest Service, Rocky Mountain Region, Species Conservation Project.

Klaus, M., R. E. Moore, and E. Vyse. 2001. Microgeographic variation in allozymes and mitochondrial DNA of *Microtus richardsoni*, the water vole, in the Beartooth Mountains of Montana and Wyoming, U.S.A. Canadian Journal of Zoology 79:1286–1295.

Klein, S. L., and C. H. Calisher. 2007. Emergence and persistence of hantaviruses. Current Topics in Microbiology and Immunology 315:217–252.

Knight, D. H., G. P. Jones, W. A. Reiners, and W. H. Romme. 2014. Mountains and Plains: The Ecology of Wyoming Landscapes. 2nd ed. Yale University Press, New Haven, CT.

Knowles, C. J., J. D. Proctor, and S. C. Forrest. 2002. Black-tailed prairie dog abundance and distribution in the Great Plains based on historic and contemporary information. Great Plains Research 12:219–254.

Knowlton, F. F. 1960. Food habits, movements, and populations of moose in the Gravelly Mountains, Montana. Journal of Wildlife Management 24:162–170.

Koch, P. L., and A. D. Barnosky. 2006. Late Quaternary extinctions: state of the debate. Annual Review of Ecology, Evolution, and Systematics 37:215–250.

Koel, T. M., P. E. Bigelow, P. D. Doepke, B. D. Ertel, and D.L. Mahony. 2005. Nonnative lake trout result in Yellowstone cutthroat trout decline and impacts to bears and anglers. Fisheries 30:10–19.

Koepfli, K.-P., K. A. Deere, G. J. Slater, C. Begg, K. Begg, L. Grassman, M. Lucherini, G. Veron, and R.K. Wayne. 2008. Multigene phylogeny of the Mustelidae: resolving relationships, tempo and biogeographic history of a mammalian adaptive radiation. BMC Biology 6:1–22.

Koepfli, K.-P., M. E. Gompper, E. Eizirik, C.-C. Ho, L. Linden, J. E. Maldonado, and R. K. Wayne. 2007. Phylogeny of the Procyonidae (Mammalia: Carnivora): molecules, morphology and the Great American Interchange. Molecular Phylogenetics and Evolution 43:1076–1095.

Koepfli, K.-P., and R. K. Wayne. 1998. Phylogenetic relationships of otters (Carnivora: Mustelidae) based on mitochondrial cytochrome b sequences. Journal of Zoology 246:401–416.

Kohn, M. J., and M. P. McKay. 2012. Paleoecology of late Pleistocene–Holocene faunas of eastern and central Wyoming, USA, with implications for LGM climate models. Palaeogeography, Palaeoclimatology, Palaeoecology 326–328:42–53.

Koprowski, J. L. 1994a. *Sciurus carolinensis.* Mammalian Species 480:1–9.

Koprowski, J. L. 1994b. *Sciurus niger.* Mammalian Species 479:1–9.

Kotliar, N. B., B. W. Baker, A. D. Whicker, and G. Plumb. 1999. A critical review of assumptions about the prairie dog as a keystone species. Environmental Management 24:177–192.

Kramer, A. 1973. Interspecific behavior and dispersion of two sympatric deer species. Journal of Wildlife Management 37:288–300.

Krohn, W. B., and C. L. Hoving. 2010. Early Maine Wildlife: Historical Accounts of Canada Lynx, Moose, Mountain Lion, White-tailed Deer, Wolverine, Wolves, and Woodland Caribou, 1603–1930. University of Maine Press, Orono.

Kruezer, M. P., and N. J. Huntley. 2003. Habitat-specific demography: evidence for source-sink population structure in a mammal, the pika. Oecologia 134:343–349.

Krumm, C. E., M. M. Conner, N. T. Hobbs, D. O. Hunter, and M.W. Miller. 2010. Mountain lions prey selectively on prion-infected mule deer. Biology Letters 6:209–211.

Krumm, R. E. 1969. An ecological study of the Audubon's cottontail on the Laramie Plains. MS thesis, University of Wyoming, Laramie.

Kunz, T. H. 1982. *Lasionycteris noctivagans.* Mammalian Species 172:1–5.

Kunz, T. H., and R. A. Martin. 1982. *Plecotus townsendii.* Mammalian Species 175:1–6.

Kunz, T. H., E. Braun de Torrez, D. Bauer, T. Lobova, and T. H. Fleming. 2011. Ecosystem services provided by bats. Annals of the New York Academy of Sciences 1223:1–38.

Kurose, N., A. V. Abramov, and R. Masuda. 2005. Comparative phylogeography between the ermine *Mustela erminea* and the least weasel *M. nivalis* of Palearctic and Nearctic regions, based on analysis of mitochondrial DNA control region sequences. Zoological Science 22:1069–1078.

Kurta, A., and R. H. Baker. 1990. *Eptesicus fuscus.* Mammalian Species 356:1–10.

Kurtén, B., and E. Anderson. 1980. Pleistocene Mammals of North America. Columbia University Press, New York.
Lack, J. B., J. E. Wilkinson, and R. A. Van Den Bussche. 2010. Range-wide population genetic structure of the pallid bat (*Antrozous pallidus*)—incongruent results from nuclear and mitochondrial DNA. Act Chiropterologica 12:401–413.
Lackey, J. A., D. G. Huckaby, and B. G. Ormiston. 1985. *Peromyscus leucopus*. Mammalian Species 247:1–10.
Lacki, M. J., and M. D. Baker. 2007. Day roosts of female fringed myotis (*Myotis thysanodes*) in xeric forests of the Pacific northwest. Journal of Mammalogy 88:967–973.
Lariviere, S. 1999. *Mustela vison*. Mammalian Species 608:1–9.
Lariviere, S., and M. Pasitschniak-Arts. 1996. *Vulpes vulpes*. Mammalian Species 537:1–11.
Lariviere, S., and L. R. Walton. 1997. *Lynx rufus*. Mammalian Species 563:1–8.
Laundré, J. W. 1994. Resource overlap between mountain goats and bighorn sheep. Great Basin Naturalist 54:114–121.
Laundré, J. W., L. Hernández, and K. B. Altendorf. 2001. Wolves, elk, and bison: reestablishing the "landscape of fear" in Yellowstone National Park, U.S.A. Canadian Journal of Zoology 79:1401–1409.
Laurence, S., D. W. Coltman, J. C. Gorrell, and A. I. Schulte-Hostedde. 2011. Genetic structure of muskrat (*Ondatra zibethicus*) and its concordance with taxonomy in North America. Journal of Heredity 102:688–696.
Lausen, C. L., I. Delisle, R. M. R. Barclay, and C. Strobeck. 2008. Beyond mtDNA: nuclear gene flow suggests taxonomic oversplitting in the little brown bat (*Myotis lucifugus*). Canadian Journal of Zoology 86:700–713.
Lawrence, I. E. 1955. An ecological study of the snowshoe hare, *Lepus americanus bairdii*, Hayden in the Medicine Bow National Forest of Wyoming. MS thesis, University of Wyoming, Laramie.
Lee, J. 1916. Diary of Rev. Jason Lee. Oregon Historical Society Quarterly 17:116–146, 397–430.
Lehman, N., A. Eisenhawer, K. Hansen, L. D. Mech, R.O. Peterson, P.J.P. Gogan, and R.K. Wayne. 1991. Introgression of coyote mitochondrial DNA into sympatric North American gray wolf populations. Evolution 45:104–119.
Lemke, T. O. 2004. Origin, expansion, and status of mountain goats in Yellowstone National Park. Wildlife Society Bulletin 32:532–541.
Leopold, A. S., S. A. Cain, C. M. Cottam, I. N. Gabrielson, and T.L. Kimball. 1963. Wildlife Management in the National Parks: The Leopold Report. National Park Service, Washington, DC.
Lillegraven, J. A., and L. M. Ostresh, Jr. 1988. Evolution of Wyoming's early Cenozoic topography and drainage patterns. National Geographic Research 4:303–327.
Lim, B. K. 1987. *Lepus townsendii*. Mammalian Species 288:1–6.
Lim, M. M., and L. J. Young. 2004. Vasopressin-dependent neural circuits underlying pair bond formation in the monogamous prairie vole. Neuroscience 125:35–45.
Lindsay, S. L. 1987. Geographic size and non-size variation in Rocky Mountain *Tamiasciurus hudsonicus*: significance in relation to Allen's rule and vicariant biogeography. Journal of Mammalogy 68:39–48.
Lindstrom, J. W., and W. A. Hubert. 2004. Mink predation on radio-tagged trout during winter in a low-gradient reach of a mountain stream, Wyoming. Western North American Naturalist 64:551–553.
Lindzey, F. G. 1971. Ecology of badgers in Curlew Valley, Utah and Idaho with emphasis on movement and activity patterns. MS thesis, Utah State University, Logan.

Lindzey, F.G. 1987. Mountain lion. Pp. 657–668. In M. Novak, J.A. Baker, M.E. Obbard, and B. Malloch (eds). Wild Furbearer Management and Conservation in North America. Ontario Trappers Association, Ontario Ministry of Natural Resources, Toronto.
Lindzey, F.G. 2003. Badger. Pp. 683–691. In G.A. Feldhamer, B.C. Thompson, and J.A. Chapman (eds). Wild Mammals of North America. 2nd ed. Johns Hopkins University Press, Baltimore, MD.
Lingle, S. 1993. Escape gaits of white-tailed deer, mule deer, and their hybrids: body configuration, biomechanics, and function. Canadian Journal of Zoology 71:708–724.
Linhart, S.B., and W.B. Robinson. 1972. Some relative carnivore densities in areas under sustained coyote control. Journal of Mammalogy 53:880–884.
Link, M.A. 1995. Kit fox (*Vulpes macrotis*) distribution in western Colorado. MS thesis, University of Northern Colorado, Greeley.
Linnaeus, C. 1758. Systema naturae. Editio decima. Impensis Laurentii Salvii, Holmiae. [In Latin]
Lobo, N., M. Duong, and J.S. Millar. 2009. Conifer-seed preferences of small mammals. Canadian Journal of Zoology 87:773–780.
Logan, K.A. 1983. Mountain lion population and habitat characteristics in the Big Horn Mountains of Wyoming. MS thesis, University of Wyoming, Laramie.
Logan, K.A. and L.L. Irwin. 1985. Mountain lion habitats in the Big Horn Mountains, Wyoming. Wildlife Society Bulletin 13:257–262.
Logan, K.A., L.L. Irwin, and R. Skinner. 1986. Characteristics of a hunted mountain lion population in Wyoming. Journal of Wildlife Management 50:648–654.
Long, B., M. Hinschberger, G. Roby, and J. Kimbal. 1980. Gros Ventre Cooperative Elk Study. Final Report 1974–1979. Wyoming Game and Fish Department and USDA Forest Service, Jackson, WY.
Long, C.A. 1963. Melanism in the Uinta ground squirrel, *Spermophilus armatus*. Transactions of the Kansas Academy of Science 66:136–137.
Long, C.A. 1965. The mammals of Wyoming. University of Kansas Publications, Museum of Natural History 14:493–758.
Long, C.A. 1972. Taxonomic revision of the North American badger, *Taxidea taxus*. Journal of Mammalogy 53:725–759.
Long, C.A. 1974. *Microsorex hoyi* and *Microsorex thompsoni*. Mammalian Species 33:1–4.
Long, C.A. 2008. The Wild Mammals of Wisconsin. Pensoft Publishers, Sofia, Bulgaria.
Long, C.A., and W.C. Kerfoot. 1963. Mammalian remains from owl-pellets in eastern Wyoming. Journal of Mammalogy 44:129–131.
Long, R.A., P. MacKay, W.J. Zielinski, and J.C. Ray, eds. 2008. Noninvasive Survey Methods for Carnivores. Island Press, Washington, DC.
Lorch, J.M., C.U. Meteyer, M.J. Behr, J.G. Boyles, P.M. Cryan, A.C. Hicks, A.E. Ballman et al. 2011. Experimental infection of bats with *Geomyces destructans* causes white-nose syndrome. Nature 480:376–378.
Lotze, J.-H., and S. Anderson. 1979. *Procyon lotor*. Mammalian Species 119:1–8.
Love, J.D., and A.C. Christianson. 1985. Geological Map of Wyoming. US Geological Survey, Denver, CO.
Luce, R.J. 2004. Species Assessment for Spotted Bat (*Euderma maculatum*) in Wyoming. Bureau of Land Management, Wyoming State Office, Cheyenne.
Luce, R.J., and D. Keinath. 2007. Spotted Bat (*Euderma maculatum*): A Technical Conservation Assessment. Prepared for USDA Forest Service, Rocky Mountain Region, Species Conservation Project.
Ludwig, D.R. 1984. *Microtus richardsoni*. Mammalian Species 223:1–6.

Luo, Z.-X. 2007. Transformation and diversification in early mammal evolution. Nature 450:1011–1019.
Luong, L.T., G.A. Vigliotti, S. Campbell, J.A. Comer, J.N. Mills, and P.J. Hudson. 2011. Dynamics of hantavirus infection in *Peromyscus leucopus* of central Pennsylvania. Vector-Borne and Zoonotic Diseases 11:1459–1464.
Lyford, M.E., S.T. Jackson, S.T. Gray, and R.G. Eddy. 2004. Validating the use of woodrat (*Neotoma*) middens for documenting natural invasions. Journal of Biogeography 31:333–342.
Lyons, S.K. 2003. A quantitative assessment of the range shifts of Pleistocene mammals. Journal of Mammalogy 84:385–402.
MacArthur, R.A., and L.C.H. Wang. 1974. Behavioral thermoregulation in the pika *Ochotona princeps*: a field study using radiotelemetry. Canadian Journal of Zoology 52:353–358.
MacCracken, J.G., and D.W. Uresk. 1984. Coyotes foods in the Black Hills, South Dakota. Journal of Wildlife Management 48:1420–1423.
MacLeod, N. 2014. The geological extinction record: history, data, biases, and testing. Geological Society of America Special Papers 505:1–28.
MacNulty, D.R., D.W. Smith, L.D. Mech, and L.E. Eberly. 2009. Body size and predatory performance in wolves: is bigger better? Journal of Animal Ecology 78:532–539.
Mager, K.J., and T.A. Nelson. 2001. Roost-site selection by eastern red bats (*Lasiurus borealis*). American Midland Naturalist 145:120–126.
Mainguy, J., K. Worley, S.D. Côté, and D.W. Coltman. 2007. Low MHC DRB class II diversity in the mountain goat: past bottlenecks and possible role of pathogens and parasites. Conservation Genetics 8:885–891.
Malaney, J.L., C.J. Conroy, L.A. Moffitt, H.D. Spoonhunter, J.L. Patton, and J.A. Cook. 2013. Phylogeography of the western jumping mouse (*Zapus princeps)* detects deep and persistent allopatry with expansion. Journal of Mammalogy 94:1016–1029.
Malaney, J.L., and J.A. Cook. 2013. Using biogeographic history to inform conservation: the case of Preble's meadow jumping mouse. Molecular Ecology 22:6000–6017.
Manley, W.F. 2002. Postglacial flooding of the Bering Land Bridge: a geospatial assessment v1. INSTAAR2002. Available at http://instaar.colorado.edu/QGISL/bering_land_bridge.
Manning, R.W. 1993. Systematics and evolutionary relationships of the long-eared myotis, *Myotis evotis* (Chiroptera: Vespertilionidae). Special Publications, The Museum, Texas Tech University 37:1–58.
Manning, R.W., and J.K. Jones, Jr. 1988. *Perognathus fasciatus*. Mammalian Species 303:1–4.
Manning, R.W., and J.K. Jones, Jr. 1989. *Myotis evotis*. Mammalian Species 329:1–5.
Maret, T.J., M. Parker, and T.E. Fannin. 1987. The effect of beaver ponds on the nonpoint source water quality of a stream in southwestern Wyoming. Water Resources 21:263–268.
Marincovich, L., Jr., and A.Y. Gladenkov. 1999. Evidence for an early opening of the Bering Strait. Nature 397:149–151.
Martin, P.S. 1984. Prehistoric overkill: the global model. Pp. 354–403. In P.S. Martin and R.G. Klein (eds). Quaternary Extinctions: A Prehistoric Revolution. University of Arizona Press, Tucson.
Martin, S.K. 1994. Feeding ecology of American martens and fishers. Pp. 297–315. In S.W. Buskirk, A.S. Harestad, M.G. Raphael, and R.A. Powell (eds). 1994. Martens, Sables and Fishers: Biology and Conservation. Cornell University Press, Ithaca, NY.
Martin, T. 2006. Early mammalian evolutionary experiments. Science 311:1109–1110.
Matthews, A.K., S.A. Neiswenter, and L.K. Ammerman. 2010. Trophic ecology of the free-tailed bats *Nyctinomops femorosaccus* and *Tadarida brasiliensis* (Chiroptera: Molossidae) in Big Bend National Park, Texas. Southwestern Naturalist 55:340–346.

Mattson, D.J. 1997. Use of ungulates by Yellowstone grizzly bears *Ursus arctos*. Biological Conservation 81:161–177.
Mattson, D.J., C.M. Gillin, S.A. Benson, and R.R. Knight. 1991. Bear feeding activity at alpine insect aggregation sites in the Yellowstone ecosystem. Canadian Journal of Zoology 69:2430–2435.
Mattson, D.J., R.R. Knight, and B.M. Blanchard. 1992. Cannibalism and predation on black bears by grizzly bears in the Yellowstone ecosystem, 1975–1990. Journal of Mammalogy 73:422–425.
Mattson, D.J., and D.P. Reinhart. 1997. Excavation of red squirrel middens by grizzly bears in the whitebark pine zone. Journal of Applied Ecology 34:926–940.
Mattson, T.A., S.W. Buskirk, and N.L. Stanton. 1996. Roost sites of the silver-haired bat (*Lasionycteris noctivagans*) in the Black Hills, South Dakota. Great Basin Naturalist 56:247–253.
Maxwell, M.H., and L.N. Brown. 1968. Ecological distribution of rodents on the high plains of eastern Wyoming. Southwestern Naturalist 13:143–158.
Mayr, E. 1942. Systematics and the Origin of Species, from the Viewpoint of a Zoologist. Columbia University Press, New York.
McAllister, J.A., and R.S. Hoffman. 1988. *Phenacomys intermedius*. Mammalian Species 305:1–8.
McCaffrey, R.E., M.C. Wallace, J.F. Kamler, and J.D. Ray. 2003. Noteworthy distributional records of the prairie vole in the Texas and Oklahoma panhandles. Southwestern Naturalist 48:717–719.
McCarty, R. 1978. *Onychomys leucogaster*. Mammalian Species 87:1–6.
McGahan, J. 1967. Quantified estimates of predation by a golden eagle population. Journal of Wildlife Management 31:496–501.
McKelvey, K.S., K.B. Aubry, N.J. Anderson, A.P. Clevenger, J.P. Copeland, K.S. Heinemeyer, R.M. Inman et al. 2014. Recovery of wolverines in the western United States: recent extirpation and colonization or range retraction and expansion? Journal of Wildlife Management 78:325–334.
McKelvey, K.S., K.B. Aubry, and Y.K. Ortega. 2000. History and distribution of lynx in the contiguous United States. Pp. 207–264. In L.F. Ruggiero, K.B. Aubry, S.W. Buskirk, G.M. Koehler, C.J. Krebs, K.S. McKelvey, and J.R. Squires. 2000. Ecology and Conservation of Lynx in the United States. University Press of Colorado, Boulder.
McKelvey, K.S., K.B. Aubry, and M.K. Schwartz. 2008. Data for rare or elusive species: the illusion of reality and a call for evidentiary standards. BioScience 58:549–555.
McKenna, M.C. 1975. Fossil mammals and early Eocene North American land continuity. Annals of the Missouri Botanical Garden 62:335–353.
McKenna, M.C., and Bell, S.K. 1997. Classification of Mammals above the Species Level. Columbia University Press, New York.
McManus, J.J. 1974. *Didelphis virginianus*. Mammalian Species 40:1–6.
McMillan, J.F. 1953. Some feeding habits of the moose in Yellowstone Park. Ecology 34:102–110.
McNamee, T. 1997. The Return of the Wolf to Yellowstone. Henry Holt, New York.
McWhirter, D. 2004. Mountain goat status and management in Wyoming. Biennial Symposium of the Northern Wild Sheep and Goat Council 14:101–113.
Meagher, M. 1986. *Bison bison*. Mammalian species 266:1–8.
Meagher, M., W.J. Quinn, and L. Stackhouse. 1992. Chlamydial-caused infectious keratoconjunctivitis in bighorn sheep of Yellowstone National Park. Journal of Wildlife Diseases 28:171–176.
Meaney, C.A. 1991. Colorado Mammal Distribution Latilong Study. Colorado Division of Wildlife, Denver.
Mech, L.D. 1974. *Canis lupus*. Mammalian Species 37:1–6.

Meece, J.K., C.E. Reynolds, P.J. Stockwell, T.A. Jenson, J.E. Christensen, and K.D. Reed. 2005. Identification of mosquito bloodmeal source by terminal restriction fragment length polymorphism profile analysis of the cytochrome B gene. Journal of Medical Entomology 42:657–667.

Melo-Ferreira, J., F.A. Seixas, E. Chang, L.S. Mills, and P.C. Alves. 2014. The hidden history of the snowshoe hare, *Lepus americanus*: extensive mitochondrial DNA introgression inferred from multilocus genetic variation. Molecular Ecology 23:4617–4630.

Melquist, W.E., P.J. Polechla, Jr., and D. Toweill. 2003. River otter. Pp. 708–734. In G.A. Feldhamer, B.C. Thompson, and J.A. Chapman (eds). Wild mammals of North America. 2nd ed. Johns Hopkins University Press, Baltimore, MD.

Menkens, G.E., Jr., and S.H. Anderson. 1989. Temporal-spatial variation in white-tailed prairie dog demography and life histories in Wyoming. Canadian Journal of Zoology 67:343–349.

Merkle, J.A., D.R. Stahler, and D.W. Smith. 2009. Interference competition between gray wolves and coyotes in Yellowstone National Park. Canadian Journal of Zoology 87:56–63.

Merriam, C.H. 1889. Description of a new genus *Phenacomys* and four new species of Arvicolinae. North American Fauna 2:27–45.

Merriam, C.H. 1898. Life zones and crop zones of the United States. USDA Division of Biological Survey Bulletin 10:1–79.

Merriam, C.H. 1918. Revision of the grizzly and big brown bears of North America (genus *Ursus*) with description of a new genus, *Vetularctos*. North American Fauna 41:1–137.

Merrill, E.H., and M.S. Boyce. 1991. Summer range and elk population dynamics in Yellowstone National Park. Pp. 263–273. In R.B. Keiter and M.S. Boyce (eds). The Greater Yellowstone Ecosystem: Redefining America's Wilderness Heritage. Yale University Press, New Haven, CT.

Merrill, E.H., T.W. Kohley, M.E. Herdendorf, W.A. Reiners, K.L. Driese, R.W. Marrs, and S.H. Anderson. 1996. Wyoming Gap Analysis: A Geographic Analysis of Biodiversity. Final Report, Wyoming Cooperative Fish and Wildlife Research Unit, University of Wyoming, Laramie, WY.

Merritt, J.F. 1981. *Clethrionomys gapperi*. Mammalian Species 146:1–9.

Mickey, A.B. 1948. A record of the shrew *Sorex nanus* for Wyoming. Journal of Mammalogy 29:294–295.

Middleton, A.D. 2012. The influence of large carnivore recovery and summer conditions on the migratory elk of Wyoming's Absaroka Mountains. PhD dissertation, University of Wyoming, Laramie.

Middleton, A.D., M.J. Kauffman, D.E. McWhirter, M.D. Jimenez, R.C. Cook, J.G. Cook, S.E. Albeke, H. Sawyer, and P.J. White. 2013. Linking anti-predator behaviour to prey demography reveals limited risk effects of an actively hunting large carnivore. Ecology Letters 16:1023–1030.

Middleton, A.D., T.A. Morrison, J.E. Fortin, C.T. Robbins, K.M. Proffitt, P.J. White, D.E. McWhirter. 2013. Grizzly bear predation links the loss of native trout to the demography of migratory elk in Yellowstone. Proceedings of the Royal Society B 280:20130870.

Miller, B., R.P. Reading, and S. Forrest. 1996. Prairie Night: Black-Footed Ferrets and the Recovery of Endangered Species. Smithsonian Institution Press, Washington, DC.

Miller, B., R. Reading, J. Hoogland, T. Clark, G. Ceballos, R. List, S. Forrest et al. 2000. The role of prairie dogs as a keystone species: response to Stapp. Conservation Biology 14:318–321.

Miller, C.R., and L.P. Waits. 2003. The history of effective population size and genetic diversity in the Yellowstone grizzly (*Ursus arctos*): implications for conservation. Proceedings of the National Academy of Sciences 100:4334–4339.

Miller, D.A., D.B. Arnett, and M.J. Lacki. 2003. Habitat management for forest-roosting bats of North America: a critical review of habitat studies. Wildlife Society Bulletin 31:30–44.
Miller, R.S. 1964. Ecology and distribution of pocket gophers (Geomyidae) in Colorado. Ecology 45:256–272.
Mills, K., K. Sato, and A. Boerger-Fields. 2006. Science, education keeps human rabies cases at bay. Reflections (University of Wyoming College of Agriculture and Natural Science):16–19.
Milner, J., C. Jones, and J.K. Jones, Jr. 1990. *Nyctinomops macrotis*. Mammalian Species 351:1–4.
Minta, S.C., K.A. Minta, and D.F. Lott. 1992. Hunting associations between badgers (*Taxidea taxus*) and coyotes (*Canis latrans*). Journal of Mammalogy 73:814–820.
Monk, R.R., and J.K. Jones, Jr. 1996. *Perognathus flavescens*. Mammalian Species 525:1–4.
Montes, C., A. Cardona, C. Jaramillo, A. Pardo, J. C. Silva, V. Valencia, C. Ayala et al. 2015. Middle Miocene closure of the Central American Seaway. Science 348:226–229.
Morales, J.C., and J.W. Bickham. 1995. Molecular systematics of the Genus *Lasiurus* (Chiroptera: Vespertilionidae) based on restriction-site maps of the mitochondrial ribosomal genes. Journal of Mammalogy 76:730–749.
Moratelli, R. 2014. Wildlife biologists are on the right track: a mammalogist's view of specimen collection. Zoologia 31:413–417.
Morrison, M.L. 2006. Wildlife–Habitat Relationships: Concepts and Applications. Island Press, Washington, DC.
Müller-Schwarze, D. 2011. The Beaver. 2nd ed. Cornell University Press, Ithaca, NY.
Mullican, T.R., and B.L. Keller. 1986. Ecology of the sagebrush vole (*Lemmiscus curtatus*) in southeastern Idaho. Canadian Journal of Zoology 64:1218–1223.
Mumford, R.E., and J.O. Whitaker, Jr. 1982. Mammals of Indiana. Indiana University Press, Bloomington.
Murie, A. 1937. Some food habits of the black bear. Journal of Mammalogy 18:238–240.
Murie, A. 1940. Fauna of the National Parks of the United States: Ecology of the Coyote in the Yellowstone. Fauna Series No. 4. Government Printing Office, Washington, DC.
Murie, A. 1944. Fauna of the National Parks of the United States: The Wolves of Mount McKinley. Fauna Series No. 5. Government Printing Office, Washington, DC.
Murie, A. 1948. Cattle on grizzly bear range. Journal of Wildlife Management 12:57–72.
Murie, A. 1960. Aquatic voles. Journal of Mammalogy 41:273–275.
Murie, A. 1961. Some food habits of the marten. Journal of Mammalogy 42:516–521.
Murie, O.J. 1934. Melanism in an Alaskan vole. Journal of Mammalogy 15:323.
Murie, O.J. 1935. Food habits of the coyote in Jackson Hole, Wyoming. USDA Circular 362:1–24.
Murie, O.J. 1951. The Elk of North America. Stackpole Books, Harrisburg, PA.
Murie, O.J. 1954. A Field Guide to Animal Tracks. Houghton Mifflin, Boston, MA.
Murie, O.J., and A. Murie. 1931. Travels of *Peromyscus*. Journal of Mammalogy 12:200–209.
Murphy, K.M., T.M. Potter, J.C. Halfpenny, K.A. Gunther, M.T. Jones, P.A. Lundberg, and N.D. Berg. 2006. Distribution of Canada lynx in Yellowstone National Park. Northwest Science 80:199–206.
Nadler, C.F., R.I. Sukernik, R.S. Hoffman, N.N. Vorontsov, C.F. Nadler, Jr., and I.I. Fomichova. 1974. Evolution in ground squirrels–I. Transferrins in Holarctic populations of *Spermophilus*. Comparative Biochemistry and Physiology A 47:663–681.
Naiman, R.J., G. Pinay, C.A. Johnston, and J. Pastor. 1994. Beaver influences on the long-term biogeochemical characteristics of boreal forest drainage networks. Ecology 75:905–921.
National Park Service. 2010. Wildlife Monitoring: Pronghorn. Grand Teton National Park, Science and Resource Management Division, Moose, WY.

National Park Service. 2011. 2011 Count of Yellowstone Pronghorn. Unpublished Report, Yellowstone Center for Resources.
Navo, K.W., and J.A. Gore. 2001. Distribution of the big free-tailed bat (*Nyctinomops macrotis*) in Colorado. Southwestern Naturalist 46:370–376.
Navo, K.W., J.A. Gore, and G.T. Skiba. 1992. Observations on the spotted bat, *Euderma maculatum*, in northwestern Colorado. Journal of Mammalogy 73:547–551.
Negus, N.C., P.J. Berger, and B.W. Brown. 1986. Microtine population dynamics in a predictable environment. Canadian Journal of Zoology 54:785–792.
Negus, N.C., and J.S. Findley. 1959. Mammals of Jackson Hole, Wyoming. Journal of Mammalogy 40:371–381.
Neiswenter, S.A., and B.R. Riddle. 2011. Landscape and climatic effects on the evolutionary diversification of the *Perognathus fasciatus* species group. Journal of Mammalogy 92:982–993.
Nelson, E.W. 1914. Description of a new subspecies of moose from Wyoming. Proceedings of the Biological Society of Washington 27:71–73.
Nelson, E.W. 1925. Status of the pronghorned antelope, 1922–1924. US Department of Agriculture Bulletin 1346:1–64.
Newmark, W.D. 1995. Extinction of mammal populations in western North American national parks. Conservation Biology 9:512–526.
Nordyke, K.A., and S.W. Buskirk. 1991. Southern red-backed vole, *Clethrionomys gapperi*, populations in relation to stand succession and old-growth character in the central Rocky Mountains. Canadian Field-Naturalist 105:330–334.
Oakley, C., and P. Riddle. 1974. Impact of a severe winter and fences on antelope mortality in southcentral Wyoming. Antelope States Workshop Proceedings 6:155–173.
Oedekoven, O.O., and F.G. Lindzey. 1987. Winter habitat-use patterns of elk, mule deer, and moose in southeastern Wyoming. Great Basin Naturalist 47:638–643.
O'Farrell, M.J., and E.H. Studier. 1980. *Myotis thysanodes*. Mammalian Species 137:1–5.
O'Gara, B.W. 2004a. Physical characteristics. Pp. 109–143. In B.W. O'Gara and J.D. Yoakum (eds). Pronghorn: Ecology and Management. University Press of Colorado, Boulder.
O'Gara, B.W. 2004b. Mortality factors. Pp. 379–408. In B.W. O'Gara and J.D. Yoakum (eds). Pronghorn: Ecology and Management. University Press of Colorado, Boulder.
O'Gara, B.W., and G. Matson. 1975. Growth and casting of horns by pronghorns and exfoliation of horns by bovids. Journal of Mammalogy 56:829–846.
Olendorff, R.R. 1976. The food habits of North American golden eagles. American Midland Naturalist 95:231–236.
Olson, T.L. 2000. Population characteristics, habitat selection patterns, and diet of swift foxes in southeast Wyoming. MS thesis, University of Wyoming, Laramie.
Olson, T.L., and F.G. Lindzey. 2002. Swift fox survival and production in southeastern Wyoming. Journal of Mammalogy 83:199–206.
Orr, R.T. 1940. The rabbits of California. Occasional Papers of the California Academy of Sciences 19:1–227.
Orr, R.T. 1954. Natural history of the pallid bat, *Antrozous pallidus*. Proceedings of the California Academy of Science, Series 4, 28:165–246.
Osborn, D.J. 1953. Age classes, reproduction, and sex ratios of Wyoming beaver. Journal of Mammalogy 34:27–44.
Osgood, W.H. 1900. Revision of the pocket mice of the genus *Perognathus*. North American Fauna 18:1–72.

O'Shea, T. J., L. E. Ellison, D. J. Neubaum, M. A. Neubaum, C. A. Reynolds, and R. A. Bowen. 2010. Recruitment in a Colorado population of big brown bats: breeding probabilities, litter size, and first-year survival. Journal of Mammalogy 91:418–428.

O'Shea, T. J., L. E. Ellison, and T. R. Stanley. 2011. Adult survival and population growth rate in Colorado big brown bats (*Eptesicus fuscus*). Journal of Mammalogy 92:433–443.

O'Shea, T. J., D. J. Neubaum, M. A. Neubaum, P. M. Cryan, L.E. Ellison, T.R. Stanley, C.E. Rupprecht, W.J. Pape, and R.A. Bowen. 2011. Bat ecology and public health surveillance for rabies in an urbanizing region of Colorado. Urban Ecosystems 14:665–697.

Owen, C. T., M. A. McGregor, G. A. Cobbs, and J. E. Alexander, Jr. 2011. Muskrat predation on a unionid mussel community: impacts of prey species composition, size and shape. Freshwater Biology 56:554–564.

Paetkau, D., L. P. Waits, P. L. Clarkson, L. Craighead, and C. Strobeck. 1997. An empirical evaluation of genetic distance statistics using microsatellite data from bear (Ursidae) populations. Genetics 147:1943–1957.

Painter, M. L., C. L. Chambers, M. Siders, R. R. Doucett, J. O. Whittaker, Jr., and D. L. Phillips. 2009. Diet of spotted bats (*Euderma maculatum*) in Arizona as indicated by fecal analysis and stable isotopes. Canadian Journal of Zoology 87:865–875.

Parkman, F. 2008. The Oregon Trail. Oxford University Press, New York.

Patterson, B. D., G. Ceballos, W. Sechrest, M. F. Tognelli, T. Brooks, L. Luna, P. Ortega, I. Salazar, and B.E. Young. 2007. Digital Distribution Maps of the Mammals of the Western Hemisphere, Version 3.0. NatureServe, Arlington, VA.

Pattie, D. L., and N. A. M. Verbeek. 1967. Alpine mammals of the Beartooth Mountains. Northwest Science 41:110–117.

Paul, G. S. 1992. The size and bulk of extinct giant land herbivores. Supplement, Journal of Vertebrate Paleontology 12:47A.

Pauley, G., and F. Lindzey. 1993. Historic and Current Distribution of White-Tailed Deer in Wyoming. Unpublished Report. Wyoming Cooperative Fish and Wildlife Research Unit, Laramie, WY.

Pauli, J. N., S. W. Buskirk, E. S. Williams, and W. H. Edwards. 2006. A plague epizootic in the black-tailed prairie dog (*Cynomys ludovicianus*). Journal of Wildlife Diseases 42:74–80.

Pauli, J. N., R. M. Stephens, and S. H. Anderson. 2006. White-Tailed Prairie Dog (*Cynomys leucurus*): A Technical Conservation Assessment. Prepared for USDA Forest Service, Rocky Mountain Region, Species Conservation Project.

Paul-Murphy, J., T. Work, D. Hunter, E. McFie, and D. Fjelline. 1994. Serologic survey and serum biochemical reference ranges of the free-ranging mountain lion (*Felis concolor*) in California. Journal of Wildlife Diseases 30:205–215.

Paulson, D. D. 1988. *Chaetodipus hispidus*. Mammalian Species 320:1–4.

Pearson, D. E., K. S. Kelvey, and L. F. Clark. 2000. Non-target effects of an introduced biological control agent on deer mouse ecology. Oecologia 122:121–128.

Peek, J. M. 2007. Habitat relationships. Pp. 351–375. In A. W. Franzman and C. C. Schwartz (eds). Ecology and Management of the North American Moose. 2nd ed. University Press of Colorado, Boulder.

Pelton, M. R. 2003. Black bear. Pp. 547–555. in G. A. Feldhamer, B. C. Thompson, and J. A. Chapman (eds). Wild Mammals of North America. 2nd ed. Johns Hopkins University Press, Baltimore, MD.

Pennisi, L. A., S. M. Holland, and T. V. Stein. 2004. Achieving bat conservation through tourism. Journal of Ecotourism 3:195–207.

Peppers, J. A., P. K. Kennedy, and M. L. Kennedy. 1996. Spatial genetic variability in the coyote (*Canis latrans*). Southwestern Naturalist 41:388–394.

Perkins, M. E., and B. P. Nash. 2002. Explosive silicic volcanism of the Yellowstone hotspot: the ash fall tuff record. Geological Society of America Bulletin 114:367–381.
Perry, R. W., and R. E. Thill. 2007. Tree roosting by male and female eastern pipistrelles in a forested landscape. Journal of Mammalogy 88:974–981.
Pettus, D., and R. R. Lechleitner. 1963. *Microsorex* in Colorado. Journal of Mammalogy 44:119.
Piaggio, A. J., and G. S. Spicer. 2000. Molecular phylogeny of the chipmunk genus *Tamias* based on the mitochondrial cytochrome oxidase subunit II gene. Journal of Mammalian Evolution 7:147–166.
Pierce, B. M., and V. C. Bleich. 2003. Mountain lion. Pp. 744–757. In G. A. Feldhamer, B. C. Thompson, and J. A. Chapman (eds). Wild Mammals of North America. 2nd ed. Johns Hopkins University Press, Baltimore, MD.
Pinter, A. J. 1986. Population dynamics and litter size of the montane vole, *Microtus montanus*. Canadian Journal of Zoology 64:1487–1490.
Pinter, A. J. 1988. Multiannual fluctuations in precipitation and population dynamics of the montane vole, *Microtus montanus*. Canadian Journal of Zoology 66:2128–2132.
Planz, J. V., E. G. Zimmerman, T. A. Spradling, and D. R. Akins. 1996. Molecular phylogeny of the *Neotoma floridana* species group. Journal of Mammalogy 77:519–535.
Plumb, G. E., and R. Sucec. 2006. A bison conservation history in the U.S. national parks. Journal of the West 45:22–28.
Pogalayen-Neuwall, I., and D. E. Toweill. 1988. *Bassariscus astutus*. Mammalian Species 327:1–8.
Polziehn, R. O., J. Hamr, F. F. Mallory, and C. Strobeck. 2000. Phylogenetic status of North American wapiti (*Cervus elaphus*) subspecies. Molecular Ecology 9:1561–1576.
Polziehn, R. O., and C. Strobeck. 2002. A phylogenetic comparison of red deer and wapiti using mitochondrial DNA. Molecular Phylogenetics and Evolution 22:342–356.
Poole, K. G., K. Stuart-Smith, and I. E. Teske. 2009. Wintering strategies by mountain goats in interior mountains. Canadian Journal of Zoology 87:273–283.
Powell, R. A. 1993. The Fisher: Life History, Ecology and Behavior. 2nd ed. University of Minnesota Press, Minneapolis.
Powell, R. A., and W. J. Zielinski. 1994. Fisher. Pp. 38–73. In L. F. Ruggiero, K. B. Aubry, S. W. Buskirk, L. J. Lyon, and W. Zielinski (eds). The Scientific Basis for Conserving Forest Carnivores: American Marten, Fisher, Lynx and Wolverine in the Western United States. USDA Forest Service, General Technical Report RM-254.
Previtali, M. A., E. M. Lehmer, J. M. C. Pearce-Duvet, J. D. Jones, C.A. Clay, B.A. Wood, P.W. Ely, S.M. Laverty, and M.D. Dearing. 2010. Roles of human disturbance, precipitation, and a pathogen on the survival and reproductive probabilities of deer mice. Ecology 9:582–592.
Priday, J., and B. Luce. 1996. Inventory of bats and bat habitat associated with caves and abandoned mines in Wyoming. Pp. 77–131. In Endangered and Nongame Bird and Mammal Investigations: Annual Completion Report. Wyoming Game and Fish Department, Biological Services Section, Nongame Program, Lander, WY.
Priday, J., and B. Luce. 1999a. Inventory of bats and bat habitat in Wyoming: completion report. Pp. 116–165. In Threatened, Endangered, and Nongame Bird and Mammal Investigations: Annual Completion Report. Wyoming Game and Fish Department, Biological Services Section, Nongame Program, Lander, WY.
Priday, J., and B. Luce. 1999b. New distributional records for spotted bat (*Euderma maculatum*) in Wyoming. Great Basin Naturalist 59:97–101.
Prothero, D. R., and T. H. Heaton. 1996. Faunal stability during the Early Oligocene climatic crash. Palaeogeography, Palaeoclimatology, Palaeoecology 127:257–283.
Proulx, G., H. N. Bryant, and P. M. Woodard, eds. 1997. *Martes*: Taxonomy, Ecology, Techniques, and Management. Provincial Museum of Alberta, Edmonton, AB.

Purcell, M. J. 2006. Pygmy rabbit (*Brachylagus idahoensis*) distribution and habitat selection in Wyoming. MS thesis, University of Wyoming, Laramie.
Quick, H. F. 1951. Notes on the ecology of weasels in Gunnison County, Colorado. Journal of Mammalogy 32:281–290.
Quimby, D. C., and D. E. Johnson. 1951. Weights and measurements of Rocky Mountain elk. Journal of Wildlife Management 15:57–62.
Radinsky, L. 1973. Are stink badgers skunks? Implications of neuroanatomy for mustelid phylogeny. Journal of Mammalogy 54:585–593.
Raesly, E. J. 2001. Progress and status of river otter reintroduction projects in the United States. Wildlife Society Bulletin 29:856–862.
Ramey, C. A., K. W. Mills, J. W. Fischer, R. G. McLean, K.A. Fagerstone, and R.M. Engeman. 2013. Graphically characterizing the movement of a rabid striped skunk epizootic across the landscape in northwestern Wyoming. EcoHealth 10:246–256.
Ramey, C. A., and D. J. Nash. 1976. Coat color polymorphism of Abert's squirrel, *Sciurus aberti*, in Colorado. Southwestern Naturalist 21:209–217.
Rausch, R. L. 1953. On the status of some arctic mammals. Arctic 6:91–148.
Rawson, T. 2001. Changing Tracks: Predators and Politics in Mount McKinley National Park. University of Alaska Press, Fairbanks.
Reed, M. P., and J. R. Choate. 1988. Noteworthy southwestern records of the prairie vole. Southwestern Naturalist 33:495–496.
Reeve, A., F. Lindzey, and S. Buskirk. 1986. Historic and Recent Distribution of the Lynx in Wyoming. Unpublished Report. Wyoming Cooperative Fishery and Wildlife Research Unit, Laramie.
Reich, L. M. 1981. *Microtus pennsylvanicus*. Mammalian Species 159:1–8.
Reichman, O. J., and E. W. Seabloom. 2002. The role of pocket gophers as subterranean ecosystem engineers. Trends in Ecology and Evolution 17:44–49.
Reimer, J. P., E. F. Baerwald, and R. M. R. Barclay. 2010. Diet of hoary (*Lasiurus cinereus*) and silver-haired (*Lasionycteris noctivagans*) bats while migrating through southwestern Alberta in late summer and autumn. American Midland Naturalist 164:230–237.
Reinhart, D. P., and D. J. Mattson. 1990. Bear use of cutthroat trout spawning streams in Yellowstone National Park. Proceedings of the 8th International Conference on Bear Research and Management 8:343–350.
Renatte, R., and S. Lariviere. 2003. Skunks. Pp. 692–707. In G. A. Feldhamer, B. C. Thompson, and J. A. Chapman (eds). Wild Mammals of North America. 2nd ed. Johns Hopkins University Press, Baltimore, MD.
Richardson, L., T. W. Clark, S. C. Forrest, and T. M. Campbell, III. 1987. Winter ecology of black-footed ferrets (*Mustela nigripes*) at Meeteetse, Wyoming. American Midland Naturalist 117:225–239.
Richardson, T. W., and S. B. Vander Wall. 2007. Yellow pine chipmunks cannot climb quaking aspens: implications for avian nest site selection. Western North American Naturalist 67:251–257.
Riddle, B. R., T. Jezkova, M. E. Eckstut, V. Olah-Hemmings, and L.N. Carraway. 2014. Cryptic divergence and revised species taxonomy within the Great Basin pocket mouse, *Perognathus parvus* (Peale, 1848), species group. Journal of Mammalogy 95:9–25.
Rideout, C. B., and R. S. Hoffmann. 1975. *Oreamnos americanus*. Mammalian Species 63:1–6.
Rieger, J. F. 1996. Body size, litter size, timing of reproduction, and juvenile survival in the Uinta ground squirrel, *Spermophilus armatus*. Oecologia 107:463–468.
Rinella, S. 2009. America Buffalo: In Search of a Lost Icon. Spiegel and Grau, New York.

Ripple, W.J., and R.L. Beschta. 2003. Wolf reintroduction, predation risk, and cottonwood recovery in Yellowstone National Park. Forest Ecology and Management 184:299–313.

Ripple, W.J., and R.L. Beschta. 2007. Restoring Yellowstone's aspen with wolves. Biological Conservation 138:514–519.

Robbins, R.L., D.E. Redfearn, and C.P. Stone. 1982. Refuges and elk management. Pp. 479–507. In J.W. Thomas and D.E. Toweill (eds). Elk of North America: Ecology and Management. Stackpole Books, Harrisburg, PA.

Roberts, N.M., and S.M. Crimmins. 2010a. Do trends in muskrat harvest indicate widespread population declines? Northeastern Naturalist 17:229–238.

Roberts, N.M., and S.M. Crimmins. 2010b. Bobcat population status and management in North America: evidence of large-scale population increase. Journal of Fish and Wildlife Management 1:169–174.

Robinson, T.J., F.F.B. Elder, and J.A. Chapman. 1983. Evolution of chromosomal variation in cottontails, genus *Sylvilagus* (Mammalia: Lagomorpha): *S. aquaticus, S. floridanus,* and *S. transitionalis.* Cytogenetic and Genome Research 35:216–222.

Robinson, T.J., and C.A. Matthee. 2005. Phylogeny and evolutionary origins of the Leporidae: a review of cytogenetics, molecular analyses and a supermatrix analysis. Mammal Review 35:231–247.

Rocke, T.E., J. Williamson, K R. Cobble, J.D. Busch, M.F. Antolin, and D.M. Wagner. 2012. Resistance to plague among black-tailed prairie dog populations. Vector-Borne and Zoonotic Diseases 12:111–116.

Rodhouse, T.J., S.A. Scott, P.C. Ormsbee, and J.M. Zinck. 2008. Field identification of *Myotis yumanensis* and *Myotis lucifugus*: a morphological evaluation. Western North American Naturalist 68:437–443.

Rogowitz, G.L. 1997. Locomotor and foraging activity of the white-tailed jackrabbit (*Lepus townsendii*). Journal of Mammalogy 78:1172–1181.

Rogowitz, G.L., and M.L. Wolfe. 1991. Intraspecific variation in life-history traits of the white-tailed jackrabbit (*Lepus townsendii*). Journal of Mammalogy 72:796–806.

Rohl, U., T. Westerhold, T.J. Bralower, and J.C. Zachos. 2007. On the duration of the Paleocene–Eocene Thermal Maximum (PETM). Geochemistry Geophysics Geosystems 8:Q12002.

Rominger, E.M., H.A. Whitlaw, D.L. Weybright, W.S. Dunn, and W.B. Ballard. 2004. The influence of mountain lion predation on bighorn sheep translocations. Journal of Wildlife Management 68:993–999.

Rompola, K.M. 2000. Small mammals of a juniper woodland and sagebrush-grassland mosaic in southwestern Wyoming. MS thesis, University of Wyoming, Laramie.

Rompola, K.M., and S.H. Anderson. 2002. Small mammal community in a juniper-woodland sagebrush-grassland mosaic in southwestern Wyoming. Intermountain Journal of Sciences 8:19–29.

Rompola, K.M., and S.H. Anderson. 2004. Habitat of three rare species of small mammals in juniper woodlands of southwestern Wyoming. Western North American Naturalist 64:86–92.

Rosatte, R., and S. Lariviere. 2003. Skunks. Pp. 692–707. In G.A. Feldhamer, B.C. Thompson, and J.A. Chapman (eds). Wild Mammals of North America. 2nd ed. Johns Hopkins University Press, Baltimore, MD.

Rothwell, R. 1979. Nest sites of red squirrels (*Tamiasciurus hudsonicus*) in the Laramie Range of southeastern Wyoming. Journal of Mammalogy 60:404–405.

Roy, M.S., E. Geffen, D. Smith, E.A. Ostrander, and R.K. Wayne. 1994. Patterns of differentiation and hybridization in North American wolflike canids, revealed by analysis of microsatellite loci. Molecular Biology and Evolution 11:553–570.

Roze, U. 2009. The North American Porcupine. 2nd ed. Cornell University Press, Ithaca, NY.
Roze, U., and L.M. Ilse. 2003. Porcupine. Pp. 371–380. In G.A. Feldhamer, B.C. Thompson, and J.A. Chapman (eds). Wild Mammals of North America. 2nd ed. Johns Hopkins University Press, Baltimore, MD.
Ruggiero, L.F., K.B. Aubry, S.W. Buskirk, G.M. Koehler, C.J. Krebs, K.S. McKelvey, and J.R. Squires. eds. 2000. Ecology and Conservation of Lynx in the United States. University Press of Colorado, Boulder.
Ruggiero, L.F., K.S. McKelvey, K.B. Aubry, J.P. Copeland, D.H. Pletscher, and M.G. Hornocker. 2007. Wolverine conservation and management. Journal of Wildlife Management 71:2145–2146.
Ruggiero, L.F., D.E. Pearson, and S.E. Henry. 1998. Characteristics of American marten den sites in Wyoming. Journal of Wildlife Management 62:663–673.
Russell, O. 1921. Journal of a Trapper; or, Nine Years in the Rocky Mountains, 1834–1843. Syms-York, Boise, ID.
Sacks, B.N., M. Moore, M .J. Statham, and H.U. Wittmers. 2011. A restricted hybrid zone between native and introduced red fox (*Vulpes vulpes*) populations suggests reproductive barriers and competitive exclusion. Molecular Ecology 20:326–341.
Salkeld, D.J., M. Salathé, P. Stapp, and J.H. Jones. 2010. Plague outbreaks in prairie dog populations explained by percolation thresholds of alternate host abundance. Proceedings of the National Academy of Sciences 107:14247–14250.
Sato, J.J., M. Wolsan, F.J. Prevosti, G. D'Elia, C. Begg, K. Begg, T. Hosoda, K.L. Campbell, and H. Suzuki. 2012. Evolutionary and biogeographic history of weasel-like carnivorans (Musteloidea). Molecular Phylogenetics and Evolution 63:745–757.
Sauer, J.R., and M.S. Boyce. 1983. Density dependence and survival of elk in northwestern Wyoming. Journal of Wildlife Management 47:31–37.
Sawyer, H., M.J. Kauffman, A.D. Middleton, T.A. Morrison, R.M. Nielson, and T.B. Wyckoff. 2012. A framework for understanding semi-permeable barrier effects on migratory ungulates Journal of Applied Ecology 50:68–78.
Sawyer, H., and C. LeBeau. 2013. Trapper's Point Wildlife Crossing Study: 2012 Progress Report. Western Ecosystems Technology, Laramie, WY.
Sawyer, H., and F.G. Lindzey. 2000. Ecology of Sympatric Mule Deer and White-Tailed Deer in Riparian Communities of Southeast Wyoming. Unpublished Report. Wyoming Cooperative Wildlife Research Unit, Laramie.
Sawyer, H., and F.G. Lindzey. 2001. Habitat selection and sexual selection of elk in northern Wyoming. Intermountain Journal of Sciences 7:70–77.
Sawyer, H., and F.G. Lindzey. 2002. A Review of Predation on Bighorn Sheep (*Ovis canadensis*). Unpublished Report. Wyoming Cooperative Wildlife Research Unit, Laramie.
Sawyer, H., F. Lindzey, and D. McWhirter. 2005. Mule deer and pronghorn migration in western Wyoming. Wildlife Society Bulletin 33:1266–1273.
Sawyer, H., F. Lindzey, D. McWhirter, and K. Andrews. 2002. Potential effects of oil and gas development on mule deer and pronghorn populations in western Wyoming. Pp. 350–365. In J. Rahm (ed.). Transactions of the Sixty-Seventh North American Wildlife and Natural Resources Conference, Washington, DC.
Sawyer, H., R.M. Nielson, F.G. Lindzey, L. Keith, J.H. Powell, and A.A. Abraham. 2007. Habitat selection of Rocky Mountain elk in a nonforested environment. Journal of Wildlife Management 71:868–874.
Sawyer, H., R.M. Nielson, F. Lindzey, and L.L. McDonald. 2006. Winter habitat selection of mule deer before and during development of a natural gas field. Journal of Wildlife Management 70:396–403.

Scheffer, V. B. 1993. The Olympic mountain goat controversy: a perspective. Conservation Biology 7:916–919.
Schmidt, C. A. 2003a. Conservation Assessment for the Western Small-Footed Myotis in the Black Hills National Forest, South Dakota and Wyoming. USDA Forest Service, Black Hills National Forest, Custer, SD.
Schmidt, C. A. 2003b. Conservation Assessment for the Long-Legged Myotis in the Black Hills National Forest, South Dakota and Wyoming. USDA Forest Service, Black Hills National Forest, Custer, SD.
Schneeweis, J. C., K. E. Severson, and L. E. Petersen. 1972. Food habits of deer in the Black Hills: northern Black Hills. Agricultural Experiment Station, South Dakota State University Bulletin 606:5–18.
Schonewald, C. 1994. *Cervus canadensis* and *C. elaphus*: North American subspecies and evaluation of clinal extremes. Acta Theriologica 39:431–452.
Schullery, P. 1996. The Yellowstone Wolf: A Guide and Sourcebook. High Plains, Worland, WY.
Schullery, P., and L. Whittlesey. 2001. Mountain goats in the Greater Yellowstone Ecosystem: a prehistoric and historical context. Western North American Naturalist 61:289–307.
Schwalm, D. L. 2012. Understanding functional connectivity in shortgrass and mixedgrass prairies using the swift fox as a model organism. PhD dissertation, Texas Tech University, Lubbock.
Schwanz, L. E., M. J. Voordouw, D. Brisson, and R. S. Ostfeld. 2011. *Borrelia burgdorferi* has minimal impact on the Lyme disease reservoir host *Peromyscus leucopus*. Vector-Borne and Zoonotic Diseases 11:117–124.
Schwartz, C. C., M. A. Haroldson, K. A. Gunther, and D. Moody. 2006. Distribution of grizzly bears in the Greater Yellowstone Ecosystem in 2004. Ursus 17:63–66.
Schwartz, C. C., M. A. Haroldson, and G. C. White. 2006. Survival of cub and yearling grizzly bears in the Greater Yellowstone Ecosystem, 1983–2001. Wildlife Monographs 161:23–31.
Schwartz, C. C., S. D. Miller, and M. A. Haroldson. 2003. Grizzly bear. Pp. 556–586. In G. A. Feldhamer, B. C. Thompson, and J. A. Chapman (eds). Wild Mammals of North America. 2nd ed. Johns Hopkins University Press, Baltimore, MD.
Schwartz, M. K., L. S. Mills, K. S. McKelvey, L. F. Ruggiero, and F.W. Allendorf. 2002. DNA reveals high dispersal synchronizing the population dynamics of Canada lynx. Nature 415:520–522.
Schwartz, M. K., K. L. Pilgrim, K. S. McKelvey, E. L. LindquistJ.J. Claar, S. Loch, and L.F. Ruggiero 2004. Hybridization between Canada lynx and bobcats: genetic results and management implications. Conservation Genetics 5:349–355.
Scott, M. D. 2004. History of pronghorns translocated from Yellowstone National Park. Proceedings of the 21st Biennial Pronghorn Workshop, Bismarck, ND.
Scurlock, B. M., and W. H. Edwards. 2010. Status of brucellosis in free-ranging elk and bison in Wyoming. Journal of Wildlife Diseases 46:442–449.
Seabloom, R. W., and T. L. Shaffer. 2005. Plains harvest mouse in North Dakota. Prairie Naturalist 37:247–250.
Sera, W. E., and C. N. Early. 2003. *Microtus montanus*. Mammalian Species 716:1–10.
Serbousek, M. R., and K. Geluso. 2009. Bats along the Republican River and its tributaries in southwestern Nebraska: distribution, abundance, and reproduction. Western North American Naturalist 69:180–185.
Seton, E. T. 1925. Lives of Game Animals. Vol. 2. Doubleday, Doran, New York.
Sexton, O. J. 1990. Replacement of fox squirrels by gray squirrels in a suburban habitat. American Midland Naturalist 124:198–205.
Shackleton, D. M. 1985. *Ovis canadensis*. Mammalian Species 230:1–9.

Sheets, R. G. 1972. A trap for capturing black-footed ferrets. American Midland Naturalist 88:461–462.
Sheffield, S. R., and C. M. King. 1994. *Mustela nivalis*. Mammalian Species 454:1–10.
Sheffield, S. R., and H. H. Thomas. 1997. *Mustela frenata*. Mammalian Species 570:1–9.
Sherburne, S. S., and J. A. Bissonette. 1994. Subnivean access point use: response to subnivean prey levels. Journal of Wildlife Management 58:400–405.
Sherwin, R. E., D. Stricklan, and D. S. Rogers. 2000. Roosting affinities of Townsend's big-eared bat (*Corynorhinus townsendii*) in northern Utah. Journal of Mammalogy 81:939–947.
Shump, K. A., Jr., and A. U. Shump. 1982a. *Lasiurus borealis*. Mammalian Species 183:1–6.
Shump, K. A., Jr., and A. U. Shump. 1982b. *Lasiurus cinereus*. Mammalian Species 185:1–5.
Siepielski, A. M. 2006. A possible role for red squirrels in structuring breeding bird communities in lodgepole pine forests. Condor 108:232–238.
Sietman, B. E., H. L. Dunn, J. K. Tucker, and D. E. Kelner. 2003. Muskrat (*Ondatra zibethicus*) predation on zebra mussels (*Dreissena polymorpha*) attached to unionid bivalves. Journal of Freshwater Biology 18:25–32.
Silva, M., J. H. Brown, and J. A. Downing. 1997. Differences in population density and energy use between birds and mammals: a macroecological perspective. Journal of Animal Ecology 66:327–340.
Simpson, M. R. 1993. *Myotis californicus*. Mammalian Species 428:1–4.
Singer, F. J., A. Harting, K. K. Symonds, and M. B. Coughenour. 1997. Density dependence, compensation, and environmental effects on elk calk mortality in Yellowstone National Park. Journal of Wildlife Management 61:12–25.
Skinner, M. P. 1927. The predatory and fur-bearing animals of the Yellowstone National Park. Roosevelt Wild Life Bulletin 4:163–281.
Skryja, D. D. 1970. Some aspects of the ecology of the least chipmunk (*Eutamias minimus operarius*) in the Laramie Mountains of southeast Wyoming. MS thesis, University of Wyoming, Laramie.
Skupski, M. P. 1995. Population ecology of the western harvest mouse, *Reithrodontomys megalotis*: a long-term perspective. Journal of Mammalogy 76:358–367.
Slade, N. A., and D. F. Balph. 1974. Population ecology of Uinta ground squirrels. Ecology 55:989–1003.
Slobodchikoff, C. N., B. S. Perla, and J. L. Verdolin. 2009. Prairie Dogs: Communication and Community in an Animal Society. Harvard University Press, Cambridge, MA.
Smith, A. T. 1974a. The distribution and dispersal of pikas: consequences of insular population structure. Ecology 55:1112–1119.
Smith, A. T. 1974b. The distribution and dispersal of pikas: influences of behavior and climate. Ecology 55:1368–1376.
Smith, A. T., and M. L. Weston. 1990. *Ochotona princeps*. Mammalian Species 352:1–8.
Smith, B. L., E. K. Cole, and D. S. Dobkin. 2004. Imperfect Pasture: A Century of Change at the National Elk Refuge in Jackson Hole, Wyoming. Grand Teton Natural History Association, Jackson, WY.
Smith, D. W., and G. Ferguson. 2005. Decade of the Wolf: Returning the Wild to Yellowstone. Lyons Press, Guilford, CT.
Smith, F. A. 1997. *Neotoma cinerea*. Mammalian Species 564:1–8.
Smith, G. W., L. C. Stoddart, and F. F. Knowlton. 2002. Long-distance movements of black-tailed jackrabbits. Journal of Wildlife Management 66:463–469.
Smith, M. E., and M. C. Belk. 1996. *Sorex monticolus*. Mammalian Species 528:1–5.
Smith, W. P. 1991. *Odocoileus virginianus*. Mammalian Species 388:1–13.
Smith, W. P. 1999. A unique observation of a fisher (*Martes pennanti*) in Grand Teton National Park. Northwestern Naturalist 80:33–34.

Smith, W. P. 2007. Ecology of *Glaucomys sabrinus*: habitat, demography, and community relations. Journal of Mammalogy 88:862–881.

Smolen, M. J., and B. L. Keller. 1987. *Microtus longicaudis*. Mammalian Species 271:1–7.

Snoke, A. W. 1993. Geologic history of Wyoming within the tectonic framework of the North American Cordillera. Pp. 3–56. In A. W. Snoke, J. R. Steidtmann, and S. M. Roberts (eds). Geology of Wyoming. Geological Survey of Wyoming Memoir No. 5, Laramie.

Snyder, M. A. 1993. Interactions between Abert's squirrel and ponderosa pine: the relationship between selective herbivory and host plant fitness. *American Naturalist* 141:866–879.

Sovada, M. A., C. C. Roy, J. B. Bright, and J. R. Gillis. 1998. Causes and rates of mortality of swift foxes in western Kansas. Journal of Wildlife Management 62:1300–1306.

Sovada, M. A., R. O. Woodward, and L. D. Igl. 2009. Historical range, current distribution, and conservation status of the swift fox, *Vulpes velox*, in North America. Canadian Field-Naturalist 123:346–367.

Spaeth, P. A., M. van Tuinen, Y. L. Chan, E. Terca, and E.A. Hadly. 2009. Phylogeography of *Microtus longicaudus* in the tectonically and glacially dynamic central Rocky Mountains. Journal of Mammalogy 90:571–584.

Sparks, D. R. 1968. Diet of black-tailed jackrabbits on sandhill rangeland in Colorado. Journal of Range Management 21:203–208.

Spencer, A. W., and D. Pettus. 1966. Habitat preferences of five sympatric species of long-tailed shrews. Ecology 47:677–683.

Springer, J. T., and J. S. Smith. 1981. Summer food habits of coyotes in central Wyoming. Great Basin Naturalist 41:449–456.

Springer, M. S., W. J. Murphy, E. Eizirik, and S. J. O'Brien. 2003. Placental mammal diversification and the Cretaceous–Tertiary boundary. Proceedings of the National Academy of Sciences 100:1056–1061.

Springer, M. S., M. J. Stanhope, O. Madsen, and W. W. de Jong. 2004. Molecules consolidate the placental mammal tree. Trends in Ecology and Evolution 19:430–438.

Squires, J. R. 2000. Food habits of Northern Goshawks nesting in south-central Wyoming. Wilson Bulletin 112:536–539.

Squires, J. R., J. P. Copeland, T. J. Ulizio, M. K. Schwartz, and L.F. Ruggiero. 2007. Sources and patterns of wolverine mortality in western Montana. Journal of Wildlife Management 71:2213–2220.

Squires, J. R., and T. Laurion. 2000. Lynx home range and movements in Montana and Wyoming: preliminary results. Pp. 337–349. In L. F. Ruggiero, K. B. Aubry, S. W. Buskirk, G. M. Koehler, C.J. Krebs, K.S. McKelvey, and J.R. Squires. (eds). Ecology and Conservation of Lynx in the United States. University Press of Colorado, Boulder.

Stahler, D. R., D. R. MacNulty, R. K. Wayne, B. vonHoldt, and D.W. Smith. 2012. The adaptive value of morphological, behavioural and life-history traits in reproductive female wolves. Journal of Animal Ecology 82:222–234.

Stahler, D. R., D. W. Smith, and D. S. Guernsey. 2006. Foraging and feeding ecology of the gray wolf (*Canis lupus*): lessons from Yellowstone National Park, Wyoming, USA. Journal of Nutrition 136:1923S–1926S.

Stalling, D. T. 1990. *Microtus ochrogaster*. Mammalian Species 355:1–9.

Stanley, T. R. 2010. Land use and small mammal predation effects on shortgrass prairie birds. Journal of Wildlife Management 74:1825–1833.

Stapp, P. 1998. A reevaluation of the role of prairie dogs in Great Plains grasslands. Conservation Biology 12:1253–1259.

Stapp, P. 1999. Size and habitat characteristics of home ranges of northern grasshopper mice (*Onychomys leucogaster*). Southwestern Naturalist 44:101–105.

Stapp, P. 2007. Rodent communities in active and inactive colonies of black-tailed prairie dogs in shortgrass steppe. Journal of Mammalogy 88:241–249.
Stapp, P., D.J. Salkeld, R.J. Eisen, R. Pappert, J. Young, L.G. Carter, K.L. Gage, D.W. Tripp, and M.F. Antolin. 2008. Exposure of small rodents to plague during epizootics in black-tailed prairie dogs. Journal of Wildlife Diseases 44:724–730.
Stapp, P., D.J. Salkeld, H.A. Franklin, J.P. Kraft, D.W. Tripp, M.F. Antolin, and K.L. Gage. 2009. Evidence for the involvement of an alternate rodent host in the dynamics of introduced plague in prairie dogs. Journal of Animal Ecology 78:807–817.
Statham, M.J., J. Murdoch, J. Janecka, K.B. Aubry, C.J. Edwards, C.D. Soulsbury, O. Berry et al. 2014. Range-wide phylogeography of the red fox reveals ancient continental divergence, minimal genomic exchange and distinct demographic histories. Molecular Ecology 23:4813–4830.
Statham, M.J., B.N. Sacks, K.B. Aubry, J.D. Perrine, and S.M. Wisely. 2012. The origin of recently established red fox populations in the United States: translocations or natural range expansions. Journal of Mammalogy 93:52–65.
Steele, M.A. 1998. *Tamiasciurus hudsonicus*. Mammalian Species 586:1–9.
Stefanich, M.R. 1995. Movements and habitat use of white-tailed deer in the northwestern Black Hills of Wyoming and South Dakota. MS thesis, University of Wyoming, Laramie.
Stephens, R.B., E.M. Anderson, S.R. Wendt, and J.K. Meece. 2014. Field identification of sympatric *Peromyscus leucopus noveboracensis* and *P. maniculatus gracilis* in Wisconsin from external measurements. American Midland Naturalist 171:139–146.
Stephens, R.M. and S.H. Anderson. 2005. Swift Fox (*Vulpes velox*): A Technical Conservation Assessment. USDA Forest Service, Rocky Mountain Region 2, Lakewood, CO. Available at http://www.fs.fed.us/r2/projects/scp/assessments/swiftfox.pdf. Accessed November 17, 2011.
Sterling, K.B. 1989. Builders of the U.S. Biological Survey, 1885–1930. Journal of Forest History 33:180–187.
Steuter, A.A., E.M. Steinhauser, G.L. Hill, P.A. Bowers, and L.L. Tieszen. 1995. Distribution and diet of bison and pocket gophers in a sandhills prairie. Ecological Applications 5:756–766.
Stewart, D.T., M. McPherson, J. Robichaud, and L. Fumagalli. 2003. Are there two species of pygmy shrews (*Sorex*)? Revisiting the question using DNA sequence data. Canadian Field-Naturalist 117:82–88.
Stoddart, L.C. 1970. A telemetric method for detecting jackrabbit mortality. Journal of Wildlife Management 34:501–507.
Stone, K.D., and J.A. Cook. 2002. Molecular evolution of Holarctic martens (genus *Martes*, Mammalia: Carnivora: Mustelidae). Molecular Phylogenetics and Evolution 24:169–179.
Storm, G.L., R.D. Andrews, R.L. Phillips, R.A. Bishop, D.B. Siniff, and J.R. Tester. 1976. Morphology, reproduction, dispersal, and mortality of midwestern red fox populations. Wildlife Monographs 49:1–82.
St-Pierre, C., J.-P. Ouellet, and M. Crête. 2006. Do competitive intraguild interactions affect space and habitat use by small carnivores in a forested landscape? Ecography 29:487–496.
St-Pierre, C., J.-P. Ouellet, F. Dufresne, A. Chaput-Bardy, and F. Hubert. 2006. Morphological and molecular discrimination of *Mustela erminea* (ermines) and *M. frenata* (long-tailed weasels) in eastern Canada. Northeastern Naturalist 13:143–152.
Streeter, R.G., and C.E. Braun. 1968. Occurrence of pine marten, *Martes americana*, (Carnivora: Mustelidae) in Colorado alpine areas. Southwestern Naturalist 13:449–451.
Streubel, D.P., and J.P. Fitzgerald. 1978a. *Spermophilus spilosoma*. Mammalian Species 101:1–4.
Streubel, D.P., and J.P. Fitzgerald. 1978b. *Spermophilus tridecemlineatus*. Mammalian Species 103:1–5.
Stromberg, M.R. 1982. New records of Wyoming bats. Bat Research News 23:42–44.

Sudman, P. D., J. K. Wickliffe, P. Horner, M J. Smolen, J.W. Bickham, and R.D. Bradley. 2006. Molecular systematics of pocket gophers of the Genus *Geomys*. Journal of Mammalogy 87:668–676.

Sullivan, E. G. 1956. Gray fox reproduction, denning, range, and weights in Alabama. Journal of Mammalogy 37:346–351.

Sullivan, J., J. R. Demboski, K. C. Bell, S. Hird, B. Sarver, N. Reid, and J.M. Good. 2014. Divergence with gene flow within the recent chipmunk radiation (*Tamias*). Heredity 113:185–194.

Sullivan, T. P., and D. S. Sullivan. 2012. Woody debris, voles, and trees: influence of habitat structures (piles and windrows) on long-tailed vole populations and feeding damage. Forest Ecology and Management 263:189–198.

Summers, C. A., and R. L. Lindner. 1978. Food habits of the black-tailed prairie dog in western South Dakota. Journal of Range Management 31:134–136.

Sundstrom, C., W. G. Hepworth, and K. L. Diem. 1973. Abundance, distribution and food habits of the pronghorn. Wyoming Game and Fish Department Bulletin 12:1–61.

Sutter, J., J. C. Munger, and D. A. Hengel. 1999. *Sorex monticolus* in shrub steppe habitat in the northern Great Basin. Great Basin Naturalist 59:102–104.

Sutton, D. A. 1992. *Tamias amoenus*. Mammalian Species 390:1–8.

Swanson, B. J., R. T. Fuhrman, and R. L. Crabtree. 2005. Elevational isolation of red fox populations in the Greater Yellowstone Ecosystem. Conservation Genetics 6:123–131.

Sweitzer, R. A., S. H. Jenkins, and J. Berger. 1997. Near-extinction of porcupines by mountain lions and consequences of ecosystem change in the Great Basin Desert. Conservation Biology 11:1407–1417.

Swenk, M. H. 1908. A preliminary review of the mammals of Nebraska. Proceedings of the Nebraska Academy of Sciences 8:61–144.

Talluto, M. V., and C. W. Benkman. 2013. Landscape-scale eco-evolutionary dynamics: selection by seed predators and fire determine a major reproductive strategy. Ecology 94:1307–1316

Thaeler, C. S., Jr. 1972. Taxonomic status of the pocket gophers, *Thomomys idahoensis*, and *Thomomys pygmaeus* (Rodentia, Geomyidae). Journal of Mammalogy 53:417–428.

Thaeler, C. S., Jr. 1980. Chromosome numbers and systematic relations in the genus *Thomomys* (Rodentia: Geomyidae). Journal of Mammalogy 61:414–422.

Thaeler, C. S., Jr., and L. L. Hinesley. 1979. *Thomomys clusius*, a rediscovered species of pocket gopher. Journal of Mammalogy 60:480–488.

Theis, N., and M. Lerdau. 2003. The evolution of function in plant secondary metabolites. International Journal of Plant Sciences 164:S93–S109.

Thiagarajan, B., Y. Bai, K. L. Gage, and J. F. Cully, Jr. 2008. Prevalence of *Yersinia pestis* in rodents and fleas associated with black-tailed prairie dogs (*Cynomys ludovicianus*) at Thunder Basin National Grassland, Wyoming. Journal of Wildlife Diseases 44:731–736.

Thies, M. L., K. Thies, and K. McBee. 1996. Organochlorine pesticide accumulation and genotoxicity in Mexican free-tailed bats from Oklahoma and New Mexico. Archives of Environmental Contamination and Toxicology 30:178–187.

Thimmayya, A. C. 2010. Habitat ecology and genetic connectivity of the pygmy rabbit (*Brachylagus idahoensis*) across southern Wyoming. MS thesis, University of Wyoming, Laramie.

Thimmayya, A. C., and S. W. Buskirk. 2012. Genetic connectivity and diversity of the pygmy rabbit (*Brachylagus idahoensis*) in southern Wyoming. Journal of Mammalogy 93:29–37.

Thoma, B. L., and W. H. Marshall. 1960. Squirrel weights and populations in a Minnesota woodlot. Journal of Mammalogy 41:272–273.

Thomas, E. M. 1954. Wyoming Furbearers: Their Life Histories and Importance. Wyoming Game and Fish Department, Bulletin No. 7, Cheyenne.

Thurber, J. M., and R. O. Peterson. 1991. Changes in body size associated with range expansion in the coyote (*Canis latrans*). Journal of Mammalogy 72:750–755.
Tileston, J. V., and R. R. Lechleitner. 1966. Some comparisons of the black-tailed and white-tailed prairie dogs in north-central Colorado. American Midland Naturalist 75:292–316.
Tomasi, T. E., and R. S. Hoffman. 1984. *Sorex preblei* in Utah and Wyoming. Journal of Mammalogy 65:708.
Trainor, A. M., T. M. Shenk, and K. R. Wilson. 2007. Microhabitat characteristics of Preble's meadow jumping mouse high-use areas. Journal of Wildlife Management 71:469–477.
Tumlison, R., and M. E. Douglas. 1992. Parsimony analysis and the phylogeny of the plecotine bats (Chiroptera: Vespertilionidae). Journal of Mammalogy 73:271–285.
Turner, R. W. 1974. Mammals of the Black Hills of South Dakota and Wyoming. University of Kansas Museum of Natural History, Miscellaneous Publications No. 60.
Tyers, D. B. 2006. Moose population history on the northern Yellowstone winter range. Alces 42:133–149.
Tyers, D. B. 2008. Absaroka-Beartooth Wilderness beaver reintroductions, 1985–present. Yellowstone Science 16(3):12–13.
Tzilkowski, W. M., and F. F. Knowlton. 1979. Man-caused mortalities of coyotes radio-marked in Grand Teton National Park. Pp. 164–168. In R. Ittner, D. R. Potter, J. K. Agee, and S. Anschell (eds). Conference Proceedings on Recreational Impact on Wildlands, October 27–29, 1978, Seattle, WA. USDA Forest Service Region 6 and USDI National Park Service.
Ubico, S. R., G. O. Maupin, K. A. Fagerstone, and R. G. McLean. 1988. A plague epizootic in the white-tailed dogs (*Cynomys leucurus*) of Meeteetse, Wyoming. Journal of Wildlife Diseases 24:399–406.
Ubico, S. R., R. G. McLean, and L. M. Cooksey. 1996. Susceptibility of selected rodent species from Colorado to *Borrelia burgdorferi*. Journal of Wildlife Diseases 32:293–299.
US Department of Agriculture. 2013. Wyoming Agricultural Statistics 2011. USDA National Agricultural Statistics Service, Wyoming Field Office, Cheyenne.
US Department of Agriculture, NRCS. 2013. The PLANTS Database, January 4. National Plant Data Team, Greensboro, NC. Available at http://plants.usda.gov.
US Fish and Wildlife Service. 1978. Endangered and threatened wildlife and plants: reclassification of the gray wolf in the United States and Mexico, with Determination of Critical Habitat in Michigan and Minnesota. Federal Register 43:9607–9615.
US Fish and Wildlife Service. 1994. Endangered and threatened wildlife and plants; proposed establishment of a nonessential experimental population of gray wolf in Yellowstone National Park in Wyoming, Idaho, and Montana. Federal Register 59:42108–42118.
US Fish and Wildlife Service. 2000. Endangered and threatened wildlife and plants: determination of threatened status for the contiguous U.S. distinct population segment of the Canada lynx and related rules. Federal Register 65:16052–16086.
US Fish and Wildlife Service. 2001. Endangered and threatened wildlife and plants: annual notice of findings on recycled petitions. Federal Register 66:1295–1300.
US Fish and Wildlife Service. 2003. Endangered and threatened wildlife and plants: Final rule to list the Columbia Basin distinct population segment of the pygmy rabbit (*Brachylagus idahoensis*) as Endangered. Federal Register 67:10388–10409.
US Fish and Wildlife Service. 2008. Endangered and threatened wildlife and plants: final rule to amend the listing for the Preble's meadow jumping mouse (*Zapus hudsonius preblei*) to specify over what portion of its range the subspecies is threatened. Federal Register 73:39790–39838.

US Fish and Wildlife Service. 2009. Endangered and threatened wildlife and plants: 12-month finding on a petition to list the black-tailed prairie dog as threatened or endangered. Federal Register 74:63343–63366.
US Fish and Wildlife Service. 2010a. Endangered and threatened wildlife and plants: 12-month finding on a petition to list the pygmy rabbit as threatened or endangered. Federal Register 75:60516–60561.
US Fish and Wildlife Service. 2010b. Endangered and threatened wildlife and plants: 12-month finding on a petition to list the North American wolverine as threatened or endangered. Federal Register 75:78030–78061.
US Fish and Wildlife Service. 2010c. Endangered and threatened wildlife and plants; reinstatement of protections for the grizzly bear in the Greater Yellowstone Ecosystem in compliance with court order. Federal Register 75:14496.
US Fish and Wildlife Service. 2011. Endangered and threatened wildlife and plants: 12-month finding on a petition to list a distinct population segment of the fisher in its United States Northern Rocky Mountain Range as endangered or threatened with critical habitat. Federal Register 76:38504–38532.
US Fish and Wildlife Service. 2012. Endangered and threatened wildlife and plants: 90-day finding on a petition to list the prairie gray fox, the plains spotted skunk, and a distinct population segment of the Mearn's eastern cottontail in East-Central Illinois and Western Indiana as endangered or threatened species. Federal Register 77:71759–71771.
US Fish and Wildlife Service. 2013. Endangered and threatened wildlife and plants: threatened status for the distinct population segment of the North American wolverine occurring in the Contiguous United States. Federal Register 78:7864–7890.
US Fish and Wildlife Service. 2014. Endangered and threatened wildlife and plants: threatened status for the distinct population segment of the North American wolverine occurring in the Contiguous United States; establishment of a nonessential experimental population of the North American wolverine in Colorado, Wyoming, and New Mexico. Federal Register 79:47522–47545.
US Fish and Wildlife Service. 2015. Endangered and threatened wildlife and plants: threatened species status for the northern long-eared bat with 4(d) rule. Federal Register 80:17974–18033.
Valdez, E.W., and P.M. Cryan. 2009. Food habits of the hoary bat (*Lasiurus cinereus*) during spring migration through Mexico. Southwestern Naturalist 54:195–200.
Van Etten, K.W., K.R. Wilson, and R.L. Crabtree. 2007. Habitat use of red foxes in Yellowstone National Park based on snow tracking and telemetry. Journal of Mammalogy 88:1498–1507.
Van Gelder, R.G. 1959. A taxonomic revision of the spotted skunks (Genus *Spilogale*). Bulletin of the American Museum of Natural History 117:229–392.
Van Manen, F.T., M.A. Haroldson, K. West, and S.C. Soileau, eds. 2013. Yellowstone Grizzly Bear Investigations: Annual Report of the Interagency Grizzly Bear Study Team. US Geological Survey, Bozeman, MT.
Van Tuinen, M., K. O'Keefe, U. Ramakrishnan, and E.A. Hadly. 2008. Fire and ice: genetic structure of the Uinta ground squirrel (*Spermophilus armatus*) across the Yellowstone hotspot. Molecular Ecology 17:1776–1788.
Van Vuren, D.H. 2001. Predation on yellow-bellied marmots (*Marmota flaviventris*). American Midland Naturalist 145:94–100.
Van Vuren, D.H., and K.B. Armitage. 1991. Duration of snow cover and its influence on life-history variation in yellow-bellied marmots. Canadian Journal of Zoology 69:1755–1758.

van Zyll de Jong, C. G. 1979. Distribution and systematic relationships of long-eared *Myotis* in western Canada. Canadian Journal of Zoology 57:987–994.
van Zyll de Jong, C. G. 1980. Systematic relationships of woodland and prairie forms of the common shrew, *Sorex cinereus cinereus* Kerr and *S. c. haydeni* Baird, in the Canadian prairie provinces. Journal of Mammalogy 61:66–75.
van Zyll de Jong, C. G. 1982. An additional morphological character useful in distinguishing two similar shrews *Sorex monticolus* and *Sorex vagrans*. Canadian Field-Naturalist 96:349–350.
van Zyll de Jong, C. G. 1983. Handbook of Canadian Mammals. 1. Marsupials and Insectivores. National Museum of Canada, Ottawa, ON.
van Zyll de Jong, C. G. 1984. Taxonomic relationships of Nearctic small-footed bats of the *Myotis leibii* group (Chiroptera: Vespertilionidae). Canadian Journal of Zoology 62:2519–2526.
van Zyll de Jong, C. G. 1985. Handbook of Canadian Mammals. 2. Bats. National Museum of Canada, Ottawa, ON.
van Zyll de Jong, C. G. 1987. A phylogenetic study of the Lutrinae (Carnivora; Mustelidae) using morphological data. Canadian Journal of Zoology 65:2536–2544.
van Zyll de Jong, C. G., and G. L. Kirkland. 1989. A morphometric analysis of the *Sorex cinereus* group in central and eastern North America. Journal of Mammalogy 70:110–122.
Vander Wall, S. B. 1997. Dispersal of singleleaf piñon pine (*Pinus monophylla*) by seed-caching rodents. Journal of Mammalogy 78:181–191.
Vaughan, T. A. 1974. Resource allocation in some sympatric, subalpine rodents. Journal of Mammalogy 55:764–795.
Vaughan, T. A. 1990. Ecology of living packrats. Pp. 14–27. In J. L. Betancourt, T. R. Van Devender, and P. S. Martin (eds). 1990. Packrat Middens: The Last 40,000 Years of Biotic Change. University of Arizona Press, Tucson.
Verts, B. J., and L. N. Carraway. 1980. *Tamias minimus*. Mammalian Species 653:1–10.
Verts, B. J., and L. N. Carraway. 1998. Land Mammals of Oregon. University of California Press, Berkeley.
Verts, B. J., and L. N. Carraway. 1999. *Thomomys talpoides*. Mammalian Species 618:1–11.
Verts, B. J., L. N. Carraway, and A. Kinlaw. 2001 *Spilogale gracilis*. Mammalian Species 674:1–10.
Verts, B. J., and G. L. Kirkland. 1988. *Perognathus parvus*. Mammalian Species 318:1–8.
Virchow, D. R., and S. E. Hygnstrom. 2002. Distribution and abundance of black-tailed prairie dogs in the Great Plains: a historical perspective. Great Plains Research 12:197–218.
VonHof, M. J., and R. M. R. Barclay. 1997. Use of tree stumps as roosts by the western long-eared bat. Journal of Wildlife Management 61:674–684.
Vucetich, J. A., D. W. Smith, and D. R. Stahler. 2005. Influence of harvest, climate, and wolf predation on Yellowstone elk, 1961–2004. Oikos 111:259–2004.
Wagner, F. H., and L. C. Stoddart. 1972. Influence of coyote predation on black-tailed jackrabbit populations in Utah. Journal of Wildlife Management 36:329–342.
Walker, D. N., and G. C. Frison. 1980. The late Pleistocene mammalian fauna from the Colby Mammoth Kill site, Wyoming. Contributions to Geology, University of Wyoming 19:69–79.
Walker, R. E. 1968. Local distribution in a population of Uinta ground squirrels. PhD dissertation, Utah State University, Logan.
Walley, W. J. 1982. Weasel skull in a short-eared owl pellet. Canadian Field-Naturalist 96:85.
Warner, R. M., and N. J. Czaplewski. 1984. *Myotis volans*. Mammalian Species 224:1–4.
Warren, E. R. 1926. A study of beaver in the Yancey region of Yellowstone National Park. Roosevelt Wild Life Annals 1:1–191.
Watkins, L. C. 1977. *Euderma maculatum*. Mammalian Species 77:1–4.

Webb, S. D. 1985. Late Cenozoic mammal dispersals between the Americas. Pp. 357–382. In F. G. Stehli and S. D. Webb (eds). The Great American Biotic Interchange. Plenum Press, New York.
Webb, S. D., and A. D. Barnosky. 1989. Faunal dynamics of Pleistocene mammals. Annual Reviews of Earth and Planetary Science 17:413–438.
Webster, W. D., and J. K. Jones. 1982. *Reithrodontomys megalotis*. Mammalian Species 167:1–5.
Weigl, P. D. 2007. The northern flying squirrel (*Glaucomys sabrinus*): a conservation challenge. Journal of Mammalogy 88:897–907.
Wells-Gosling, N., and L. R. Heaney. 1984. *Glaucomys sabrinus*. Mammalian Species 229:1–8.
Wengeler, W. R., D. A. Kelt, and M. L. Johnson. 2010. Ecological consequences of invasive lake trout on river otters in Yellowstone National Park. Biological Conservation 143:1144–1153.
Western Bat Working Group. 1998. Western bat species: regional priority matrix. Available at http://www.wbwg.org/speciesinfo/species_matrix/spp_matrix.pdf. Accessed April 10, 2012.
Western Bat Working Group. 2005. *Corynorhinus townsendii*. Available at http://www.wbwg.org/speciesinfo/species_accounts/vespertilonidae/coto.pdf. Accessed August 14, 2013.
Weyandt, S. E., and R. A. Van Den Bussche. 2007. Phylogeographic structuring and volant mammals: the case of the pallid bat (*Antrozous pallidus*). Journal of Biogeography 34:1233–1245.
Whicker, A. D. and J. K. Detling. 1988. Ecological consequences of prairie dog disturbances. BioScience 38:778–785.
Whitaker, J. O., Jr. 1972. *Zapus hudsonius*. Mammalian Species 11:1–7.
Whitaker, J. O., Jr. 2004. *Sorex cinereus*. Mammalian Species 743:1–9.
White, J. A., and K. Geluso. 2012. Seasonal link between food hoarding and burrow use in a nonhibernating rodent. Journal of Mammalogy 93:149–160.
White, P. J., T. O. Lemke, D. B. Tyers, and J. A. Fuller. 2008. Initial effects of reintroduced wolves *Canis lupus* on bighorn sheep *Ovis canadensis* dynamics in Yellowstone National Park. Wildlife Biology 14:138–146.
Whittlesey, L., and P. Schullery. 2011. How many wolves were in the Yellowstone area in the 1870s? Yellowstone Science 19(1):23–28.
Wilkins, K. T. 1986. *Reithrodontomys montanus*. Mammalian Species 257:1–5.
Wilkins, K. T. 1989. *Tadarida brasiliensis*. Mammalian Species 331:1–10.
Williams, B. W., and K. T. Scribner. 2010. Effects of multiple founder populations on spatial genetic structure of reintroduced American martens. Molecular Ecology 19:227–240.
Williams, E. S. 2005. Chronic wasting disease. Veterinary Pathology 42:530–549.
Williams, E. S., K. Mills, D. R. Kwiatkowski, E. T. Thorne, and A. Boerger-Fields. 1994. Plague in a black-footed ferret (*Mustela nigripes*). Journal of Wildlife Diseases 30:581–585.
Williams, E. S., E. T. Thorne, M. J. G. Appel, and D. W. Belitsky. 1988. Canine distemper in black-footed ferrets (*Mustela nigripes*) from Wyoming. Journal of Wildlife Diseases 24:385–398.
Williams, J. W., and S. T. Jackson. 2007. Novel climates, no-analog communities, and ecological surprises. Frontiers in Ecology and the Environment 5:475–482.
Williams, O. 1959. Food habits of the deer mouse. Journal of Mammalogy 40:415–419.
Willingham, A. 2008. Emerging factors associated with the decline of a gray fox population and multi-scale land cover associations of mesopredators in the Chicago metropolitan area. Electronic thesis or dissertation, Ohio State University, Columbus. Available at https://etd.ohiolink.edu/.
Willner, G. R., G. A. Feldhamer, E. E. Zucker, and J. A. Chapman. 1980. *Ondatra zibethicus*. Mammalian Species 141:1–8.

Wilson, D.E. and D.M. Reeder, eds. 2005. Mammal Species of the World: A Taxonomic and Geographic Reference. 3rd ed. Johns Hopkins University Press, Baltimore, MD.
Wilson, G.M., R.A. Den Bussche, K. McBee, L.A. Johnson, and C.A. Jones. 2005. Intraspecific phylogeography of red squirrels (*Tamiasciurus hudsonicus*) in the central Rocky Mountain region of North America. Genetica 125:141–154.
Wilson, G.P. 2014. Mammalian extinction, survival, and recovery dynamics across the Cretaceous–Paleogene boundary in northeastern Montana, USA. Geological Society of America Special Paper 503:365–392.
Wilson, M.C. 1996. Late Quaternary vertebrates and the opening of the ice-free corridor, with special reference to the genus *Bison*. Quaternary International 32:97–105.
Winhold, L., A. Kurta, and R. Foster. 2008. Long-term change in an assemblage of North American bats: are eastern red bats declining? Acta Chiropterologica 10:359–366.
Wolf, E.C., D.J. Cooper, and N.T. Hobbs. 2007. Hydrologic regime and herbivory stabilize an alternative state in Yellowstone National Park. Ecological Applications 17:1572–1587.
Wolf Management Committee. 1991. Reintroduction and Management of Wolves in Yellowstone National Park and the Central Idaho Wilderness Area: A Report to the United States Congress. Wolf Management Committee, Washington, DC.
Woodburne, M.O. 2010. The Great American Biotic Interchange: dispersals, tectonics, climate, sea level and holding pens. Journal of Mammalian Evolution 17:245–264.
Woods, C.A. 1973. *Erethizon dorsatum*. Mammalian Species 29:1–6.
Woolf, A., T. O'Shea, and D.L. Gilbert. 1970. Movements and behavior of bighorn sheep on summer ranges in Yellowstone National Park. Journal of Wildlife Management 34:446–450.
Wu, S., W. Wu, F. Zhang, J. Ye, X. Ni, J. Sun, S.V. Edwards, J. Meng, and C.L. Organ. 2012. Molecular and paleontological evidence for a post-Cretaceous origin of rodents. PLOS ONE 7:e46445.
Wyoming Game and Fish Commission. 2011. Furbearing Animal Hunting or Trapping Seasons. Wyoming Game and Fish Commission, Cheyenne.
Wyoming Game and Fish Department. 1982. A Summary of Red Fox Statistics. Unpublished Planning Report, December. Wyoming Game and Fish Department, Cheyenne.
Wyoming Game and Fish Department. 2009. Annual Report. Wyoming Game and Fish Department, Cheyenne.
Wyoming Game and Fish Department. 2010a. State Wildlife Action Plan 2010. Unpublished Report. Wyoming Game and Fish Department, Cheyenne.
Wyoming Game and Fish Department. 2010b. Annual Report. Wyoming Game and Fish Department, Cheyenne.
Wyoming Game and Fish Department. 2011. Annual Report. Wyoming Game and Fish Department, Cheyenne.
Yates, T.L., and D.J. Schmidly. 1978. *Scalopus aquaticus*. Mammalian Species 105:1–4.
Yoakum, J.D. 2004a. Distribution and abundance. Pp. 75–105. In B.W. O'Gara, and J.D. Yoakum (eds). Pronghorn: Ecology and Management. University Press of Colorado, Boulder.
Yoakum, J.D. 2004b. Foraging ecology, diet studies, and nutrient values. Pp. 447–502. In B.W. O'Gara and J.D. Yoakum (eds). Pronghorn: Ecology and Management. University Press of Colorado, Boulder.
Zahratka, J.L., and T.M. Shenk. 2008. Population estimates of snowshoe hares in the southern Rocky Mountains. Journal of Wildlife Management 72:906–912.
Zanazzi, A., M.J. Kohn, B.J. MacFadden, and D.O. Terry, Jr. 2007. Large temperature drop across the Eocene–Oligocene transition in central North America. Nature 445:639–642.
Zegers, D.A. 1984. *Spermophilus elegans*. Mammalian Species 214:1–7.

Ziegler, P.E., S.E. Wade, J. Oliver, S.L. Schaaf, and H.O. Mohammed. 2009. Factors associated with the risk of *Giardia* species infection in two sympatric species of *Peromyscus* mice. Journal of Parasitology 95:1062–1065.

Zielinski, W.J., R.L. Truex, F.V. Schlexer, L.A. Campbell, and C. Carroll. 2005. Historical and contemporary distributions of carnivores in forests of the Sierra Nevada, California, USA. Journal of Biogeography 32:1385–1407.

Zimmer, J.E. 2004. Pronghorn productivity and fawn survival in central and southeastern Wyoming. MS thesis, University of Wyoming, Laramie.

Zimmerman, G., P. Stapp, and B. Van Horne. 1996. Seasonal variation in the diet of great horned owls (*Bubo virginianus*) on shortgrass prairie. American Midland Naturalist 96:149–156.

Zinck, J.M., D.A. Duffield, and P.C. Ormsbee. 2004. Primers for identification and polymorphism assessment of vespertilionid bats in the Pacific Northwest. Molecular Ecology Notes 4:239–242.

PHOTO CREDITS

Copyright of all images is held by the photographers noted below unless otherwise stated.

J. Scott Altenbach ©, Bat Conservation International, pp. 108, 112
Alex Badyaev, pp. 81, 84, 87, 93
Francis and Janice Bergquist ©, pp. 173, 189, 206, 262, 310, 316, 345, 351
Jim Brandenburg/Minden Pictures, p. 308
Matthias Breiter/Minden Pictures, p. 335
Milo Burcham, p. 213
Beth Buskirk, p. 200
Alan Carey/Photo Researchers, p. 264
Mark Chappell, pp. 208, 235
Barbara Clauson, p. 231
E. R. Degginger/Photo Researchers, p. 95
Michael Durham, p. 191
Michael Durham/Minden Pictures, pp. 199
Michael Durham ©/Minden Pictures, Bat Conservation International, pp. 105, 106, 118, 122, 127, 131, 135
Robert J. Erwin/Photo Researchers, p. 202
Gary Fairfield ©, p. 301
Tim Fitzharris/Minden Pictures, pp. 332
Keith Geluso, pp. 116, 137
Bob Gibbons/Minden Pictures, p. 144
Mark Gocke, pp. 153 bottom, 303
Greater Yellowstone Wolverine Program, p. 313 bottom
Bob Gress ©, p. 229
Steve Griffith/Murie Center Archives, p. 15
Cassie Hartman, pp. 177, 355

Dan and Cindy Hartman, pp. 151, 156, 180, 280 bottom, 294, 367
Gordon and Kathy Illg ©/Animals Animals, p. 298
Gerald C. Kelley/Photo Researchers, p. 341
Steve Kilpatrick, p. 369
Stephen Kraseman/Photo Researchers, pp. 224
Thomas and Pat Leeson/Photo Researchers, pp. 277, 328
Travis Livieri, p. 23 top
J. Michael Lockhart, p. 305
Charles Long, pp. 16, 24
Craig K. Lorenz/Photo Researchers, pp. 251, 257
Tom McHugh/Photo Researchers, pp. 209, 215, 222
Katie Mortenson, p. 363
Rolf Nussbaumer/npl/Minden Pictures, pp. 149, 187
Gregory Ochocki/Photo Researchers, p. 325
Craig Okraska/Hayden-Wing Assoc., p. 239
Michael Patrikeev/Wild Nature Images/www.wildnatureimages.org, pp. 76, 160, 233
Michael Quinton/Minden Pictures, pp. 246, 293, 318
Joe Riis, p. 63
Michael Rose/FLPA/Minden Pictures, p. 145
Royal Ontario Museum, with permission, p. 98 bottom
L. Lee Rue/FLPA/Minden Pictures, p. 289
Chas. & Elizabeth Schwartz Trust ©/Animals Animals, pp. 158, 204
Wendy Shattil/Bob Rozinski ©, pp. 147, 162, 164, 168, 171, 175, 193, 196, 211, 217, 220, 228, 236, 244, 260, 266, 274, 280 top, 284, 286, 312, 321, 323, 348, 360
Stouffer Productions ©/Animals Animals, p. 90
Merlin D. Tuttle ©, Bat Conservation International, www.batcon.org, pp. 110, 112, 114, 120, 124, 126, 129, 133
Tom Vezo/Minden Pictures, p. 153 top
Jim Witham, pp. 255, 256, 270
Wyoming Game and Fish Department, p. 23 bottom, 338
Wyoming Natural Diversity Database, p. 243
Wyoming State Archives, Department of State Parks and Cultural Resources, p. 56

INDEX

Page numbers in bold denote taxon accounts.